Mathematics Study Resources

Volume 25

Series Editors

Kolja Knauer, Departament de Matemàtiques i Informàtica, Universitat de Barcelona, Barcelona, Spain

Elijah Liflyand, Department of Mathematics, Bar-Ilan University, Ramat-Gan, Israel

This series comprises direct translations of successful foreign language titles, especially from the German language.

Powered by advances in automated translation, these books draw on global teaching excellence to provide students and lecturers with diverse materials for teaching and study.

Thomas Richter · Henry von Wahl ·
Thomas Wick

Introduction to Numerical Mathematics

Theory, Practice and Numerous Examples

 Springer

Thomas Richter
Fakultät für Mathematik
Universität Magdeburg
Magdeburg, Germany

Henry von Wahl
Fakultät für Mathematik
Friedrich-Schiller-Universität Jena
Jena, Germany

Thomas Wick
Institut für Angewandte Mathematik
Leibniz Universität Hannover
Hanover, Germany

ISSN 2731-3824   ISSN 2731-3832   (electronic)
Mathematics Study Resources
ISBN 978-3-662-72545-0   ISBN 978-3-662-72546-7   (eBook)
https://doi.org/10.1007/978-3-662-72546-7

This book is a translation of the original German edition "Einführung in die Numerische Mathematik," 2nd edition, by Thomas Richter et al., published by Springer-Verlag GmbH, DE in 2024. The translation was done with the help of an artificial intelligence machine translation tool. A subsequent human revision was done primarily in terms of content, so that the book will read stylistically differently from a conventional translation. Springer Nature works continuously to further the development of tools for the production of books and on the related technologies to support the authors.

Translation from the German language edition: "Einführung in die Numerische Mathematik" by Thomas Richter et al., © Der/die Herausgeber bzw. der/die Autor(en), exklusiv lizenziert an Springer-Verlag GmbH, DE, ein Teil von Springer Nature 2024. Published by Springer Berlin Heidelberg. All Rights Reserved.

Responsible Editor: Iris Ruhmann

This Springer imprint is published by the registered company Springer-Verlag GmbH, DE, part of Springer Nature.
The registered company address is: Heidelberger Platz 3, 14197 Berlin, Germany

If disposing of this product, please recycle the paper.

Preface

Accompanying the lecture *Introduction to Numerical Mathematics (Numerik 0)* in the summer term 2012 at the University of Heidelberg, the first version of these lecture notes was written. The contents of these lecture notes are kept elementary and can be understood with basic knowledge of calculus and linear algebra. However, numerical mathematics differs significantly from these lectures. In calculus and linear algebra, structural investigations are made on various problems: Is a function integrable? Does the solution to a system exist? How can vector spaces be characterized? And so on.

In numerical mathematics, the calculation of specific solutions using approximate methods is central. In addition to the design of approximation methods, numerical mathematics is particularly concerned with the analysis of these methods in terms of accuracy, computational effort (efficiency), and robustness. These methods then form the basis of scientific computing, i.e., the *mathematical modeling* of real problems from various disciplines, their *discretization* and *approximation* using numerical methods, and the subsequent implementation in *algorithms* for conducting *computer simulations*.

It turns out that many mathematical problems cannot—or only very laboriously—be really solved. And even if a simple solution exists, for example, $x = \sqrt{2}$ is one of the two roots of $f(x) = x^2 - 2$, this solution cannot be represented exactly on a calculator or computer. Numerical algorithms are generally implemented on a computer, and errors, which, for example, arise from rounding, must be included in the analysis. Of course, numerical mathematics is mathematically exact and sound. However, we cannot assume that mathematical tasks, such as the calculation of the sum of two numbers $x + y$, can also be realized exactly on the computer. This inaccuracy in the realization must be taken into account in numerical analysis.

These lecture notes are divided into two major parts: the numerical consideration of problems of linear algebra and the investigation of problems of calculus. We aim to establish a connection between problems, mathematical analysis, and numerical methods.

Heidelberg, Germany   Thomas Richter
August 2012   Thomas Wick

For the lectures *Introduction to Numerical Mathematics (Numerik 0)* in the summer term 2014 at the University of Heidelberg and *Numerical Mathematics* in the winter term 2015/2016 at the University of Erlangen, these lecture notes were revised in some parts, supplemented and corrected and especially expanded by numerous concrete examples. In particular, however, countless minor errors could be eliminated through the intensive correction of all tutors.

Heidelberg and Erlangen, Germany   Thomas Richter
March 2014 and October 2015

For the lecture *Numerical Modeling for STEEM (MAP502)* in the winter term 2016/2017 at École Polytechnique, Institut Polytechnique de Paris, parts of the chapter *Root determination* were translated into English, further elaborated, and then translated back into the German version.

Paris, France   Thomas Wick
November 2016

First Edition of the Book

For the first edition of the book, the material was further revised and completed. Many minor errors could be eliminated. Completely new and a special concern are application-related examples, so-called *excursions* from various areas of research, technology, and everyday life, in which numerical mathematics is not visible at first glance, but plays a significant role.

Magdeburg, Germany and Paris, France   Thomas Richter
February 2017   Thomas Wick

Second Edition of the Book

For the second edition, numerous errors were corrected, for which we are very grateful. Material on neural networks was added, and another chapter on the basics of numerical analysis was designed. Due to the topicality of Numerical Mathematics in many applications, we are very pleased to include another author, so that Python code snippets better illustrate the path of scientific computing from mathematical tasks, over algorithm design via pseudo-code to actual implementation.

♣ Supplementary Python programs in Jupyter notebooks, we have summarized on GitHub under github.com/sn-code-inside/EinfNumMath-RWW for most sections of this book. The corresponding places are marked with the Python logo.

In addition to its use at our home universities, the book was also used internationally, for example, in the course MAP 502 of the École Polytechnique, as well as in the context of the DAAD-funded project PeCCC (Peruvian Competence Center for Scientific Computing) at various universities in Latin America. In the online teaching of the years 2020–2021, the book has been a valuable support to us.

Magdeburg, Germany	Thomas Richter
Vienna, Austria and Jena, Germany	Henry von Wahl
Hanover, Germany	Thomas Wick
April 2024	

English Version of the Book

Along with the second German edition, we decided to translate the book into English due to our international collaborations and various schools on Numerical Mathematics that we are being asked to deliver. The translation was made with the help of an artificial intelligence translation tool and then revised by ourselves. We further made a small number of corrections, fixed some typos, and added a few new plots. Nevertheless, this English version corresponds in general to the second edition of the German book. We thank Springer for giving us the opportunity to publish this translated edition.

Magdeburg, Germany	Thomas Richter
Jena, Germany	Henry von Wahl
Hanover, Germany	Thomas Wick

Competing Interests The authors have no competing interests to declare that are relevant to the content of this manuscript.

Contents

Introduction

In numerical mathematics, methods to "solve" mathematical problems and tasks are designed and analyzed. Numerical mathematics is closely linked with other branches of mathematics and computer science, and always needs to keep applications from in chemistry, physics, medicine, engineering, or economics in mind. The tasks we try to solve often originate from outside mathematics and come from a wide variety of areas. We place particular emphasis on these applications in this book and mark the sections as *excursions*.

The usual path from problem to solution is long and can be divided into the following steps:

1. *Mathematical modeling.* The underlying problem is mathematically formulated, and equations are developed (e.g., based on physical laws) that describe the problem. The result of the mathematical model can be a system of linear equations, a differential equation, or a function whose roots are sought.
2. *Analysis of the model.* The mathematical model must be examined for its properties: Does a solution exist? Is this solution unique? Does the solution continuously depend on the data? Here, all sub-areas of mathematics are used, from analysis, linear algebra, statistics, group theory, functional analysis, to the theory of partial differential equations.
3. *Development and analysis of numerical methods.* A numerical solution or approximation is developed for the model. The mathematical task is algorithmically prepared in mathematical notation, then formulated in pseudo-code, and then implemented in a programming language. Many mathematical models (such as differential equations) cannot be solved exactly, and often a solution, even if it exists, cannot be specified. The solution of a differential equation is a function $f : [a, b] \to \mathbb{R}$. To describe it exactly, the

T. Richter et al., *Introduction to Numerical Mathematics*, Mathematics Study Resources 25, https://doi.org/10.1007/978-3-662-72546-7_1

function value would have to be determined at the infinitely many points of the interval $[a, b]$. However, any computer realization will only be able to perform a finite number of steps. The problem must first be *discretized*, i.e., reduced to a finite-dimensional task. In addition to algorithm development, numerical mathematics deals with the analysis of numerical methods, i.e., the examination of convergence and approximation errors, as the discretized solution increasingly approaches the infinite-dimensional solution (by adding more points in $[a, b]$).

4. *Implementation.* The numerical method must be implemented on a computer. An *efficient implementation* requires good knowledge of a programming language (for example, C++, MATLAB, Python, Octave, Java, Julia, Fortran, etc.). An *efficient program* requires the development of special algorithms. To utilize modern computer architectures, the method must be modified, for example, for the use of parallel computers. At the same time, there is interest in developing *robust algorithms*. That is, for changes in geometry, interval length, or various problem parameters, the algorithm should always operate similarly reliably.

5. *Evaluation.* The numerical results must be evaluated. This includes a graphical or tabular representation of the results, as well as in technical applications, for example, the evaluation of forces that are only indirectly given.

6. *Interpretation.* The quality of the results must be assessed based on plausibility analyses. Such analyses are either made in comparison to known (analytical) solutions or experimental data; known as validation. Purely numerical analyses in the absence of analytical or experimental data using computer-aided convergence analyses are also possible, known as verification, and we provide the corresponding tools in this book.

The aspects mentioned above must not be seen separately from each other. An efficient algorithm is worthless if the underlying problem has no well-defined solution at all, and a numerical method for an application problem is worthless if the computer does not come to a solution in a foreseeable time.

Furthermore, the above list is usually not processed sequentially; the evaluation of the results can lead to a review of the underlying mathematical models and modified numerical methods. We are thus dealing with a constant interplay of modeling, analysis, numerical approximation, and application.

Ultimately, with the steps discussed above, we lay the foundation for *Scientific Computing*, which has established itself as the third pillar alongside experiment and theory. Numerical mathematics has traits of an experimental science, in which we experiment with algorithms and computer programs, taking advantage of clearly defined mathematical structures.

1.1 Concepts of Numerical Mathematics

Since the goals and procedures in numerical mathematics differ significantly from other disciplines, in this section, we introduce the essential concepts that are characteristic and which we will encounter again and again in all subsequent chapters of this book. In the following, we discuss seven central concepts using examples and definitions.

1.1.1 Approximation

Typically, numerical mathematics involves the computation of mathematical problems and the provision of specific solutions. For many problems, we know that a solution exists, but we do not know a method to calculate this solution. An example is the search for the roots of a general polynomial

$$p(x) = a_0 + a_1 x + a_2 x^2 + a_3 x^3 + \ldots + a_n x^n.$$

The Fundamental Theorem of Algebra [5] states that every (non-constant) polynomial has at least one (possibly complex) root. However, for polynomials of degree $n \geq 5$, there is no analytical solution formula. We know that a solution exists, but we do not know a method to calculate it. This is where numerical mathematics comes into play. We will not be able to solve this problem *exactly*, but we can develop methods to *approximate* a solution.

Approximations are also central in other areas of mathematics. The *Weierstrass Approximation Theorem* states that any continuous function on an interval $I = [a, b]$ can be approximated arbitrarily well by a polynomial. This concept of approximation is closely related to the concept of *convergence*: There is a sequence of polynomials $p_n \in P_n$, which converges to a given function $f \in C[a, b]$:

$$\max_{x \in [a,b]} \| f(x) - p_n(x) \| \to 0 \quad (n \to \infty)$$

In numerical mathematics, we will essentially replace the concept of a solution with the concept of a sufficiently accurate approximation. We consider a problem, for example, the approximation of a continuous function by a polynomial, as solved if we manage to determine a polynomial $p_n \in P_n$ such that the error between $f(x)$ and $p_n(x)$ in $I = [a, b]$ is sufficiently small. What does sufficiently mean? We cannot answer this question. The step from the Weierstrass Approximation Theorem to the specific indication of an approximating polynomial is still long. If the solution must be specified concretely, it is far from sufficient to know that such a solution exists. Instead, we must find ways to construct this solution.

If mathematical problems cannot be solved analytically or if the analytical solution cannot be calculated in finite time, the solution must be **approximated**. In many cases, we will accept a **numerical approximation** as a solution if it is sufficiently accurate.

A simple example is the number π. There are numerous formulas for its approximation. Leibniz published the following series for the approximation of π (the series was known much earlier and was used by the Indian mathematician Madhava):

$$\frac{\pi}{4} = 1 - \frac{1}{3} + \frac{1}{5} - \frac{1}{7} + \frac{1}{9} \pm \cdots \tag{1.1}$$

It is based on a series expansion of $y = \arctan(x)$ as well as the relationship $\arctan(1) = \pi/4$. If we want to use this series to approximate π, we must stop the calculation at some point. We define the finite sum Π_n

$$\pi \approx \Pi_n := 4 \left(\sum_{k=1}^{n} \frac{(-1)^{k+1}}{2k - 1} \right) \tag{1.2}$$

as an approximation.

1.1.2 Convergence

Analysis provides us with methods for investigating the *convergence* of the sequence (1.2). It holds

$$\Pi_n \to \pi \quad (n \to \infty),$$

since it is an alternating series whose terms form a null sequence. Convergence is a central concept in analysis and plays a vital role in numerical mathematics.

Convergence is a qualitative term. The convergence of a sequence $(a_n)_{n\in\mathbb{N}}$ towards a limit $a_n \to a$ for $n \to \infty$ states that for every positive value $\epsilon > 0$ there is an index $n_0 \in \mathbb{N}$ such that all further sequence elements a_n for $n \geq n_0$ are closer to the limit a than ϵ, so $|a - a_n| < \epsilon$ for all $n \geq n_0$. The limit a is often the sought-after solution of the problem in the context of numerical mathematics. The sequence $(a_n)_{n\in\mathbb{N}}$ is the result of a numerical approximation procedure.

The Leibniz series converges to the desired solution, the number π. Although this statement is important, it offers little practical help in numerical mathematics.

1.1.3 Errors

How can we determine the number π with an error of 1%? At what point can we stop the calculation of the series? It holds

$$\Pi_1 = 4, \ \Pi_2 \approx 2.67, \ \Pi_3 \approx 3.47, \ \Pi_4 \approx 2.90, \dots, \Pi_{31} \approx 3.17, \ \Pi_{32} \approx 3.11, \dots$$

and finally, for the relative error, the estimate

$$\frac{|\Pi_{32} - \pi|}{\pi} \approx 0.99\%.$$

This result was not foreseeable, and the *error estimate* is only possible here because we already know the number π we are looking for.

> Numerical mathematics can be considered the mathematics of **errors**. Numerical mathematics is not wrong, inexact, or imprecise. However, in the analysis of numerical methods, we must take into account **errors**. When we solve problems from applications, measurement errors occur. When we solve problems with the computer, calculation errors occur, for example, due to a finite representation of numbers. When we approximate solutions, an error occurs when a theoretically infinite process is terminated after a finite number of steps.

1.1.4 Error Estimates

In general, in numerical mathematics, we naturally try to solve problems whose solution we do not yet know. The term *error estimate* explains another critical concept of numerical mathematics. If we have created a numerical approximation Π_n for a problem with (unknown) solution π, we must ensure that the error between the approximation and the exact solution is small. Such an *error estimator* should be computable. We are looking for a computable bound for the error, i.e., a formula that guarantees

$$|\Pi_n - \pi| \le E_n.$$

> An **error estimator** is a (computable) formula that provides an estimate for the error, which, for example, arises from the termination of an infinite process after a finite number of steps. An **error estimator** can also be a computable bound for the error that arises from rounding on the computer or is given by faulty data. A goal of **error estimators** is always to assess the quality of numerical approximations. **Error estimators** and **error control** are essential features of numerical mathematics.

Using the example of the Leibniz series, we can derive a corresponding error estimator by finding an idea in analysis: The Leibniz series is a so-called *alternating series*, whose summands form a *monotonically decreasing null sequence*. It holds:

$$\Pi_n = 4 \sum_{k=1}^{n} (-1)^{n+1} a_k, \quad a_k := \frac{1}{2k-1}$$

For such a series, it always holds that the limit lies between two consecutive partial sums. Applied to this example, this means that

$$\Pi_{2n} \leq \pi \leq \Pi_{2n+1}.$$

From this estimation, we can derive an error estimator. It holds

$$0 \leq \pi - \Pi_{2n} \leq \Pi_{2n+1} - \Pi_{2n} = \frac{4}{2n+1} \tag{1.3}$$

and to ensure an error of 1%, it must hold for $n \in \mathbb{N}$:

$$\frac{4}{2n+1} < \frac{1}{100} \quad \Leftrightarrow \quad n > 100$$

For $\Pi_{2n} = \Pi_{200}$, we can therefore guarantee that we do not exceed an error of 1% (without knowing π). It holds

$$\Pi_{200} - \pi \approx 0.0050 < 0.01 = 1\%.$$

This error estimator is not sharp. This means that the desired accuracy is achieved much earlier, already for $n = 50$ it holds

$$\Pi_{101} - \pi \approx 0.0099 < 0.01 = 1\%.$$

But the estimate is *robust*, i.e., it guarantees us a corresponding approximation quality, even if we do not yet know the desired result.

1.1.5 Convergence Speed

A modified series for the approximation of π is

$$\pi \approx \Pi_n := \sqrt{12} \sum_{k=1}^{n} \frac{(-3)^{-k+1}}{2k-1} \tag{1.4}$$

For the corresponding partial sums Π_n it holds:

$$\Pi_1 \approx 3.46, \quad \Pi_2 \approx 3.08, \quad \Pi_3 \approx 3.16, \quad \Pi_4 \approx 3.14, \ldots$$

After only four steps, an error of less than 1% is reached. This approximation seems much more powerful and faster than the previous one in (1.2) and better suited for a numerical approximation.

> To assess and compare different methods, the **convergence speed** is introduced: Does a sequence $(a_n)_{n\in\mathbb{N}}$ converge faster or slower than a second sequence $(b_n)_{n\in\mathbb{N}}$ towards the same limit? Often, we use the term **convergence order** for the classification of different methods: Does the error halve in each step of the approximation? Does a method converge linearly or quadratically?

1.1.6   Efficiency

In numerical mathematics, we do not just design any algorithms for solving problems, but these must also be *efficient*. This is a requirement that is not posed in many areas of mathematics. But we now set ourselves the task of approximating the number π to 10 significant digits as efficiently as possible. Using the error estimation (1.3), we can derive a relationship between the desired relative error $\epsilon = 10^{-10}$ and the number of steps $2n_\epsilon$ for the first Leibniz series:

$$\frac{|\pi - \Pi_{2n_\epsilon}|}{\pi} < \frac{1}{(2n_\epsilon - 1)\pi} \overset{!}{<} \epsilon \quad \Leftrightarrow \quad n_\epsilon > \frac{1 + \pi\epsilon}{2\epsilon\pi}$$

Roughly speaking, to gain one digit, we have to perform 10 times more steps. To determine the first 10 digits of π, we have to perform about $10^{10} = 10\,000\,000\,000$ steps with the Leibniz series. Even modern computers take a while to compute this task. So this approach may be *robust*, but it is certainly not *efficient*. The modified Leibniz series is also the series of an alternating null sequence, and we derive a corresponding error estimate:

$$0 \leq \pi - \Pi_{2n} \leq \Pi_{2n+1} - \Pi_{2n} = \sqrt{12}\frac{(-3)^{-2n}}{4n + 1}$$

Now, in a strong simplification, we have

$$\frac{|\pi - \Pi_{2n_\epsilon}|}{\pi} < \sqrt{12}\frac{(-3)^{-2n}}{(4n + 1)\pi} \overset{!}{<} \epsilon \quad \Leftrightarrow \quad n_\epsilon > \frac{\log\left(\frac{4\pi\epsilon}{\sqrt{12}}\right)}{2\log\left(\frac{1}{3}\right)}. \tag{1.5}$$

For $\epsilon = 10^{-10}$ we get

$$n_\epsilon > 10.$$

The modified sequence, therefore, seems to converge much faster.

We call a numerical method **efficient** if it solves or approximates a mathematical problem "resource-saving". Important "resources" include not only *computing time* but also *memory usage*. The term **efficiency** is usually a relative term. A method is **more efficient** or **less efficient** than an alternative method. For most mathematical problems, it is not known which method is the **most efficient**.

Srinivasa Ramanujan has a series for approximating π:

$$\Pi_n := \frac{9801}{2\sqrt{2}} \left(\sum_{k=0}^{n} \frac{(4k)!(1103 + 26390k)}{(k!)^4 396^{4k}} \right)^{-1} \to \pi \quad (n \to \infty).$$

For the first three approximations, we have:

$$\Pi_0 \approx \mathbf{3.141592}730013305660313996$$
$$\Pi_1 \approx \mathbf{3.14159265358979}3877998905$$
$$\Pi_2 \approx \mathbf{3.14159265358979323846264}9$$

The correct digits are underlined. About eight correct digits are added in each iteration. Instead of 10 000 000 000 steps, only three steps are necessary here. Each step in this procedure is, of course, much more complex. Instead of a division and an addition in the Leibniz series, two factorials, a power, and some multiplications have to be performed. But even if we take into account this greater effort, this procedure is by far *more efficient*.

1.1.7 Stability

To explain the concept of *stability*, we turn to another critical problem in numerical mathematics, namely the calculation of eigenvalues of a matrix $A \in \mathbb{R}^{n \times n}$. These are given as roots of the characteristic polynomial:

$$\det(\lambda I - A) = 0$$

For example, let the matrix $A \in \mathbb{R}^{5 \times 5}$ have the eigenvalues $1, 2, 3, 4, 5$. The characteristic polynomial, therefore, has the form

$$p(\lambda) = \det(\lambda I - A) = \prod_{i=1}^{5}(\lambda - i) = \lambda^5 - 15\lambda^4 + 85\lambda^3 - 225\lambda^2 + 274\lambda - 120.$$

We must assume that a computer will not be able to determine the polynomial exactly. The coefficients will be provided with a (small) error. We consider a small perturbation ϵ and the polynomial

Table 1.1 The roots of a perturbed polynomial for the calculation of the eigenvalues of a matrix. Even with a relative error of 0.01%, the error in the eigenvalues is almost 5%. With a perturbation of the polynomial coefficient of only 0.08%, the polynomial already has complex roots

ϵ	λ_1	λ_2	λ_3	λ_4	λ_5	rel. error
0	1	2	3	4	5	0
$1 \cdot 10^{-4}$	1.00	1.97	3.13	3.82	5.08	4.5%
$2 \cdot 10^{-4}$	1.00	1.95	3.38	3.52	5.14	12%
$3 \cdot 10^{-4}$	1.00	1.93	$3.43 + 0.3\mathrm{i}$	$3.43 - 0.3\mathrm{i}$	5.21	18%

$$p_\epsilon(\lambda) = \lambda^5 - 15(1 + \epsilon)\lambda^4 + 85(1 - \epsilon)\lambda^3 - 225(1 + \epsilon)\lambda^2 + 274(1 - \epsilon)\lambda - 120(1 + \epsilon).$$

For different perturbations, we determine the roots of this polynomial in Table 1.1 (we only give the first three significant digits) and compute the maximum relative error in each case.

At $\epsilon = 10^{-4}$, i.e., a relative error of the coefficients of only 0.01%, the error in the eigenvalues is already almost 5%. The error is amplified by a factor of 500. With an error of only 0.03%, complex eigenvalues already occur. An essential structural property of, for example, symmetric matrices is no longer preserved. The calculation of eigenvalues via the characteristic polynomial is not a *numerically stable* algorithm.

The *stability* is probably the most important and also the challenging concept in numerical mathematics. There is no uniform definition for this. However, Hadamard gave a common explanation for the well-posedness of a mathematical problem in 1921 and demanded: existence, uniqueness, and continuous dependence of the solution on the problem data. The latter point is a form of stability: How strongly does the solution change when the input data are changed? Mathematical problems that violate one of the three Hadamard criteria are called *ill-posed*. But we must immediately question this: The complex roots of a polynomial depend continuously on the coefficients, thus fulfilling the stability criterion according to Hadamard. This apparent contradiction is again an example of the interplay between qualitative and quantitative statements, which are so crucial in numerics. Stability, according to Hadamard, is a qualitative term. In numerics, however, we are primarily interested in quantitative relationships. It is not enough for us to know that a continuous dependence exists; we want to know how strongly the perturbation affects the error. The concept of *stability* is always applied when, in numerical mathematics, the general thought patterns no longer apply:

- In analysis, we prove the convergence of sequences to their limit $a_n \to a$. But how good is the approximation for *small n*? We will call numerical methods *stable* if they also provide *meaningful results* for small n, i.e., far away from convergence. What is *meaningful* must be distinguished on a case-by-case basis.

- Simple mathematical laws such as the distributive law no longer apply on the computer. Unavoidable rounding errors lead, depending on the calculation path, to different results. We will call numerical methods stable if they are robust despite erroneous data and erroneous computer implementation.

> The term **stability** describes the behavior of numerical methods in their practical implementation. Is the method **robust** against errors? How do convergent processes behave when they terminate after a finite number of iterations?

All the terms introduced here as *concepts* have in common that, unlike in most areas of mathematics, they cannot be clearly defined. We speak of large and small errors, but we cannot give clear boundaries for when an error is large or small. An exception is convergence, which is well known from analysis and already represents one of the essential concepts there. Its definition and idea fulfill the same purpose in numerical mathematics. The vagueness of many statements in numerical mathematics does not make the latter easier than other disciplines, just because possibly less accurate results are acceptable. On the contrary, since we know that errors constantly occur, it is essential to describe these in definitions, theorems, and proofs clearly. Moreover, in practice (in computer simulations), it is even crucial to control the occurring errors to avoid an unwanted termination of the algorithm, incorrect results, or to improve efficiency. Numerous examples in this book demonstrate the necessity for careful investigation of the *errors* in numerical mathematics.

1.2   Definition of an Algorithm

The result of a numerical task is generally a number or a finite set of numbers. Examples of numerical tasks include calculating the roots of a function, evaluating integrals, finding the derivative of a function at a point, and solving more complex problems such as differential equations. To solve numerical problems, we will learn about different methods. We first define:

Definition 1.1 (*Numerical Method, Algorithm*) A numerical method is a procedure for systematically solving or approximating a mathematical problem. A numerical method is called *direct* if the solution can be calculated exactly, except for rounding errors. A method is called *approximative* if the solution can only be approximated. A method is called *iterative* if the approximation is gradually improved by repeatedly executing a procedure.

Examples of direct solution methods are the factorization formula for calculating the roots of quadratic polynomials (see Chap. 8) or the Gaussian elimination algorithm (Chap. 3) for solving linear systems. Approximative methods must be used, for example, to determine complicated integrals or to calculate the roots of general functions. Often, an exact solution to problems is not appropriate. If, for example, the input data are fraught with significant measurement errors, an exact solution is not necessary.

Polynomial Evaluation and Horner's Method

We consider an example and its analysis of a direct method in detail:

Example 1.2 (Polynomial evaluation) Let

$$p(x) = a_0 + a_1 x + \cdots + a_n x^n$$

be a given polynomial. Here, $n \in \mathbb{N}$ is very large. We evaluate $p(x)$ at a point $\xi \in \mathbb{R}$ using the trivial method:

Algorithm 1.3: Polynomial Evaluation

> **Input:** Polynomial $p(x) = a_0 + a_1 x + \cdots + a_n x^n$, evaluation point $\xi \in \mathbb{R}$.
> 1 **for** $i = 1$ **to** n **do**
> 2 $y_i = a_i \xi^i$
> 3 $p = p + y_i$
> **Result:** The evaluated polynomial $p = p(\xi)$.

We calculate the *effort* for polynomial evaluation: In step 3 of the algorithm, i multiplications are necessary to calculate the y_i, in total

$$0 + 1 + 2 + \cdots + n = \frac{n(n + 1)}{2} = \frac{n^2}{2} + \frac{n}{2}.$$

In step 4, another n additions are necessary. The total effort of the algorithm is therefore $n^2/2 + n/2$ multiplications and n additions. We combine an addition and a multiplication into one *elementary operation* and obtain the effort of the trivial polynomial evaluation as

$$A_1(n) = \frac{n^2}{2} + \frac{n}{2}$$

elementary operations. We now rewrite the polynomial using the distributive law

$$p(x) = a_0 + x(a_1 + x(a_2 + \cdots + x(a_{n-1} + xa_n)\ldots)) \tag{1.6}$$

and derive a second algorithm from this:

Algorithm 1.4: Horner's Method

> **Input:** Polynomial $p(x) = a_0 + a_1 x + \cdots + a_n x^n$, evaluation point $\xi \in \mathbb{R}$.
> 1 $p = a_n$
> 2 **for** $i = n - 1$ **to** 0 **do**
> 3 $p = a_i + \xi \cdot p$
> **Result:** The evaluated polynomial $p = p(\xi)$.

Each of the n steps of the method requires one multiplication and one addition, so the effort is

$$A_2(n) = n$$

elementary operations. The *Horner's method* requires significantly fewer operations for the same task. Consider a polynomial of degree $n = 1000$ and compare $A_1(1000) \approx 500\,000$ and $A_2(1000) = 1000$. ◀

Example 1.5 (Horner's Method) We calculate the value of the polynomial $p(x) = 3x^5 - 2x^4 + 1$ at the point $\xi = 2$. The coefficients are $a_5 = 3, a_4 = -2, a_3 = a_2 = a_1 = 0, a_0 = 1$. Then we get the scheme:

$$
\begin{array}{c|cccccc}
i & 5 & 4 & 3 & 2 & 1 & 0 \\
a_i & 3 & -2 & 0 & 0 & & 1 \\
\hline
p & 3 & 4 & 8 & 16 & 32 & \underline{65}
\end{array}
$$

Thus, $p(2) = 65$. In Python, we can implement Horner's method as follows:

♣ **Implementation 1.6: Horner's Method**

```python
def horner(coeffs, x):
    n = len(coeffs)
    p = coeffs[-1]
    for i in range(n - 1, 0, -1):
        p = coeffs[i - 1] + x * p
    return p
```

This code is available in the jupyter notebooks respository.[1] We apply the function to the same coefficients

```python
horner([1, 0, 0, 0, -2 ,3], 2)
```

and thus obtain the output

```
65
```
◀

Horner's method can be naturally generalized to calculate the derivatives of the polynomial. Thus, we obtain the following algorithm for calculating the derivatives $p^{(k)}(\xi)$ of a polynomial:

---

[1] See the subdirectory EN in the repository https://github.com/sn-code-inside/EinfNumMath-RWW.

Algorithm 1.7: Complete Horner's Method

Input: polynomial $p(x) = a_0^{(0)} + a_1^{(0)}x + \cdots + a_n^{(0)}x^n$, evaluation point $\xi \in \mathbb{R}$.

1 $a_n^{(1)} = a_n^{(0)}, a_n^{(2)} = a_n^{(0)}, \ldots, a_n^{(n+1)} = a_n^{(0)}$
2 **for** $k = 0$ **to** $n - 1$ **do**
3   **for** $j = n - 1$ **to** k **do**
4          $a_j^{(k+1)} = a_j^{(k)} + a_{j+1}^{(k+1)}\xi$
5      $p^{(k)}(\xi) = k! \cdot a_k^{(k+1)}$

Result: The evaluations of the polynomial and its derivatives $p^{(k)}(\xi)$, for $k = 0, \ldots, n - 1$

This algorithm can be illustrated using the following scheme:

i	n	$n-1$	$n-2$	$\cdots$	2	1	0
	$a_n^{(0)}$	$a_{n-1}^{(0)}$	$a_{n-2}^{(0)}$	$\cdots$	$a_2^{(0)}$	$a_1^{(0)}$	$a_0^{(0)}$
$p_i(\xi)$	$a_n^{(1)}$	$a_{n-1}^{(1)}$	$a_{n-2}^{(1)}$	$\cdots$	$a_2^{(1)}$	$a_1^{(1)}$	$a_0^{(1)}$
$p_i'(\xi)$	$a_n^{(2)}$	$a_{n-1}^{(2)}$	$a_{n-2}^{(2)}$	$\cdots$	$a_2^{(2)}$	$a_1^{(2)}$	
$\vdots$	$\vdots$	$\vdots$	$\vdots$	$\ddots$			
$p_i^{(n-2)}(\xi)$	$a_n^{(n-1)}$	$a_{n-1}^{(n-1)}$	$a_{n-2}^{(n-1)}$				
$p_i^{(n-1)}(\xi)$	$a_n^{(n)}$	$a_{n-1}^{(n)}$					
$p_i^{(n)}(\xi)$	$a_n^{(n+1)}$						

We continue the above example:

Example 1.8 (Complete Horner's Method) We calculate the values of the derivatives of the polynomial $p(x) = 3x^5 - 2x^4 + 1$ at the point $\xi = 2$. For this, we obtain the following scheme:

i	5	4	3	2	1	0	
a_i	3	-2	0	0	0	1	
p_i	3	4	8	16	32	65	$\Rightarrow p(\xi) = 0! \cdot 65 = 65$
p_i'	3	10	28	72	176		$\Rightarrow p'(\xi) = 1! \cdot 176 = 176$
p_i''	3	16	60	192			$\Rightarrow p''(\xi) = 2! \cdot 192 = 386$
$p_i^{(3)}$	3	22	104				$\Rightarrow p^{(3)}(\xi) = 3! \cdot 104 = 624$
$p_i^{(4)}$	3	28					$\Rightarrow p^{(4)}(\xi) = 4! \cdot 28 = 672$
$p_i^{(5)}$	3						$\Rightarrow p^{(5)}(\xi) = 5! \cdot 3 = 360$

In Python, we can implement the complete Horner's method using the `factorial` function from the built-in library `math` and the `array` data structure from the library `numpy` to store the resulting matrix.

♣ Implementation 1.9: Complete Horner's Method

```python
import numpy as np
from math import factorial

def horner_complete(coeffs, x):
    n = len(coeffs)
    p = np.zeros(n)
    a = np.zeros((n + 1, n))
    a[:, -1] = coeffs[-1]
    a[0, :] = coeffs

    for k in range(n):
        for j in range(n - 1, k, -1):
            a[k + 1, j - 1] = a[k, j - 1] + a[k + 1, j] * x
        p[k] = factorial(k) * a[k + 1, k]
    return p
```

The output is now an `array`, in which the values of the i-th derivative are stored in the i-th entry of the array. It is important to note that in Python, indices always start at 0.

Applying this code to the same coefficients

```python
horner_complete([1, 0, 0, 0, -2, 3], 2)
```

gives the output

```python
array([ 65., 176., 384., 624., 672., 360.])
```

so an `array` with the function value and the values of all derivatives.   ◀

1.2.1   Numerical Effort and Landau Symbols

In the following, we introduce a general notation for describing the effort of a method.

Definition 1.10 (*Numerical Effort in Elementary Operations*)   The *effort* of a numerical method is the number of necessary elementary operations. An *elementary operation* is composed of one addition as well as one multiplication.

Usually, the effort of a method depends on the problem size. The problem size $N \in \mathbb{N}$ is defined from problem to problem. When solving a linear system $Ax = b$ with a matrix $A \in \mathbb{R}^{N \times N}$, the size of the matrix N is the problem size. When evaluating a polynomial $p(x) = a_n x^n + a_{n-1} x^{n-1} + \cdots + a_0$ at a point x, the problem size is the degree n of the polynomial. For an easier notation, we define:

Definition 1.11 (*Landau Symbols*) *(i)* Let $g(n)$ be a function with $g \to \infty$ for $n \to \infty$. Then $f \in O(g)$ if and only if

$$\limsup_{n \to \infty} \left| \frac{f(n)}{g(n)} \right| < \infty,$$

and $f \in o(g)$ if and only if

$$\lim_{n \to \infty} \left| \frac{f(n)}{g(n)} \right| = 0.$$

(ii) Let $g(h)$ be a function with $g(h) \to 0$ for $h \to 0$. We define $f \in O(g)$ and $f \in o(g)$ as above.

Simply put: $f \in O(g)$ if f converges to ∞ at most as fast as g, and $f \in o(g)$ if g goes faster than f to ∞. Correspondingly, for $g \to 0$, $f \in O(g)$ if f converges to zero at least as fast as g, and $f \in o(g)$ if f converges to zero faster than g. The effort of a method can be characterized more easily using the Landau symbols.

Example 1.12 (Landau Symbols) *(i)* In the case of the trivial polynomial evaluation in Algorithm 1.3, the effort $A_1(n)$ depends on the polynomial size n

$$A_1(n) \in O(n^2)$$

and in the case of Horner's method from Algorithm 1.4

$$A_2(n) \in O(n).$$

We say: The effort of the trivial polynomial evaluation grows quadratically; the effort of Horner's method grows linearly. Furthermore, with the Landau symbols, convergence concepts can be quantified and compared. We will return to this point in the corresponding chapters.

(ii) A numerical method provides for the parameter $h > 0$ the approximation $\bar{a} \approx a(h)$ with $a(h) \to \bar{a}$ for $h \to 0$. We say that the method *converges linearly* if

$$\|\bar{a} - a(h)\| = O(h)$$

and converges quadratically in the case

$$\|\bar{a} - a(h)\| = O(h^2).$$

Generally, we speak of the convergence of order α if

$$\|\bar{a} - a(h)\| = O(h^\alpha).$$

In the case $\|\bar{a} - a(h)\| = o(h)$ we call the convergence *superlinear*.   ◀

Remark 1.13 (Landau Symbols) The symbol $O(g)$ describes a set of functions. It is

$$f \in O(g),$$

if $f(n)/g(n) \leq C$ remains bounded for $n \geq n_0$. Often, we will use the equality sign when using the Landau symbols, i.e., we write

$$f = O(g).$$

This does not mean actual equality. The implication

$$f_1 = O(g), \quad f_2 = O(g) \quad \Rightarrow \quad f_1 = f_2$$

is generally wrong. $f = O(g)$ is just a notation for $f \in O(g)$ which is often handy. We can hereby easily calculate with the Landau symbols and write, for example,

$$f(n) = \frac{1}{1 - \frac{1}{n}} = 1 + \frac{1}{n} + O\left(\frac{1}{n^2}\right).$$

The equality holds up to an error of maximum order $O(n^{-2})$. That this is really the case is easy to check with the Taylor expansion of $1/(1 - x)$ for $x = 1/n$.   ◄

1.3   Errors, Error Amplification and Conditioning

Numerical solutions are often fraught with errors. Sources of errors are numerous: The input may be tainted with a measurement error, an approximate method will not calculate exactly, but only approximately, some tasks can only be solved using a computer or calculator, so the solution is accompanied by *rounding errors*. Other sources of errors include discretization errors, interpolation errors, iteration errors, regularization errors, implementation errors (bugs), model errors, data errors, and uncertainties. In numerical mathematics, we always have to assume that inputs, intermediate results, and results are subject to errors. An "error-free calculation" is actually only possible with a lot of luck. This does not make numerical mathematics "wrong mathematics". On the contrary, it is part of numerical mathematics to identify errors and to control their propagation and amplification as much as possible. We define:

Definition 1.14 *(Error)* Let $\tilde{x} \in \mathbb{R}$ be the approximation of a quantity $x \in \mathbb{R}$. With $|\delta x| = |\tilde{x} - x|$ we denote the *absolute error* and with $|\delta x|/|x|$ the *relative error*.

Usually, the consideration of the relative error is of greater importance: An absolute measurement error of 100 m is small when trying to determine the distance between the Earth and the Sun, but large when the distance between the cafeteria and the mathematics building is to be measured.

In the analysis of numerical methods, errors, especially the propagation of errors, play a crucial role. We consider an example:

Example 1.15 (Size Determination) Thomas wants to determine his height h, but does not have a tape measure available. However, he has a clock, a ball, and he paid good attention in physics class. To solve the task, he has two ideas:

Method 1   Thomas drops the ball from head height and measures the time t_0 until the ball hits the ground. The height he calculates from the formula for free fall,

$$y(t) = h - \frac{1}{2}gt^2 \quad \Rightarrow \quad h = \frac{1}{2}gt_0^2,$$

with the gravitational constant $g = 9.81\,m/s^2$.

Method 2   The ball is thrown $2\,m$ above the head and we measure the time until the ball has landed back on the ground. The distance $s = 2\,m$ is covered in $t' = \sqrt{2s/g}$ seconds for which the ball requires a starting speed of $v_0 = gt' = \sqrt{2\,sg} \approx 6.26\,m/s$. For the trajectory it holds:

$$y(t) = h + v_0 t - \frac{1}{2}gt^2 \quad \Rightarrow \quad h = \frac{1}{2}gt_0^2 - v_0 t_0$$

The second method is chosen because the time t_0, which the ball needs to fall in the first method, is very small. Large measurement errors are suspected. In Fig. 1.1, we sketch the procedure in both variants.

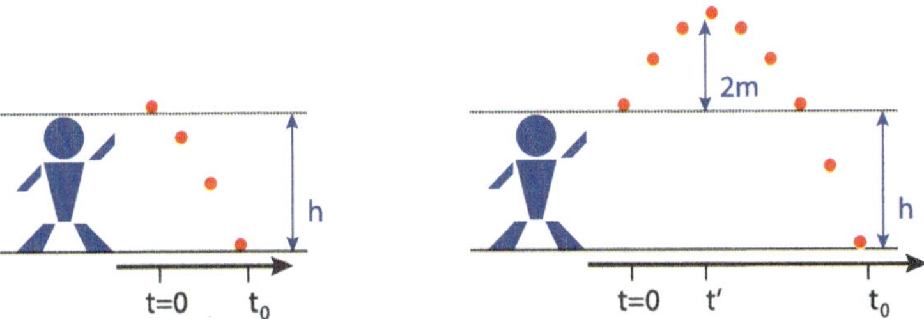

Fig. 1.1 Example 1.15. Two experiments to determine one's own size. Left: A ball is dropped from head height, and we measure the time t_0 until it lands on the ground. Right: To minimize measurement errors, the ball is first thrown 2 m into the air. We again measure the time t_0 until the ball lands on the ground, which is a bit longer. The height h can be derived from the formula for free fall, or for the parabolic trajectory

We first carry out Algorithm 1 and measure 5 times (exact solution $h = 1.80\,\text{m}$ and $t_0 \approx 0.606\,\text{s}$).

Measurement	0.58 s	0.61 s	0.62 s	0.54 s	0.64 s	0.598 s
Measurement error (rel.)	4%	0.5%	2%	10%	6%	1%
Size	1.65 m	1.82 m	1.89 m	1.43 m	2.01 m	1.76 m
Error (abs.)	0.15 m	0.02 m	0.09 m	0.37 m	0.21 m	0.04 m
Error (rel.)	8%	1 %	5%	21%	12%	2%

In the last column, the average of all measurement results was chosen. We now carry out Algorithm 2 and measure 5 times (exact solution $h = 1.80\,\text{m}$ and $t_0 \approx 1.519\,\text{s}$). In the last column, the average of all measurement results is considered again:

Measurement	1.60 s	1.48 s	1.35 s	1.53 s	1.45 s	1.482 s
Measurement error (rel.)	5%	2.5%	11%	$< 1\%$	5%	2%
Size	2.53 m	1.47 m	0.48 m	1.90 m	1.23 m	1.49 m
Error (abs.)	0.73 m	0.33 m	1.32 m	0.10 m	0.57 m	0.31 m
Error (rel.)	40%	18%	73%	6%	32%	17%

◀

Despite initial doubts, Algorithm 1 proves to be better, with smaller errors in the result. Possible sources of errors are the measurement error in the determination of time t_0 and in Algorithm 2, the accuracy in reaching the height of $2\,m$. In Algorithm 1, the measurement error leads to about twice the relative error in the size. In Algorithm 2, even small errors $\leq 1\%$ in the measurements lead to significantly larger errors in the result. The still small error of 5% in the first measurement leads to a result error of about 40%. Even when considering the average of all measured values, the determined size of $1.49\,m$ is not a good approximation. The results of the two methods also show that the relative measurement error of each goes into the relative error in the result by a fixed factor. In method 1, the error is approximately doubled; in method 2, it is approximately multiplied by 8.

We will now analytically investigate this relationship. We first consider a scalar task, i.e., a function $f : \mathbb{R} \to \mathbb{R}$, $f(x) = y$, with the input x and the output y. The disturbed input $\tilde{x} = x + \delta x$ yields a disturbed result $\tilde{y} = y + \delta y$, i.e., $\tilde{y} = f(\tilde{x})$. It holds

$$\delta y = \tilde{y} - y = f(\tilde{x}) - f(x) = f(x + \delta x) - f(x) \tag{1.7}$$

and after expanding with $\delta x \cdot \delta x^{-1}$ and Taylor expansion (we require that the function $f : \mathbb{R} \to \mathbb{R}$ is differentiable), it holds

$$\delta y = \frac{f(x + \delta x) - f(x)}{\delta x} \cdot \delta x = f'(x)\delta x + o(|\delta x|). \tag{1.8}$$

The expression $o(|\delta x|)$ is one of the Landau symbols from Definition 1.11 and means that the error goes to zero faster than linearly in δx. The disturbed input δx is thus essentially amplified or attenuated by the factor $f'(x)$, depending on whether $|f'(x)| > 1$ or $|f'(x)| < 1$. We get the relative error amplification by dividing by $y = f(x)$ and expanding with $x \cdot x^{-1}$

$$\frac{\delta y}{y} = \frac{f'(x)}{f(x)}\delta x + o(|\delta x|) = f'(x)\frac{x}{f(x)} \cdot \frac{\delta x}{x} + o(|\delta x|)$$

We call

$$\kappa := \left| f'(x)\frac{x}{f(x)} \right|$$

the *condition number* of the task, and it describes the relative error amplification in relation to an input size.

Definition 1.16 (*Condition Number*) For the continuously differentiable function $f : x \mapsto y$ the *condition number* is given by

$$\kappa(x) := \left| f'(x)\frac{x}{f(x)} \right|.$$

A task is called *ill-conditioned* for an input x if $|\kappa(x)| \gg 1$, otherwise *well-conditioned*. In the case $|\kappa(x)| < 1$, we speak of *error damping*, otherwise of *error amplification*.

We continue the example of experimental size determination:

Example 1.17 (Size Determination, Continuation of Example 1.15)

Method 1   To perform the method $h(t_0)$, the measured time t_0 is required as input. For the condition number, it holds

$$\kappa_{h,t_0} := \frac{\partial h(t_0)}{\partial t_0}\frac{t_0}{h(t_0)} = g t_0 \frac{t_0}{\frac{1}{2}g t_0^2} = 2.$$

A relative error in the input $\delta t_0/t_0$ can therefore cause a twice as large relative error in the output.

Method 2   We determine the condition number in relation to the time t_0:

$$\kappa_{h,t_0} = (g t_0 - v_0)\frac{t_0}{\frac{1}{2}g t_0^2 - v_0 t_0} = 2\frac{g t_0 - v_0}{g t_0 - 2 v_0}$$

We know that the ball, with exact throw and without measurement error, is in transit for $t_0 \approx 1.519\,s$ at a starting speed of $v_0 \approx 6.26\,m/s$. This results in:

$$\kappa_{h,t_0} \approx \kappa_{h,v_0} \approx 8$$

Errors in the inputs $\delta t_0/t_0$ and $\delta v_0/v_0$ are amplified by a factor of 8. The error amplifications predicted by the condition numbers are presented in the tables for Example 1.15. ◄

The definition of the condition number can be extended to higher-dimensional tasks $A : x \mapsto y$ with $x \in \mathbb{R}^n$ and $y \in \mathbb{R}^m$. The input x is associated with the error

$$\delta x = \delta_j \cdot e_j \in \mathbb{R}^n,$$

where $e_j = (0, \ldots, 0, 1, 0, \ldots, 0) \in \mathbb{R}^n$ is the j-th unit vector and $\delta_j \geq 0$. For the disturbed solution $\tilde{y} = A(\tilde{x})$, the following applies in component $i = 1, \ldots, m$:

$$\delta y_i = \tilde{y}_i - y_i = A_i(\tilde{x}) - A_i(x) = A_i(x + \delta x_j) - A_i(x)$$

As above, we expand with $\delta x_j \cdot \delta x_j^{-1}$ and divide by $y_i = A_i(x)$

$$\frac{\delta y_i}{y_i} = \frac{A_i(x + \delta x_j) - A_i(x)}{\delta x_j} \frac{x_j}{A_i(x)} \frac{\delta x_j}{x_j}.$$

With the Taylor expansion of $A_i(\cdot)$ in the direction of x_j, we get

$$\left| \frac{\delta y_i}{y_i} \right| = \left| \frac{\partial A_i(x)}{\partial x_j} \frac{x_j}{A_i(x)} \right| \cdot \left| \frac{\delta x_j}{x_j} \right| + o(\|\delta x\|).$$

We define:

Definition 1.18 (*Condition Number of Vector-Valued Mappings*) Let $A : \mathbb{R}^n \to \mathbb{R}^m$ be continuously differentiable. We call

$$\kappa_{i,j} := \left| \frac{\partial A_i(x)}{\partial x_j} \frac{x_j}{A_i(x)} \right|, \quad i = 1, \ldots, n, \ j = 1, \ldots, m$$

the *condition numbers* of the function $A(\cdot)$.

1.3.1 Conditioning of Basic Operations

Rounding errors occur in every elementary operation. Therefore, the conditioning of the basic operations, from which all algorithms are built, is of crucial importance:

Example 1.19 (Conditioning of Basic Operations)

1. Addition, subtraction: $A(x, y) = x + y$:

$$\kappa_{A.x} = \left| \frac{x}{x + y} \right| = \left| \frac{1}{1 + \frac{y}{x}} \right|$$

In the case $x \approx -y$, the condition number of the addition ($x \approx y$ for subtraction) can become arbitrarily large. An example with four-digit calculation:

$$x = 1.021, \quad y = -1.019 \quad \Rightarrow \quad x + y = 0.002$$

Now let $\tilde{y} = 1.020$ be disturbed. The relative error in y is very small: $|1.019 - 1.020|/|1.019| \leq 0.1\%$. We get the disturbed result

$$x + \tilde{y} = 0.001,$$

and an error of 100%. The enormous error amplification in the addition of numbers with approximately the same magnitude is called *cancellation*. Because, if the first significant digits of x and y match in the calculation of $x - y$, these are cancelled out in the result: all the first digits are zero and the number becomes smaller in magnitude. Only very few non-zero digits remain to represent the accuracy. Cancellation typically occurs when the result of a numerical operation compared to the input values has a very small magnitude. Small relative errors in the input amplify to large relative errors in the result.

2. Multiplication: $A(x, y) = x \cdot y$. The multiplication of two numbers is always well-conditioned:

$$\kappa_{A,x} = \left| y \frac{x}{xy} \right| = 1$$

3. Division: $A(x, y) = x/y$. The same applies to division:

$$\kappa_{A,x} = \left| \frac{1}{y} \frac{x}{\frac{x}{y}} \right| = 1, \quad \kappa_{A,y} = \left| \frac{x}{y^2} \frac{y}{\frac{x}{y}} \right| = 1$$

4. Square root: $A(x) = \sqrt{x}$:

$$\kappa_{A,x} = \left| \frac{1}{2\sqrt{x}} \frac{x}{\sqrt{x}} \right| = \frac{1}{2}$$

An error in the input is even reduced in the result.

◀

1.4 Representation of Numbers and Its Importance in Numerics

1.4.1 Representations of Numbers

The analysis of errors and error propagation plays a central role in numerical mathematics. Errors occur in many ways, even without inaccurate input values. Often, numerical algorithms are very complex, consisting of many operations. When approximating with the computer, *rounding errors* inevitably occur. Since the memory space in the computer or the number of digits on the calculator is limited, rounding errors already occur in the mere representation of a number. Even if there is a simple solution to the numerical task, such as

$$x^2 = 2, \quad \Leftrightarrow x = \pm\sqrt{2},$$

with $x = \pm\sqrt{2}$, these cannot be represented exactly on a computer:

$$\sqrt{2} = 1.41421356237309504880\ldots$$

The obvious reason for a mandatory representation error is the limited memory of a computer. Another reason lies in efficiency. A computer cannot efficiently compute with arbitrarily long numbers. The reason is the limited data bandwidth (that is, the 8-bit, 16-bit, 32-bit, or 64-bit of the processors). Operations with numbers in longer representation must be composed, similar to the written multiplication or addition from school. Computers store numbers rounded and in binary representation, i.e., in base 2.

$$\text{rd}(x) = \pm \sum_{i=-n_1}^{n_2} a_i 2^i, \quad a_i \in \{0, 1\}. \tag{1.9}$$

The accuracy of the representation and the rounding error depend on the number of digits, i.e., on n_1 and n_2. The *fixed-point representation* of the number in the binary system is:

$$\text{rd}(x) = [a_{n_2} a_{n_2-1} \ldots a_1 a_0 . a_{-1} a_{-2} \ldots a_{-n_1}]_2 \tag{1.10}$$

More practical is the *floating-point representation* of numbers. For this, the binary representation is normalized, and a common exponent is introduced:

$$\text{rd}(x) = \pm \left(\sum_{i=-n_1-n_2}^{0} a_{i+n_2} 2^i \right) 2^{n_2}, \quad a_i \in \{0, 1\}. \tag{1.11}$$

The leading term (the a_i) is called the *mantissa* and we denote it by M, the *exponent* is denoted by E. The exponent can be a positive or negative number. To simplify, the exponent E is written as $E = e - b$ with a positive number $e \in \mathbb{N}$ and the bias $b \in \mathbb{N}$. The bias value b is chosen in a specific number representation and thus determines the largest possible negative exponent. The variable exponent part e is itself stored in binary format. What remains is to choose the number of binary digits for the mantissa and exponent. In addition, there is a bit for the sign $S \in \{+, -\}$. The floating-point representation in the binary system is:

$$\text{rd}(x) = S \cdot [m_0.m_{-1}m_{-2}\ldots m_{-\#m}]_2 \cdot 2^{[e_{\#e}\ldots e_1 e_0]_2 - b} \tag{1.12}$$

The mantissa is usually normalized with $m_0 = 1$ to $M \in [1, 2)$. This means that the leading digit does not need to be explicitly stored. On the computer, the bit sequence remains

$$\text{rd}(x) = [Se_{\#e}\ldots e_1 e_0 m_{-1}m_{-2}\ldots m_{-\#m}]_2, \tag{1.13}$$

For the interpretation, we need to know the number of bits in the mantissa and exponent, as well as the bias.

On modern computers, the *IEEE-754* format [57][2] for storing numbers has been established. The basic idea of the format is the binary floating-point representation according to (1.12), but several tricks are used to make the most optimal use of the available bits for mantissa and exponent. For understanding numerical mathematics, these are not essential. However, it is important to realize that only a limited accuracy is available (given by the number of digits in the mantissa) and that there are the largest and the smallest number that can be represented (determined by the number of digits in the exponent and by the bias b).

In Table 1.2, various floating-point representations are summarized, which are either historical or refer to current formats. Currently, the IEEE format is used almost exclusively. So far, two representations were common, *single-precision* (in C++ float) and *double-precision* (in C++ double). The arithmetic units of modern computers internally use an increased accuracy for performing elementary operations (80 bits in modern Intel CPUs). Rounding is only done after the result is calculated.

Historically, different number representations were chosen in computing systems. While the first computers still had processor-integrated arithmetic units for floating-point arithmetic, the so-called *floating-point processing unit* (FPU), this initially disappeared again from the usual computers and was only present in special computers. In the form of *coprocessors*, an FPU could be retrofitted (e.g., the Intel 8087 to the Intel 8086). The "486" was the first processor for home computers with an integrated FPU.

Today, floating-point calculations can be efficiently outsourced to graphics cards. The processors of the graphics cards, the *graphics processing units* (GPU) are specifically designed for such calculations (e.g., fast calculation of light refractions and reflections, mapping of patterns onto 3D surfaces). Special plug-in cards, which contain several GPUs, are used in high-performance systems. These accelerator cards were initially developed for computer graphics applications, and today they play a major role in *machine learning* and *artificial intelligence*. In both cases, speed is more important than accuracy. Therefore, calculations with *half-precision* numbers have become established.

Example 1.20 (Floating Point Representation) We assume a four-digit mantissa and four digits in the exponent with bias $2^{4-1} - 1 = 7$, so $\#m = \#e = 4$.

- The number $x = -96$ has a negative sign, so $S = 1$. The binary representation of 96 according to (1.9) is

$$96_{10} = 1 \cdot 64 + 1 \cdot 32 + 0 \cdot 16 + 0 \cdot 8 + 0 \cdot 4 + 0 \cdot 2 + 0 \cdot 1 = 1100000_2,$$

and normalized according to (1.11) and (1.12), we get

---

[2] See also the Wikipedia page on the IEEE-754 format https://de.wikipedia.org/wiki/IEEE_754.

Table 1.2 IEEE 754 format in single and double precision as well as floating point formats in current and historical hardware. Modern accelerator hardware has various modes and can (partially simultaneously) calculate with different number models

	Size	Sign	Exponent	Mantissa	Bias
Single precision	32 Bit	1 Bit	8 Bit	23+1 Bit	127
Double precision	64 Bit	1 Bit	11 Bit	52+1 Bit	1023
Zuse Z1 (1938)	24 Bit	1 Bit	7 Bit	15 Bit	–
IBM 704 (1954)	36 Bit	1 Bit	8 Bit	27 Bit	128
i8087 Coprocessor (1980)	First use of IEEE (single + double)				
Intel 486 (1989)	First integrated FPU in standard PC				
Nvidia Tesla A100 (2020) FP 32	32 Bit	1 Bit	8 Bit	23 Bit	127
Nvidia Tesla A100 (2020) TF 32	19 Bit	1 Bit	8 Bit	10 Bit	127
Nvidia Tesla A100 (2020) FP 16	16 Bit	1 Bit	5 Bit	10 Bit	15

$$96_{10} = 1.1000_2 \cdot 2^{6_{10}} = 1.1_2 \cdot 2^{13_{10} - 7_{10}} = 1.1_2 \cdot 2^{1101_2 - b}.$$

As a floating point representation in the format according to (1.13) we get

$$x = [111011000]_2,$$

where $s = 1$, $e = 1101_2$ and $m = 1000_2$.

- The number $x = -384$ again has a negative sign and $S = 1$. The binary representation of 384 is

$$384_{10} = 1 \cdot 256 + 1 \cdot 128 + 0 \cdot 64 + 0 \cdot 32 + 0 \cdot 16 + 0 \cdot 8 + 0 \cdot 4 + 0 \cdot 2 + 0 \cdot 1$$
$$= [110000000]_2,$$

which, in normalized form, gets

$$384_{10} = 1.1000 \cdot 2^{8_{10}} = 1.1000 \cdot 2^{15_{10}-7_{10}} = 1.1000_2 \cdot 2^{1111_2-b}.$$

All bits in the exponent are 1. In the IEEE-754 format, however, this special exponent $e = 1111_2$ is reserved for storing *NaN* (not a number). The number 384 is too large to be stored in this number format. Instead, $-\infty$ or binary $[111110000]_2$ is stored.

- The number $x = 1/3 = 0.33333\ldots$ is positive, so $S = 0$. The binary representation of $1/3$ is

$$\frac{1}{3} = 0.01010101\ldots_2.$$

Normalized with a four-digit mantissa, we get

$$\frac{1}{3} \approx 1.0101_2 \cdot 2^{-2} = 1.0101_2 \cdot 2^{5_{10}-7_{10}} = 1.0101_2 \cdot 2^{0101_2-b},$$

and hence, the binary representation is 001010101_2. The number is rounded. If we transform it back to the decimal system, we get

$$1.0101_2 \cdot 2^{0101_2-7} = 1.3125 \cdot 2^{-2} = 0.328125$$

with the relative error

$$\left| \frac{\frac{1}{3} - 1.0101_2 \cdot 2^{0101_2-b}}{\frac{1}{3}} \right| \approx 0.016.$$

- The number $x = \sqrt{2} \approx 1.4142135623\ldots$ is positive, so $S = 0$. Rounded, we have:

$$\sqrt{2} \approx 1.0111_2$$

This number is already normalized. With bias $b = 7$, the exponent $e = 0 = 7 - 7$

$$\sqrt{2} \approx 1.0111_2 \cdot 2^{0111_2-b},$$

thus $\sqrt{2} \approx 001110111_2$. Reverting to the decimal representation, we obtain

$$1.0111_2 \cdot 2^{0111_2-b} = 1.4375$$

with the relative representation error

$$\left| \frac{\sqrt{2} - 1.4375}{\sqrt{2}} \right| \approx 0.016.$$

- Finally, we consider the number $x = -0.003$. With $S = 1$, it holds:

$$0.003_{10} \approx 0.00000000110001_2$$

and normalized

$$0.003_{10} \approx 1.1001_2 \cdot 2^{-9} = 1.1001_2 \cdot 2^{-2-7}.$$

The exponent $-9 = -2 - b$ is too small and cannot be represented in this format (i.e., four digits for the exponent). The IEEE-754 provides, as an extension, the *denormalized numbers* for numbers that are particularly close to zero. Here, all bits (except for the sign) are used to store the mantissa, and the exponent is always 2^{1-b}. Details can be found in [57]. ◄

Example 1.20 has shown that a number can be too large (or too negative) to be represented in a given floating point convention. At the same time, a number can be too close to zero to be distinguished from zero in the floating-point representation. These important cases receive their own definition.

Definition 1.21 (*Overflow and Underflow*) *Overflow* is the situation when the absolute value of a number is larger than the largest representable machine number. *Underflow* is the situation when a non-zero number is rounded down to zero. The interval $[-u_{min}, u_{min}] \subset \mathbb{R}$ of numbers that are too small to be represented as non-zero floating point numbers is called the *underflow gap*.

The occurrence of an *overflow* or *underflow* essentially depends on the number of digits in the exponent.

The inaccuracy in the number representation must always be taken into account in numerical mathematics because it leads to the fact that even the simplest calculations cannot be carried out exactly, or that the result cannot be stored exactly. The exact definition of the IEEE-754 format is mostly irrelevant. Also, the choice of the binary system is not decisive. Instead, we give a simplified interpretation:

Definition 1.22 (*Number Representation with n-Digit Accuracy*) For a number $x \in \mathbb{R}$, we denote by the *number representation with n digits accuracy* the approximation

$$\mathrm{rd}(x) = \mathrm{sgn}(x)0.x_1x_2\ldots x_n \cdot 10^E,$$

where $x_1, \ldots, x_n \in \{0, 1, 2, 3, 4, 5, 6, 7, 8, 9\}$ are the n significant digits, $\mathrm{sgn}(x)$ is the sign of x and the exponent $E \in \mathbb{N}$ is given as

$$E = \begin{cases} \lceil \log_{10}(|x|) \rceil & x \neq 0 \\ 0 & x = 0 \end{cases}$$

Alternatively, we write a number with n digits of accuracy as a sequence of n contiguous digits with any number of leading or trailing zeros. This means, for example, that the number $x = 100/3$ with three-digit accuracy has the simplified representation

$$\mathrm{rd}\left(\frac{1}{3}\right) = 33.3 = 0.333 \cdot 10^2.$$

The speed of light is given with five-digit accuracy as

$$\mathrm{rd}(299\,792\,458)\,\mathrm{m\,s}^{-1} = 299\,790\,000\,\mathrm{m\,s}^{-1} = 0.29979 \cdot 10^9\,\mathrm{m\,s}^{-1},$$

and finally, the Planck constant is given with 4-digit accuracy as

$$\mathrm{rd}(6.626\,070\,15 \cdot 10^{-34})\,\mathrm{Js}$$
$$= 0.000\,000\,000\,000\,000\,000\,000\,000\,000\,000\,000\,000\,000\,662\,6\,\mathrm{Js}$$
$$= 0.6626 \cdot 10^{-35}\,\mathrm{Js}.$$

1.4.2   Machine Precision

The limited number of digits of the mantissa inevitably results in an error when performing numerical algorithms. The relative error that can occur in the computer representation $\tilde{x}$ of a number $x \in \mathbb{R}$,

$$\left| \frac{\mathrm{rd}(x) - x}{x} \right|,$$

is bounded by the so-called *machine precision*:

Definition 1.23 (*Machine Precision*) The *machine precision* eps is the maximum relative rounding error of the number representation and is determined as

$$\mathrm{eps} := \inf\{x > 0 : \mathrm{rd}(1 + x) > 1\}.$$

Machine precision plays a major role in numerical mathematics. It is less an estimate for the maximum relative error that can occur in each calculation, but rather a measure of the error that we must expect in each calculation step. The machine precision does not apply uniformly throughout the entire representable number space. For example, if denormalized numbers are provided, the accuracy deteriorates for x close to zero. However, Definition 1.23 is generally sufficient to describe the essential effects.

Since rounding errors inevitably occur, fundamental mathematical laws such as the associative law

$$(a + b) + c = a + (b + c), \quad (a \cdot b) \cdot c = a \cdot (b \cdot c)$$

or the distributive law

$$a \cdot (b + c) = ab + ac, \quad (a + b) \cdot c = ac + bc$$

no longer apply on computers. Even a simple comparison of numbers is often not possible, the query

```
1 if 3.8/10 == 0.38:
```

can deliver the wrong result due to rounding and must be replaced by queries of the type

```
1 if np.abs(3.8/10 - 0.38) < eps:
```

(the command np.abs(...) in the Python library numpy stands for the absolute value of a number). The machine precision can be determined in Python with

```
1 import numpy
2 print(numpy.finfo(float).eps)
```

We obtain the smallest and largest representable numbers with

```
1 import sys
2 sys.float_info.min
3 sys.float_info.max
```

These values depend on the operating system and not the software used. On our computer, we get

```
2.220446049250313e-16      # eps (machine precision)
2.2250738585072014e-308    # min (smallest representable number)
1.7976931348623157e+308    # max (largest representable number)
```

Machine precision also plays an important role in determining the termination criteria of iterative algorithms. For example, we consider the error estimator for the Leibniz series from Sect. 1.1.6, see (1.5). We want to stop the iteration when the error is smaller than a chosen tolerance $\epsilon > 0$. That is, in the case of the Leibniz series, we stop the iteration when

$$\frac{|\Pi_{2n+1} - \Pi_{2n}|}{\pi} < \epsilon$$

applies. We must always choose $\epsilon \gg$ eps. Because the requirement $\epsilon <$ eps we will never be able to fulfill, since the representation error of the solution can already be larger.

Remark 1.24 Machine precision has nothing to do with the smallest number that can be represented on a computer. This is determined by the number of digits in the exponent. The machine precision is determined by the number of digits in the mantissa.   ◄

1.4.3 Computing Speed—Floating Point Operations Per Second

Most numerical algorithms consist at their core of the repeated execution of basic operations. The effort of algorithms is measured in the number of necessary elementary operations (depending on the problem size). The runtime of an algorithm depends on the performance of the computer. This is measured in *FLOPS*, i.e., *floating point operations per second*.

In Table 1.3, we summarize the achievable FLOPS for different computer systems. The performance has increased rapidly over the years, with a factor of about 1000 being reached every ten years. This increase in performance is not just due to higher clock speed but also due to more efficient handling. Early computers had 8-bit registers that could be processed in each clock cycle (this is the MHz specification). Current hardware has 64-bit registers available. In addition, there is a more efficient execution of commands through so-called *pipelining*: The usual sequence in the processor includes *1. read memory*, *2. process data* and *3. store result*. Pipelining allows the processor to read the memory for the next operation already while the current one is still being processed. This has significantly reduced the number of necessary processor cycles per computing operation. The combination of Intel 8086 with FPU 8087 required about 150 cycles per floating point operation at a clock rate of 8 MHz and 50,000 FLOPS. The 486 needed only 50 cycles per operation at 66 MHz and about 1,000,000 FLOPS. The Pentium III is at about five cycles per floating point operation.

In recent years, the increase in efficiency has been largely due to a very high degree of parallelization. A Core i7 processor can perform several operations simultaneously. An Nvidia L40 GPU achieves its performance with almost 20,000 computing cores. To be able to use this performance efficiently, the algorithms must be adapted accordingly so that all cores are utilized. If an algorithm cannot exploit this special architecture, the performance drops back to about one GigaFLOP, i.e., to the level of the Pentium III from 2001. If the 20,000 cores are optimally utilized, a single graphics card of the L40 type achieves the same performance class as the K computer, which was the fastest computer in the world in 2011. This comparison becomes even more impressive when considering the acquisition costs per FLOP or the energy consumption per FLOP. As first supercomputer, *Frontier* reached a performance of one exaFLOP.

The fastest computer in 2025, *El Capitan*, combines about 10 000 000 cores on CPUs with more than 600 000 000 stream processors on accelerator cards. A current overview is provided by the *Top 500 list of supercomputers*.[3]

---

[3] https://www.top500.org List of the 500 fastest computers in the world.

Table 1.3 Floating point speed and acquisition costs of some current and historical computers (HD = Heidelberg). Not all computers or accelerator cards achieve their performance at the same precision

CPU	Year	FLOPS	Price	Energy consumption
Zuse Z3	1941	0.3	Single pieces	4 kW
IBM 704	1955	$5 \cdot 10^3$	$>10\,000\,000$ €	75 kW
Intel 8086 + 8087	1980	$50 \cdot 10^3$	$2\,000$ €	200 W
i486	1991	$1.4 \cdot 10^6$	$2\,000$ €	200 W
IPhone 4 s	2011	$100 \cdot 10^6$	500 €	10 W
2 x Pentium III	2001	$800 \cdot 10^6$	$2\,000$ €	200 W
Core i7	2011	$100 \cdot 10^9$	$1\,000$ €	200 W
Helics I (Parallel computer in HD)	2002	$1 \cdot 10^{12}$	$1\,000\,000$ €	40 kW
NVIDIA GTX 580 (GPU)	2010	$1 \cdot 10^{12}$	500 €	250 W
K computer	2011	$10 \cdot 10^{15}$	$500\,000\,000$ €	12\,000 kW
Frontier	2022	$1.1 \cdot 10^{18}$	$600\,000\,000$ €	22\,300 kW
NVIDIA L40 (GPU)	2022	$1.2 \cdot 10^{15}$	$8\,000$ €	300 W
El capitan	2024	$2.7 \cdot 10^{18}$	$600\,000\,000$ €	30\,000 kW

In parallel high-performance computers, power consumption also plays a significant role. *El Capitan* consumes as much electricity as $100\,000$ private individuals. Each hour of use results in an electricity bill of several thousand euros. New designs, especially smaller layouts of the circuit boards, can constantly reduce the power consumption per FLOP.

1.5 Rounding Error Analysis and Stability of Numerical Algorithms

When designing numerical algorithms for a task, there are often various options, which can differ, for example, in the order of the procedure steps. Different algorithms for the same task can lead to different results due to rounding errors.

Example 1.25 (Distributive Law) We consider the task $A(x, y, z) = x \cdot z - y \cdot z = (x - y)z$ and two different algorithms:

$$a_1^{(1)} := x \cdot z, \qquad a_1^{(2)} := x - y,$$
$$a_2^{(1)} := y \cdot z, \qquad a^{(2)} := a_1^{(2)} \cdot z,$$
$$a^{(1)} := a_1^{(1)} - a_2^{(1)}.$$

Let $x = 0.519$, $y = 0.521$, $z = 0.941$. In four-digit arithmetic, we get:

$$a_1^{(1)} := \mathrm{rd}(x \cdot z) = 0.4884, \qquad a_1^{(2)} := \mathrm{rd}(x - y) = -0.002,$$
$$a_2^{(1)} := \mathrm{rd}(y \cdot z) = 0.4903, \qquad a_2^{(2)} := \mathrm{rd}(a_1^{(2)} \cdot z) = -0.001882,$$
$$a_3^{(1)} := \mathrm{rd}(a_1^{(1)} - a_2^{(1)}) = -0.0019.$$

With $A(x, y, z) = -0.001882$ the relative errors are:

$$\left| \frac{-0.0001882 - a_3^{(1)}}{0.0001882} \right| \approx 0.01, \qquad \left| \frac{-0.0001882 - a_2^{(2)}}{0.0001882} \right| = 0.$$

The distributive law does not hold on the computer. ◀

The stability depends crucially on the design of the algorithm. Input errors or rounding errors that occur in individual steps are amplified in subsequent steps of the algorithm. We now analyze both methods in detail and assume that in each of the elementary steps (here, only addition and multiplication), a relative rounding error ϵ with $|\epsilon| \leq$ eps occurs, i.e.,

$$\mathrm{rd}(x + y) = (x + y)(1 + \epsilon), \quad \mathrm{rd}(x \cdot y) = (x \cdot y)(1 + \epsilon).$$

To analyze a given algorithm, we track the rounding errors that occur in each step and their accumulation:

Example 1.26 (Stability of the Distributive Law) We first calculate the condition numbers of the task:

$$\kappa_{A,x} = \left| \frac{1}{1 - \frac{y}{x}} \right|, \quad \kappa_{A,y} = \left| \frac{1}{1 - \frac{x}{y}} \right|, \quad \kappa_{A,z} = 1.$$

For $x \approx y$, the task is ill-conditioned. We start with Algorithm 1 and estimate in each step the rounding errors. In addition, we consider input errors (or representation errors) of x, y, and z. We always only consider first-order error terms and combine all further terms with the Landau symbols (for small ϵ):

$$\begin{aligned}
a_1 &= x(1 + \epsilon_x)z(1 + \epsilon_z)(1 + \epsilon_1) = xz(1 + \epsilon_x + \epsilon_z + \epsilon_1 + O(\mathrm{eps}^2)) \\
&= xz(1 + 3\epsilon_1 + O(\mathrm{eps}^2)) \\
a_2 &= y(1 + \epsilon_y)z(1 + \epsilon_z)(1 + \epsilon_2) = yz(1 + \epsilon_y + \epsilon_z + \epsilon_2 + O(\mathrm{eps}^2)) \\
&= yz(1 + 3\epsilon_2 + O(\mathrm{eps}^2)) \\
a_3 &= (xz(1 + 3\epsilon_1) - yz(1 + 3\epsilon_2) + O(\mathrm{eps}^2))(1 + \epsilon_3) \\
&= (xz - yz)(1 + \epsilon_3) + 3xz\epsilon_1 - 3yz\epsilon_2 + O(\mathrm{eps}^2)
\end{aligned}$$

We determine the relative error:

$$\left|\frac{a_3 - (xz - yz)}{xz - yz}\right| = \frac{|(xz - yz)\epsilon_3 + 3xz\epsilon_1 - 3yz\epsilon_2|}{|xz - yz|} \le \text{eps} + 3\frac{|x| + |y|}{|x - y|}\,\text{eps}$$

The error amplification of this algorithm can become large for $x \approx y$, which corresponds approximately (factor 3) to the conditioning of the task. We therefore call the algorithm stable.

We now consider Algorithm 2:

$$\begin{aligned}
a_1 &= (x(1 + \epsilon_x) - y(1 + \epsilon_y))(1 + \epsilon_1)\\
&= (x - y)(1 + \epsilon_1) + x\epsilon_x - y\epsilon_y + O(\text{eps}^2)\\
a_2 &= z(1 + \epsilon_z)\big((x - y)(1 + \epsilon_1) + x\epsilon_x - y\epsilon_y + O(\text{eps}^2)\big)(1 + \epsilon_2)\\
&= z(x - y)(1 + \epsilon_1 + \epsilon_2 + \epsilon_z + O(\text{eps})^2) + zx\epsilon_x - zy\epsilon_y + O(\text{eps}^2)
\end{aligned}$$

For the relative error, to first order, we have:

$$\begin{aligned}
\left|\frac{a_2 - (xz - yz)}{xz - yz}\right| &= \frac{|z(x - y)(\epsilon_1 + \epsilon_2 + \epsilon_z) + zx\epsilon_x - zy\epsilon_y|}{|xz - yz|}\\
&\le 3\,\text{eps} + \frac{|x| + |y|}{|x - y|}\,\text{eps}
\end{aligned}$$

The error amplification can become large for $x \approx y$. However, the amplification factor is smaller than in Algorithm 1. In particular, it is noticeable that this second algorithm does not exhibit any error amplification with error-free input data. (This is the case $\epsilon_x = \epsilon_y = \epsilon_z = 0$.)

Both algorithms are stable; the second one has better stability properties than the first. ◄

We call both algorithms stable, although one has significantly better stability properties. The concept of stability is therefore often used to compare different algorithms. We define:

Definition 1.27 (*Stability*) A numerical algorithm for solving a task is called *stable* if the accumulated rounding errors during execution do not exceed the unavoidable error given by the condition of the task.

We note: For a poorly conditioned problem, there is no stable algorithm with a lower error propagation than determined by the conditioning. For well-conditioned problems, however, arbitrarily unstable methods can exist.

The essential difference between the two algorithms from the example is the order of operations. In Algorithm 1, the last step is a subtraction, whose poor condition we have got to know learned under the term *cancellation*. Accumulated rounding errors at the beginning of the procedure are significantly amplified here. In Algorithm 2, the rounding errors that occur at the beginning are not further amplified by the final multiplication. From the analyzed example, we derive the rule:

In the design of numerical algorithms, poorly conditioned operations should be performed at the beginning.

1.6   Excursus: The Tax Function in Income Tax Calculation

We present in the following a first excursion, which provides quite a surprising insight into which everyday areas numerics can be found. We illustrate Horner's method as well as the concrete application of rounding error calculations (Sect. 1.3) using a practical example from German tax legislation. Income tax rates can be described by polynomials to specify them in tax tables then. In the versions of the Income Tax Act (EstG §32a) valid until 2009, it is explicitly noted (see paragraph (3) in EstG §32a), that the polynomials are to be evaluated using Horner's method.[4] However, since the year 2010, paragraphs (2)–(4) have been removed from the law. Nonetheless, let's look at the current values (year 2018). The official income tax rate is already given in the form (1.6) [24] (EstG §32a Income Tax Rate, Paragraph 1, as of 2023)[5][6]:

$$
E(x) = \begin{cases}
0 \,\text{€} & x \leq 10\,908 \,\text{€} \\
(979.18x + 1\,400)x \,\text{€} & 10\,909 \,\text{€} \leq x - 10\,908 \leq 15\,999 \,\text{€} \\
(192.59x + 2\,397)x \,\text{€} + 966.53 \,\text{€} & 16\,000 \,\text{€} \leq x - 15\,999 \leq 62\,809 \,\text{€} \\
0.42x \,\text{€} - 9\,972.98 \,\text{€} & 62\,810 \,\text{€} \leq x \leq 277\,825 \,\text{€} \\
0.45x \,\text{€} - 18\,307.73 \,\text{€} & 277\,826 \,\text{€} \leq x.
\end{cases}
$$

$$(1.14)$$

In the footnotes, y and z are used as other designations for the variables, depending on which range the income falls into. This piecewise-defined function $E(x)$ is called the tax function. This is illustrated in Fig. 1.2.

We immediately recognize the form of the Horner polynomials (1.6):

---

[4] EstG §32a, paragraph 3 [24] (year 2002): "The calculation steps necessary for the calculation of the tariff income tax are to be carried out in the order that results from Horner's method. In doing so, the intermediate results arising from the multiplications for each further calculation step are to be set with three decimal places; the subsequent decimal places are to be omitted. The resulting tax amount is to be rounded down to the nearest full Euro amount."

[5] EstG §32a Income Tax Rate, Paragraph 1, Footnote 1: The income tax is based on the taxable income rounded down to the nearest full Euro.

[6] (3) The quantity "y" is one ten-thousandth of the part of the taxable income that exceeds the basic tax-free allowance, rounded down to the nearest full Euro. (4) The quantity "z" is one ten-thousandth of the part of the taxable income that exceeds 15,999 Euros, rounded down to the nearest full Euro. (5) The quantity "x" is the taxable income rounded down to the nearest full Euro. (6) The resulting tax amount is to be rounded down to the nearest full Euro.

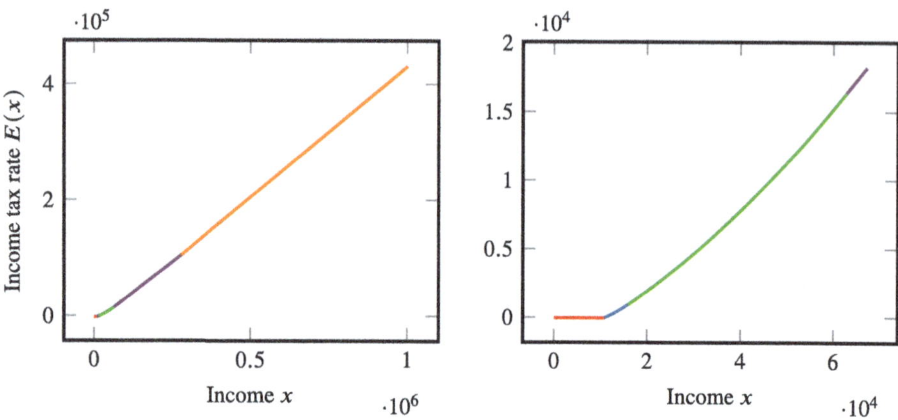

Fig. 1.2 Tax function $E(x)$. Left: entire tax function. Right: restriction to income up to 70 000€

$$p(x) = a_0 + x(a_1 + xa_2)$$

for example with $a_0 = 0$, $a_1 = 1\,400$, $a_2 = 979.18$.

We assume an entry-level salary for an engineer with $x = 48\,000$ € gross annual income (Bachelor (Uni/TH), as of 2022[7]). This means we evaluate the tax function at the point $\xi = 48000$ €. This falls into group 3. First, $x = \frac{48\,000 - 13\,770}{10\,000} = 3.423$ (in the official calculation, three decimal places are set). Horner's method now reads:

i	2	1	0
a_i	223.76	2 397.00	11 939.57
s_i	223.76	3 162.90	11 766.00 $= s(\xi)$

In the year 2023, the engineer has to pay $s(48\,000) = 11\,766$ EUR in income tax.

We now calculate the top tax rate at the point $\xi = 48\,000$ EUR, which indicates what proportion of each additional Euro earned must be paid to taxes.

The top tax rate curve is a marginal tax rate and mathematically given by the first derivative $s'(x)$ of the tax function $s(x)$. As we can easily see in (1.14) for the 5th income range, the top tax rate of $s'(x) = 0.45$, i.e., 45%, since $s(x) = 0.45x - 16\,164.53$. To calculate the top tax rate for our engineer with $\xi = 48\,000$ EUR income, caution is required when deriving since the function $s(z)$ is given in z, but we are interested in x. Accordingly, the chain rule must be applied, since $s(z) := s(z(x))$ and $s'(z(x)) \cdot z'(x)$.

Then, for the middle income range (3), see (1.14), Horner's method extended by the 1st derivative applies with the representation from Table 1.4. To get the top tax rate, we still need to include the derivative of the inner function $z'(x)$:

---

[7] https://www.academics.de/ratgeber/ingenieur-gehalt.

Table 1.4 Complete Horner scheme to calculate the tax function and its derivative in the income range (3) from (1.14)

i	2	1	0
a_i	223.76	2 397.00	11 939.57
s_i	223.76	3 162.90	$11\,766.00 = s(\xi)$
s_i'	223.76	3 928.80	

$$s'(\xi) = a_1^2 \cdot z'(x) = 3928.8 \cdot \frac{1}{10\,000},$$

since $z(x) = \frac{x - 13\,770}{10\,000}$. The top tax rate at an income of 48 000 EUR is therefore 0.39, i.e., 39% of each additional Euro must be paid in taxes. In comparison, the overall average tax rate, which is defined by $d(x) = s(x)/x$, is $d(x) = \frac{11\,766}{48\,000} = 0.25$, i.e., 25%.

Part I

Numerical Methods of Linear Algebra

In linear algebra, the structure of linear mappings $T : V \to W$ between finite dimensional vector spaces V and W is studied. In *numerical linear algebra*, we deal with some practical questions of linear algebra. These can be roughly divided into three main areas: *linear systems, orthogonalization methods*, and *eigenvalue problems*. First, consider linear systems $Ax = b$. The focus of the investigation is on the solution of linear systems with real square matrices $A \in \mathbb{R}^{n \times n}$. The well-known Gaussian elimination method is the starting point. However, we will quickly find that this simple method is neither very stable nor very efficient. If the systems become large and consist of, for example, $n \gg 1{,}000{,}000$ equations, the Gaussian elimination method is no longer feasible, and we must investigate alternatives.

Furthermore, we investigate the problem of orthogonalization: A generating system $\{v_1, \ldots, v_n\} \subset V$ shall be transformed into an orthogonal system with respect to a scalar product $(.\,,.)_V : V \times V \to \mathbb{R}$:

$$(v_i, v_j) = 0, \forall i \neq j.$$

In linear algebra, the *Gram-Schmidt orthogonalization* method was introduced for this purpose. Again, we see that this method is straightforward and efficient, but it has numerical stability problems. A numerically computed *orthogonal system* will no longer be orthogonal due to rounding errors.

Finally, we will study methods for calculating eigenvalues of linear mappings, i.e., finding complex numbers $\lambda \in \mathbb{C}$ with

$$A\omega = \lambda\omega,$$

where $\omega \in \mathbb{C}^n$ is a corresponding eigenvector. We will see that the numerical calculation of the eigenvalues as roots of the characteristic polynomial $\chi(\lambda) := \det(A - \lambda I) = 0$ is not stable. Instead, we will learn about iterative methods to approximate the eigenvalues.

Most problems of linear algebra occur as subproblems of other methods. Large linear systems must be solved for the discretization of differential equations, but also for the approximation of functions, for example. Orthogonalization methods are needed in the solution of linear systems with fixed-point iterations, but also in the design of methods for numerical quadrature (the approximation of integrals).

We will regularly refer back to the *concepts* introduced in the introduction and characterize our results based on them.

Fundamentals of Linear Algebra

We first collect some definitions and basic results. For a detailed introduction to linear algebra, see, for example, [67, 88]. Let V always be a finite-dimensional vector space over the field $\mathbb{K}$. We usually consider the space of real-valued vectors $V = \mathbb{R}^n$.

Definition 2.1 (*Basis and Dimension*) A subset $B = \{v_1, \ldots, v_n\} \subset V$ of a vector space over $\mathbb{K}$ is called a *basis*, if every element $v \in V$ can be uniquely represented as a linear combination of the *basis vectors* B:

$$v = \sum_{i=1}^{n} \alpha_i v_i, \quad \alpha_i \in \mathbb{K}.$$

The number of vectors n in the basis is called the *dimension of the vector space*.

The unique representability of each $v \in V$ by basis vectors allows us to identify the vector space V with the vector space of coefficient vectors $\alpha \in \mathbb{K}^n$.

Definition 2.2 (*Norm*) A mapping $\| \cdot \| : V \to \mathbb{R}_+$ is called a *norm*, if it has the following three properties:

1. positive definiteness: $\quad \|x\| \geq 0 \ \ \forall x \in V, \quad \|x\| = 0 \ \Rightarrow \ x = 0,$
2. linearity: $\qquad\qquad\quad \|\alpha x\| = |\alpha| \, \|x\| \qquad\qquad \forall x \in V, \ \alpha \in \mathbb{K},$
3. triangle inequality: $\quad \|x + y\| \leq \|x\| + \|y\| \qquad\qquad \forall x, y \in V.$

A vector space with a norm is called a *normed space*. In $\mathbb{K}^n$, frequently used norms are the *maximum norm* $\| \cdot \|_\infty$, the *Euclidean norm* $\| \cdot \|_2$ and the l_1-*norm* $\| \cdot \|_1$

T. Richter et al., *Introduction to Numerical Mathematics*, Mathematics Study Resources 25, https://doi.org/10.1007/978-3-662-72546-7_2

$$\|x\|_\infty := \max_{i=1,\ldots,n} |x_i|, \quad \|x\|_2 := \left(\sum_{i=1}^n |x_i|^2\right)^{\frac{1}{2}}, \quad \|x\|_1 := \sum_{i=1}^n |x_i|.$$

In the vector space $\mathbb{R}^n$ as well as in all finite-dimensional vector spaces, all norms are equivalent.

Theorem 2.3 (Norm Equivalence) *For any two norms* $\|\cdot\|$ *and* $\|\cdot\|'$ *in the finite-dimensional vector space* V, *there exists a constant* $c > 0$, *such that:*

$$\frac{1}{c}\|x\| \leq \|x\|' \leq c\|x\| \quad \forall x \in V.$$

Norms are important for the concept of convergence, and the theorem states that a sequence $x_n \to x$, which converges with respect to a norm $\|\cdot\|$, also converges with respect to any other norm $\|\cdot\|'$. This relationship is typical for finite-dimensional spaces and does not hold in (infinite-dimensional) function spaces. For example, consider the sequence of functions $f_n(x) = \exp(-nx^2)$ in the L^2-norm

$$\|f_n - 0\|_{L^2([-1,1])} := \left(\int_{-1}^1 |f_n(x) - 0|^2 \, dx\right)^{\frac{1}{2}} \xrightarrow{n\to\infty} 0.$$

However, using the maximum norm, it holds

$$\|f_n - 0\|_\infty := \max_{x\in[-1,1]} |f_n(x) - 0| \xrightarrow{n\to\infty} 0.$$

In finite-dimensional vector spaces, all norms are equivalent, but the constant c that appears here can depend on the dimension n of the vector space. This becomes important when considering convergence statements for $n \to \infty$. In the limit, the equivalence no longer holds.

Definition 2.4 (*Scalar Product*) A mapping $(\cdot, \cdot) : V \times V \to \mathbb{K}$ is called a *scalar product*, if it has the following properties:

1. definiteness: $\qquad\qquad\quad (x, x) > 0 \qquad\qquad\qquad\quad \forall x \in V,\ x \neq 0,$

$\qquad\qquad\qquad\qquad\qquad (x, x) = 0 \quad \Rightarrow \quad x = 0$

2. linearity: $\qquad\qquad\qquad (x, \alpha y + z) = \alpha(x, y) + (x, z) \quad \forall x, y, z \in V,\ \alpha \in \mathbb{K},$

3. conjugate symmetry: $\qquad (x, y) = \overline{(y, x)} \qquad\qquad\qquad \forall x, y \in V,$

where $\bar{z}$ is the complex conjugation of $z \in \mathbb{C}$.

In real spaces, symmetry $(x, y) = (y, x)$ holds. Furthermore, from 2. and 3. it follows that $(\alpha x, y) = \bar{\alpha}(x, y)$. An example is the *Euclidean scalar product* for vectors $x, y \in \mathbb{R}^n$

$$(x, y)_2 = x^T y = \sum_{i=1}^{n} x_i y_i.$$

Vector spaces with a scalar product are called *pre-Hilbert spaces*. Complex vector spaces with a scalar product are also called *unitary spaces*, in the real case we speak of *Euclidean spaces*.

Scalar products are closely linked to norms:

Theorem 2.5 (Induced Norm) *Let V be a vector space with a scalar product. Then by*

$$\|x\| = \sqrt{(x, x)}, \quad x \in V,$$

the induced norm *is given on V.*

A pre-Hilbert space is thus always also a normed space. If the pre-Hilbert space V is complete with respect to the norm induced by the scalar product, i.e., a Banach space (see, e.g., [95]), then V is called a *Hilbert space*.

The Euclidean norm is the norm induced by the Euclidean scalar product:

$$\|x\|_2 = (x, x)_2^{\frac{1}{2}} = \left(\sum_{i=1}^{n} x_i^2 \right)^{\frac{1}{2}}$$

Some important theorems apply to pairs of scalar products and induced norms:

Theorem 2.6 *Let V be a vector space with scalar product $(\cdot, \cdot)$ and induced norm $\|\cdot\|$. Then the* Cauchy-Schwarz inequality *holds:*

$$|(x, y)| \leq \|x\| \, \|y\| \quad \forall x, y \in V,$$

as well as the parallelogram identity*:*

$$\|x + y\|^2 + \|x - y\|^2 = 2\|x\|^2 + 2\|y\|^2 \quad \forall x, y \in V.$$

We further define:

Definition 2.7 (*Orthogonality*) Two vectors $x, y \in V$ of a pre-Hilbert space are called *orthogonal*, if $(x, y) = 0$.

The introduction of a norm mapping on V allows us to define the distance $\|x - y\|$. The existence of a scalar product will enable us to assign an angle to two elements. It holds in the Euclidean scalar product

$$\cos(\sphericalangle(x, y)) = \frac{(x, y)_2}{\|x\|_2 \|y\|_2},$$

and from this follows $x \perp y$, if $(x, y)_2 = 0$. This concept of angle and orthogonality can be transferred to any space with a scalar product.

One of the tasks of numerical linear algebra is the orthogonalization (or orthonormalization) of given systems of vectors:

Definition 2.8 (*Orthonormal Basis*) A basis $B = \{v_1, \ldots, v_n\}$ of V is called *orthogonal basis* with respect to the scalar product $(\cdot, \cdot)$, if:

$$(v_i, v_j) = 0 \quad \forall i \neq j,$$

and *orthonormal basis*, if:

$$(v_i, v_j) = \delta_{ij}$$

with the Kronecker symbol

$$\delta_{ij} = \begin{cases} 1 & i = j \\ 0 & i \neq j \end{cases}.$$

With different scalar products, different orthonormal bases exist for the same vector space V. Orthogonality then generally does not coincide with the geometric concept of orthogonality in Euclidean space:

Example 2.9 (Scalar Products and Orthonormal Basis) Let $V = \mathbb{R}^2$. By

$$(x, y)_2 := x_1 y_1 + x_2 y_2, \quad (x, y)_\omega := 2 x_1 y_1 + x_2 y_2,$$

two different scalar products are given. The first is the Euclidean scalar product. The scalar product properties of the second are easy to check. By

$$x_1 = \frac{1}{\sqrt{2}}(1, 1)^T, \quad x_2 = \frac{1}{\sqrt{2}}(-1, 1)^T,$$

an orthonormal basis with respect to $(\cdot, \cdot)_2$ is given. However, it holds:

$$(x_1, x_2)_\omega = \frac{1}{2}(-2 + 1) = \frac{-1}{\sqrt{2}} \neq 0.$$

An orthonormal basis with respect to $(\cdot, \cdot)_\omega$ is given, e.g., through

$$x_1 = \frac{1}{\sqrt{3}}(1, 1)^T, \quad x_2 = \frac{1}{2}(-1, 2)^T.$$

◀

Orthonormal bases are needed for numerous numerical methods, in Gaussian quadrature (Sect. 10.4), in Gaussian approximation Sect. 9.5.1 of functions, and, e.g., for the QR factorization (Chap. 4) of a matrix.

If a linearly independent set of vectors $\{a_1, \ldots, a_n\}$ of a vector space is given, we can use the Gram-Schmidt orthogonalization method to create an orthogonal or even orthonormal system:

Theorem 2.10 (Gram-Schmidt Orthonormalization) *Let $\{a_1, \ldots, a_n\}$ be a linearly independent subset of a vector space V, and let $(\cdot, \cdot)$ be a scalar product on V with induced norm $\|\cdot\|$. The algorithm*

$$(i) \quad q_1 := \frac{a_1}{\|a_1\|},$$

$$(ii) \quad i = 2, \ldots, n: \quad \tilde{q}_i := a_i - \sum_{j=1}^{i-1}(a_i, q_j)q_j, \quad q_i := \frac{\tilde{q}_i}{\|\tilde{q}_i\|},$$

generates an orthonormal system $\{q_1, \ldots, q_n\}$ in V. It also holds that

$$(q_i, a_j) = 0 \quad \forall 1 \leq j < i \leq n.$$

If $\{a_1, \ldots, a_n\}$ is a basis of V, then $\{q_1, \ldots, q_n\}$ is an orthonormal basis of V.

Proof We prove this by induction. For $i = 1$, $a_1 \neq 0$, since $a_1, \ldots, a_n$ are linearly independent. For $i = 2$, we have

$$(\tilde{q}_2, q_1) = (a_2, q_1) - (a_2, q_1)\underbrace{(q_1, q_1)}_{=1} = 0.$$

From the linear independence of a_2 and q_1, it follows that $\tilde{q}_2 \neq 0$. Now let $(q_j, q_k) = \delta_{jk}$ for $k, j < i$. Then for any $k < i$, we have

$$(\tilde{q}_i, q_k) = (a_i, q_k) - \sum_{j=1}^{i-1}(a_i, q_j)\underbrace{(q_j, q_k)}_{=\delta_{jk}} = (a_i, q_k) - (a_i, q_k) = 0.$$

Since $\text{span}\{q_1, \ldots, q_j\} = \text{span}\{a_1, \ldots, a_j\}$, it follows that $(q_i, a_j) = 0$ for $j < i$. $\qquad\square$

The process is named after the mathematicians Gram and Schmidt, but was already known to Laplace and Cauchy.

We will consider the vector space of all $\mathbb{R}^{n \times m}$ matrices. This vector space is also finite-dimensional, and in principle, we can identify the vector space of $n \times m$ matrices with the vector space of (nm) vectors. Of particular interest to us is the vector space of square $\mathbb{R}^{n \times n}$ matrices. We define:

Definition 2.11 (*Eigenvalues, Eigenvectors*) The *eigenvalues* $\lambda \in \mathbb{C}$ of a real[1] matrix $A \in \mathbb{R}^{n \times n}$ are defined as the roots of the characteristic polynomial:

$$\det(A - \lambda I) = 0$$

The set of all eigenvalues of a matrix A is called the *spectrum* of A

$$\sigma(A) := \{\lambda \in \mathbb{C}, \ \lambda \text{ is an eigenvalue of } A\}.$$

The *spectral radius* $\rho : \mathbb{R}^{n \times n} \to \mathbb{R}_+$ is the largest eigenvalue in absolute value:

$$\rho(A) := \max\{|\lambda|, \ \lambda \in \sigma(A)\}$$

An element $w \in \mathbb{R}^n \setminus \{0\}$ is called *eigenvector* for the eigenvalue λ, if it holds:

$$Aw = \lambda w.$$

When investigating linear systems $Ax = b$, the question arises whether such a system is solvable at all and whether the solution is unique. We summarize:

Theorem 2.12 (Regular Matrix) *For a square matrix $A \in \mathbb{R}^{n \times n}$, the following statements are equivalent:*

1. *The matrix A is regular.*
2. *The transposed matrix A^T is regular.*
3. *The inverse A^{-1} is regular.*
4. *The linear system $Ax = b$ is uniquely solvable for every $b \in \mathbb{R}^n$.*
5. *It holds that $\det(A) \neq 0$.*
6. *All eigenvalues of A are non-zero.*

We further define:

Definition 2.13 (*Positive Definiteness*)   A matrix $A \in \mathbb{K}^{n \times n}$ is called *positive definite*, if

$$(Ax, x) > 0 \ \ \forall x \neq 0, \ \ x \in \mathbb{K}^n.$$

It is called *positive semidefinite*, if

$$(Ax, x) \geq 0 \ \ \forall x \in \mathbb{K}^n.$$

---

[1] In this book, we mostly limit ourselves to real matrices. However, the concept of eigenvalues can be directly applied to matrices with complex entries. Matrices with real entries can still have complex eigenvalues.

Correspondingly, the matrix is called *negative definite*, if

$$(Ax, x) < 0 \quad \forall x \neq 0, \quad x \in \mathbb{K}^n,$$

and *negative semidefinite*, if

$$(Ax, x) \leq 0 \quad \forall x \in \mathbb{K}^n.$$

If none of these inequalities apply, the matrix is *indefinite*.

Example 2.14 (Definiteness of Matrices) The matrix

$$A = \begin{pmatrix} 1 & 4 & 1 \\ 0 & 2 & 1 \\ 0 & 0 & 3 \end{pmatrix}$$

is positive definite. The matrix

$$A = \begin{pmatrix} -1 & 0 & 0 \\ 0 & 1 & 0 \\ 0 & 0 & 1 \end{pmatrix}$$

is indefinite.   ◄

Conversely, we can read from the defining property of the positive definiteness of a matrix: If A is positive definite and hermitian, then $(A \cdot , \cdot)$ is a scalar product.

Theorem 2.15 (Positive Definite Matrices) *Let $A \in \mathbb{R}^{n \times n}$ be a symmetric matrix. Then A is positive definite if and only if all (real) eigenvalues of A are positive. If A is symmetric positive definite, then all diagonal elements of A are positive, i.e., strictly greater than zero, and the element with the largest absolute value is on the diagonal.*

Proof *(i)* Let A be a symmetric matrix with an orthonormal basis of eigenvectors $w_1, \ldots, w_n$. A is positive definite. Then, for an arbitrary eigenvector w_i with eigenvalue λ_i:

$$0 < (Aw_i, w_i) = \lambda_i (w_i, w_i) = \lambda_i$$

Conversely, let all λ_i be positive. For $x = \sum_{i=1}^{n} \alpha_i w_i$ with $x \neq 0$ it holds:

$$(Ax, x) = \sum_{i,j} (\lambda_i \alpha_i \omega_i, \alpha_j \omega_j) = \sum_i \lambda_i \alpha_i^2 > 0$$

(ii) Let A be a real, positive definite matrix. Let e_i be the i-th unit vector. Then it holds:

$$0 < (Ae_i, e_i) = a_{ii}.$$

That is, all diagonal elements are positive.

(iii) Now we choose $x = e_i - \text{sign}(a_{ij})e_j$. We assume that $a_{ji} = a_{ij}$ is the element with the largest absolute value of the matrix. Then it holds:

$$0 < (Ax, x) = a_{ii} - \text{sign}(a_{ij})(a_{ij} + a_{ji}) + a_{jj} = a_{ii} + a_{jj} - 2|a_{ij}| \leq 0.$$

From this contradiction follows the last statement of the theorem: The element with the largest absolute value must be on the diagonal. $\qquad\square$

For norms on the space of matrices, we define further structural properties.

Definition 2.16 *(Matrix Norms)* A norm $\|\cdot\| : \mathbb{R}^{n\times m} \to \mathbb{R}_+$, for all $m, n \in \mathbb{N}$ is called *matrix norm*, if it is submultiplicative i.e.:

$$\|AB\| \leq \|A\|\,\|B\| \quad \forall A \in \mathbb{R}^{k\times l},\ B \in \mathbb{R}^{l\times m}.$$

It is called *compatible* with a vector norm $\|\cdot\| : \mathbb{R}^m \to \mathbb{R}_+$, if

$$\|Ax\| \leq \|A\|\,\|x\| \quad \forall A \in \mathbb{R}^{n\times m},\ x \in \mathbb{R}^m$$

holds. A matrix norm $\|\cdot\| : \mathbb{R}^{n\times m} \to \mathbb{R}_+$ is called *induced* by a vector norm $\|\cdot\| : \mathbb{R}^m \to \mathbb{R}_+$, if

$$\|A\| := \sup_{x\neq 0} \frac{\|Ax\|}{\|x\|}.$$

Matrix norms induced by a vector norm are also called *natural matrix norms*.

For an induced matrix norm, it always holds:

$$\|A\| := \sup_{x\neq 0} \frac{\|Ax\|}{\|x\|} = \sup_{\|x\|=1} \|Ax\| = \sup_{\|x\|\leq 1} \|Ax\|$$

It is easy to prove that every matrix norm induced by a vector norm is also compatible with it. Compatible with the Euclidean norm is also the *Frobenius norm*

$$\|A\|_F := \left(\sum_{i,j=1}^{n} a_{ij}^2 \right)^{\frac{1}{2}}, \tag{2.1}$$

which is not induced by a vector norm. For general norms on the vector space of matrices, it does not necessarily hold that $\|I\| = 1$, where $I \in \mathbb{R}^{n\times n}$ is the identity matrix. However, this relationship holds for every matrix norm induced by a vector norm. We summarize the essential induced matrix norms in the following theorem.

Theorem 2.17 (Induced Matrix Norms) *Let $A \in \mathbb{R}^{n \times m}$. The matrix norms induced by the Euclidean vector norm, the maximum norm and the l_1-norm are the* spectral norm $\| \cdot \|_2$, *the maximum row sum $\| \cdot \|_\infty$, and the maximum column sum $\| \cdot \|_1$:*

$$\|A\|_2 = \sqrt{\rho(A^T A)}, \quad \|A\|_\infty = \max_{i=1,\dots,n} \sum_{j=1}^{m} |a_{ij}|,$$

$$\|A\|_1 = \max_{j=1,\dots,m} \sum_{i=1}^{n} |a_{ij}|.$$

Proof *(i)* We have:

$$\|A\|_2^2 = \sup_{x \neq 0} \frac{\|Ax\|_2^2}{\|x\|_2^2} = \sup_{x \neq 0} \frac{(Ax, Ax)}{\|x\|_2^2} = \sup_{x \neq 0} \frac{(A^T Ax, x)}{\|x\|_2^2}.$$

The matrix $A^T A$ is symmetric and consequently only has real eigenvalues. It possesses an orthonormal basis $\omega_i \in \mathbb{R}^m$, $i = 1, \dots, m$, of eigenvectors with eigenvalues $\lambda_i \geq 0$. All eigenvalues λ_i are greater than or equal to zero, because:

$$\lambda_i = \lambda_i (\omega_i, \omega_i) = (A^T A\omega_i, \omega_i) = (A\omega_i, A\omega_i) = \|A\omega_i\|^2 \geq 0 \tag{2.2}$$

Let $x \in \mathbb{R}^m$ be arbitrary with basis representation $x = \sum_i \alpha_i \omega_i$. Then, due to $(\omega_i, \omega_j)_2 = \delta_{ij}$, the relation $\|x\|_2^2 = \sum_i \alpha_i^2$ with $\alpha_i \in \mathbb{R}$ holds:

$$\|A\|_2^2 = \sup_{|\alpha| \neq 0} \frac{(\sum_i \alpha_i A^T A\omega_i, \sum_i \alpha_i \omega_i)}{\sum_i \alpha_i^2} = \sup_{|\alpha| \neq 0} \frac{(\sum_i \alpha_i \lambda_i \omega_i, \sum_i \alpha_i \omega_i)}{\sum_i \alpha_i^2}$$

$$= \sup_{|\alpha| \neq 0} \frac{\sum_i \lambda_i \alpha_i^2}{\sum_i \alpha_i^2} \leq \max_i \lambda_i,$$

where

$$|\alpha| = \max_i |\alpha_i|.$$

Now, conversely, let λ_k be the largest eigenvalue. Then, due to (2.2) with $\alpha_i = \delta_{ki}$:

$$0 \leq \max_i \lambda_i = \lambda_k = \sum_i \lambda_i \alpha_i^2 = \sum_{i,j} (\lambda_i \alpha_i \omega_i, \alpha_j \omega_j) = (A^T Ax, x) = \|Ax\|_2^2. \tag{2.3}$$

Therefore, $\max_i \lambda_i \leq \|A\|_2^2$.

(ii) We demonstrate the result exemplarily for the maximum norm:

$$\|Ax\|_\infty = \sup_{\|x\|_\infty = 1} \left(\max_i \sum_{j=1}^{m} a_{ij} x_j \right)$$

This sum with $\|x\|_\infty = 1$ reaches its maximum if $|x_j| = 1$ and if the sign of x_j is chosen such that $a_{ij}x_j \geq 0$ for all $j = 1, \ldots, m$. Then:

$$\|Ax\|_\infty = \max_i \sum_{j=1}^m |a_{ij}|$$

$\square$

As a side result, we obtain from (2.3) that every eigenvalue is bounded in absolute value by the spectral norm of the matrix A. It even holds with any matrix norm and a compatible vector norm for an eigenvalue λ with corresponding eigenvector $w \in \mathbb{R}^n$ of A, that

$$|\lambda| = \frac{\|\lambda w\|}{\|w\|} = \frac{\|Aw\|}{\|w\|} \leq \frac{\|A\| \, \|w\|}{\|w\|} = \|A\|.$$

A simple bound for the largest eigenvalue in absolute value is thus obtained by analyzing any (compatible) matrix norms.

From Theorem 2.17, we further conclude that for symmetric matrices the $\| \cdot \|_2$-norm coincides with the spectral radius of the matrix itself, hence the name spectral norm.

Theorem 2.18 *Let $A \in \mathbb{R}^{m \times n}$ be an arbitrary rectangular matrix. Then:*

$$Rank(A) = Rank(A^T) = Rank(AA^T) = Rank(A^T A),$$

where $Rank(A) := dim(range(A))$ denotes the (column) rank of the matrix A, i.e., the maximum number of linearly independent column vectors.

The proof can be found, for example, in [67, 88].

3

Solving systems of linear equations is one of the most essential numerical tasks. Often, problems are not directly given in the form of a linear system, but arise in the course of modeling or the discretization and approximation of more complex problems, e.g., of differential equations or optimization problems. Such systems of linear equations can be very large. Large means that many million[1] to billions[2] unknowns must be solved. In this section, we will deal exclusively with real-valued matrices $A \in \mathbb{R}^{n \times n}$. Methods to solve systems of linear equations are classified into *direct methods*, which calculate the solution directly and up to rounding error influences exactly, and *iterative methods*, which approximate the solution by a fixed point iteration. In this chapter, we first deal with direct methods. Iterative methods are then the subject of Sect. 3.7.

As an introductory example, we consider a system with floating-point numbers

$$\begin{pmatrix} 0.988 & 0.960 \\ 0.992 & 0.963 \end{pmatrix} \begin{pmatrix} x \\ y \end{pmatrix} = \begin{pmatrix} 0.084 \\ 0.087 \end{pmatrix}$$

with the solution $(x, y)^T = (3, -3)^T$. We determine the solution by Gaussian elimination and round to three-digit accuracy after each calculation:

---

[1] For example, in the course of the European space mission Gaia, a system with 600 million (i.e., $6 \cdot 10^8$) unknowns must be solved [9].

[2] In 2016, on a so-called *Petascale computer*, i.e., a parallel computer with more than 10^{15} FLOPS, linear systems with up to $6 \cdot 10^{11}$ (0.6 trillion) unknowns were solved [55], and in [64] with 1.036×10^{12} unknowns per second.

© The Author(s), under exclusive license to Springer-Verlag GmbH, DE, part of Springer Nature 2026

T. Richter et al., *Introduction to Numerical Mathematics*, Mathematics Study Resources 25, https://doi.org/10.1007/978-3-662-72546-7_3

$$
\begin{pmatrix} 0.988 & 0.960 & 0.084 \\ 0.992 & 0.963 & 0.087 \end{pmatrix} \times 0.988/0.992
$$

$$
\begin{pmatrix} 0.988 & 0.960 & 0.084 \\ 0.988 & 0.959 & 0.0866 \end{pmatrix} \downarrow -
$$

$$
\begin{pmatrix} 0.988 & 0.959 & 0.084 \\ 0 & 0.001 & -0.0026 \end{pmatrix},
$$

Using Gaussian elimination, we have transformed the matrix A into a triangular shape. The right-hand side b was modified accordingly. The resulting triangular system can now be very easily solved by *backward substitution* (with three-digit calculation):

$$
0.001 y = -0.0026 \qquad\qquad \Rightarrow y = -2.6
$$
$$
0.988 x = 0.087 - 0.959 \cdot (-2.6) \approx 2.58 \quad \Rightarrow x = 2.61.
$$

So we get $(x, y) = (2.61, -2.60)$. The relative error of the numerical solution is thus more than 10%. The numerical task of solving a system either seems to be very poorly conditioned (see Chap. 1), or the elimination method is numerically highly unstable and not well suited. We will investigate the question of conditioning and stability in the following section.

3.1   Stability Analysis for Linear Systems

For a square regular matrix $A \in \mathbb{R}^{n \times n}$ and a vector $b \in \mathbb{R}^n$ we consider the linear system

$$
Ax = b.
$$

Due to numerical errors, rounding errors, input errors, or measurement inaccuracies, both A and b are only disturbed:

$$
\tilde{A}\tilde{x} = \tilde{b}
$$

Here, $\tilde{A} = A + \delta A$ and $\tilde{b} = b + \delta b$ with the perturbations δA and δb.

We now come to the core statement of this section and want to consider the error amplification when solving linear systems. Errors can occur both in the matrix A and in the right-hand side b. We first consider perturbations of the right-hand side.

Theorem 3.1 (Pertubation of the Right-Hand Side) *Let $A \in \mathbb{R}^{n \times n}$ be a regular matrix, $b \in \mathbb{R}^n$. Let $x \in \mathbb{R}^n$ be the solution of the linear system $Ax = b$. Let δb be a perturbation of the right-hand side $\tilde{b} = b + \delta b$ and $\tilde{x}$ the solution of the disturbed system $A\tilde{x} = \tilde{b}$. Furthermore, let $\| \cdot \|$ be a matrix norm compatible with a vector norm $\| \cdot \|$. Then it holds:*

$$\frac{\|\delta x\|}{\|x\|} \le \text{cond}(A)\frac{\|\delta b\|}{\|b\|},$$

with the condition number *of the matrix*

$$\text{cond}(A) = \|A\| \cdot \|A^{-1}\|.$$

Proof Let $\|\cdot\|$ be an arbitrary matrix norm with compatible vector norm $\|\cdot\|$. For the solution $x \in \mathbb{R}^n$ and perturbed solution $\tilde{x} \in \mathbb{R}^n$, we have

$$\tilde{x} - x = A^{-1}(A\tilde{x} - Ax) = A^{-1}(\tilde{b} - b) = A^{-1}\delta b.$$

Therefore,

$$\frac{\|\delta x\|}{\|x\|} \le \|A^{-1}\|\frac{\|\delta b\|}{\|x\|} \cdot \frac{\|b\|}{\|b\|} = \|A^{-1}\|\frac{\|\delta b\|}{\|b\|} \cdot \frac{\|Ax\|}{\|x\|} \le \underbrace{\|A\| \cdot \|A^{-1}\|}_{=:\text{cond}(A)} \frac{\|\delta b\|}{\|b\|}$$

$\square$

Remark 3.2 (*Condition Number of a Matrix*) The *condition number* of a matrix plays a crucial role in numerical linear algebra. For instance, considering the conditioning of the matrix-vector multiplication $y = Ax$, we obtain for the perturbed input $\tilde{x} = x + \delta x$ the estimate

$$\frac{\|\delta y\|}{\|y\|} \le \text{cond}(A)\frac{\|\delta x\|}{\|x\|}.$$

The condition number of a matrix depends on the chosen norm. However, since all matrix norms in $\mathbb{R}^{n \times n}$ are equivalent, all condition concepts are equivalent. With Theorem 2.17, we infer for symmetric matrices for the special case $\text{cond}_2(A)$

$$\text{cond}_2(A) = \|A\|_2 \cdot \|A^{-1}\|_2 = \frac{\max\{|\lambda|,\ \lambda \text{ eigenvalue of } A\}}{\min\{|\lambda|,\ \lambda \text{ eigenvalue of } A\}}.$$

$\blacklozenge$

We now consider the case where the matrix A of a linear system is perturbed by δA. The first question is whether the perturbed matrix $\tilde{A} = A + \delta A$ is still regular.

Lemma 3.3 *Let $\|\cdot\|$ be a matrix norm induced by the vector norm. Further, let $B \in \mathbb{R}^{n \times n}$ be a matrix with $\|B\| < 1$. Then, the matrix $I + B$ is regular, and the estimate*

$$\|(I + B)^{-1}\| \le \frac{1}{1 - \|B\|}$$

holds.

Proof We have

$$\|(I + B)x\| \geq \|x\| - \|Bx\| \geq (1 - \|B\|)\|x\|.$$

Since $1 - \|B\| > 0$, $I + B$ is an injective mapping. Therefore, $I + B$ is a regular matrix. Further, we have:

$$1 = \|I\| = \|(I + B)(I + B)^{-1}\| = \|(I + B)^{-1} + B(I + B)^{-1}\|$$
$$\geq \|(I + B)^{-1}\| - \|B\| \, \|(I + B)^{-1}\| = \|(I + B)^{-1}\|(1 - \|B\|) > 0.$$

$\square$

With this lemma, we can address the perturbation of the matrix in the following theorem.

Theorem 3.4 (Matrix Perturbation) *Let $A \in \mathbb{R}^{n \times n}$ be a regular matrix, $b \in \mathbb{R}^n$. Further, let $x \in \mathbb{R}^n$ be the solution of the linear system $Ax = b$ and let $\tilde{A} = A + \delta A$ be a perturbed matrix with $\|\delta A\| \leq \|A^{-1}\|^{-1}$. For the perturbed solution $\tilde{x} = x + \delta x$ of $\tilde{A}\tilde{x} = b$ we have:*

$$\frac{\|\delta x\|}{\|x\|} \leq \frac{\text{cond}(A)}{1 - \text{cond}(A)\|\delta A\|/\|A\|} \frac{\|\delta A\|}{\|A\|}.$$

Proof For the solution x as well as the perturbed solution $\tilde{x}$ and the error $\delta x := \tilde{x} - x$ it holds

$$\begin{aligned} (A + \delta A)\tilde{x} &= b \\ (A + \delta A)x &= b + \delta Ax \end{aligned} \quad \Rightarrow \quad \delta x = -[A + \delta A]^{-1}\delta Ax.$$

According to the assumption $\|A^{-1}\delta A\| \leq \|A^{-1}\| \, \|\delta A\| < 1$, it follows with Lemma 3.3:

$$\|\delta x\| \leq \|[I + A^{-1}\delta A]^{-1}A^{-1}\delta A\| \, \|x\| \leq \frac{\|A^{-1}\|}{1 - \|A^{-1}\delta A\|}\|\delta A\| \, \|x\|$$
$$\leq \frac{\text{cond}_2(A)}{1 - \|A^{-1}\|\|\delta A\|} \frac{\|\delta A\|}{\|A\|}\|x\|$$

The result is obtained by extending with $\|A\|/\|A\|$. $\square$

These two perturbation theorems can be easily combined to simultaneously estimate the perturbation due to the right-hand side and the matrix.

Theorem 3.5 (Perturbation Theorem for Linear Systems) *Let $A \in \mathbb{R}^{n \times n}$ be a regular matrix, $b \in \mathbb{R}^n$ the right-hand side. Furthermore, let $x \in \mathbb{R}^n$ be the solution of the linear system $Ax = b$. For the solution $\tilde{x} \in \mathbb{R}^n$ of the perturbed system $\tilde{A}\tilde{x} = \tilde{b}$ with perturbations $\delta b = \tilde{b} - b$ and $\delta A = \tilde{A} - A$, under the condition*

$$\|\delta A\| < \frac{1}{\|A^{-1}\|},$$

the following estimate holds

$$\frac{\|\delta x\|}{\|x\|} \leq \frac{\text{cond}(A)}{1 - \text{cond}(A)\|\delta A\|/\|A\|} \left(\frac{\|\delta b\|}{\|b\|} + \frac{\|\delta A\|}{\|A\|} \right),$$

where the condition number *is defined as*

$$\text{cond}(A) = \|A\| \, \|A^{-1}\|.$$

Proof We combine the statements of Theorems 3.1 and 3.4. For this, let x be the solution of $Ax = b$, $\tilde{x}$ the perturbed solution $\tilde{A}\tilde{x} = \tilde{b}$, and $\hat{x}$ the solution to the perturbed right-hand side $A\hat{x} = \tilde{b}$. Then it holds:

$$\|x - \tilde{x}\| \leq \|x - \hat{x}\| + \|\hat{x} - \tilde{x}\| \leq \text{cond}(A)\frac{\|\delta b\|}{\|b\|}\|x\| + \frac{\text{cond}(A)}{1 - \text{cond}(A)\frac{\|\delta A\|}{\|A\|}} \frac{\|\delta A\|}{\|A\|}\|x\|.$$

Note $\|\delta A\| < \|A^{-1}\|^{-1}$. Hereby,

$$0 \leq \text{cond}(A)\frac{\|\delta A\|}{\|A\|} < \|A\| \, \|A^{-1}\| \frac{1}{\|A\| \, \|A^{-1}\|} = 1,$$

and it holds

$$\text{cond}(A) \leq \frac{\text{cond}(A)}{1 - \text{cond}(A)\frac{\|\delta.A\|}{\|A\|}}.$$

This proves the statement. $\square$

With this result, we return to the introductory example from Sect. 3:

$$A = \begin{pmatrix} 0.988 \ 0.959 \\ 0.992 \ 0.963 \end{pmatrix} \quad \Rightarrow \quad A^{-1} \approx \begin{pmatrix} 8302 & -8267 \\ -8552 & 8517 \end{pmatrix}$$

In the maximum row sum norm $\| \cdot \|_\infty$, it holds:

$$\|A\|_\infty = 1.955, \quad \|A^{-1}\|_\infty \approx 17069, \quad \text{cond}_\infty(A) \approx 33370.$$

Solving a linear system with the matrix A is therefore extremely poorly conditioned if the matrix has a large condition number. This results in the enormous rounding error in the example at the beginning of the chapter. We note: With a large condition number, the influence of rounding errors is considerable. The large error is inherent in the task and not necessarily due to a stability problem of the method.

3.2   The Gaussian Elimination Method and the LU Decomposition

The most important method for solving a linear system $Ax = b$ with a square matrix $A \in \mathbb{R}^{n \times n}$ is the Gaussian elimination method. The matrix $(A|b)$, extended by the right-hand side b, is first brought to triangular form by stepwise elimination. In the ith step, a multiple of the ith row is added to the rows $j > i$ to eliminate the elements below the diagonal, i.e., a_{ji}, to zero:

$$\begin{pmatrix} * & * & * & \cdots & * & | & * \\ * & * & * & & * & | & * \\ * & * & * & & * & | & * \\ \vdots & & \ddots & & \vdots & | & \vdots \\ * & * & * & \cdots & * & | & * \end{pmatrix} \rightarrow \begin{pmatrix} * & * & * & \cdots & * & | & * \\ 0 & * & * & & * & | & * \\ 0 & * & * & & * & | & * \\ \vdots & & \ddots & & \vdots & | & \vdots \\ 0 & * & * & \cdots & * & | & * \end{pmatrix} \rightarrow \cdots \rightarrow \begin{pmatrix} * & * & * & \cdots & * & | & * \\ 0 & * & * & & * & | & * \\ 0 & 0 & * & & * & | & * \\ \vdots & & & \ddots & \vdots & | & \vdots \\ 0 & 0 & \cdots & 0 & * & | & * \end{pmatrix}$$

Subsequently, the backward elimination follows. Starting with $i = n, n - 1, \ldots,$ we add multiples of the ith row to all rows $j < i$, this time to bring the elements above the diagonal a_{ji} to zero:

$$\begin{pmatrix} * & * & * & \cdots & * & | & * \\ 0 & * & * & & * & | & * \\ \vdots & \ddots & \ddots & & \vdots & | & \vdots \\ 0 & & 0 & * & * & | & * \\ 0 & \cdots & 0 & 0 & * & | & * \end{pmatrix} \rightarrow \begin{pmatrix} * & * & \cdots & * & 0 & | & * \\ 0 & * & \cdots & * & 0 & | & * \\ \vdots & \ddots & \ddots & & \vdots & | & \vdots \\ 0 & & 0 & * & 0 & | & * \\ 0 & \cdots & 0 & 0 & * & | & * \end{pmatrix} \rightarrow \cdots \rightarrow \begin{pmatrix} * & 0 & & \cdots & 0 & | & * \\ 0 & * & 0 & & 0 & | & * \\ \vdots & \ddots & \ddots & \ddots & \vdots & | & \vdots \\ 0 & & & * & 0 & | & * \\ 0 & 0 & \cdots & 0 & * & | & * \end{pmatrix}$$

The result in the left $n \times n$ block is a diagonal matrix, and the result can be read off after division by the diagonal.

The LU factorization is based on Gaussian elimination. The main difference is that the transformation of the matrix and the solution of the linear system are split into two steps. First, we compute a multiplicative factorization $A = LU$ and afterwards we solve the actual linear system using L and U.

In the following, we describe the factorization of the matrix into the two factors L and U in detail. The algorithm is iterative, and in the ith step, we aim to eliminate the elements a_{ki} for $k > i$. The entries $a_{kl}^{(i-1)}$ of $A^{(i-1)}$ get new values $a_{kl}^{(i)}$ of $A^{(i)}$ in the following manner:

$$a_{kl}^{(i)} = a_{kl}^{(i-1)} - \frac{a_{ki}^{(i-1)}}{a_{ii}^{(i-1)}} a_{il}^{(i-1)} \quad k, l = i + 1, \ldots, n$$

We write this step compactly in the form of a matrix-vector multiplication.

Theorem 3.6 (Elimination Step of the LU Factorization)   *Let $A^{(i-1)} \in \mathbb{R}^{n \times n}$ be a matrix with partial rectangular shape up to column i, i.e.,*

$$a_{kl}^{(i-1)} = 0 \; for \; l = 1, \ldots, i-1 \; and \; k = l+1, \ldots, n,$$

and with pivot element $a_{ii}^{(i-1)} \neq 0.$ *Then the matrix*

$$A^{(i)} = F^{(i)} A^{(i-1)}$$

has a partial rectangular shape with level i, i.e.,

$$a_{kl}^{(i)} = 0 \; for \; l = 1, \ldots, i \; and \; k = l+1, \ldots, n,$$

if $F^{(i)}$ *is the* Frobenius matrix

$$F^{(i)} := \begin{pmatrix} 1 & & & & & \\ & \ddots & & & & \\ & & 1 & & & \\ & & -g_{i+1}^{(i)} & \ddots & & \\ & & \vdots & & \ddots & \\ & & -g_{n}^{(i)} & & & 1 \end{pmatrix}, \quad g_k^{(i)} := \frac{a_{ki}^{(i-1)}}{a_{ii}^{(i-1)}}.$$

Proof The proof is demonstrated with element-by-element calculation. First, $a_{ii}^{(i-1)} \neq 0$, so that the $g_k^{(i-1)}$ and thus all of $F^{(i)}$ are well defined. If we calculate the matrix-matrix product element by element and use the special form of the Frobenius matrix, we get

$$[F^{(i)} A^{(i-1)}]_{kl} = \sum_{j=1}^{n} F_{kj}^{(i)} a_{jl}^{(i-1)}$$

$$= a_{kl}^{(i-1)} - \begin{cases} 0 & k \leq i \\ g_k^{(i)} a_{il}^{(i-1)} & k > i \end{cases}$$

$$= a_{kl}^{(i-1)} - \begin{cases} 0 & k \leq i \\ \dfrac{a_{ki}^{(i-1)}}{a_{ii}^{(i-1)}} a_{il}^{(i-1)} & k > i \end{cases}$$

$$= \begin{cases} a_{kl}^{(i-1)} & k \leq i \\ 0 & k > i, \; l \leq i \\ a_{kl}^{(i-1)} - \dfrac{a_{ki}^{(i-1)} a_{il}^{(i-1)}}{a_{ii}^{(i-1)}} & k > i, \; l > i. \end{cases}$$

Multiplication by $F^{(i)}$ from the left leaves the first i rows unchanged. Likewise, the first $i-1$ columns remain unchanged. The ith column below the diagonal is set to zero, so that the resulting matrix has a partial rectangular form in the first i columns. The remaining block $k, l = i+1, \ldots, n$ receives new values. $\square$

The transformation of the matrix A to upper right triangular shape is now achieved by repeated application of this theorem, i.e., after multiplication with $F^{(1)}, F^{(2)}, \ldots, F^{(n-1)}$

$$U := A^{(n-1)} = F^{(n-1)} F^{(n-2)} \cdots F^{(1)} A. \tag{3.1}$$

If all *pivot elements* of the matrices $A^{(i)} = F^{(i)} F^{(i-1)} \cdots F^{(1)} A$ appearing in the course of the sequence are non-zero, i.e., $a_{ii}^{(i-1)} \neq 0$, then the Frobenius matrices $F^{(i)}$ are well-defined and invertible such that Theorem 3.6 is applicable and shows that U has upper right triangular shape. To identify the matrix L, we look more closely at the Frobenius matrices.

Theorem 3.7 (Frobenius Matrix) *Let $F^{(i)} \in \mathbb{R}^{n \times n}$ be a Frobenius matrix of the form*

$$F^{(i)} := \begin{pmatrix} 1 & & & & & \\ & \ddots & & & & \\ & & 1 & & & \\ & & -g_{i+1} & \ddots & & \\ & & \vdots & & \ddots & \\ & & -g_n & & & 1 \end{pmatrix},$$

where $g_i \in \mathbb{R}$. Each Frobenius matrix $F^{(i)} \in \mathbb{R}^{n \times n}$ is regular, and its inverse is given by:

$$[F^{(i)}]^{-1} := \begin{pmatrix} 1 & & & & & \\ & \ddots & & & & \\ & & 1 & & & \\ & & g_{i+1} & \ddots & & \\ & & \vdots & & \ddots & \\ & & g_n & & & 1 \end{pmatrix}.$$

For two Frobenius matrices $F^{(i_1)}$ and $F^{(i_2)}$ with $i_1 < i_2$, the following holds:

$$F^{(i_1)} F^{(i_2)} = F^{(i_1)} + F^{(i_2)} - I = \begin{pmatrix} 1 & & & & & & \\ & \ddots & & & & & \\ & & 1 & & & & \\ & & -g_{i_1+1}^{(i_1)} & 1 & & & \\ & & \vdots & & \ddots & & \\ & & \vdots & & 1 & & \\ & & \vdots & & -g_{i_2+1}^{(i_2)} & \ddots & \\ & & \vdots & & \vdots & & \ddots \\ & & -g_n^{(i_1)} & & -g_n^{(i_2)} & & 1 \end{pmatrix}.$$

Proof This follows by component-wise calculation of the products $F^{(i)}[F^{(i)}]^{-1}$ and $F^{(i_1)}F^{(i_2)}$. We exemplarily show the second part. To this end, we first introduce the matrices $\hat{F}^{(i)} = F^{(i)} - I$ and write the product as

$$F^{(i_1)}F^{(i_2)} = I + \hat{F}^{(i_1)} + \hat{F}^{(i_2)} + \hat{F}^{(i_1)}\hat{F}^{(i_2)}.$$

For the product, we obtain by component-wise calculation using the special structure of $\hat{F}^{(i_1)}$ and $\hat{F}^{(i_2)}$ for $i_1 < i_2$

$$[\hat{F}^{(i_1)}\hat{F}^{(i_2)}]_{kl} = \sum_{r=1}^{n} \hat{F}^{(i_1)}_{kr}\hat{F}^{(i_2)}_{rl} = 0,$$

as $F^{(i_1)}_{kr} \neq 0$ is only possible for $r = i_1$ and $F^{(i_2)}_{rl} \neq 0$ may only hold for $r > l = i_2$. $\square$

The product of Frobenius matrices is not commutative. That is, the second part of the theorem does not hold in the case $i_1 > i_2$. However, the theorem can easily be generalized to the situation $1 \leq i_1 < i_2 < \cdots < i_r < n$, and it holds

$$\prod_{j=1}^{r} F^{(i_j)} = \sum_{j=1}^{r} F^{(i_j)} - (r - 1)I. \tag{3.2}$$

We now continue with (3.1) and multiply successively with the inverses of $F^{(n-1)}$, $F^{(n-2)}$, ... and obtain

$$U := A^{(n-1)} = F^{(n-1)}F^{(n-2)} \cdots F^{(1)}A$$
$$\Leftrightarrow \quad [F^{(1)}]^{-1}[F^{(2)}]^{-1} \cdots [F^{(n-1)}]^{-1}u = A. \tag{3.3}$$

The factors $[F^{(i)}]^{-1}$ are according to Theorem 3.7 all Frobenius matrices and with (3.2) it follows that the product

$$L := \prod_{j=1}^{n-1}[F^{(j)}]^{-1} = \sum_{j=1}^{n-1}[F^{(j)}]^{-1} - (n - 2)I \tag{3.4}$$

is a regular matrix and more precisely a lower left triangular matrix with ones on the diagonal. Thus, $A = LU$ and we have derived the sought factorization of a matrix A into a lower left triangular matrix (with ones on the diagonal) and an upper right triangular matrix.

Theorem 3.8 (LU Factorization) *Let $A \in \mathbb{R}^{n \times n}$ be a square, regular matrix. Assume that all pivot elements $a_{ii}^{(i-1)}$ occurring during the elimination are non-zero. Then the uniquely determined LU factorization into a right upper regular triangular matrix $u \in \mathbb{R}^{n \times n}$ and a* *left lower regular triangular matrix $L \in \mathbb{R}^{n \times n}$ with diagonal entries 1 exists. The effort to perform the LU factorization is*

$$\frac{1}{3}n^3 + O(n)$$

elementary operations, i.e., multiplications and additions.

Proof *(i) Uniqueness.* Assume there exist two LU factorizations

$$A = L_1 U_1 = L_2 U_2 \quad \Leftrightarrow \quad L_2^{-1} L_1 = U_2 U_1^{-1}.$$

The product of triangular matrices is again a triangular matrix, so both products must be diagonal matrices. The product $L_2^{-1} L_1$ has only ones on the diagonal, so it follows

$$L_2^{-1} L_1 = U_2 U_1^{-1} = I,$$

and thus $L_1 = L_2$ and $U_1 = U_2$.

(ii) Feasibility. Each step of the elimination is feasible as long as division by $a_{ii}^{(i-1)} = 0$ is not required. The matrix F is regular by construction, and thus the inverse L exists.

(iii) Effort. In the ith elimination step

$$A^{(i)} = F^{(i)} A^{(i-1)}$$

initially $n - i$ arithmetic operations are required to calculate $g_j^{(i)}$ for $j = i + 1, \ldots, n$. The matrix-matrix multiplication only affects all elements a_{kl} with $k > i$ and $l > i$. It holds

$$a_{kl}^{(i)} = a_{kl}^{(i-1)} - g_k^{(i)} a_{il}^{(i-1)}, \quad k, l = i + 1, \ldots, n.$$

For this, $(n - i)^2$ arithmetic operations for $g_k^{(i)} a_{il}^{(i-1)}$ and $(n - i)$ subsequent subtractions of the type $a_{kl}^{(i-1)} - g_k^{(i)} a_{il}^{(i-1)}$ are needed. In total, the effort in the $n - 1$ steps sums up to

$$N_{LU}(n) = \sum_{i=1}^{n-1} \left(n - i + (n - i)^2 \right) = \sum_{i=1}^{n-1} i + i^2.$$

With the known sum formula, it follows:

$$N_{LU}(n) = \frac{n^3}{3} - \frac{n}{3}$$

$\square$

The complexity of the LU factorization is therefore $O(n^3)$. If we double the matrix size n, the effort increases asymptotically by a factor of 8. If we increase the size of the matrix tenfold, the effort increases by a factor of $1\,000$. Once the matrix A is decomposed, linear systems $Ax = b$ can easily be solved afterwards:

Algorithm 3.9: Solving Linear Systems with the LU Factorization

> **Input:** $A \in \mathbb{R}^{n \times n}$ a regular matrix (whose LU factorization exists), $b \in \mathbb{R}^n$ a right-hand side.
> 1 Create LU factorization $A = LU$
> 2 Forward substitution: $Ly = b$
> 3 Backward substitution: $Ux = y$
> **Result:** The solution $x \in \mathbb{R}^n$ of the system $Ax = b$.

If we substitute $y = Ux$ into the equation $Ly = b$, we see with $Ly = LUx = Ax = b$ that the two steps yield the solution of the linear system. The solution steps in lines 2 and 3 of the algorithm are called *forward substitution* and *backward substitution*. Solving systems with triangular matrices is easily done by direct algorithms:

Algorithm 3.10: Backward Substitution

> **Input:** $U \in \mathbb{R}^{n \times n}$ a regular upper right triangular matrix, $b \in \mathbb{R}^n$ a right-hand side.
> 1 Set $x_n = u_{nn}^{-1} b_n$.
> 2 **for** $i = n - 1$ **to** *1* **do**
> 3 $\quad x_i = u_{ii}^{-1} \left(b_i - \sum_{j=i+1}^{n} u_{ij} x_j \right)$
> **Result:** The solution $x \in \mathbb{R}^n$ of the upper triangular system $Ux = b$.

In Python, the backward substitution can then be implemented together with the library numpy as follows and is available in the Jupyter notebooks.[3] Here, u is a numpy.array of size $n \times n$ and b is a numpy.array of length n.

♣ Implementation 3.11: Backward Substitution

```python
def backward(U, b):
    n = b.shape[0]
    x = np.empty_like(b)
    x[n - 1] = b[n - 1] / U[n - 1, n - 1]
    for i in range(n - 2, -1, -1):
        xr = 0
        for j in range(i + 1, n):
            xr += U[i, j] * x[j]
        x[i] = (b[i] - xr) / U[i, i]
    return x
```

It holds:

Theorem 3.12 (Backward substitution) *Let $U \in \mathbb{R}^{n \times n}$ be an upper right triangular matrix with $u_{ii} \neq 0$. Then the matrix U is regular, and the backward substitution requires*

---

[3] github.com/sn-code-inside/EinfNumMath-RWW.

$$N_U(n) = \frac{n^2}{2} + O(n)$$

elementary operations.

Proof It holds $\det(u) = \prod u_{ii} \neq 0$. Thus, the matrix U is regular.

Each step of the backward substitution consists of additions, multiplications, and divisions by the diagonal elements. With $u_{ii} \neq 0$, each step is feasible.

To calculate x_i, $n - i$ multiplications and additions are necessary. In addition, there is one division per step. This results in

$$n + \sum_{i=1}^{n-1}(n - i) = n + (n - 1)n - \frac{(n - 1)n}{2} = \frac{n^2}{2} + \frac{n}{2}$$

operations. $\square$

The forward elimination follows the backward substitution of Algorithm 3.10 and, according to Theorem 3.12, it requires $O(n^2)$ operations. Thus, the actual solution of a linear system is far less complex than the creation of the factorization. In many applications, such as the discretization of parabolic differential equations, many systems with different right-hand sides, but identical matrices, must be solved consecutively. In this case, it is advisable to create the factorization just once and then apply it repeatedly.

Remark 3.13 (*Practical Aspects*) The matrix L is a lower left triangular matrix with ones on the diagonal. Therefore, the known diagonal elements do not need to be stored. Similarly, the zero elements of the matrices $A^{(i)}$ below the diagonal do not need to be stored. L and U can be memorized in the same square matrix, if we write step i of the algorithm as:

$$\tilde{A}^{(i)} = \begin{pmatrix}
a_{11} & a_{12} & a_{13} & \cdots & & \cdots & \cdots & a_{1n} \\
\mathbf{l_{21}} & a_{22}^{(1)} & a_{23}^{(1)} & & & & & \vdots \\
\mathbf{l_{31}} & \mathbf{l_{32}} & a_{33}^{(2)} & & \ddots & & & \vdots \\
\vdots & & \ddots & \ddots & & & & \vdots \\
\vdots & & & & \mathbf{l_{i+1,i}} & a_{i+1,i+1}^{(i)} & \cdots & a_{i+1,n}^{(i)} \\
\vdots & & & & \vdots & & \ddots & \vdots \\
\mathbf{l_{n1}} & \mathbf{l_{n2}} & \cdots & \mathbf{l_{n,i}} & a_{n,i+1}^{(i)} & \cdots & & a_{nn}^{(i)}
\end{pmatrix}$$

The bold printed values are the entries of L. The values above the line do not change during the process and already form the entries L and U. $\blacklozenge$

Since the LU factorization *in place* has a special significance, we provide the corresponding algorithm here.

Algorithm 3.14: LU Factorization Without Additional Storage

> **Input:** $A = (a_{ij})_{ij} \in \mathbb{R}^{n \times n}$ a regular matrix (whose LU factorization exists).
> 1 **for** $i = 1$ **to** n **do**
> 2   **for** $k = i + 1$ **to** n **do**
> 3 $a_{ki} = a_{ki}/a_{ii}$
> 4   **for** $j = i + 1$ **to** n **do**
> 5 $a_{kj} = a_{kj} - a_{ki}a_{ij}$;
> **Result:** The matrix A overwritten with its LU factorization.

A Python implementation of the LU factorization without additional memory, using `numpy.arrays` to store the matrices, could look like this:

✿ Implementation 3.15: LU Factorization Without Additional Memory

```python
def LU_simple(A):
    assert (A.shape[0] == A.shape[1]), 'The matrix is not square'
    n = A.shape[0]
    for i in range(0, n):
        for k in range(i + 1, n):
            A[k, i] = A[k, i] / A[i, i]
            for j in range(i + 1, n):
                A[k, j] = A[k, j] - A[k, i] * A[i, j]
    return None
```

We note again that we no longer have access to the original matrix, as it has been overwritten. Further, we need to remember that the L matrix has only ones on the diagonal, which we do not store.

3.2.1 LU Factorization with Pivoting

The element $a_{ii}^{(i-1)}$ is called the *pivot element*. To perform the LU factorization, it must always hold $a_{ii}^{(i-1)} \neq 0$. However, even for regular matrices, this is not necessarily the case. As an example, we consider

$$A := \begin{pmatrix} 1 & 4 & 2 \\ 2 & 8 & 1 \\ 1 & 2 & 1 \end{pmatrix}.$$

In the first step of creating the LU factorization, $a_{11}^{(0)} = 1$ and it holds:

$$A^{(1)} = F^{(1)}A = \begin{pmatrix} 1 & 0 & 0 \\ -2 & 1 & 0 \\ -1 & 0 & 1 \end{pmatrix} \begin{pmatrix} 1 & 4 & 2 \\ 2 & 8 & 1 \\ 1 & 2 & 1 \end{pmatrix} = \begin{pmatrix} 1 & 4 & 2 \\ 0 & 0 & -3 \\ 0 & -2 & -1 \end{pmatrix}$$

At this point, the algorithm stops, because $a_{22}^{(1)} = 0$. However, we could continue the algorithm with the choice of $a_{32}^{(i)} = -2$ as a new pivot element. This is done systematically by introducing a *permutation*. In the i-th step of the process, a suitable pivot element a_{ki} in the i-th column with $k \geq i$ is first sought. The k-th and i-th rows are swapped, and the LU factorization can be continued. The exchanging of the k-th and i-th row is done by multiplying with a permutation matrix:

$$P^{ki} := \begin{pmatrix} 1 & & & & & & & & & \\ & \ddots & & & & & & & & \\ & & 1 & & & & & & & \\ & & & 0\,0\,\ldots\,0\,1 & & & & & \\ & & & 0\,1 \qquad\quad 0 & & & & \\ & & & \vdots \quad \ddots \quad \vdots & & & & \\ & & & 0 \qquad\quad 1\,0 & & & & \\ & & & 1\,0\,\cdots\,0\,0 & & & & \\ & & & & 1 & & & & \\ & & & & & \ddots & & \\ & & & & & & 1 \end{pmatrix}$$

It holds that $P_{jj}^{ki} = I$ for $j \neq k$ and $j \neq i$ as well as $P_{ki}^{ki} = P_{ik}^{ki} = 1$, all other elements are zero. We summarize some properties of P:

Theorem 3.16 (Permutation Matrices) *Let $P = P^{ki}$ be the pivot matrix with $P_{jj}^{ki} = 1$ for $j \neq k, i$ and $P_{ki}^{ki} = P_{ik}^{ki} = 1$. The product $P^{ki} A$ from the left swaps the k and ith row of A, the product $A P^{ki}$ from the right swaps the k and ith column. It holds:*

$$P^2 = I \text{ and thus } P^{-1} = P.$$

Proof Exercise. $\qquad\qquad\qquad\qquad\qquad\qquad\qquad\qquad\qquad\qquad\qquad\qquad\qquad\quad$ □

Algorithm 3.17: Pivot Search in Step $(i - 1) \mapsto (i)$

> **Input:** $A^{(i-1)} = (a_{kl}^{(i-1)})_{kl} \in \mathbb{R}^{n \times n}$ a regular matrix.
> 1. Search for index $j \geq i$, such that $|a_{ji}| = \max_{l \geq i} |a_{li}|$.
> 2. Set $P^{(i)} := P^{ji}$.
> **Result:** Permutation matrix $P^{(i)}$

Subsequently, we determine $A^{(i)}$ as

$$A^{(i)} = F^{(i)} P^{(i)} A^{(i-1)}.$$

Pivoting ensures that all elements $g_k^{(i)} = a_{ki}^{(i-1)}/a_{ii}^{(i-1)}$ of $F^{(i)}$ are bounded in magnitude by 1. In total, we obtain the factorization:

$$U = A^{(n-1)} = F^{(n-1)} P^{(n-1)} \cdots F^{(1)} P^{(1)} A = \Big(\prod_{i=n-1}^{1} F^{(i)} P^{(i)} \Big) A. \tag{3.5}$$

The pivot matrices do not commute with A or the $F^{(i)}$. Therefore, a transition to the LU factorization is not straightforward. We define the matrices

$$\tilde{P}^{(i+1)} := P^{(n-1)} \cdots P^{(i+1)}, \quad \tilde{P}^{(n)} := I,$$

with $\tilde{P}^{(i)} \tilde{P}^{(i)} = 1$ and insert this product into (3.5):

$$\begin{aligned}
U &= \Big(\prod_{i=n-1}^{1} F^{(i)} \tilde{P}^{(i+1)} \tilde{P}^{(i+1)} P^{(i)} \Big) A \\
&= \Big(\prod_{i=n-1}^{1} F^{(i)} \tilde{P}^{(i+1)} \tilde{P}^{(i)} \Big) A = \underbrace{\tilde{P}^{(n)}}_{=I} \Big(\prod_{i=n-1}^{1} \tilde{P}^{(i+1)} F^{(i)} \tilde{P}^{(i+1)} \Big) \tilde{P}^{(1)} A
\end{aligned} \tag{3.6}$$

We define

$$\tilde{F}^{(i)} := \tilde{P}^{(i+1)} F^{(i)} \tilde{P}^{(i+1)} = P^{(n-1)} \cdots P^{(i+1)} F^{(i)} P^{(i+1)} \cdots P^{(n-1)}. \tag{3.7}$$

The matrix $\tilde{F}^{(i)}$ is obtained by multiple row and column swaps of $F^{(i)}$. Only rows and columns $j > i$ are swapped. The matrix $\tilde{F}^{(i)}$ has the same sparsity structure as $F^{(i)}$ and, in particular, only ones on the diagonal. That is, it is again a Frobenius matrix, and Theorem 3.7 still applies. Only the entries in the i-th column below the diagonal are permuted. The same applies to its inverse

$$\tilde{L}^{(i)} := [\tilde{F}^{(i)}]^{-1} = P^{(n-1)} \cdots P^{(i+1)} [F^{(i)}]^{-1} P^{(i+1)} \cdots P^{(n-1)}. \tag{3.8}$$

We can, therefore, write the LU factorization with pivoting as

$$U = \underbrace{\tilde{F}^{(n-1)} \tilde{F}^{(n-2)} \cdots \tilde{F}^{(1)}}_{=: \tilde{F}} \underbrace{\tilde{P}^{(1)}}_{=: \tilde{P}} A.$$

Since all $\tilde{F}^{(i)}$ are Frobenius matrices, $\tilde{F}$ is a regular lower triangular matrix with ones on the diagonal, and with $\tilde{L} = \tilde{F}^{-1}$, we get

$$\tilde{L} U = \tilde{P} A.$$

When creating the LU factorization, not only the $A^{(i)}$, but in addition, the previously calculated $L^{(i)}$ must be permuted.

Theorem 3.18 (LU factorization with pivoting) *Let $A \in \mathbb{R}^{n \times n}$ be a regular matrix. Then, there always exists an LU factorization with pivoting*

$$PA = LU,$$

where P is a product of pivot matrices, L is a lower triangular matrix with diagonal entries 1, and U is an upper right triangular matrix. The LU factorization without pivoting $P = I$ is unique if it exists.

Proof Exercise.   □

With pivoting, we ensure that the LU factorization exists for every regular matrix A. On the other hand, the stability of the factorization can be improved by suitable pivoting. By choosing a pivot element a_{ki} with maximum relative magnitude (relative to the row), the risk of cancellation can be reduced.

Example 3.19 (*LU Factorization Without Pivoting*) Let

$$A = \begin{pmatrix} 2.3 & 1.8 & 1.0 \\ 1.4 & 1.1 & -0.7 \\ 0.8 & 4.3 & 2.1 \end{pmatrix}, \quad b = \begin{pmatrix} 1.2 \\ -2.1 \\ 0.6 \end{pmatrix},$$

be given. The solution of the linear system $Ax = b$ is given by (five-digit accuracy):

$$x \approx \begin{pmatrix} 0.34995 \\ -0.98023 \\ 2.1595 \end{pmatrix}$$

For the matrix A, it holds $\mathrm{cond}_\infty(A) = \|A\|_\infty \|A^{-1}\|_\infty \approx 7.2 \cdot 1.2 \approx 8.7$. The task is well-conditioned; the condition number 8.7 suggests an amplification of the error by about one digit. We first create the LU factorization (three-digit calculation). We write the entries of L in bold in the resulting matrix:

$$F^{(1)} = \begin{pmatrix} 1 & 0 & 0 \\ -\frac{1.4}{2.3} & 1 & 0 \\ -\frac{0.8}{2.3} & 0 & 1 \end{pmatrix} \approx \begin{pmatrix} 1 & 0 & 0 \\ -0.609 & 1 & 0 \\ -0.348 & 0 & 1 \end{pmatrix},$$

$$[L^{(1)}, A^{(1)}] \approx \begin{pmatrix} 2.3 & 1.8 & 1.0 \\ \mathbf{0.609} & 0.0038 & -1.31 \\ \mathbf{0.348} & 3.67 & 1.75 \end{pmatrix}.$$

In the second step, it holds:

$$F^{(2)} = \begin{pmatrix} 1 & 0 & 0 \\ 0 & 1 & 0 \\ 0 & -\frac{3.67}{0.0038} & 1 \end{pmatrix} \approx \begin{pmatrix} 1 & 0 & 0 \\ 0 & 1 & 0 \\ 0 & -966 & 1 \end{pmatrix},$$

$$[L^{(2)}L^{(1)}, A^{(2)}] \approx \begin{pmatrix} 2.3 & 1.8 & 1.0 \\ \mathbf{0.609} & 0.0038 & -1.31 \\ \mathbf{0.348} & \mathbf{966} & 1270 \end{pmatrix}.$$

The LU factorization results in

$$L = \begin{pmatrix} 1 & 0 & 0 \\ 0.609 & 1 & 0 \\ 0.348 & 966 & 1 \end{pmatrix}, \quad U = \begin{pmatrix} 2.3 & 1.8 & 1.0 \\ 0 & 0.0038 & -1.31 \\ 0 & 0 & 1270 \end{pmatrix}.$$

We next solve the system by forward and backward substitution:

$$A\tilde{x} = L \underbrace{U\tilde{x}}_{=y} = b$$

Initially, it holds

$$y_1 = 1.2, \qquad y_2 = -2.1 - 0.609 \cdot 1.2 \approx -2.83,$$
$$y_3 = 0.6 - 0.348 \cdot 1.2 + 966 \cdot 2.83 \approx 2730.$$

And finally:

$$\tilde{x}_3 = \frac{2730}{1270} \approx 2.15,$$
$$\tilde{x}_2 = \frac{-2.83 + 1.31 \cdot 2.15}{0.0038} \approx -3.55,$$
$$\tilde{x}_1 = \frac{1.2 + 1.8 \cdot 3.55 - 1 \cdot 2.15}{2.3} \approx 2.37.$$

For the solution $\tilde{x}$, it holds:

$$\tilde{x} = \begin{pmatrix} 2.37 \\ -3.55 \\ 2.15 \end{pmatrix}, \quad \frac{\|\tilde{x} - x\|_2}{\|x\|_2} \approx 1.4,$$

i.e., a relative error of 140%, although we only considered rounding errors and no disturbed input yet. To use the implementation of the LU factorization for solving the linear system in Python, we still need to implement the forward substitution. We assume the case that the LU factorization was stored instead of the original matrix A, and the ones on the diagonal must be considered separately.

♣ Implementation 3.20: Forward Substitution

```
1 def forward(L, b):
2     x = np.zeros_like(b)
3     for i in range(0, b.shape[0]):
4         xr = 0
5         for j in range(0, i):
6             xr += L[i, j] * x[j]
7         x[i] = (b[i] - xr)
8     return x
```

To solve the system, we create the matrix and the right-hand side as a `numpy.array`. To see rounding effects, we store the entries as `half` floating-point numbers.

```
1 import numpy as np
2 A = np.array([[2.3, 1.8, 1], [1.4, 1.1, -0.7], [0.8, 4.3, 2.1]], dtype=np.half)
3 b = np.array([1.2, -2.1, 0.6], dtype=np.half)
```

The simple LU factorization without additional memory then yields

```
1 LU_simple(A)
2 print('Modified A =\n', A)

  Modified A =
   [[ 2.301e+00  1.800e+00  1.000e+00]
   [ 6.089e-01  3.906e-03 -1.309e+00]
   [ 3.477e-01  9.410e+02  1.233e+03]]
```

By forward and backward substitution, we obtain

```
1 y = forward(A, b)
2 print('y = ', y)
3 x = backward(A, y)
4 print('x = ', x)

  y =  [ 1.200e+00 -2.830e+00  2.664e+03]
  x =  [ 0.365 -1.    2.16 ]
```

To check the error, we compare the solution with that computed by the linear algebra library numpy [51] using `double` floating point numbers:

```
1 A = np.array([[2.3, 1.8, 1], [1.4, 1.1, -0.7], [0.8, 4.3, 2.1]], dtype=np.double)
2 b = np.array([1.2, -2.1, 0.6], dtype=np.double)
3 x_np = np.linalg.solve(A, b)
4
5 print('x_np = ', x_np)
6 print('Relative error:', np.linalg.norm(x - x_np) / np.linalg.norm(x_np))

  x_np =  [ 0.34994583 -0.98022752  2.15953413]
  Relative error: 0.0103672221141273248
```

So even with the higher accuracy of `half`, compared with three significant digits, we still have a relative error of 1%.   ◀

This negative example shows the poor stability of the LU factorization for solving systems of linear equations. In the second step, a value close to 0 was chosen as the pivot element with $a_{22}^{(1)} = 0.0038$. This results in values of very different magnitudes in the matrices L and U. This has a negative effect on further stability.

Example 3.21 (*LU Factorization With Pivoting*) We continue the example in step 2 and first search for the pivot element:

$$[L^{(1)}, A^{(1)}] = \begin{pmatrix} 2.3 & 1.8 & 1.0 \\ \mathbf{0.609} & 0.0038 & -1.31 \\ \mathbf{0.348} & \boxed{3.67} & 1.75 \end{pmatrix}, \quad P^{(2)} = \begin{pmatrix} 1 & 0 & 0 \\ 0 & 0 & 1 \\ 0 & 1 & 0 \end{pmatrix}.$$

Pivoting yields results in:

$$[\tilde{L}^{(1)}, \tilde{A}^{(1)}] = \begin{pmatrix} 2.3 & 1.8 & 1.0 \\ \mathbf{0.348} & 3.67 & 1.75 \\ \mathbf{0.609} & 0.0038 & -1.31 \end{pmatrix}.$$

The next elimination step gives

$$F^{(2)} = \begin{pmatrix} 1 & 0 & 0 \\ 0 & 1 & 0 \\ 0 & -\frac{0.0038}{3.67} & 1 \end{pmatrix} \approx \begin{pmatrix} 1 & 0 & 0 \\ 0 & 1 & 0 \\ 0 & -0.00104 & 1 \end{pmatrix},$$

$$[L^{(2)}\tilde{L}^{(1)}, \tilde{A}^{(2)}] \approx \begin{pmatrix} 2.3 & 1.8 & 1.0 \\ \mathbf{0.348} & 3.67 & 1.75 \\ \mathbf{0.609} & \mathbf{0.00104} & -1.31 \end{pmatrix}.$$

We obtain the factorization $\tilde{L}U = PA$ as

$$\tilde{L} := \begin{pmatrix} 1 & 0 & 0 \\ 0.348 & 1 & 0 \\ 0.609 & 0.00104 & 1 \end{pmatrix}, \quad U := \begin{pmatrix} 2.3 & 1.8 & 1.0 \\ 0 & 3.67 & 1.75 \\ 0 & 0 & -1.31 \end{pmatrix}, \quad P := \begin{pmatrix} 1 & 0 & 0 \\ 0 & 0 & 1 \\ 0 & 1 & 0 \end{pmatrix}.$$

We solve the linear system in the form:

$$PAx = \tilde{L}\underbrace{Ux}_{=y} = Pb.$$

First, for the right side $\tilde{b} = Pb = (1.2, 0.6, -2.1)^T$ forward substitution in $\tilde{L}y = \tilde{b}$ yields

$$y_1 = 1.2,$$
$$y_2 = 0.6 - 0.348 \cdot 1.2 \approx 0.182,$$
$$y_3 = -2.1 - 0.609 \cdot 1.2 - 0.00104 \cdot 0.182 \approx -2.83.$$

As an approximation $\tilde{x}$, we obtain

$$\tilde{x} = \begin{pmatrix} 0.350 \\ -0.980 \\ 2.160 \end{pmatrix}$$

with a relative error

$$\frac{\|\tilde{x} - x\|_2}{\|x\|_2} \approx 0.0002,$$

thus only 0.02% instead of 140%.

In Python, we will not store the pivot matrices, but directly swap both the A parts and the L parts of the matrices occurring during the process. To solve the system afterwards by forward and backward substitution, the right-hand side b must then be permuted. For this, it is sufficient to store the pivoting indices in each step:

♣ Implementation 3.22: LU Factorization With Pivoting

```python
def LU_pivot(A):
    n = A.shape[0]
    pivot = []
    for i in range(0, n):
        k = i
        for j in range(i, n):
            if abs(A[j, i]) > abs(A[k, i]):
                k = j
        A[[i, k], :] = A[[k, i], :]
        pivot.append([i, k])
        for k in range(i + 1, n):
            A[k, i] = A[k, i] / A[i, i]
            for j in range(i + 1, n):
                A[k, j] = A[k, j] - A[k, i] * A[i, j]
    return pivot
```

If we apply this to the same example, we get with `half` floating-point numbers

```python
A = np.array([[2.3, 1.8, 1], [1.4, 1.1, -0.7], [0.8, 4.3, 2.1]], dtype=np.half)
b = np.array([1.2, -2.1, 0.6], dtype=np.half)

pivot = LU_pivot(A)
print('Modified A =\n', A)

for p in pivot:
    b[p] = b[[p[1], p[0]]]
print('P b = ', b)

y = forward(A, b)
print('y = ', y)
x = backward(A, y)
print('x = ', x)
```

```
Modified A =
[[ 2.301e+00  1.800e+00  1.000e+00]
 [ 3.477e-01  3.676e+00  1.752e+00]
 [ 6.089e-01  1.062e-03 -1.311e+00]]
P b =  [ 1.2  0.6 -2.1]
y =  [ 1.2     0.1829 -2.83  ]
x =  [ 0.3494 -0.98    2.16  ]
```

Table 3.1 Computing time to create the LU factorization of a matrix $A \in \mathbb{R}^{n \times n}$ on a hypothetical computer with 10 GigaFLOPS at peak performance

n	Operations ($\approx \frac{1}{3}n^3$)	Time LU factorization
100	$300 \cdot 10^3$	30 µs
1 000	$300 \cdot 10^6$	30 ms
10 000	$300 \cdot 10^9$	30 s
100 000	$300 \cdot 10^{12}$	10 h
1 000 000	$300 \cdot 10^{15}$	1 year

The LU factorization with pivoting then gives the relative error

```
1 np.linalg.norm(x - x_np) / np.linalg.norm(x_np)
```

```
1 0.00036962585275811158
```

only 0.037% instead of 1%.   ◀

The examples show that the calculated LU factorization in practical application is, of course, not a real factorization, but only an approximation of the matrix $A \approx LU$ due to rounding errors. One can easily check by calculating LU and determining the error $A - LU$.

The LU factorization is one of the most important direct methods for solving linear systems. However, the effort to calculate the LU factorization increases with the third order $O(n^3)$. Even on modern computers, the runtime for large systems quickly exceeds a reasonable limit.

In the following sections, we discuss how special structural properties of the matrix A can be exploited to speed up the LU factorization in exceptional cases significantly or to improve the stability of the factorization.

3.2.2   LU Factorization for Diagonally Dominant Matrices

Theorem 3.18 states that the LU factorization for any regular matrix with pivoting is possible. On the other hand, there are many matrices for which the LU factorization exists without pivoting. However, it is generally not apparent from the matrix A whether the factorization without pivoting is possible. Positive examples are positive definite or *diagonally dominant matrices*.

Definition 3.23 (*Diagonal Dominance*) A matrix $A \in \mathbb{R}^{n \times n}$ is called *diagonally dominant*, if

$$|a_{ii}| \geq \sum_{j \neq i} |a_{ij}|, \quad i = 1, \ldots, n.$$

A diagonally dominant matrix has the largest element in magnitude on the diagonal, and for regular matrices, the diagonal elements are non-zero.

Theorem 3.24 (LU Factorization of Diagonally Dominant Matrices) *Let $A \in \mathbb{R}^{n \times n}$ be a regular and diagonally dominant matrix. Then the LU factorization without pivoting is feasible and all occurring pivot elements $a_{ii}^{(i-1)}$ are non-zero.*

Proof We prove by induction and show that all submatrices $A_{k,l>i}^{(i)}$ are again diagonally dominant. For a diagonally dominant matrix, it holds

$$|a_{11}| \geq \sum_{j>1} |a_{1j}| \geq 0,$$

and, since A is regular, it further necessarily holds $|a_{11}| > 0$. The first step of the LU factorization is thus feasible.

We now show that the resulting matrix $\tilde{A}$ after an elimination step has a diagonally dominant submatrix $\tilde{A}_{i,j>1}$. For the entries $\tilde{a}_{ij}$, it holds:

$$\tilde{a}_{ij} = a_{ij} - \frac{a_{i1}a_{1j}}{a_{11}}, \quad i, j = 2, \ldots, n.$$

For the submatrix, we get

$$i = 2, \ldots, n : \quad \sum_{j=2,\, j \neq i}^{n} |\tilde{a}_{ij}| \leq \underbrace{\sum_{j=1,\, j \neq i}^{n} |a_{ij}| - |a_{i1}|}_{\leq |a_{ii}|} + \frac{|a_{i1}|}{|a_{11}|} \underbrace{\sum_{j=2}^{n} |a_{1j}|}_{\leq |a_{11}|} - \frac{|a_{i1}|}{|a_{11}|}|a_{1i}|$$

$$\leq |a_{ii}| - \frac{|a_{i1}|}{|a_{11}|}|a_{1i}| \leq |a_{ii} - \frac{a_{i1}}{a_{11}}a_{1i}| = |\tilde{a}_{ii}|,$$

and the resulting matrix is diagonally dominant. $\qquad\square$

The diagonal dominance of a matrix is easy to check and, therefore, a good criterion to estimate the necessity of pivoting.

3.3   The Cholesky Factorization for Symmetric Positive Definite Matrices

Another important class of matrices is that of positive definite matrices, see Definition 2.13 and Theorem 2.15. It turns out that for symmetric positive definite matrices $A \in \mathbb{R}^{n \times n}$, a symmetric factorization $A = \tilde{L}\tilde{L}^T$ into a lower triangular matrix $\tilde{L}$ can be created without pivoting. This decomposition is called *Cholesky factorization* and the effort to create the Cholesky factorization is just half as large as the effort to create the LU factorization.

Theorem 3.25 (LU Factorization of a Positive Definite Matrix) *Let $A \in \mathbb{R}^{n \times n}$ be a symmetric positive definite matrix. Then there exists a unique LU factorization without pivoting.*

Proof We proceed similarly to the proof of Theorem 3.24 (diagonally dominant matrices), and use induction to show that each step results in a symmetric positive definite submatrix.

Let A be a symmetric positive definite matrix. A step of the LU factorization is feasible, since, according to Theorem 2.15, $a_{11} > 0$ holds. We show that the submatrix $\tilde{A}_{i,j>1}$ after an elimination step is symmetric positive definite. Due to the symmetry of A, it holds:

$$\tilde{a}_{ij} = a_{ij} - \frac{a_{1j}a_{i1}}{a_{11}} = a_{ji} - \frac{a_{1i}a_{j1}}{a_{11}} = \tilde{a}_{ji},$$

i.e., $\tilde{A}_{i,j>1}$ is symmetric.

Let $x \in \mathbb{R}^n$ be a vector $x = (x_1, \tilde{x})$ with $\tilde{x} = (x_2, \ldots, x_n) \in \mathbb{R}^{n-1}$ arbitrary. The entry x_1 will be specified in the course of the proof. Due to the positive definiteness of A, it holds

$$0 < (Ax, x) = \sum_{ij} a_{ij}x_i x_j = a_{11}x_1^2 + 2x_1 \sum_{j=2}^{n} a_{1j}x_j + \sum_{i,j=2}^{n} a_{ij}x_i x_j$$

$$= a_{11}x_1^2 + 2x_1 \sum_{j=2}^{n} a_{1j}x_j + \sum_{i,j=2}^{n} \underbrace{\left(a_{ij} - \frac{a_{1j}a_{i1}}{a_{11}} \right)}_{=\tilde{a}_{ij}} x_i x_j + \sum_{i,j=2}^{n} \frac{a_{1j}a_{i1}}{a_{11}} x_i x_j$$

$$= a_{11} \left(x_1^2 + 2x_1 \frac{1}{a_{11}} \sum_{j=2}^{n} a_{1j}x_j + \frac{1}{a_{11}^2} \sum_{i,j=2}^{n} a_{1j}a_{i1}x_i x_j \right) + \sum_{i,j=2}^{n} \tilde{a}_{ij}x_i x_j$$

$$= a_{11} \left(x_1 + \frac{1}{a_{11}} \sum_{j=2}^{n} a_{1j}x_j \right)^2 + (\tilde{A}_{i,j>1}\tilde{x}, \tilde{x})$$

The positive definiteness of $\tilde{A}_{i,j>1}$ thus follows with the choice

$$x_1 = -\frac{1}{a_{11}} \sum_{j=2}^{n} a_{1j} x_j.$$

$\square$

For a symmetric positive definite matrix A, the LU factorization can always be performed without pivoting. Only positive pivot elements $a_{ii}^{(i-1)}$ occur. That is, the matrix U has only positive diagonal elements $u_{ii} > 0$. Let $D \in \mathbb{R}^{n \times n}$ be the diagonal matrix with $d_{ii} = u_{ii} > 0$. Then we have $A = LU = LD\tilde{U}$ with an upper right triangular matrix $\tilde{U} = D^{-1}U$, which only has ones on the diagonal. Since A is symmetric, it follows:

$$A = LU = LD\tilde{U} = \tilde{U}^T DL^T = A^T$$

Due to the uniqueness of the LU factorization, it must be $L = \tilde{U}^T$ and $u = DL^T$. Since D only has positive diagonal entries, the matrix $\sqrt{D}$ exists and we write

$$A = LU = LD\tilde{U} = LD^{\frac{1}{2}}D^{\frac{1}{2}}\tilde{U} = \underbrace{LD^{\frac{1}{2}}}_{=:\tilde{L}} D^{-\frac{1}{2}}u.$$

Since $L = \tilde{U}^T$, it follows that $\tilde{L}^T = D^{\frac{1}{2}}\tilde{U} = D^{-\frac{1}{2}}U$, so $A = \tilde{L}\tilde{L}^T$.

Theorem 3.26 (Cholesky Factorization) *Let $A \in \mathbb{R}^{n \times n}$ be a symmetric positive definite matrix. Then the Cholesky factorization into a lower left triangular matrix $\tilde{L}$*

$$A = \tilde{L}\tilde{L}^T$$

exists, and it can be computed without pivoting in

$$\frac{n^3}{6} + O(n^2)$$

elementary operations.

Instead of a proof, we provide an efficient algorithm for the direct computation of the Cholesky factorization. Here, the necessary effort can easily be deduced.

Algorithm 3.27: Direct Computation of the Cholesky Factorization

Input: $A \in \mathbb{R}^{n \times n}$ symmetric positive definite.
1 **for** $j = 1$ **to** n **do**
2      $l_{11} = \sqrt{a_{11}}$, or $l_{jj} = \sqrt{a_{jj} - \sum_{k=1}^{j-1} l_{jk}^2}$.
3   **for** $i = j + 1$ **to** n **do**
4          $l_{ij} = l_{jj}^{-1} \left(a_{ij} - \sum_{k=1}^{j-1} l_{ik} l_{jk} \right)$.
Result: The lower triangular matrix of the Cholesky factorization $L = (l_{ij})$, $j \le i$.

The Python implementation is

♣ Implementation 3.28: Direct Computation of the Cholesky Factorization

```python
def cholesky(A):
    n = A.shape[0]
    L = np.zeros_like(A)

    for j in range(0, n):
        l = 0
        for k in range(0, j):
            l += L[j, k]**2
        L[j, j] = np.sqrt(A[j, j] - l)
        for i in range(j + 1, n):
            l = 0
            for k in range(j):
                l += L[i, k] * L[j, k]
            L[i, j] = (A[i, j] - l) / L[j, j]
    return L
```

The algorithm can be derived iteratively from the relationship $\tilde{L}\tilde{L}^T = A$. It holds:

$$a_{ij} = \sum_{k=1}^{\min\{i,j\}} l_{ik} l_{jk}$$

We proceed column by column for $j = 1, 2, \ldots, n$. That is, $\tilde{L}$ is known for all columns up to $j - 1$. Then, in column j initially for the diagonal element:

$$a_{jj} = \sum_{k=1}^{j} l_{jk}^2 \quad \Rightarrow \quad l_{jj} = \sqrt{a_{jj} - \sum_{k=1}^{j-1} l_{jk}^2}$$

If the jth diagonal element l_{jj} is known, then for $i > j$, it holds that

$$a_{ij} = \sum_{k=1}^{j} l_{ik} l_{jk} \quad \Rightarrow \quad l_{ij} l_{jj} = a_{ij} - \sum_{k=1}^{j-1} l_{ik} l_{jk}$$

$$\Rightarrow \quad l_{ij} = l_{jj}^{-1} \left(a_{ij} - \sum_{k=1}^{j-1} l_{ik} l_{jk} \right).$$

To solve the linear system with the Cholesky factorization, the algorithm for forward and backward substitution must be adapted, as $\tilde{L}$ has arbitrary (positive) entries on the diagonal and not necessarily ones.

3.4 Sparse Matrices and Band Matrices

The effort to create the LU or Cholesky factorization grows cubically with the size of the matrix. In many application problems, *sparse matrices* occur:

Definition 3.29 (*Sparse Matrix*) A matrix $A \in \mathbb{R}^{n \times n}$ is called *sparse matrix*, if it only has $O(n)$ non-zero entries. The *sparsity pattern* $B \subset \{1, \ldots, n\}^2$ of A is the set of all index pairs (i, j) with $a_{ij} \neq 0$.

Other, less strict definitions of sparse matrices require that the matrix has $O(n \log(n))$ or even $O(n\sqrt{n})$ or simply $o(n^2)$ non-zero entries.

Examples of sparse matrices are a *tridiagonal matrices* of the form

$$A \in \mathbb{R}^{n \times n}, \quad a_{ij} = 0 \quad \forall |i - j| > 1.$$

A generalization of tridiagonal systems is a banded matrix.

Definition 3.30 (*Banded Matrix*) A matrix $A \in \mathbb{R}^{n \times n}$ is called a *banded matrix* with *bandwidth* $m \in \mathbb{N}$, if

$$a_{ij} = 0 \quad \forall |i - j| > m.$$

A banded matrix has at most $n(2m + 1)$ non-zero entries.

It turns out that the LU factorization of a banded matrix is again a banded matrix and can therefore be performed efficiently.

Theorem 3.31 (LU Factorization of a Banded Matrix) *Let $A \in \mathbb{R}^{n \times n}$ be a banded matrix with bandwidth m. The LU factorization (without permutation)*

$$LU = A$$

is again a banded matrix, i.e.,

$$l_{ij} = u_{ij} = 0 \quad \forall |i - j| > m,$$

and can be performed in

$$O(nm^2)$$

operations.

Proof We show inductively that the resulting elimination matrices $\tilde{A}_{i, j>1}$ are again banded matrices. It holds:

$$\tilde{a}_{ij} = a_{ij} - \frac{a_{1j}a_{i1}}{a_{11}}, \quad i, j = 2, \ldots, n.$$

Let $|i - j| > m$. Then $a_{ij} = 0$ and $a_{1j} = a_{i1} = 0$, since $1 \le i, j$.

At most m^2 elementary operations must be executed in step i instead of $(n - i)^2$ operations. The Frobenius matrix has a maximum of m non-trivial elements to be computed. In total, the effort is

$$\sum_{i=1}^{n}(m + m^2) = nm^2 + O(nm).$$

$\square$

For tridiagonal matrices, the LU factorization can be performed very efficiently, and the algorithm has its own name.

Theorem 3.32 (Thomas Algorithm) *Let $A \in \mathbb{R}^{n \times n}$ be a regular tridiagonal matrix. The matrices L and U of the LU factorization of A are again tridiagonal and can be created in $O(n)$ operations.*

Proof This follows directly from Theorem 3.31.

$\square$

The difference between performing an LU factorization for a dense matrix and a banded matrix is substantial. To discretize the Laplace equation with finite differences, linear equation systems with the so-called *model matrix* have to be solved:

$$A = \begin{pmatrix} A_m & -I_m & 0 & \cdots & 0 \\ -I_m & A_m & -I_m & & \vdots \\ 0 & -I_m & \ddots & \ddots & 0 \\ \vdots & & \ddots & \ddots & -I_m \\ 0 & \cdots & 0 & -I_m & A_m \end{pmatrix}, \quad A_m = \left. \begin{pmatrix} 4 & -1 & 0 & \cdots & 0 \\ -1 & 4 & -1 & & \vdots \\ 0 & -1 & \ddots & \ddots & 0 \\ \vdots & & \ddots & \ddots & -1 \\ 0 & \cdots & 0 & -1 & 4 \end{pmatrix} \right\} m$$

Table 3.2 Computing time for creating the LU factorization of a banded matrix $A \in \mathbb{R}^{n \times n}$ with bandwidth $m = \sqrt{n}$ on a computer with 10 GigaFLOPS

n	Operations	Time (band structure)	Time (dense)
100	10^5	1 µs	30 µs
1 000	10^6	100 µs	30 ms
10 000	10^8	10 ms	30 s
100 000	10^{10}	1 s	10 h
1 000 000	10^{12}	2 min	1 year

The matrix $A \in \mathbb{R}^{n \times n}$ (for a square number n) has the bandwidth $m = \sqrt{n}$. In Table 3.2, we give the necessary computing times for creating the LU factorization on a hypothetical computer with 10 GigaFLOPS. Compare with Table 3.1.

The model matrix is a banded matrix with bandwidth $m = \sqrt{n}$, which has at most four non-zero entries in each row besides the diagonal element, namely the entries $a_{i,i\pm1}$ and $a_{i,i\pm m}$. For sparse matrices, the question arises whether the LU factorization, i.e., the matrices L and U, have the same sparse pattern. This is generally not the case, and the matrices L and U can be dense. In the case of the model matrix, L and U are nearly dense banded matrices with bandwidth m.

The effort to calculate the LU factorization of a sparse matrix depends significantly on the sorting, i.e., the pivot of the matrix. We consider a simple example for this:

Example 3.33 (*LU Factorization of a Sparse Matrix*) We consider the two matrices:

$$
A_1 := \begin{pmatrix} 1 & 2 & 3 & 4 \\ 2 & 1 & 0 & 0 \\ 3 & 0 & 1 & 0 \\ 4 & 0 & 0 & 1 \end{pmatrix}, \quad
A_2 := \begin{pmatrix} 1 & 0 & 0 & 4 \\ 0 & 1 & 0 & 3 \\ 0 & 0 & 1 & 2 \\ 4 & 3 & 2 & 1 \end{pmatrix}.
$$

The two matrices are obtained by simultaneously swapping the first and fourth rows and columns, as well as the second and third rows and columns. It holds:

$$
P^T A_1 P = \begin{pmatrix} 0 & 0 & 0 & 1 \\ 0 & 0 & 1 & 0 \\ 0 & 1 & 0 & 0 \\ 1 & 0 & 0 & 0 \end{pmatrix}
\begin{pmatrix} 1 & 2 & 3 & 4 \\ 2 & 1 & 0 & 0 \\ 3 & 0 & 1 & 0 \\ 4 & 0 & 0 & 1 \end{pmatrix}
\begin{pmatrix} 0 & 0 & 0 & 1 \\ 0 & 0 & 1 & 0 \\ 0 & 1 & 0 & 0 \\ 1 & 0 & 0 & 0 \end{pmatrix}
= \begin{pmatrix} 1 & 0 & 0 & 4 \\ 0 & 1 & 0 & 3 \\ 0 & 0 & 1 & 2 \\ 4 & 3 & 2 & 1 \end{pmatrix} = A_2
$$

The LU factorization of the matrix A_1 (with Pivoting) is:

$$L_1 = \begin{pmatrix} 1 & 0 & 0 & 0 \\ 2 & 1 & 0 & 0 \\ 3 & 2 & 1 & 0 \\ 4 & \frac{8}{3} & 1 & 1 \end{pmatrix}, \quad U_1 = \begin{pmatrix} 1 & 2 & 3 & 4 \\ 0 & -3 & -6 & -8 \\ 0 & 0 & 4 & 4 \\ 0 & 0 & 0 & \frac{7}{3} \end{pmatrix},$$

and for the matrix A_2, we get (again with Pivoting)

$$L_2 = \begin{pmatrix} 1 & 0 & 0 & 0 \\ 0 & 1 & 0 & 0 \\ 0 & 0 & 1 & 0 \\ 4 & 3 & 2 & 1 \end{pmatrix}, \quad U_2 = \begin{pmatrix} 1 & 0 & 0 & 4 \\ 0 & 1 & 0 & 3 \\ 0 & 0 & 1 & 2 \\ 0 & 0 & 0 & -28 \end{pmatrix}.$$

Although both matrices describe the same linear system apart from row and column swaps, the LU factorizations have a completely different sparsity pattern: The LU factorization of matrix A_1 is dense. In contrast, the LU factorization of matrix A_2 manages with the same sparse pattern as the matrix A_2 itself.

This example can be generalized to the corresponding $n \times n$ matrix. In case 1, n^2 entries are necessary to store the LU factorization, in case 2, only $3n$. For $n \gg 1\,000$, this difference is crucial. The number of operations required differs even more significantly in both cases. ◄

3.4.1 Sparse Matrices in Practice

Sparse matrices often occur in applications, such as in the discretization of differential equations, but also in the calculation of cubic splines. In the Excursions 3.8 and 5.6, we will encounter sparse matrices of the model matrix type. For the LU factorization to benefit from the sparse structure, the matrix must be appropriately permuted (also referred to as "sorted"). If zero entries in A are overwritten in the LU factorization, this is referred to as *fill-ins*. Current improvements in the LU factorization are less based on the factorization itself but on the development of efficient sorting methods to reduce the *fill-ins*. We know that the LU factorization of a banded matrix with bandwidth m is again a banded matrix with bandwidth m and can be performed in $O(nm^2)$ operations. One idea for sorting the matrix A is to arrange the entries in such a way that the sorted matrix is a banded matrix with the thinnest possible bandwidth.

3.4.2 The Cuthill-McKee Sorting Algorithm

A classic method is the *Cuthill-McKee algorithm*, named after Elizabeth Cuthill and James McKee. This creates a sorting of the n indices, i.e., a bijection

$$f : \{1, 2, \dots, n\} \to \{1, 2, \dots, n\},$$

so that connected indices i and j, i.e., indices for matrix entries $a_{ij} \neq 0$, are close together. To derive the method, we need some terminology. Let $A \in \mathbb{R}^{n \times n}$ be a sparse matrix with sparsity pattern

$$B(A) = \{(i, j) \in \{1, \dots, n\} \times \{1, \dots, n\} : a_{ij} \neq 0\}.$$

For simplicity, we assume that B is symmetric. From $(i, j) \in B$ it follows that $(j, i) \in B$. However, the matrix itself does not have to be symmetric. For an index $i \in \{1, \dots, n\}$ let $\mathcal{N}(i)$ be the number of all indices connected with i:

$$\mathcal{N}(i) := \{j \in \{1, \dots n\} : (i, j) \in B\}.$$

And for an arbitrary set $N \subset \{1, \dots, n\}$ we denote with $\#N$ the set of elements in N, i.e., with $\#\mathcal{N}(i)$ the number of neighbors of i.

The algorithm gradually fills an index list $I = (i_1, i_2, \dots)$, until all indices $1, \dots, n$ are treated. We start with the index $I = (i_1)$, which has the smallest number of neighbors $\#\mathcal{N}(i_1) \leq \#\mathcal{N}(j), \forall j$. Subsequently, we add the neighbors $j \in \mathcal{N}(i_1)$ of i_1 in the order of their respective number of neighbors. In this way, the algorithm moves from neighbor to neighbor and processes the sparsity structure of the matrix until all indices have been added to the list.

In the precise definition of the algorithm, some special cases need to be considered so that it actually terminates and finds all indices:

Algorithm 3.34: Cuthill-McKee Algorithm

Input: $A \in \mathbb{R}^{n \times n}$ a sparse matrix with symmetric sparsity pattern $B \in (i, j)$
1 Initialize empty index list $I = [\ ]$ and an empty queue Q.
2 **for** $k = 1$ **to** n **do**
3   **if** $\#I = n$ **then**
4   Stop
5   **else if** $\#I = k - 1 < n$ **then**
6   Determine index element $i_k \notin I$: $\#\mathcal{N}(i_k) \leq \#\mathcal{N}(j) \ \forall\, l \in \{1, \dots, n\} \setminus I$.
7   Add i_k to the end of the list I.
8   Add i_k to the queue Q.
9   Select the first index i_k in the queue I.
10     $N_k := \mathcal{N}(i_k) \setminus I$
11   **if** $\#N_k > 0$ **then**
12   Sort $N_k = [s_1^k, s_2^k, \dots]$ according to $\#\mathcal{N}(s_i^k) \leq \#\mathcal{N}(s_{i+1}^k)$
13   Add $[s_1^k, s_2^k, \dots]$ sorted to the end of the list I.
14   Add $s_1, s_2, \dots$ to the end of the queue Q.
15   Remove i_k from the queue.
Result: A sorting of the indices of the variables with which the sorted matrix has a thinner bandwidth.

In Python, this can be implemented using the `scipy` library, which implements sparse matrices, as follows:

♣ Implementation 3.35: Cuthill-McKee Algorithm

```python
def cuthill_mckee(A):
    n = A.shape[0]
    row, col = A.nonzero()
    N = [(l, len(l)) for l in [list(row[col == i]) for i in range(n)]]

    I, Q = [], []
    R = [i for i in range(n)]

    for k in range(n):
        if len(I) == n:
            break
        elif len(I) == k:
            i = R[np.argmin(np.array([N[i][1] for i in R]))]
            I.append(i)
            Q.append(i)
            R.remove(i)

        i = Q[0]
        neighbours = [n for n in N[i][0] if n not in I]
        neighbours_sort = sorted(neighbours, key=lambda i: N[i][1])
        for ik in neighbours_sort:
            I.append(ik)
            Q.append(ik)
            R.remove(ik)

        Q.pop(0)

    data, row, col = [], [], []
    for key in A.todok().keys():
        data.append(A[key])
        row.append(I.index(key[0]))
        col.append(I.index(key[1]))

    return sp.sparse.csr_matrix((data, (row, col)), shape=(n, n))
```

To illustrate this, let us consider an example:

Example 3.36 (*Cuthill-McKee*) Let $A \in \mathbb{R}^{8 \times 8}$ be given with the following sparsity pattern:

$$
A := \begin{pmatrix}
* & * & & & & & & * \\
* & * & & & * & * & & * \\
& & * & & * & & * & * \\
& & & * & & & & \\
& * & * & & * & & * & \\
& * & & & & * & & \\
& & * & & * & & * & \\
* & * & * & & & & & *
\end{pmatrix}
$$

Index	$N(i)$	$\#N(i)$
1	$1, 2, 8$	3
2	$1, 2, 5, 6, 8$	5
3	$3, 5, 7, 8$	4
4	4	1
5	$2, 3, 5, 7$	4
6	$2, 6$	2
7	$3, 5, 7$	3
8	$1, 2, 3, 8$	4

- *Step $k = 1$*: (1) The index list I is empty. The index 4 has only one neighbor (itself), i.e.,

$$
i_1 = 4, \quad I = (4).
$$

(2) For the index 4, $N_1 = \mathcal{N}(4) \setminus I = \emptyset$, i.e., continue with $k = 2$.

- *Step* $k = 2$: (1) The index list has only $k - 1 = 1$ element. Of the remaining indices, index 6 has the minimal number of two neighbors, i.e.,

$$i_2 = 6, \quad I = (4, 6).$$

(2) It holds $N_2 = \mathcal{N}(6) \setminus I = \{2, 6\} \setminus \{4, 6\} = \{2\}$.
(3) A single element is naturally sorted, i.e.,

$$i_3 = 2, \quad I = (4, 6, 2).$$

- *Step* $k = 3$: (1) This step does not apply, as I already has three elements, it is $i_3 = 2$.
(2) It holds $N_3 = \mathcal{N}(2) \setminus I = \{1, 2, 5, 6, 8\} \setminus \{4, 6, 2\} = \{1, 5, 8\}$.
(3) It holds $\#N(1) = 3$, $\#N(5) = 4$ and $\#N(8) = 4$, i.e., we add in sorted order:

$$i_4 = 1, \ i_5 = 5, \ i_6 = 8, \quad I = (4, 6, 2, 1, 5, 8).$$

- *Step* $k = 4$: (1) I has more than three elements.
(2) $i_4 = 1$. It is $N_4 = \mathcal{N}(1) \setminus I = \{1, 2, 8\} \setminus \{4, 6, 2, 1, 5, 8\} = \emptyset$. Therefore, continue with $k = 5$
- *Step* $k = 5$. (1) I has enough elements.
(2) $i_5 = 5$. It is $N_5 = \mathcal{N}(5) \setminus I = \{2, 3, 5, 7\} \setminus \{4, 6, 2, 1, 5, 8\} = \{3, 7\}$.
(3) It holds $\#\mathcal{N}(3) = 4$ and $\#\mathcal{N}(7) = 3$, i.e., index 7 is added first:

$$i_7 = 7, \ i_8 = 3, \quad I = (4, 6, 2, 1, 5, 8, 7, 3)$$

- *Step* $k = 6$: (0) Stop, since $\#I = 8$.

For $I_0 = (1, 2, 3, 4, 5, 6, 7, 8)$, $(I_0)_j :\mapsto I_j$ we get the sorting:

$$1 \to 4, \ 2 \to 6, \ 3 \to 2, \ 4 \to 1, \ 5 \to 5, \ 6 \to 8, \ 7 \to 7, \ 8 \to 3$$

We create the sorted matrix $\tilde{A}$ according to $\tilde{a}_{kj} = a_{i_k i_j}$:

$$
A :=
\begin{array}{c}
\begin{array}{cccccccc}
4 & 6 & 2 & 1 & 5 & 8 & 3 & 7
\end{array} \\
\left(
\begin{array}{cccccccc}
* &   &   &   &   &   &   &   \\
  & * & * &   &   &   &   &   \\
  & * & * & * & * & * &   &   \\
  &   & * & * &   & * &   &   \\
  &   & * &   & * &   & * & * \\
  &   & * & * &   & * &   & * \\
  &   &   &   & * &   & * & * \\
  &   &   &   & * & * & * & *
\end{array}
\right)
\begin{array}{c}
4 \\ 6 \\ 2 \\ 1 \\ 5 \\ 8 \\ 7 \\ 3
\end{array}
\end{array}
$$

The sorted matrix is a banded matrix with bandwidth $m = 3$.

Many methods for sorting a matrix are based on graph theory. The Cuthill-McKee algorithm seeks a permutation of the sparsity pattern so that closely adjacent indices stand close together in the order. Other methods try to decompose the graph spanned by the sparsity pattern as much as possible into subgraphs. These subgraphs correspond to blocks in the matrix A. The subsequent LU factorization generates dense but small blocks. The individual blocks are only weakly coupled. An advantage of this sorting is the possibility of using efficient parallelization methods for the individual blocks.

In Fig. 3.1, we show the sparsity structure of a matrix $A \in \mathbb{R}^{1\,089 \times 1\,089}$ before and after corresponding sorting. In a grid of $1089 \cdot 1089$ points (i, j), a point is colored black if a_{ij} belongs to a matrix entry not equal to zero. Even if the first impression suggests something different, the same number of black points is present in all three figures. The matrix has a symmetric sparsity structure, i.e., if (i, j) is black, then (j, i) is likewise black and there are $9\,409$ entries not equal to zero (which is less than 1% of all entries). The LU factorization of the unsorted matrix is almost dense with about $1\,000\,000$ entries and its assembly requires about $400 \cdot 10^6$ operations. The Cuthill-McKee algorithm generates a banded matrix with very thin bandwidth $m \approx 70$. To store the LU factorization, only $150,000$ entries are necessary and the calculation requires about $5 \cdot 10^6$ operations, just $1/80$ of the effort. Finally, we show for comparison the sorting with a so-called *multi-frontal method*. Here, the matrix is divided into individual blocks. For the calculation of the LU factorization, $3 \cdot 10^6$ operations are necessary ($1/133$ of the effort). Details can be found in [45].

Example 3.37 (*Cholesky Factorization for Banded Matrices*) From Theorem 3.31, we know that the LU factorization of a banded matrix produces two banded matrices with the same bandwidth. It follows immediately that the same statement applies to the Cholesky factorization. We can exploit this in an implementation to avoid computational effort at points where we know that the L matrix must have a zero entry.

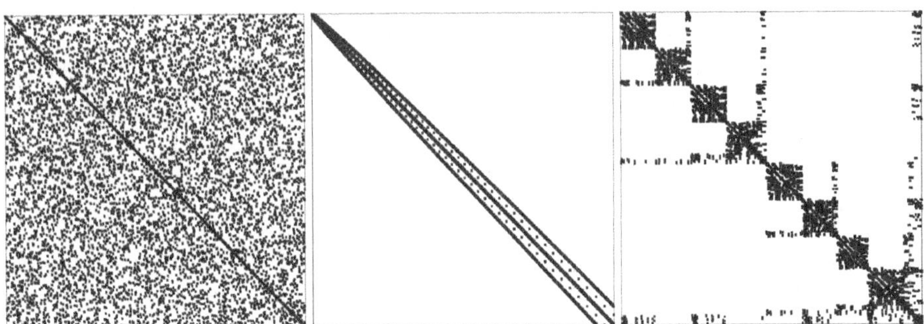

Fig. 3.1 Sparsity structure of a sparse matrix $A \in \mathbb{R}^{1089 \times 1089}$ with a total of $9,409$ non-zero entries. Left: Before sorting; Middle: Cuthill-McKee algorithm; Right: After sorting with a multi-frontal method from [45]

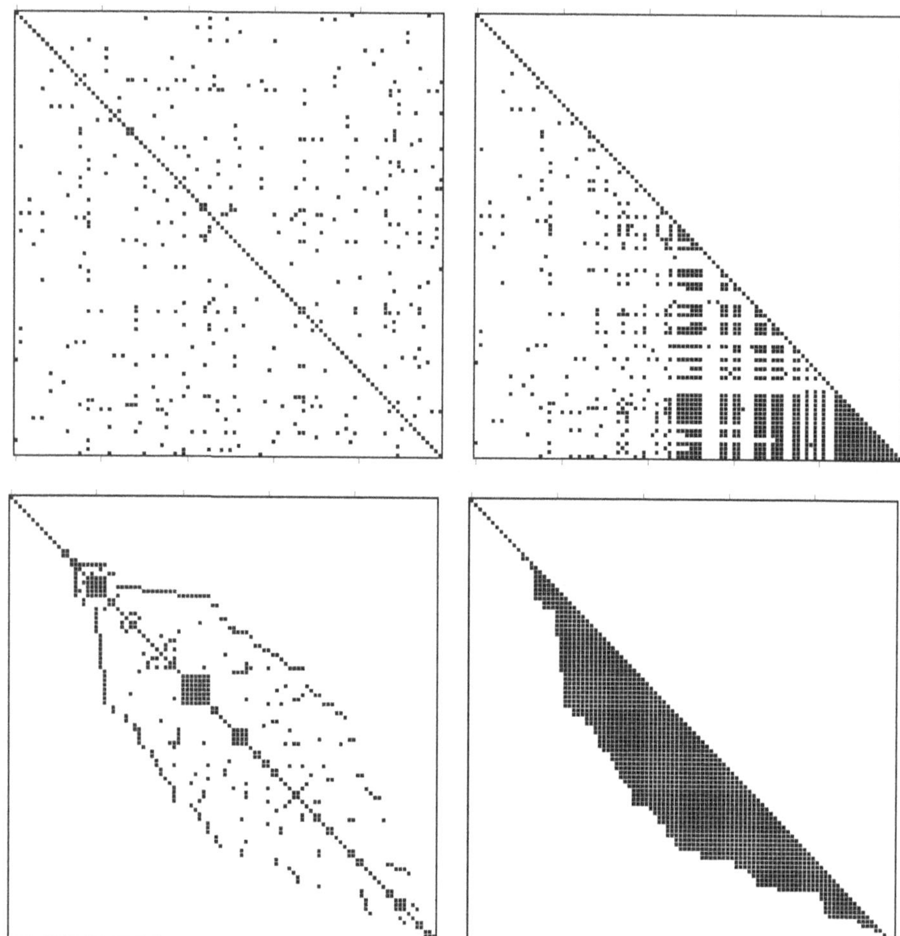

Fig. 3.2 Top left: Sparsity structure of a sparse symmetric positive definite matrix. Top right: Sparsity structure of the Cholesky factorization. Bottom left: Sparsity structure of the matrix after application of the Cuthill-McKee algorithm. Bottom right: Sparsity structure of the Cholesky factorization of the reordered matrix

With the `scipy` library, sparse matrices with random sparsity structure can be generated. If we construct such a matrix B, then we know that $A = I + BB^T$ is symmetric and positive definite. The sparsity structure of such a matrix is given in the top right of Fig. 3.2. If we apply the Cholesky factorization directly to it, we obtain a nearly dense matrix L. The sparsity structure of this is shown in the top right of Fig. 3.2.

With the Cuthill-McKee algorithm, we can reorder the matrix A to obtain a banded matrix with as small a bandwidth as possible. The sparsity structure that results for our matrix can be seen in the bottom left of Fig. 3.2. If we apply the Cholesky factorization to this, we see that we again have a banded matrix. The number of non-zero entries of this lower left triangular matrix is larger than that of the Cholesky factorization of the original matrix. Still, the factorization is more efficient to calculate and the band structure can now be exploited in the forward and backward substitution, so that these only have the effort $O(pn)$, where p is the bandwidth.   ◀

3.5   Excursus: LU Factorization on Parallel Computers

In Sect. 1.4 of the introduction, we already discussed the performance of modern computers. The increase in performance results from three factors:

- Increase in clock rate, i.e., the number of operations that a computer can perform per second,
- Increase in efficiency, i.e., a reduction in cycles that the processor needs per operation (e.g., the multiplication of two numbers),
- Use of many parallel cores.

In the 20 years from the Intel 486 in 1991 to the Intel Core i7 in 2011, the clock speed has increased from 50 MHz to about 3 GHz by a factor of 60. However, the increase in performance in *FLOPS (floating point operations per second)* is 70,000-fold, see Table 1.3. A significant part of this performance increase can be attributed to a better chip design, as fewer cycles are needed per operation. In addition, modern processors have several *cores*, so they can perform several operations simultaneously *in parallel*. The Core i7 can perform up to eight calculations simultaneously. However, this does not happen automatically, i.e., from the hardware side. Each numerical method must be implemented explicitly for such parallel computers. A usual *sequential* algorithm will not utilize this potential, i.e., the processor can only be utilized to about 10%. With graphics chips (GPU) with thousands of cores, this imbalance is even greater. If the numerical methods are not adapted to the new hardware, a large part of the potential performance is lost.

Here, using the example of the LU factorization of a dense matrix, we want to introduce some basic concepts of *parallel computing*, and otherwise refer to the literature [79].

Concepts of Parallel Computing

First, we clarify what is meant by parallel computing or a parallel computer. From [4], we take a first definition:

Definition 3.38 (*Parallel Computer*) A *parallel computer* is a collection of computing units (processors) that can quickly solve large problems through coordinated cooperation.

Parallelism can refer to the data and the algorithms:

- Modern processors have several (usually 4–16) *cores*, i.e., independent operating units. Independent algorithms run in parallel and access a common memory. This model is called *multiple instruction, single data (MISD)* and the parallel use of several cores is referred to as *multi-core parallelization.*
- Some computers have multiple processors, each of which has several cores. The processors continue to access a common memory area, so these models again fall into the *multiple instruction, single data* category. However, the processors each have their own local memory, the so-called *cache*. Access to data from the cache is very fast. This leads to the exchange of data between multiple processors being slower than working with their own data.
- Large parallel computers, so-called *clusters,* consist of networks of individual computers. On each computer, algorithms run in parallel, each of which can only access the local memory. So there is also a parallelization of the data. This model is called *multiple instruction, multiple data (MIMD)*. Each of the computers themselves can be of the MISD type and have several processors with several cores. To access data from other computers, they must be exchanged. This data exchange is much slower than accessing data in one's own memory.
- Modern graphics cards use a special form of parallelization. The same algorithm runs in parallel on different data. This model is called *single instruction, multiple data (SIMD)*.

Efficient use of parallel computers varies depending on the type of algorithm. When parallelizing algorithms, it is not enough to only consider the computing operations. The computer requires additional data for computation. If they cannot be read quickly enough from memory or the results cannot be written quickly enough into memory, then the processor has nothing to do.

Remark 3.39 (*Computing Speed and Bandwidth*) A modern multi-core CPU theoretically achieves a performance of more than 100 GigaFLOPS, i.e., more than $100\,000\,000\,000 = 10^{11}$ calculations per second. We consider the sum of two vectors

$$z = x + y, \quad z_i = x_i + y_i \text{ for } i = 1, \ldots, N,$$

with $N = 10^{10}$. Theoretically, the processor should be able to handle this task in 0.1 seconds. To add the entries $x_i + y_i$, x_i, y_i must be copied from memory to the CPU and z_i back into memory. In total, $3 \cdot 10^{10}$ floating-point numbers must be written from memory to the CPU or back. At double precision, see Sect. 1.4, these are $8 \cdot 3 \cdot 10^{10} = 24 \cdot 10^{10}$ bytes. A modern computer achieves a data rate of about 30 gigabytes per second, so that it can transfer $30 \cdot 10^9$ bytes between CPU and memory each second. In this example, about 8 seconds are added to the 0.1 seconds of computing time for data transfer. The efficiency of some operations is bounded by the memory, which makes the use of parallel computers significantly more challenging.   ◆

In light of this discussion, we will always assume throughout this section that the transfer of data involves an effort that may be much greater than the actual computational effort.

The goal of parallelization is to reduce runtimes by simultaneous use of $P \in \mathbb{N}$ parallel *processes*. As an extension of Definition 1.10, we define:

Definition 3.40 (*Parallel Effort, Speedup, Overhead and Efficiency*) To solve a problem of size $n \in \mathbb{N}$, an *optimal sequential algorithm* with the *effort* $T(n)$ is given. (Here, the effort can be measured either in elementary operations or in runtime.) We denote the effort of the parallel algorithm with $P \in \mathbb{N}$ parallel processes by $T_P(n)$. The *parallel speedup* $S_P(n)$, the *parallel efficiency* $E_P(n)$ and the *parallel overhead* $O_P(n)$ are given as

$$S_P(n) = \frac{T(n)}{T_P(n)}, \quad E_P(n) = \frac{S_P(n)}{P}, \quad O_P(n) = PT_P(n) - T(n).$$

We consider an example for this.

Example 3.41 (*Parallel Addition of Two Vectors*) The task is to compute $x = y + z$ with $x, y, z \in \mathbb{R}^n$. The sequential algorithm is given as

Algorithm 3.42: Sequential Addition of Two Vectors

```
1  for i = 1 to n do
2       x_i = y_i + z_i
```

and has a runtime of $T(n) = n$, measured in elementary operations. We now consider a parallel variant of the procedure, distributed over $P \in \mathbb{N}$ processes, which can access the same memory area. We assume that P is a divisor of n.

Algorithm 3.43: Parallel Addition of Two Vectors

```
1  n_p = n/P
   /* For each process p=1,...,P:                                    */
2  for i = p * n_p + 1 to (p + 1) * n_p do
3       x_i = y_i + z_i
```

The actual work, i.e., the addition of the vectors, is optimally divided among the P processes. Nevertheless, this algorithm has a small parallel overhead, the calculation of the elements per process, $n_p = n/P$, as well as the calculation of the boundaries for each process, $p * n_p$ and $(p + 1) * n_p$. Thus, we have:

$$T_P(n) = \frac{n}{P} + 3, \quad O_P(n) = PT_P(n) - T(n) = (n - n) + 3P = 3P.$$

The parallel speedup is given by

$$S_P(n) = \frac{T(n)}{T_P(n)} = \frac{n}{\frac{n}{P} + 3} = \frac{P}{1 + 3\frac{P}{n}},$$

and the parallel efficiency is given by

$$E_P(n) = \frac{S_P(n)}{P} = \frac{1}{1 + 3\frac{P}{n}}. \tag{3.9}$$

For the efficiency, $0 < E_P(n) < 1$ holds. For fixed P and $n \to \infty$, it follows that $E_P(n) \to 1$. Conversely, for fixed n and $P \to n$, it follows that $E_P(n) \to 1/4$. In Fig. 3.3, the parallel speedup for $P = 1, \ldots, 500$ processes is shown for $n = 1\,0000$ and $n = 10\,000$. Depending on the problem size, the speedup and thus the parallel efficiency decrease significantly. For a small number of parallel processes, this parallel procedure will be very efficient. On the eight cores of a Core i7, almost optimal speedup can be expected. Note that we did not consider the access to memory, which, on a shared memory computer, would significantly alter the picture and lower the efficiency. ◄

This simple example shows that the analysis of parallel algorithms is not trivial. The parallel efficiency depends on the number of processors and the problem size. For a more detailed analysis of parallel procedures, two further terms are introduced:

Definition 3.44 (*Scalability*) The *strong scalability* of a parallel procedure refers to the behavior of the parallel runtime with an increasing number of processes P and a fixed problem size n. The *weak scalability* of a parallel procedure refers to the behavior of the parallel runtime with an increasing number of processes P and a fixed problem size per process n/P.

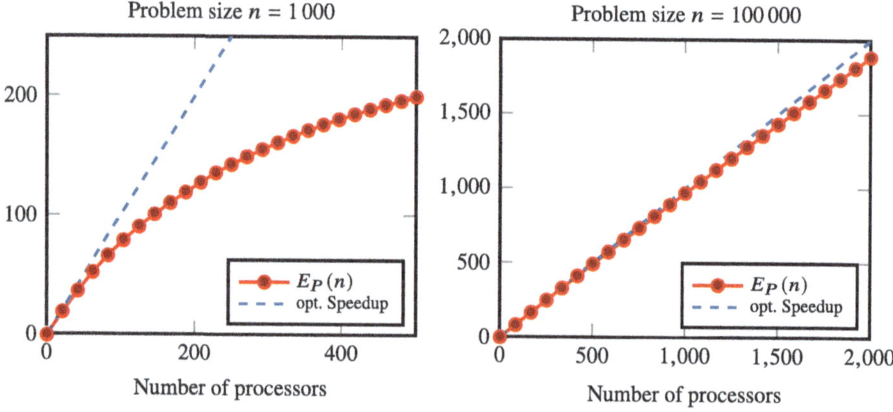

Fig. 3.3 Theoretical parallel speedup for the addition of two vectors for an increasing number of processes. The achieved efficiency increase strongly depends on the problem size

We have already seen that the *strong scalability* worsens as the number of processes increases. To analyze the *weak scalability*, we introduce the size $c = n/P$ and obtain from (3.9)

$$E_P(n) = \frac{1}{1 + \frac{3}{c}} \text{ for } c = \frac{n}{P} \text{ fixed.}$$

For $n \to \infty$ while simultaneously increasing the number of processes with $P = c/n$, it follows for the efficiency that $E_P(n) \to 1/(1 + 3/c)$. If we keep the problem size per process fixed, for example, at $c = 100$, then the parallel efficiency always remains $E_P(n) \approx 0.97$.

The addition of two vectors is a straightforward procedure to parallelize, as the individual threads can work independently of each other, each on their own data. Problems usually arise when data needs to be exchanged between threads. To illustrate this, let's consider a second simple example:

Example 3.45 *(Parallel Dot Product)* Let $x, y \in \mathbb{R}^n$. The task is to compute the dot product $r = (x, y)$. A sequential algorithm is given by

```
1  r = 0
2  for i = 1 to n do
3      r = r + x_i y_i
```

and again we get a linear runtime $T(n) = n$. For a parallel algorithm, we now imagine that the vectors $x, y \in \mathbb{R}^n$ are already divided among P processes. Each process calculates the partial product, and then the result is collected on the process with id $p = 1$.

Algorithm 3.46: Parallel Dot Product

```
     /* On the process with id p ∈ {1,...,P}                              */
1  n_p = n/P
2  r_p = 0
3  for i = p n_p + 1 to (p + 1) n_p do
4      r_p = r_p + x_i y_i
5  if p = 1 then
6      Set r = r_1
7      for i = 2 to P do
8          Receive r_i from process i
9          r = r + r_i
10     for i = 2 to P do
11         Send result r to process i
12 else
13     Send partial result r_p to process 1
14     Receive total result r from process 1
```

The first phase of the parallel algorithm is comparable to the addition of two vectors. The parallel runtime is n/P. In the second phase of the algorithm, the results are added. This happens on the process with number 1. This process receives the results from the processes $p = 2, \ldots, P$ in sequence, adds them, and then sends them. The effort for this second step is

$$T_{sum}(n, P) = 2c_k P + P,$$

where c_k describes how long the communication, i.e., the exchange of a value between the processes, takes. The total effort is

$$T(n, P) = \frac{n}{P} + 2c_k P + P,$$

and a *parallel efficiency* of

$$E_P(n) = \frac{T(n)}{PT_P(n)} = \frac{n}{n + 2c_k P^2 + P^2} = \frac{1}{1 + \frac{2c_k+1}{n} P^2}.$$

For the analysis, we examine the weak scalability, thus choosing $c = n/P$ fixed. Then

$$E_P(n) = \frac{1}{1 + \frac{2c_k+1}{c} P}. \tag{3.10}$$

In Fig. 3.4, we show the efficiency curve for $c = 10\,000$. We thus assume that each process has a subproblem of size $10\,000$. Assuming a shared memory computer, we choose $c_k = 10$: Accessing a number from memory is thus 10 times as expensive as a single calculation. If we assume that the processes are distributed on a large cluster, we choose $c_k = 1\,000$, the transfer rate is significantly slower.   ◄

This figure shows that the trivial implementation of the dot product is not suitable for parallelization. For each value that must be transferred to memory, just one calculation is necessary. On systems with many parallel processes and a poor network, the dot product can hardly be parallelized efficiently and must, as far as possible, be avoided.

Algorithm 3.46 is not optimal. The collection of distributed data, as it happens in the second step of the calculation, is a parallel operation that appears in many algorithms. It is called *all-to-one reduction* and can be performed more efficiently. The idea is not to employ just one thread: Initially, each thread with number $p = 2k$ forms the sum of itself and its neighbor, i.e., $r_{2k} = r_{2k} + r_{2k+1}$. Subsequently, all threads with number $p = 4k$ form the sums $r_{4k} = r_{4k} + r_{4k+2}$. This is done recursively in $\log_2(P)$ steps until the first thread has the result, see [79].

Algorithm 3.47: Communication — All to Process 1

```
    /* For  p = 1,..., P = 2^k                                      */
 1  for i = 1 to k − 1 do
 2      r = ⌈p/2^i⌉
 3      s = p mod 2^i
 4      if s = 1 then
 5          Receive from p + 2^{i−1}
 6      else
 7          Send to s
```

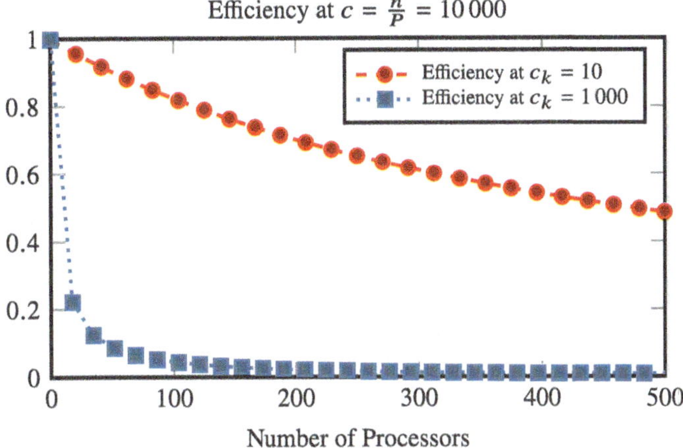

Fig. 3.4 Weak scalability for the parallel dot product (non-optimal algorithm) at fixed problem size per thread $c = n/P = 10\,000$. The case $c_k = 10$ corresponds to a system with high-speed access to the memory, e.g., a *shared memory system* with many cores and shared main memory, while the choice $c_k = 1\,000$ corresponds to a system with slow memory connection, such as a *cluster* with an interconnecting network

Process 1 collects the information from all other processes in $k = \log_2(P)$ steps. Using this method, the parallel runtime of the dot product can be reduced to

$$T(n, P) = \frac{n}{P} + 2c_k \log_2(P) + \log_2(P).$$

For a fixed problem size per thread, this leads to

$$E_P(n) = \frac{1}{1 + \frac{(2c_k+1)}{c} \log_2(P)},$$

and significantly better parallel efficiencies. In the case $c_k = 1\,000$, however, the parallel efficiency for $c = 10\,000$ decreases very quickly, see Fig. 3.5.

Parallel LU Decomposition

These two simple examples show that the analysis of parallel algorithms is very complex and even a small carelessness (like a too simple summation) can prevent efficient parallelization. We now want to turn to the actual task, the development of a parallel LU factorization. Let $A \in \mathbb{R}^{n \times n}$ be a (dense) regular matrix. We again assume that the communication of values between the processes is "expensive", i.e., $c_k > 1$. We consider the LU-factorization $A = LU$ without pivoting (see Sect. 3.2.2) and want to parallelize the algorithm of the LU *factorization in place*, see Algorithm 3.14:

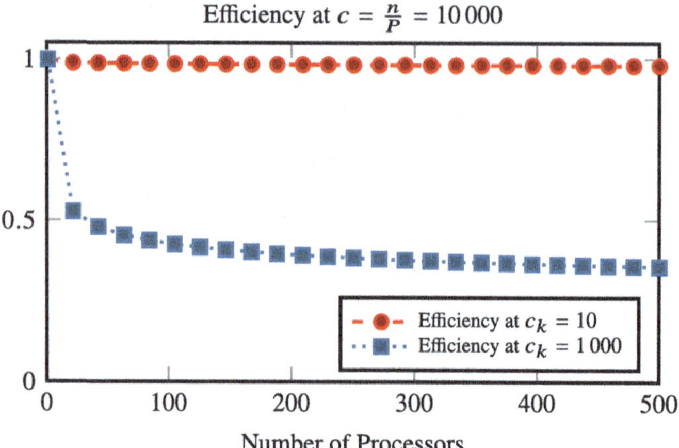

Efficiency at $c = \frac{n}{P} = 10\,000$

Fig. 3.5 Weak scalability for the parallel dot product (optimal logarithmic algorithm) for a fixed problem size per thread $c = n/P = 10\,000$ and for $c_k = 10$ (fast memory access) and $c_k = 1\,000$ (slow memory access)

Algorithm 3.48: Sequential LU Factorization of a Matrix A

Input: $A = (a_{ij})_{ij} \in \mathbb{R}^{n \times n}$ a regular matrix (whose LU factorization exists)
1 **for** $i = 1$ **to** n **do**
2   **for** $k = i + 1$ **to** n **do**
3          $a_{ki} = a_{ki}/a_{ii}$
4   **for** $j = i + 1$ **to** n **do**
5              $a_{kj} = a_{kj} - a_{ki} \cdot a_{ij}$

In Theorem 3.8, we have determined the sequential effort of the LU factorization as

$$T(n) = \frac{1}{3}n^3 + O(n^2) \tag{3.11}$$

For the parallelization, we assume that the matrix $A \in \mathbb{R}^{n \times n}$ is distributed row-wise to the P processes. For this, let $n_p = n/P \in \mathbb{N}$ be an integer. We choose the factorization initially so that process 1 has the rows $1, \ldots, n_p$, process 2 has $n_p + 1, \ldots, 2n_p$ and so on. For this, we define the index sets

$$p_s := (p-1) \cdot n_p + 1, \quad p_e := p \cdot n_p, \quad N(p) = \{p_s, \ldots, p_e\}, \quad p = 1, \ldots, P.$$

A row $i \in \{1, \ldots, N\}$ is in index set $N(\lceil i/n_p \rceil)$.

$$
A := \begin{pmatrix}
a_{1,1} & a_{1,2} & \cdots\cdots\cdots\cdots & a_{1,n} \\
\vdots & \vdots & & \vdots \\
a_{n_p,1} & a_{n_p,2} & \cdots\cdots\cdots\cdots & a_{n_p,n} \\
\hline
a_{n_p+1,1} & a_{n_p+1,2} & \cdots\cdots\cdots\cdots & a_{n_p+1,n} \\
\vdots & \vdots & & \vdots \\
a_{2n_p,1} & a_{2n_p,2} & \cdots\cdots\cdots\cdots & a_{2n_p,n} \\
\hline
\vdots & \vdots & & \vdots \\
\vdots & \vdots & & \vdots \\
\hline
a_{(p-1)\cdot n_p+1,1} & a_{(p-1)n_p+1,2} & \cdots\cdots\cdots\cdots & a_{(p-1)n_p+1,n} \\
\vdots & \vdots & & \vdots \\
a_{n,1} & a_{n,2} & \cdots\cdots\cdots\cdots & a_{n,n}
\end{pmatrix}
$$

The rows below the diagonal can now be processed in parallel in each process. To perform steps 3 and 5 of Algorithm 3.48, the processes must know the pivot row i. If this row is not stored in the corresponding process, the elements $a_{i,j}$ for $j \geq i$ must first be transferred.

Algorithm 3.49: Parallel LU Factorization of a Matrix A

> **Input:** $A = (a_{ij})_{ij} \in \mathbb{R}^{n \times n}$ a regular matrix (whose LU factorization exists).
> Process p has rows $a_{i,j}$ with $i \in \mathcal{N}(p)$.
> ```
> /* On process  p:                                               */
> ```
> 1 $p_s = (p-1) \cdot n_p + 1,\ p_e = p \cdot n_p$
> 2   **for** $i = 1$ **to** n **do**
> ```
>        /* Exchange pivot row                                    */
> ```
> 3   **if** $\lceil i/n_p \rceil = p$ **then**
> 4   Send $a_{i,i}, \ldots, a_{i,n}$ to processes $p+1, \ldots, P$
> 5   **else if** $\lceil i/n_p \rceil < p$ **then**
> 6   Receive $a_{i,i}, \ldots, a_{i,n}$ from process p
> ```
>        /* Eliminate                                             */
> ```
> 7   **for** $k = \min\{i+1, p_s\}$ **to** $\max\{n, p_e\}$ **do**
> 8 $a_{ki} = a_{ki}/a_{ii}$
> 9   **for** $j = i+1$ **to** n **do**
> 10 $a_{kj} = a_{kj} - a_{ki} \cdot a_{ij}$

The actual elimination loop $7{-}10$ is performed on each process p for the rows $i \in \mathcal{N}(p)$. In step i, each process performs at most $(n-i) \cdot n_p$ operations. The additional parallel work involves the transfer of the pivot row, i.e., the transfer of $n-i$ entries to the processes $q = p+1, \ldots, P$. Here, the active process p has the most work, as it sends the row to all other processes (which spend most of the time waiting for this). The distribution of the values to the $P - p$ processes can be done with the optimal method from Algorithm 3.47 in $\lceil \log_2(P-p) \rceil$ steps. We summarize the effort of the parallel method in step i:

$$T_i(n, P) = c_k \log_2\left(P - \lceil i/n_p \rceil\right) \cdot (n - i) + (n - i)\min\{n_p, n - i\}$$

The minimum comes into play because in the last n_p steps, no process has a full block to process anymore. The logarithmic term describes the effort in communication. We roughly approximate it as $\log_2(P - \lceil i/n_p \rceil) \approx \log_2(P)$, sum over i from 1 to n and obtain with the known sum formulas

$$T(n, P) = \sum_{i=1}^{n} T_i(n, P) \approx \sum_{i=1}^{n} \left(c_k(n - i)\log_2(P) + (n - i)\min\{n_p, n - i\}\right)$$

$$= c_k \frac{n^2 \log_2(P)}{2} + \frac{n^3}{2P} + O(n^2) + O(n^3/P^2).$$

The *parallel speedup* gets

$$S_P(n) = \frac{T(n)}{T(n, P)} \approx \frac{1}{\frac{3}{2P} + \frac{3c_k \log_2(P)}{2n}}$$

and the *parallel efficiency* is calculated as

$$E_P(n) = \frac{S_P(n)}{P} = \frac{1}{\frac{3}{2} + \frac{3c_k P \log_2(P)}{2n}} = \frac{1}{\frac{3}{2} + \frac{3c_k \log_2(P)}{2n_p}}, \tag{3.12}$$

where the problem size per process is $n_p = n/P$. For the cases $c_k = 10$ and $c_k = 1\,000$ at $n_p = 1\,000$, we show the achievable parallel efficiency in Fig. 3.6. For $c_k = 10$, the LU factorization is highly parallelizable. However, for $c_k = 1\,000$, the efficiency drops very quickly. Note that $n_p = 1\,000$ for $p = 500$ processors means that $n = p \cdot n_p = 500\,000$. So we are already dealing with a matrix of size $A \in \mathbb{R}^{500\,000 \times 500\,000}$ with $2.5 \cdot 10^{11}$ entries. For double precision, already $2\,000$ gigabytes of main memory are needed for storage. On a *cluster* of 500 individual computers, each with only 4 gigabytes of memory, this is possible, but the slow communication between the individual nodes makes the execution inefficient.

The problem with the previous algorithm is the poor distribution of work across the P processes. Once row n_P is passed, the first process has no work left. After row $2n_p$, the second process is idle. This is due to the row-wise partitioning of the data. This is referred to as poor *load balancing*. A better distribution is achieved when the rows are alternately assigned to the P processes. This results in the index sets

$$\mathcal{N}(p) = \{p, p + P, p + 2P, \ldots, p + (n_p - 1)P\}, \quad p = 1, \ldots, P.$$

Communication now always takes place with all other processes.

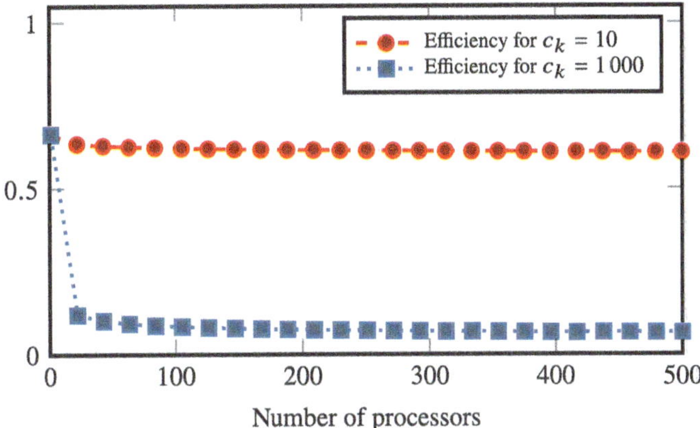

Fig. 3.6 Weak scalability for parallel LU factorization with logarithmic communication and fixed problem size per processor of $n_p = 1\,000$

Algorithm 3.50: Parallel LU Factorization of a Matrix A with Alternating Distribution

> **Input:** $A = (a_{ij})_{ij} \in \mathbb{R}^{n \times n}$ a regular matrix (whose LU factorization exists).
> Process p has rows $a_{i,j}$ with $i \in \mathcal{N}(p)$
>   `/* On process p:`   `*/`
> 1 **for** $i = 1$ **to** n **do**
>   `/* Exchange pivot row`   `*/`
> 2   **if** $i \bmod P = p$ **then**
> 3   Send $a_{i,i}, \ldots, a_{i,n}$ to other processes
> 4   **else if** $i \bmod P \neq p$ **then**
> 5   Receive $a_{i,i}, \ldots, a_{i,n}$ from process $i \bmod P$
>   `/* Eliminate`   `*/`
> 6   **for** $k = i + 1$ **to** n **do**
> 7   **if** $k \bmod P = p$ **then**
> 8 $a_{ki} = a_{ki}/a_{ii}$
> 9   **for** $j = i + 1$ **to** n **do**
> 10 $a_{kj} = a_{kj} - a_{ki} \cdot a_{ij}$

To analyze the runtime, we again separate the effort for the communication and the actual computational effort. Since each process in step i still has to eliminate $n_p - \lceil i/P \rceil$ rows of length $n - i$, the approximation is

$$T_i(n, P) \approx c_k \log_2(P)(n - i) + \left(n_p - \frac{i}{P}\right)(n - i),$$

and summed up

$$T(n, P) = \sum_{i=1}^{n} \left(c_k \log_2(P)(n - i) + \frac{(n - i)^2}{P} \right)$$

$$= \frac{c_k n(n + 1) \log_2(P)}{2} + \frac{n(n + 1)(2n + 1)}{6P}.$$

This results in the *parallel efficiency* (we skip the *speedup*)

$$E(n, P) = \frac{1}{\frac{c_k P \log_2(P)}{2n} + 1} = \frac{1}{1 + c_k \frac{\log_2(P)}{n_p}}.$$

The comparison with (3.12) shows a 50% better speedup. In particular, we achieve the efficiency of the sequential algorithm for small P. We cannot avoid a drop in efficiency for slow data exchange rates.

Good parallelization of numerical methods is very complex. Problems come into focus that usually don't need to be considered in mathematics: An approach can be good on a specific type of computer, but fail on another one. For further studies, we refer to the extensive literature, both on theoretical concepts of parallel computing as well as on practical applications and tips for implementation [79, 85].

3.6   Iterative Refinement

The numerically performed LU factorization $A = LU$ is only an approximation due to rounding errors, and the thus approximated solution $LU\tilde{x} = b$ is afflicted with an error $x - \tilde{x}$. The *residual* provides an idea to the error.

Definition 3.51 (*Residual of a Linear System*) Let $\tilde{x} \in \mathbb{R}^n$ be the approximation to the solution $x \in \mathbb{R}^n$ of $Ax = b$. By

$$r(\tilde{x}) := b - A\tilde{x},$$

we denote the *residual*.

For the exact solution $x \in \mathbb{R}^n$ of $Ax = b$, it holds $r(x) = 0$. The more accurate our solution is, the smaller the residual.

Theorem 3.52 (Error Estimate for Linear Systems) *Let $A \in \mathbb{R}^{n \times n}$ be a regular matrix and $\tilde{x} \in \mathbb{R}^n$ the approximation to the solution $x \in \mathbb{R}^n$ of $Ax = b$. Then we have*

$$\frac{\|x - \tilde{x}\|}{\|x\|} \leq \operatorname{cond}(A) \frac{\|(\tilde{x})\|}{\|b\|}.$$

Proof We have

$$x - \tilde{x} = A^{-1}(b - A\tilde{x}) = A^{-1}(\tilde{x}) \quad \Rightarrow \quad \|x - \tilde{x}\| \leq \|A^{-1}\|\,\|(\tilde{x})\|.$$

Dividing by $\|b\| = \|Ax\| \leq \|A\|\,\|x\|$ yields the desired result

$$\frac{\|x - \tilde{x}\|}{\|x\|} \leq \|A\|\,\|A^{-1}\|\,\frac{\|(\tilde{x})\|}{\|b\|}.$$

$\square$

Assume that the residual $r(\tilde{x})$ can be computed without rounding error. Furthermore, we assume that we can solve the *residual equation*

$$Aw = r(\tilde{x}) = b - A\tilde{x}$$

exactly for $w \in \mathbb{R}^n$. Then $\tilde{x} + w$

$$\tilde{x} + w = \tilde{x} + A^{-1}(b - A\tilde{x}) = \tilde{x} + x - \tilde{x} = x$$

is the exact solution of the system. This process is called *residual correction* or *iterative refinement*. The assumption that the residual equation can be solved without rounding error is, of course, not realistic (then the original system could be solved exactly). However, the principle of iterative refinement can be exploited to perform the complex parts of the calculation, i.e., the factorization of the matrix $A = LU$, with reduced accuracy. This requires less memory and is much faster on many computer systems (such as GPU accelerators). The calculation of the residual $r(\tilde{x})$ as well as the forward and backward substitution is then performed with high accuracy.

Iterative refinement can be performed in several steps.

Algorithm 3.53: Iterative Refinement

Input: $A \in \mathbb{R}^{n \times n}$ a regular matrix, $b \in \mathbb{R}^n$, number of iterations N_{max}.

1 Create $LU = PA$ // single precision
2 $r^{(1)} = b$
3 $x^{(1)} = 0$
4 **for** $i = 1, \ldots, N_{max}$ **do**
5      $Ly^{(i)} = Pr^{(i)}$ // double precision
6      $Rw^{(i)} = y^{(i)}$ // double precision
7      $x^{(i+1)} = x^{(i)} + w^{(i)}$ // double precision
8      $r^{(i+1)} = b - Ax^{(i+1)}$ // double precision

Result: Approximate solution $x \in \mathbb{R}^n$ of $Ax = b$.

The advantage of iterative refinement lies in the multiple use of the assembled LU factorization. The calculation of the LU factorization in the first line of the procedure requires $O(n^3)$ operations. In contrast, forward and backward substitution, as well as the calculation of the residual, only need $O(n^2)$ operations. This means that even when using higher precision, the effort in step $5-8$ of the procedure is small compared to step 1.

Theorem 3.54 *(Defect Correction) Let $A \in \mathbb{R}^{n \times n}$ be a regular matrix and $b \in \mathbb{R}^n$. Furthermore, let $\epsilon > 0$ be the sufficiently small error tolerance of single precision and $0 < \alpha < \epsilon$ be the error tolerance of high precision (for simplicity, $\alpha = \epsilon^2$). We define the iteration*

$$x^{(0)} = 0, \quad x^{(i)} = x^{(i-1)} + (LU)^{-1}(b - Ax^{(i-1)}),$$

where the LU factorization of the matrix A is calculated with only single precision $O(\epsilon)$ and all further calculations are performed with high precision $O(\alpha)$. Then, for the exact solution $x = A^{-1}b$, the following asymptotic error estimate holds

$$\frac{\|x^{(i)} - x\|}{\|x\|} = c(A)O(\epsilon)\frac{\|x^{(i-1)} - x\|}{\|x\|} + O(\alpha),$$

where $c(A) > 0$ is a constant that depends on the condition number of the matrix A.

Proof We write a step of the iterative refinement in compact form and indicate the respective accuracies:

$$x^{(i)} = \underbrace{x^{(i-1)} + \overbrace{(LU)^{-1}}^{\text{low}}\ \overbrace{(b - Ax^{(i-1)})}^{\text{high}}}_{\text{high}}$$

With the estimates for the relative errors of the LU factorization, the matrix-vector multiplication, and the solution of linear systems, we obtain with the low accuracy $\epsilon > 0$ and the high accuracy $\epsilon > \alpha > 0$

$$x^{(i)} = \left(1 + O(\alpha)\right)\left(x^{(i-1)} + \left(I + c(A)O(\epsilon)\right)A^{-1}\left(1 + c(A)O(\alpha)\right)\left(b - Ax^{(i-1)}\right)\right),$$

where $c(A)$ depends on the condition number of the matrix A, see Theorem 3.5. Multiplying out and using $b = Ax$, we get

$$x^{(i)} = x^{(i-1)} + \left(1 + c(A)O(\epsilon) + c(A)^2 O(\epsilon\alpha)\right)\left(x - x^{(i-1)}\right) + O(\alpha)x.$$

With this we obtain

$$x - x^{(i)} = \left(c(A)O(\epsilon) + c(A)^2 O(\epsilon\alpha)\right)(x - x^{(i-1)}) + O(\alpha)x.$$

The result follows by taking the absolute value and dividing by $\|x\|$. In the case $c(A)\epsilon < 1$, the quadratic term is smaller and can be asymptotically neglected. $\square$

Each iterative refinement step improves the error by the factor $c(A)O(\epsilon)$ until the threshold $O(\alpha)$ is reached. The benefit of iterative refinement lies on the one hand in the reduced memory requirement when L and U are calculated with low accuracy, but also in the faster execution. This is particularly relevant on modern accelerator hardware, where calculations can be performed in double precision, in single, and, on so-called *tensor cores*, in half precision. However, iterative refinement reaches its limits with very poorly conditioned matrices, as convergence necessitates $\epsilon c(A) < 1$.

Example 3.55 (*Iterative Refinement*) We continue the Python part of Example 3.21 and use `np.half` precision for the LU factorization and `np.single` precision for the stored vectors.

To implement iterative refinement with `numpy.arrays`, we need to be careful to change the entries of the vectors and not overwrite the vectors with new ones.

♣ **Implementation 3.56: Iterative Refinement**

```python
def iterative_refinement(A, b, n=3, precision2=np.half):
    A_2 = A.astype(precision2)
    pivot = LU_pivot(A_2)
    r = b.copy()
    x = np.zeros_like(b)
    for i in range(n):
        for p in pivot:
            r[p] = r[[p[1], p[0]]]
        y = forward(A_2, r)
        w = backward(A_2, y)
        x[:] += w
        r[:] = b - A.dot(x)
    return x
```

If we apply this implementation with one iterative refinement to the problem from Example 3.21, we get

```python
A = np.array([[2.3, 1.8, 1],[1.4, 1.1, -0.7],[0.8, 4.3, 2.1]], dtype=np.single)
b = np.array([1.2, -2.1, 0.6], dtype=np.single)
x = post_iteration(A_single, b_single, n=2, precision2=np.half)
print(np.linalg.norm(x - x_np) / np.linalg.norm(x_np))
```

    2.4034927140331146e-07

a relative error of 0.000024%. Thus, the solution is about three orders of magnitude better compared to the simple application of the LU factorization with pivoting with `half` floating point numbers.   ◀

3.7 Iterative Methods for Solving Linear Systems

In Sect. 3.6, we introduced the method of iterative refinement as an approach to reduce the influence of errors through successive iterations. With the perturbed LU factorization $A \approx \tilde{L}\tilde{U}$, we defined the iteration

$$x^{k+1} = x^k + \tilde{U}^{-1}\tilde{L}^{-1}(b - Ax^k), \quad k = 0, 1, 2, \ldots$$

Although $\tilde{L}$ and $\tilde{U}$ are not exact, this iteration converges (if the residual $r^k := b - Ax^k$ can be calculated exactly), see Theorem 3.54. With the perturbed LU factorization $\tilde{L}\tilde{U} \approx A$, an iterative refinement step can be written compactly as

$$x^{(i+1)} = x^{(i)} + Cr(\tilde{x}^{(i)}), \quad C := \tilde{U}^{-1}\tilde{L}^{-1}.$$

Here, the matrix C is an approximation to the inverse of A. The question arises whether iterative refinement also forms a convergent method when $\tilde{C}$ is a coarser approximation of the inverse $\tilde{C} \approx A^{-1}$. One could consider approximations that, although further from A^{-1}, are more efficient, e.g., requiring $O(n^2)$ operations to compute.

Definition 3.57 (*Iterative Methods to Solve Linear Systems*) Let $A \in \mathbb{R}^{n \times n}$ as well as $b \in \mathbb{R}^n$ and $C \in \mathbb{R}^{n \times n}$. For an arbitrary initial value $x^0 \in \mathbb{R}^n$, iterate for $k = 1, 2, \ldots$

$$x^k = x^{k-1} + C(b - Ax^{k-1}). \tag{3.13}$$

Alternatively, we introduce the notations $B := I - CA$ and $c := Cb$. Then

$$x^k = (I - CA)x^{k-1} + Cb = Bx^{k-1} + c. \tag{3.14}$$

Due to the construction, it is easy to see that the map $g(x) = Bx + c = x + C(b - Ax)$ gives a fixed point iteration with the solution of $Ax = b$ as a fixed point, because for $Ax = b$, it is

$$g(x) = x + C \underbrace{(b - Ax)}_{=0} = x.$$

Iterative methods are known from analysis and are briefly recapitulated in Chap. 7. Methods of type (3.14) are called *stationary iterations*, since the method rule $x^k = Bx^{k-1} + c$, in particular B and c, always remain the same and do not depend on the current approximation x^k.

The convergence analysis of iterative methods of type (3.13) is quite simple. If we subtract the exact solution $x = A^{-1}b$ from both sides, we get

$$x^k - x = x^{k-1} - x + C(b - Ax^{k-1}) = (I - CA)(x - x^{k-1})$$

$$\Rightarrow \quad \|x^k - x\| \leq \|I - CA\| \cdot \|x - x^{k-1}\|$$

and the method converges if the *iteration matrix* $I - CA$ satisfies $\|I - CA\| = \rho < 1$. However, there is the apparent dilemma that, depending on the used matrix norm $\|\cdot\|$, different convergence results are predicted. For example, for the matrix

$$B = \begin{pmatrix} 0.4 & -0.6 \\ 0.5 & 0.5 \end{pmatrix},$$

it holds $\|B\|_2 \approx 0.78$ but $\|B\|_1 = 1.1$ and $\|B\|_\infty = 1$, compare Theorem 2.17. This seems to be a contradiction. The fixed point iteration is always the same, regardless of the considered norm, and the sequence x^k either converges or it does not. The problem lies in the sharpness of the estimate. We noticed that $\|I - CA\| < 1$ is a sufficient condition, but it is not a necessary condition in every norm.

Lemma 3.58 (Matrix Norm and Spectral Radius) *For any arbitrary matrix $B \in \mathbb{R}^{n \times n}$ and any $\epsilon > 0$, there exists a natural matrix norm (i.e., a matrix norm induced by a vector norm) $\| \cdot \|_\epsilon$, such that:*

$$\rho(B) \leq \|B\|_\epsilon \leq \rho(B) + \epsilon$$

Proof For symmetric matrices, the statement is clear, since

$$\rho(B) = \|B\|_2$$

holds with the matrix norm induced by the Euclidean norm. For the general case, we refer to [75]. $\square$

In the following, we will therefore always work with the spectral radius as the norm. With this lemma, we show the fundamental result about general linear iterative methods:

Theorem 3.59 (Iterative Method to Solve Linear Systems) *Iteration (3.13) converges for every initial value $x^0 \in \mathbb{R}^n$ exactly to the solution $x \in \mathbb{R}^n$ of $Ax = b$, if and only if $\rho := \rho(B) < 1$. The asymptotic convergence factor is then given as*

$$\lim_{k \to \infty} \left(\frac{\|x^k - x\|}{\|x^0 - x\|} \right)^{1/k} = \rho(B).$$

Proof In step *(i)* of the proof, we will show that $\rho(B) < 1$ is sufficient for convergence $x^k \to x$. Step *(ii)* shows that the condition is also necessary, and finally, step *(iii)* gives the convergence estimate.

We define the error $e^k := x^k - x$, and with the fixed point property $x = Bx + c$ we have

$$e^k = x^k - x = Bx^{k-1} + c - (Bx + c) = Be^{k-1}.$$

Accordingly, it holds

$$e^k = B^k e^0 = B^k(x^0 - x). \tag{3.15}$$

(i) Lemma 3.58 states that for every $\epsilon > 0$ a natural matrix norm $\| \cdot \|_\epsilon$ exists with

$$\rho(B) \leq \|B\|_\epsilon \leq \rho(B) + \epsilon.$$

Assumed $\rho(B) < 1$. Then, there exists an $\epsilon > 0$ such that

$$\|B\|_\epsilon \leq \rho(B) + \epsilon < 1.$$

From (3.15) we immediately obtain for $k \to \infty$ in the corresponding induced vector norm $\|\cdot\|_\epsilon$

$$\|e^k\|_\epsilon = \|B^k e^0\|_\epsilon \leq \|B^k\|_\epsilon \|e^0\|_\epsilon \leq \|B\|_\epsilon^k \|e^0\|_\epsilon \to 0 \quad (k \to \infty).$$

Since all norms are equivalent in $\mathbb{R}^n$, $x^k \to x$ converges for $k \to \infty$.

(ii) The iteration is convergent $x^k \to x$. Let $w \in \mathbb{R}^n$ be an eigenvector to the largest eigenvalue $\lambda_{\max}$ of $B = I - CA$ in absolute value. Then for $x^0 = x + w$, i.e., $w = x^0 - x = -e^0$ with (3.15):

$$\lambda_{\max}^k w = B^k w = -B^k e^0 = -e^k$$

Since the iteration is convergent by assumption (for every initial value), it follows

$$-e^k = \lambda_{\max}^k w \to 0 \quad (k \to \infty),$$

and necessarily

$$\rho(B) = |\lambda_{\max}| < 1.$$

(iii) Error estimate: Due to the equivalence of all norms, there exist constants m, M with $m \leq M$, such that

$$m\|x\| \leq \|x\|_\epsilon \leq M\|x\|, \quad x \in \mathbb{R}^n.$$

Therefore, it holds

$$\|e^k\| \leq \frac{1}{m}\|e^k\|_\epsilon = \frac{1}{m}\|B^k e^0\|_\epsilon \leq \frac{1}{m}\|B\|_\epsilon^k \|e^0\|_\epsilon \leq \frac{M}{m}(\rho(B) + \epsilon)^k \|e^0\|.$$

Because

$$\left(\frac{M}{m}\right)^{1/k} \to 1, \quad (k \to \infty)$$

it follows that

$$\lim_{k \to \infty} \left(\frac{\|e^k\|}{\|e^0\|}\right)^{1/k} \leq \rho(B) + \epsilon.$$

Since $\epsilon > 0$ can be chosen arbitrarily small, this proves the claim. $\qquad\square$

Theorem 3.59 lays the theoretical foundation for general iterative methods. For the spectral radius $\rho := \rho(B) = \rho(I - CA)$, it must hold that $\rho < 1$. The goal of this section is the construction of iteration matrices C, which

1. are as close as possible to the inverse $C \approx A^{-1}$, so that $\rho(I - CA) \ll 1$,
2. allows a relatively simple calculation of the iteration $x^k = Bx^{k-1} + c$.

The first requirement is intuitive, and $C = A^{-1}$ represents the optimal matrix in this sense. The second condition describes the effort to perform the iteration. If we choose $C = \tilde{U}^{-1}\tilde{L}^{-1}$, each step means a forward and backward substitution, i.e., an effort of the order $O(n^2)$. In addition, there are the $O(n^3)$ operations for the one-time creation of the factorization. The choice $C = I$ allows a very efficient iteration with $O(n)$ operations in each step, without additional effort to set up the iteration matrix C. However, the unit matrix I is generally a very poor approximation of A^{-1}. The two demands lie on opposite ends of the scale and represent extreme positions, and they appear to contradict each other. In the following, we will discuss several simple choices that lie in between.

Definition 3.60 (*Richardson Iteration*) To solve $Ax = b$, let $x^0 \in \mathbb{R}^n$ be an arbitrary initial value. Iterate for $k = 1, 2, \ldots$

$$x^k = x^{k-1} + \omega(b - Ax^{k-1})$$

with a relaxation parameter $\omega > 0$.

To construct further simple iterative methods, we split the matrix A additively into $A = L + D + U$ with

$$A = \underbrace{\begin{pmatrix} 0 & & \cdots & 0 \\ a_{21} & \ddots & & \\ \vdots & \ddots & \ddots & \\ a_{n1} & \cdots & a_{n,n-1} & 0 \end{pmatrix}}_{=:L} + \underbrace{\begin{pmatrix} a_{11} & & \cdots & 0 \\ & \ddots & & \\ & & \ddots & \\ 0 & \cdots & & a_{nn} \end{pmatrix}}_{=:D} + \underbrace{\begin{pmatrix} 0 & a_{12} & \cdots & a_{1n} \\ & \ddots & \ddots & \vdots \\ & & \ddots & a_{n-1,n} \\ 0 & \cdots & & 0 \end{pmatrix}}_{=:U}.$$

This additive splitting should not be confused with the multiplicative LU factorization. If the matrix A is known, the additive splitting into L, D, U can be specified without further calculations. We define the two most important iterative methods.

Definition 3.61 (*Jacobi Method*) To solve $Ax = b$, let $x^0 \in \mathbb{R}^n$ be an arbitrary initial value. Iterate for $k = 1, 2, \ldots$

$$x^k = x^{k-1} + D^{-1}(b - Ax^{k-1}),$$

which using the *Jacobi iteration matrix* $J := -D^{-1}(L + U)$ corresponds to

$$x^k = Jx^{k-1} + D^{-1}b.$$

Definition 3.62 (*Gauss-Seidel Method*) To solve $Ax = b$, let $x^0 \in \mathbb{R}^n$ be an arbitrary initial value. Iterate for $k = 1, 2, \ldots$

$$x^k = x^{k-1} + (D + L)^{-1}(b - Ax^{k-1}),$$

which, using the *Gauss-Seidel iteration matrix* $H := -(D + L)^{-1}U$ corresponds to

$$x^k = Hx^{k-1} + (D + L)^{-1}b.$$

Theorem 3.63 (Implementation of the Jacobi and Gauss-Seidel Methods) *A step of the Jacobi or Gauss-Seidel method can be performed in* $n^2 + O(n)$ *operations.*

Instead of a proof, we give the two algorithms realizing Jacobi and Gauss-Seidel iterations.

Algorithm 3.64: Jacobi Method

Input: $A \in \mathbb{R}^{n \times n}$, $b \in \mathbb{R}^n$, $x^0 \in \mathbb{R}^n$ and a tolerance $\epsilon > 0$.
1 **for** $k = 0, 1, \dots$ **do**
2   **for** $i = 1, \dots, n$ **do**
3       $x_i^{k+1} = \frac{1}{a_{ii}}\left(b_i - \sum_{j=1, j\neq i}^{n} a_{ij}x_j^k\right)$
4   **if** $\|b - Ax^{k+1}\| < \epsilon$ **then**
5   Stop
  Result: Approximate solution $x = x^{k+1}$ of $Ax = b$.

Algorithm 3.65: Gauss-Seidel Method

Input: $A \in \mathbb{R}^{n \times n}$, $b \in \mathbb{R}^n$, $x^0 \in \mathbb{R}^n$ and a tolerance $\epsilon > 0$.
1 **for** $k = 0, 1, \dots$ **do**
2   **for** $i = 1, \dots, n$ **do**
3       $x_i^{k+1} = \frac{1}{a_{ii}}\left(b_i - \sum_{j<i} a_{ij}x_j^{k+1} - \sum_{j>i} a_{ij}x_j^k\right)$
4   **if** $\|b - Ax^{k+1}\| < \epsilon$ **then**
5   Stop
  Result: Approximate solution $x = x^{k+1}$ of $Ax = b$.

Proof Exercise.   □

Each step of these methods requires $O(n^2)$ operations, which is the same order of magnitude as the iterative refinement method with the LU factorization. However, with Jacobi and Gauss-Seidel, worse convergence factors are to be expected (we still do not know if these methods converge at all). The advantage of simple iteration methods only becomes apparent with sparse matrices. If a matrix $A \in \mathbb{R}^{n \times n}$ only has $O(n)$ entries, each iteration step only requires $O(n)$ operations. For a sparse matrix, the LU factorization $A = LU$ can consist of dense matrices L and u. Then, forward and backward substitution still require $O(n^2)$ arithmetic operations.

3.7.1   Convergence Criteria for the Jacobi and Gauss-Seidel Methods

To investigate convergence, according to Theorem 3.59, the spectral radius of the iteration matrices $J = -D^{-1}(L + U)$ and $H := -(D + L)^{-1}U$ must be examined. In individual cases, this investigation is not straightforward, and for a matrix A, the corresponding spectral radius can be difficult to determine. Therefore, in this section, we derive simple, verifiable criteria to make a statement about the convergence of the two methods.

Theorem 3.66 (Strong Row Sum Criterion) *If the row sums of the matrix $A \in \mathbb{R}^{n \times n}$ satisfy* strict diagonal dominance,

$$|a_{jj}| > \sum_{k=1, k \neq j}^{n} |a_{jk}|, \quad j = 1, \ldots, n,$$

then $\rho(J) < 1$ and $\rho(H) < 1$ and Jacobi and Gauss-Seidel methods converge.

Proof We prove the theorem for both methods simultaneously. Let λ_J be an eigenvalue of J and λ_H an eigenvalue of H. We denote by v_J and v_H the corresponding normalized eigenvectors $\|v_J\|_\infty = \|v_H\|_\infty = 1$. Then, we have

$$\lambda_J v_J = J v_J = -D^{-1}(L + U)v_J$$

for the Jacobi and Gauss-Seidel methods, and we get

$$\lambda_H v_H = H v_H = -(D + L)^{-1}U v_H \quad \Leftrightarrow \quad (D + L)\lambda_H v_H = -U v_H.$$

Hence,

$$\lambda_H v_H = -D^{-1}\big(U + L\lambda_H\big)v_H.$$

For the Jacobi method, with $\|v_J\|_\infty = 1$, we get

$$|\lambda_J| \leq \|D^{-1}(L + U)v_J\|_\infty \leq \|D^{-1}(L + U)\|_\infty \leq \max_{j=1,\ldots,n} \left\{ \frac{1}{|a_{jj}|} \sum_{k=1, k \neq j}^{n} |a_{jk}| \right\} < 1,$$

i.e., all eigenvalues are less than 1 in absolute value and the Jacobi method converges. For the Gauss-Seidel method, we get

$$|\lambda_H| \leq \|D^{-1}(\lambda_H L + U)\|_\infty \|v_H\|_\infty = \|D^{-1}(\lambda_H L + U)\|_\infty$$

$$\leq \max_{j=1,\ldots,n} \left\{ \frac{1}{|a_{jj}|} \left(\sum_{k<j} |\lambda_H||a_{jk}| + \sum_{k>j} |a_{jk}| \right) \right\}. \qquad (3.16)$$

We show $|\lambda_H| < 1$ using a contradiction argument. Suppose $|\lambda_H| \geq 1$. Then we have

$$\max_{j=1,\dots,n}\left\{\frac{1}{|a_{jj}|}\left(\sum_{k<j}|\lambda_H||a_{jk}|+\sum_{k>j}|a_{jk}|\right)\right\}$$

$$=|\lambda_H|\max_{j=1,\dots,n}\left\{\frac{1}{|a_{jj}|}\left(\sum_{k<j}|a_{jk}|+\frac{1}{|\lambda_H|}\sum_{k>j}|a_{jk}|\right)\right\}$$

$$\le|\lambda_H|\max_{j=1,\dots,n}\left\{\frac{1}{|a_{jj}|}\left(\sum_{k<j}|a_{jk}|+\sum_{k>j}|a_{jk}|\right)\right\}=|\lambda_H|\underbrace{\|D^{-1}(L+U)\|_\infty}_{<1}.$$

Now (3.16) and the strict diagonal dominance lead to a contradiction:

$$|\lambda_H|\le|\lambda_H|\|D^{-1}(L+U)\|_\infty<|\lambda_H|$$

It necessarily follows that $|\lambda_H|<1$. $\qquad\qquad\qquad\qquad\qquad\Box$

This criterion is easy to check and allows an immediate assessment of whether the Jacobi and Gauss-Seidel methods converge for a given matrix. However, it turns out that strict diagonal dominance is too strong a requirement. As an example, we take the model matrix of the Laplace operator, which appears in the context of a minimal surface approximation in Sect. 3.8. The matrix has the form

$$A=\begin{pmatrix}B&-I&&\\-I&B&\ddots&\\&\ddots&\ddots&-I\\&&-I&B\end{pmatrix},\quad B=\begin{pmatrix}4&-1&&\\-1&4&\ddots&\\&\ddots&\ddots&-1\\&&-1&4\end{pmatrix},\quad I=\begin{pmatrix}1&&&\\&\ddots&&\\&&\ddots&\\&&&1\end{pmatrix}\qquad(3.17)$$

and is only diagonally dominant (see Definition 3.23), but not strictly diagonally dominant. However, it turns out that both the Jacobi and Gauss-Seidel methods still converge. To weaken the requirements, we define

Definition 3.67 (*Irreducibility*) A matrix $A\in\mathbb{R}^{n\times n}$ is called irreducible if there is no permutation matrix P such that the matrix A can be decomposed into a block triangular matrix by rows and columns

$$PAP^T=\begin{pmatrix}\tilde{A}_{11}&0\\\tilde{A}_{21}&\tilde{A}_{22}\end{pmatrix}$$

with the matrices $\tilde{A}_{11}\in\mathbb{R}^{p\times p}$, $\tilde{A}_{22}\in\mathbb{R}^{q\times q}$, $\tilde{A}_{21}\in\mathbb{R}^{q\times p}$, with $p,q>0,\ p+q=n$.

Whether a given matrix $A\in\mathbb{R}^{n\times n}$ is irreducible cannot be determined directly. Often, the following equivalent criterion, which is easier to check, helps:

Theorem 3.68 (Irreducibility) *A matrix $A \in \mathbb{R}^{n \times n}$ is irreducible if and only if the associated directed graph*

$$\mathcal{G}(A) = \left\{ Nodes\colon \{1, 2, \ldots, n\}, Edges\colon \{(i, j),\ a_{ij} \neq 0\} \right\}$$

is connected. That is: For any two nodes i *and* j *there exists a path* $(i, i_1) = (i_0, i_1)$, $(i_1, i_2), \ldots, (i_{m-1}, i_m) = (i_{m-1}, j)$ *with edge* $(i_{k-1}, i_k) \in \mathcal{G}(A)$.

For the proof, we refer to [75]. According to this alternative characterization, irreducibility means that for any two indices i, j there exists a path $i = i_0 \mapsto i_1 \mapsto \ldots \mapsto i_m = j$ such that $a_{i_{k-1}, i_k} \neq 0$. Intuitively, this corresponds to alternating jumps in rows and columns of the matrix A, where only entries not equal to zero may be hit.

Theorem 3.69 (Weak Row Sum Criterion) *The matrix $A \in \mathbb{R}^{n \times n}$ is irreducible and satisfies the* weak row sum criterion, *i.e., the matrix A is diagonally dominant*

$$|a_{jj}| \geq \sum_{k=1, k \neq j}^{n} |a_{jk}|, \quad j = 1, \ldots, n,$$

and in at least one row $r \in \{1, \ldots, n\}$ there is strong diagonal dominance

$$|a_{rr}| > \sum_{k=1, k \neq r}^{n} |a_{rk}|.$$

Then A is regular and $\rho(J) < 1$ or $\rho(H) < 1$. That is, Jacobi and Gauss-Seidel methods converge.

Proof *(i)* Feasibility of the methods: Due to the irreducibility of A, it is necessary that

$$\sum_{k=1}^{n} |a_{jk}| > 0, \quad j = 1, \ldots, n.$$

Because of the assumed diagonal dominance, it follows that $a_{jj} \neq 0$ for $j = 1, 2, \ldots, n$.
(ii) Show $\rho(J) \leq 1$ and $\rho(H) \leq 1$: The proof of Theorem 3.66 can be redone using "$\leq$" instead of "$<$". It remains to show that no eigenvalue has the absolute value one.
(iii) Proof that $|\lambda| < 1$: Assume there is an eigenvalue $\lambda \in \sigma(J)$ with $|\lambda| = 1$. Let $v \in \mathbb{C}^n$ be the associated normalized eigenvector with $\|v\|_\infty = 1$. In particular, let $|v_s| = \|v\|_\infty = 1$ for a $s \in \{1, \ldots, n\}$. Then, using the structure of the iteration matrix (here comes the Jacobi case)

$$J = -D^{-1}(L + U) = \begin{pmatrix} 0 & \frac{-a_{12}}{a_{11}} & \frac{-a_{13}}{a_{11}} & \cdots & \frac{-a_{1n}}{a_{11}} \\ \frac{-a_{21}}{a_{22}} & 0 & \frac{-a_{23}}{a_{22}} & \cdots & \frac{-a_{2n}}{a_{22}} \\ \vdots & & \ddots & & \vdots \\ \frac{-a_{n1}}{a_{nn}} & \cdots & & \cdots & 0 \end{pmatrix},$$

the following estimate holds

$$|v_i| = \underbrace{|\lambda|}_{=1} |v_i| = |(Av)_i| \leq \sum_{k \neq i} \frac{|a_{ik}|}{|a_{ii}|} |v_k| \leq \sum_{k \neq i} \frac{|a_{ik}|}{|a_{ii}|} \leq 1, \quad i = 1, 2, \ldots, n. \tag{3.18}$$

Due to the irreducibility, for any two indices s, r there always exists a chain of indices $i_1 \mapsto i_2 \mapsto \ldots \mapsto i_m$ with $i_m \in \{1, \ldots, n\}$, such that

$$a_{s,i_1} \neq 0, a_{i_1,i_2} \neq 0, \ldots, a_{i_m,r} \neq 0.$$

By repeatedly applying (3.18), we get the contradiction (namely that $|\lambda| = 1$):

$$|v_r| = |\lambda v_r| \leq \frac{1}{|a_{rr}|} \sum_{k \neq r} |a_{rk}| \|v\|_\infty < \|v\|_\infty \quad \text{(strict DD in one row)},$$

$$|v_{i_m}| = |\lambda v_{i_m}| \leq \frac{1}{|a_{i_m,i_m}|} \left[\sum_{k \neq i_m, r} |a_{i_m,k}| \|v\|_\infty + |a_{i_m,r}| \, |v_r| \right] < \|v\|_\infty,$$

$$\vdots$$

$$|v_{i_1}| = |\lambda v_{i_1}| \leq \frac{1}{|a_{i_1,i_1}|} \left[\sum_{k \neq i_1,i_1} |a_{i_1,k}| \|v\|_\infty + |a_{i_1,i_2}| \, |v_{i_2}| \right] < \|v\|_\infty,$$

$$\|v\|_\infty = |\lambda v_s| \leq \frac{1}{|a_{ss}|} \left[\sum_{k \neq s,i_1} |a_{s,k}| \|v\|_\infty + |a_{s,i_1}| \, |v_{i_1}| \right] < \|v\|_\infty.$$

Therefore, $\rho(J) < 1$ must hold.

With a similar procedure, $\rho(H) < 1$ is shown. Again, the structure of the iteration matrix H as well as the reverse triangle inequality ($|a - b| \geq ||a| - |b||$) are used to obtain (3.18). The remaining argumentation then proceeds similarly. $\qquad\square$

This theorem can be applied to the model matrix to demonstrate the convergence of Jacobi and Gauss-Seidel methods.

Example 3.70 (*Jacobi and Gauss-Seidel Methods for the 1D-Model Matrix*) We consider
the linear system $Ax = b$ with the model matrix[4] $A \in \mathbb{R}^{n \times n}$

$$A = \begin{pmatrix} 2 & -1 & & & \\ -1 & 2 & -1 & & \\ & \ddots & \ddots & \ddots & \\ & & -1 & 2 & -1 \\ & & & -1 & 2 \end{pmatrix} \tag{3.19}$$

and the right-hand side $b \in \mathbb{R}^n$ with $b = (1, \dots, 1)^T$. For i, j with $i < j$ it holds

$$a_{i,i} \neq 0 \rightarrow a_{i,i+1} \neq 0 \rightarrow a_{i+1,i+2} \neq 0 \rightarrow a_{i+2,i+3} \neq 0 \rightarrow \cdots \rightarrow a_{j-1,j} \neq 0.$$

The matrix is therefore irreducible. Furthermore, it is diagonally dominant and strongly
diagonally dominant in the first and last row. The Jacobi and Gauss-Seidel methods converge. Experimentally, we determine for the problem size $n = 10 \cdot 2^k$ for $k = 2, 3, \dots$ the
number of necessary iteration steps as well as the computation time (Core i7 processor
2011) to approximate the system with an error tolerance $\|x^k - x\| < 10^{-4}$:   These results

Matrix size	Jacobi		Gauss-Seidel	
	Steps	Time [s]	Steps	Time [s]
80	9453	0.02	4727	0.01
160	37232	0.13	18617	0.06
320	147775	0.92	73888	0.43
640	588794	7.35	294398	3.55
1280	2149551	58	1074776	29
2560	8590461	466	4295231	233

demonstrate that the Gauss-Seidel method requires half as many steps as the Jacobi method.
Furthermore, the number of iterations increases with the matrix size n. Doubling the matrix
size increases the iteration count by a factor of 4. The computing time even increases by a
factor of 8, as each step is more costly.

We note that the Jacobi and Gauss-Seidel methods were efficiently programmed, taking
advantage of the sparse occupancy structure. Nevertheless, the time effort for approximation
with a given accuracy increases with the third order in the problem size n.   ◄

3.7.2 Relaxation Methods: The SOR Method

The previous example shows that Jacobi and Gauss-Seidel methods converge very slowly.
For the model matrix $A \in \mathbb{R}^{n \times n}$, the number of necessary iterations (to achieve a given

---

[4] In (3.17) we already defined a *model matrix* resulting from the discretization of the 2d Laplace
problem. Here, we consider a variant of it that results from the discretization of the 1d Laplace
problem.

accuracy) increases quadratically, i.e., $O(n^2)$. Although each step is very simple and can be performed extremely efficiently in $O(n)$ operations, these methods are not superior to the direct ones. Often, for example, with tridiagonal systems, direct solvers with a cost of $O(n)$ are unbeatably fast.

The SOR method is an advancement of the Gauss-Seidel iteration by introducing a *relaxation parameter* $\omega > 0$. The i-th element is calculated according to Theorem 3.63 as

$$x_i^{k,\mathrm{GS}} = \frac{1}{a_{ii}} \left(b_i - \sum_{j<i} a_{ij} x_j^{k,\mathrm{GS}} - \sum_{j>i} a_{ij} x_j^{k-1} \right), \quad i = 1, \ldots, n.$$

To determine the SOR solution, we do not directly use this approximation $x_i^{k,\mathrm{GS}}$, but introduce a relaxation parameter $\omega > 0$ and define

$$x_i^{k,\mathrm{SOR}} = \omega x_i^{k,\mathrm{GS}} + (1 - \omega) x_i^{k-1}, \quad i = 1, \ldots, n$$

as a weighted average between the Gauss-Seidel iteration and the old approximation. This relaxation parameter ω can now be used to significantly influence the convergence properties of the iteration. In the case $\omega = 1$, the Gauss-Seidel method is obtained. In the case $\omega < 1$, we speak of *under-relaxation*, in the case $\omega > 1$ of *over-relaxation*. "SOR" stands for *successive over relaxation*, thus using relaxation parameters $\omega > 1$. "Successive" (i.e., step-by-step) means that the relaxation is applied to each index. The relaxation can be efficiently integrated into the iteration in index notation for $i = 1, \ldots, n$

$$x_i^{k,\mathrm{SOR}} = \omega \frac{1}{a_{ii}} \left(b_i - \sum_{j<i} a_{ij} x_j^{k,\mathrm{SOR}} - \sum_{j>i} a_{ij} x_j^{k-1} \right) + (1 - \omega) x_i^{k-1}.$$

In vector notation, we have

$$x^{k,\mathrm{SOR}} = \omega D^{-1} (b - L x^{k,\mathrm{SOR}} - U x^{k-1}) + (1 - \omega) x^{k-1}.$$

Separating the terms according to $x^{k,\mathrm{SOR}}$ and x^{k-1} yields

$$(D + \omega L) x^{k,\mathrm{SOR}} = \omega b + [(1 - \omega) D - \omega U] x^{k-1},$$

thus the iteration

$$x^{k,\mathrm{SOR}} = H_\omega x^{k-1} + \omega [D + \omega L]^{-1} b, \quad H_\omega := [D + \omega L]^{-1} [(1 - \omega) D - \omega U].$$

This method with the iteration matrix H_ω fits into the scheme of general iterative methods, and according to Theorem 3.59, the convergence of the method depends on the condition $\rho_\omega := \rho(H_\omega) < 1$.

The difficulty in implementing the SOR method is the determination of good relaxation parameters, so that the matrix H_ω has the smallest possible spectral radius.

Theorem 3.71 (Relaxation Parameter of the SOR Method) *Let $A \in \mathbb{R}^{n \times n}$ with regular diagonal part $D \in \mathbb{R}^{n \times n}$. Then it holds*

$$\rho(H_\omega) \geq |\omega - 1|, \quad \omega \in \mathbb{R}.$$

For $\rho(H_\omega) < 1$, $\omega \in (0, 2)$ must therefore hold.

Proof We use the matrix representation of the iteration

$$H_\omega = [D + \omega L]^{-1}[(1 - \omega)D - \omega U] = (I + w \underbrace{D^{-1}L}_{=:\tilde{L}})^{-1} \underbrace{D^{-1}D}_{=I}[(1 - \omega)I - \omega \underbrace{D^{-1}U}_{=:\tilde{U}}].$$

The matrices $\tilde{L}$ and $\tilde{U}$ are strict triangular matrices with zeros on the diagonal. That is, $\det(I + \omega\tilde{L}) = 1$ and $\det((1 - \omega)I - \omega\tilde{U}) = (1 - \omega)^n$. For the determinant of H_ω, it holds

$$\det(H_\omega) = (1 - \omega)^n.$$

For the eigenvalues λ_i of H_ω it therefore follows

$$\prod_{i=1}^{n} \lambda_i = \det(H_\omega) = (1 - \omega)^n$$

$$\Rightarrow \quad \rho(H_\omega) = \max_{1 \leq i \leq n} |\lambda_i| \geq \left(\prod_{i=1}^{n} |\lambda_i|\right)^{\frac{1}{n}} = |1 - \omega|.$$

The last estimate uses the fact that the geometric mean of n numbers is smaller than the maximum. $\qquad\square$

This theorem provides a first estimate for the choice of the relaxation parameter, but does not yet help in determining an optimal value. For the important class of positive definite matrices, it can be shown that the SOR method converges for all values $\omega \in (0, 2)$.

For symmetric positive definite matrices, the following relationship can be shown for the matrix of the Jacobi J and Gauss-Seidel method H_1:

$$\rho(J)^2 = \rho(H_1) \tag{3.20}$$

One step of the Gauss-Seidel iteration leads to the same error reduction as two steps of the Jacobi iteration. This result is exactly reflected in Example 3.70. Furthermore, for these matrices, a relationship between the eigenvalues of the matrix H_ω and the eigenvalues of the matrix J can be derived. Suppose $\rho_J := \rho(J) < 1$. Then the spectral radius of the SOR matrix is given by

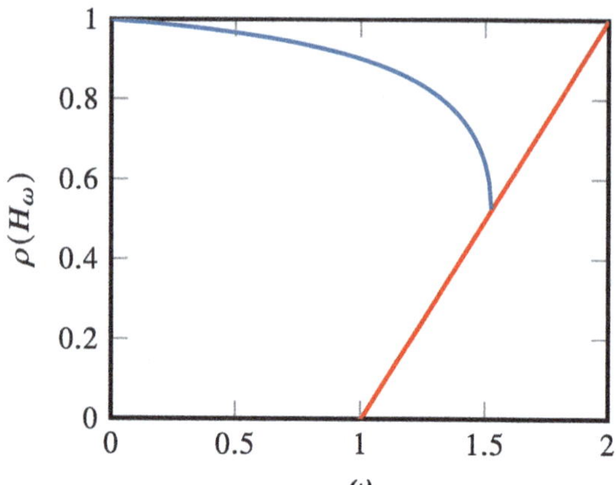

Fig. 3.7 Optimal relaxation parameter ω_{opt} as a function of the spectral radius

$$
\rho(H_\omega) = \begin{cases} \omega - 1 & \omega \geq \omega_{\text{opt}}, \\ \frac{1}{4}\left(\rho_J\omega + \sqrt{\rho_J^2\omega^2 - 4(\omega - 1)}\right)^2 & \omega \leq \omega_{\text{opt}}. \end{cases}
$$

If the spectral radius of the matrix J is known, the optimal parameter ω_{opt} can be found as the intersection of these two functions, see Fig. 3.7. It holds

$$
\omega_{\text{opt}} = \frac{2(1 - \sqrt{1 - \rho_J^2})}{\rho_J^2}. \tag{3.21}
$$

Example 3.72 (*Solving the Model Matrix Problem with the SOR Method*) We once again consider the simplified model matrix from Example 3.70. For the matrix $J = -D^{-1}(L + U)$ we have

$$
J = \begin{pmatrix} 0 & \frac{1}{2} & & & \\ \frac{1}{2} & 0 & \frac{1}{2} & & \\ & \ddots & \ddots & \ddots & \\ & & \frac{1}{2} & 0 & \frac{1}{2} \\ & & & \frac{1}{2} & 0 \end{pmatrix}.
$$

First, we determine the eigenvalues λ_i and eigenvectors w_i for $i = 1, \ldots, n$ of this matrix. The eigenvalues are given by

$$
w^i = (w_k^i)_{k=1,\ldots,n}, \quad w_k^i = \sin\left(\frac{\pi i k}{n + 1}\right),
$$

and, using the trigonometric identity $\sin(x \pm y) = \sin(x)\cos(y) \pm \cos(x)\sin(y)$, we have

$$(Jw^i)_k = \frac{1}{2}w^i_{k-1} + \frac{1}{2}w^i_{k+1} = \frac{1}{2}\left(\sin\left(\frac{\pi i(k-1)}{n+1}\right) + \sin\left(\frac{\pi i(k+1)}{n+1}\right)\right)$$

$$= \frac{1}{2}\sin\left(\frac{\pi i k}{n+1}\right)\left(\cos\left(\frac{\pi i}{n+1}\right) + \cos\left(\frac{\pi i}{n+1}\right)\right) = w^i_k\cos\left(\frac{\pi i}{n+1}\right).$$

Note that this equation is also valid for the first and last row, i.e., for $i = 1$ and $i = n$. It is $\lambda_i = \cos(\pi i/(n+1))$ and the largest eigenvalue of J is taken for $i = 1$ and $i = n$. Using the series expansion of the cosine, we have

$$\lambda_{\max} = \lambda_1 = \cos\left(\frac{\pi}{n+1}\right) = 1 - \frac{\pi^2}{2(n+1)^2} + O\left(\frac{1}{(n+1)^4}\right).$$

The largest eigenvalue approaches 1 quadratically as $n \to \infty$. From this, we determine the optimal relaxation parameters for Example 3.70:

n	$\lambda_{\max}(J)$	ω_{opt}
320	0.9999521084	1.980616162
640	0.9999879897	1.990245664
1280	0.9999969927	1.995107064

Finally, we compare the SOR method with the optimal relaxation parameter to Jacobi and Gauss-Seidel iteration:

Matrix size	Jacobi		Gauss-Seidel		SOR	
	Steps	Time [s]	Steps	Time [s]	Steps	Time [s]
320	147 775	0.92	73 888	0.43	486	$\ll 1$
640	588 794	7.35	294 398	3.55	1 034	0.02
1 280	2 149 551	58	1 074 776	29	1 937	0.05
2 560					4 127	0.22
5 120	Too complex				8 251	0.90
10 240					16 500	3.56

The number of necessary steps in the SOR method only increases linearly with the problem size. This is a significant improvement compared to the quadratic increase in the Jacobi and Gauss-Seidel methods. Since the effort of a step in the SOR method is comparable to that of Jacobi and Gauss-Seidel, the SOR method leads to a total effort of only $O(n^2)$ operations. However, this positive result only applies if the optimal SOR parameter is known.

◀

3.7.3 Practical Aspects

In deriving iterative methods, we started from the representation

$$x^{k+1} = x^k + C^{-1}(b - Ax^k) \tag{3.22}$$

Here it gets obvious that these are indeed fixed point mappings, as the residual $r(x) = b - Ax$ disappears in the case $Ax^k = b$ and thus there is no further update of the approximation. Convergence occurs when $\rho = \rho(I - C^{-1}A) < 1$. In general, it holds

$$\|x - x^k\| \le \rho\|x - x^{k-1}\| \le \rho^k\|x - x^0\|. \tag{3.23}$$

For $k \to \infty$, x^k converges to the exact solution of the linear system. However, these methods are not used as exact solvers. Instead, the iteration is stopped after sufficient reduction of the error. In principle, based on the estimate (3.23), the number of necessary steps to reach a tolerance ϵ_{tol} can be determined, but the rate ρ is generally not available.

Remark 3.73 (*Error Reduction*) When performing iterative solution methods, tolerances relative to the initial error are often required as stopping criteria. The iteration is stopped if:

$$\|x^k - x\| \le \epsilon_{\text{tol}} \|x^0 - x\|$$

As a practically feasible criterion, the unknown errors are replaced by the residuals:

$$\|b - Ax^k\| \le \epsilon_{\text{tol}} \|b - Ax^0\|$$

As $b - Ax^k = A(x - x^k)$, it then follows that

$$\|x^k - x\| \le \|A^{-1}\|\|b - Ax^k\| \le \epsilon_{\text{tol}} \|A^{-1}\|\|b - Ax^0\|$$
$$\le \epsilon_{\text{tol}} \cdot \text{cond}(A)\|x^0 - x\|.$$

Therefore, there can be a significant misestimation of the error.   ◆

Example 3.74 (*Linear System with the Model Matrix*) Let $A \in \mathbb{R}^{n \times n}$ with $n = m^2$ be the model matrix

$$A = \begin{pmatrix} B & -I & & \\ -I & B & \ddots & \\ & \ddots & \ddots & -I \\ & & -I & B \end{pmatrix}, \quad B = \begin{pmatrix} 4 & -1 & & \\ -1 & 4 & \ddots & \\ & \ddots & \ddots & -1 \\ & & -1 & 4 \end{pmatrix}, \quad I = \begin{pmatrix} 1 & & \\ & \ddots & \\ & & 1 \end{pmatrix}$$

with $B, I \in \mathbb{R}^{m \times m}$. The matrix A is a band matrix with bandwidth $2m$. Furthermore, the matrix A is symmetric positive definite. It is irreducible and diagonally dominant and fulfills the strong row sum criterion at the edges of the blocks. All previously discussed methods can be applied to the matrix A. The largest and smallest eigenvalues and spectral condition of A are calculated as

$$\lambda_{\min} = \frac{2\pi^2}{n} + O(n^{-2}), \quad \lambda_{\max} = 8 - \frac{2\pi^2}{n} + O(n^{-2}), \quad \kappa \approx \frac{4}{\pi^2}n \approx n.$$

Direct Methods

The LU factorization can be performed (since A is positive definite) without permutation. According to Theorem 3.31, the effort for this is $N_{LU} = nm^2 = 4n^2$ operations. The solution is then given exactly, except for rounding error influences. Alternatively, the Cholesky factorization of A can be created. The effort for this is $N_{LL} = 2n^2$. LU or Cholesky factorization are densely populated band matrices. The subsequent solving with forward and backward elimination requires another $2nm$ operations. We summarize the necessary operations in the following table:

$n = m^2$	N_{LU}	N_{LL}
100	$5 \cdot 10^4$	$2 \cdot 10^4$
10000	$5 \cdot 10^9$	$2 \cdot 10^9$
1000000	$5 \cdot 10^{12}$	$2 \cdot 10^{12}$

With efficient implementation on modern hardware, the system with 10 000 unknowns can be solved in a few seconds, the largest system in a few hours. Fundamental for the efficient application of direct methods is a pre-sorting of the sparsely populated systems, see Sect. 3.4.

Simple Iterative Methods

We first estimate the spectral radii for Jacobi and Gauss-Seidel methods. With an eigenvalue λ and eigenvector w of A it holds

$$\begin{aligned}
Aw = \lambda w \quad &\Rightarrow \quad Dw + (L + U)w = \lambda w \\
&\Rightarrow \quad -D^{-1}(L + U)w = -D^{-1}(\lambda I - D)w.
\end{aligned}$$

That is, due to $D_{ii} = 4$ it holds

$$Jw = \frac{\lambda - 4}{4}w$$

and the eigenvalues of J lie between

$$\lambda_{\min}(J) = \frac{1}{4}(\lambda_{\min}(A) - 4) \approx -1 + \frac{\pi^2}{2n},$$

$$\lambda_{\max}(J) = \frac{1}{4}(\lambda_{\max}(A) - 4) \approx 1 - \frac{\pi^2}{2n}.$$

Minimum and maximum eigenvalues are each close to -1 and 1 respectively. For the convergence factor, it holds

$$\rho_J := \rho(J) = 1 - \frac{\pi^2}{2n}$$

and with (3.20) it follows for the Gauss-Seidel method that

$$\rho_H := \rho_J^2 \approx 1 - \frac{\pi^2}{n}.$$

To reduce the error by a given factor $\epsilon > 0$, $t \in \mathbb{N}$ steps are required

$$\rho_J^{t_J} = \epsilon \quad \Rightarrow \quad t_J = \frac{\log(\epsilon)}{\log(\rho)} \approx \frac{2}{\pi^2} \log(\epsilon)n, \quad t_H \approx \frac{1}{\pi^2} \log(\epsilon)n.$$

Each step has the cost of a matrix-vector product, i.e., in the given case $5n$ operations. Finally, using (3.21), we determine the optimal SOR parameter as

$$\omega_{\mathrm{opt}} \approx 2 - 2\frac{\pi}{\sqrt{n}}.$$

Then it holds

$$\rho_\omega = \omega_{\mathrm{opt}} - 1 \approx 1 - 2\frac{\pi}{\sqrt{n}}$$

and we obtain

$$t_\omega \approx \frac{\log(\epsilon)}{2\pi} \sqrt{n}.$$

The cost of the SOR method corresponds to that of the Gauss-Seidel method with an additional relaxation, i.e., $6n$ operations per step. We summarize for the three methods the convergence factor, number of necessary steps, and total cost. In each case, $\epsilon = 10^{-4}$ is chosen:

$n = m^2$	Jacobi		Gauss-Seidel		SOR		
	t_J	N_J	t_H	N_H	ω_{opt}	t_J	N_J
100	180	10^5	90	$5 \cdot 10^4$	1.53	9	10^5
10000	18500	10^9	9300	$5 \cdot 10^8$	1.94	142	10^7
1000000	1866600	10^{13}	933000	$5 \cdot 10^{12}$	1.99	1460	10^9

While the largest system can be solved in a few seconds with the optimal SOR method, Jacobi and Gauss-Seidel methods require several hours.

We will continue this test case in Sect. 6.2, compare Examples 6.9 and 6.13. There, we include further iterative methods that will be derived in a completely different way. ◄

3.8   Excursus: Numerical Approximation of the Minimal Surface Problem

The *minimal surface problem* is one of the classical problems of analysis. Given is a closed curve $\Gamma \subset \mathbb{R}^3$, the task is to find a smooth surface $\mathcal{F} \subset \mathbb{R}^3$, bounded by the curve Γ, and having the minimal surface area among all surfaces. Minimal surfaces have many applications and often correspond physically to the principle of *energy minimization*. Soap bubbles between a wire ring are approximations of minimal surfaces. In architecture, minimal surfaces are models for roof constructions such as the one of the Munich Olympic Stadium, which we show in Fig. 3.8, or the main station in Stuttgart. For details on the theory and for more precise definitions, we refer to the literature [7, 60].

Modeling

For *modeling*, i.e., the mathematical description of the problem, we first look for a simple way to describe surfaces in space. We initially restrict ourselves to curves Γ and surfaces $\mathcal{F}$, which can be written as the graph of a function $f : \mathbb{R}^2 \to \mathbb{R}^3$ over $[0, 1]^2$

Fig. 3.8 The Munich Olympic Stadium

$$\mathcal{F}(f) = \left\{ \begin{pmatrix} x \\ y \\ f(x, y) \end{pmatrix}, \ 0 \leq x, y \leq 1 \right\}$$

Hereby, a *parametrization* of the surface is given. This excludes general surfaces and especially those that are intertwined and cannot be represented as a graph at all.

Definition 3.75 (*Simplified Minimal Surface Problem*) Let $\Omega = (0, 1)^2$ and $\Gamma = \partial\Omega$ be the boundary of the domain. Further, let $g : \Gamma \to \mathbb{R}$ be a smooth representation of the boundary curve. Then,

$$\Gamma(g) = \left\{ \begin{pmatrix} x \\ y \\ g(x, y) \end{pmatrix}, \ (x, y) \in \Gamma \right\} \subset \mathbb{R}^3$$

is the curved boundary of the 3d surface. We denote the space of permissible parametrizations by

$$X^g := \{\phi \in C(\Omega \cup \Gamma), \ \phi = g \text{ on } \Gamma\}.$$

We are looking for the parametrization $f \in X^g$, such that the associated surface $\mathcal{F}(f)$ has the smallest surface area among all surfaces with a parametrization in X^g:

$$A(\mathcal{F}(f)) \leq A(\mathcal{F}(h)) \quad \forall h \in X^g$$

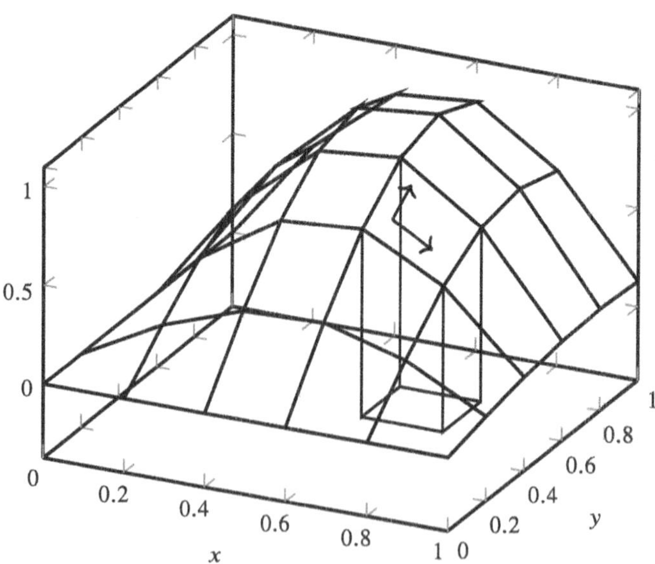

Fig. 3.9 Calculation of the area of the surface for a parametrically given surface $f : \Omega \mapsto \mathcal{F}$. The tangential vectors $\mathbf{t}_x$ and $\mathbf{t}_y$ span the infinitesimal surface elements

In order to be able to solve the minimal surface problem, we first need to calculate the area of the surface for a given parametrization.

Theorem 3.76 (Area of a parametrized Surface) *Let $f \in C^1(\bar{\Omega})$ be the parametrization of a surface $\mathcal{F}$. Then, the area of the surface $\mathcal{F}$ is given by*

$$A(\mathcal{F}) = \int_0^1 \int_0^1 \sqrt{1 + f_x(x, y)^2 + f_y(x, y)^2}\, dx\, dy,$$

with the partial derivatives $f_x(x, y) = \frac{\partial f}{\partial x}(x, y)$ and $f_y(x, y) = \frac{\partial f}{\partial y}(x, y)$.

Proof For a formal derivation, we refer to the literature [7]. Here, we refer to Fig. 3.9. Let $(x, y) \in \Omega$ be a point on the base surface and by dx and dy an infinitesimal rectangular surface element $\Delta_2 := (x, x + dx) \times (y, y + dy) \subset \Omega$ is given. This surface element Δ_2 is mapped by f onto a surface $\Delta(f) \subset \Omega$ with the two tangential vectors

$$\mathbf{t}_x(x, y) = \begin{pmatrix} 1 \\ 0 \\ f_x(x, y) \end{pmatrix}, \quad \mathbf{t}_y(x, y) = \begin{pmatrix} 0 \\ 1 \\ f_y(x, y) \end{pmatrix},$$

such that it holds

$$\frac{|\Delta(f)|}{|\Delta_2|} \to \|\mathbf{t}_x \times \mathbf{t}_y\| \text{ for } |\Delta_2| \to 0.$$

The result follows with

$$\|\mathbf{t}_x \times \mathbf{t}_y\| = \left\| \begin{pmatrix} -f_x(x, y) \\ -f_y(x, y) \\ 1 \end{pmatrix} \right\| = \sqrt{1 + f_x(x, y)^2 + f_y(x, y)^2}.$$

$\square$

The functions in the space $\mathcal{X}^g$ must be at least once continuously differentiable, i.e.,

$$\mathcal{X}^g := \{\phi \in C^1(\Omega) \cap C(\bar{\Omega}),\ \phi = g \text{ on } \Gamma\}.$$

We start by reformulating the minimal surface problem as a *variational problem*, see [60]. Suppose that $f \in \mathcal{X}^g$ is the parameterization of a minimal surface. Then it holds

$$A(\mathcal{F}(f)) \le A(\mathcal{F}(h)) \quad \forall h \in \mathcal{X}^g. \tag{3.24}$$

The difference between the two parametrizations f and g lies in the space

$$\delta f := h - f \in \mathcal{X}^0 := \{\phi \in C^1(\Omega) \cap C(\bar{\Omega}),\ \phi = 0 \text{ on } \Gamma\}$$

as all functions in $\mathcal{X}^g$ satisfy the boundary values. We can thus write (3.24) in the form

$$A(\mathcal{F}(f)) \leq A(\mathcal{F}(f + \delta f)) \quad \forall \delta f \in \mathcal{X}^0.$$

Theorem 3.77 (Minimal Surface as a Variational Problem) *Let $f \in \mathcal{X}^g$ be a solution of the minimal surface problem, which is twice continuously differentiable, i.e., $f \in C^2(\Omega)$. Then f is given as a solution of the* partial differential equation

$$- f_{xx}(x, y)\left(1 + f_y(x, y)^2\right) - f_{yy}(x, y)\left(1 + f_x(x, y)^2\right)$$
$$+ 2 f_{xy}(x, y) f_x(x, y) f_y(x, y) = 0$$

in the domain $\Omega = (0, 1)^2$, satsifying the boundary values

$$f(x, y) = g(x, y) \text{ on } \Gamma = \partial\Omega.$$

Proof Let $f \in \mathcal{X}^g$ be the parameterization of a minimal surface. Further, let $\phi \in \mathcal{X}^0$ be arbitrary, but fixed. Then it holds

$$A(\mathcal{F}(f)) \leq A(\mathcal{F}(f + s\phi)) =: a(s) \quad \forall s \in \mathbb{R}.$$

The function $a(s)$ thus has a (local) minimum at the point $s = 0$. It therefore holds for all $\phi \in \mathcal{X}^0$ that

$$a'(s)\Big|_{s=0} = 0.$$

We calculate the derivative using the formula from Theorem 3.76:

$$a'(0) = \frac{\partial}{\partial f} A(\mathcal{F}(f + s\phi))\Big|_{s=0}$$
$$= \int_0^1 \int_0^1 \frac{2(f_x + s\phi_x)\phi_x + 2(f_y + s\phi_y)f_y\phi_y}{2\sqrt{1 + f_x^2 + f_y^2}} \, dx\, dy\Big|_{s=0}$$
$$= \int_0^1 \int_0^1 \frac{f_x\phi_x + f_y\phi_y}{\sqrt{1 + f_x^2 + f_y^2}} \, dx\, dy$$

For simplicity, we have omitted the arguments "x" and "y" of the functions $f(x, y)$ and $\phi(x, y)$. With

$$\gamma(x, y) := \sqrt{1 + f_x(x, y)^2 + f_y(x, y)^2}$$

it holds by partial integration

$$a'(0) = -\int_0^1 \int_0^1 \left(\frac{\partial}{\partial x} \frac{f_x}{\gamma} + \frac{\partial}{\partial y} \frac{f_y}{\gamma}\right) \cdot \phi \, dx\, dy + \underbrace{\int_0^1 \frac{f_x\phi}{\gamma}\Big|_0^1 dy + \int_0^1 \frac{f_y\phi}{\gamma}\Big|_0^1 dx}_{=0}.$$

The boundary terms vanish, since the functions $\phi \in X^0$ are zero on Γ. Further, it holds (the same equality is given for the y-derivative)

$$\frac{\partial}{\partial x} \frac{f_x}{\gamma} = \frac{f_{xx}\gamma - f_x \gamma_x}{\gamma^2} \text{ with } \gamma_x = \frac{f_x f_{xx} + f_y f_{xy}}{\gamma},$$

i.e., with $\gamma^2 = 1 + f_x^2 + f_y^2$

$$\frac{\partial}{\partial x} \frac{f_x}{\gamma} + \frac{\partial}{\partial y} \frac{f_y}{\gamma} = \frac{f_{xx}}{\gamma^3}\left(1 + f_y^2\right) + \frac{f_{yy}}{\gamma^3}\left(1 + f_x^2\right) - \frac{2 f_x f_y f_{xy}}{\gamma^3}.$$

Finally, the derivative $a'(0)$ is given by

$$a'(0) = \int_0^1 \int_0^1 \frac{-f_{xx}(1 + f_y^2) - f_{yy}(1 + f_x^2) + 2 f_{xy} f_x f_y}{\gamma^3} \cdot \phi \, dx \, dy \overset{!}{=} 0.$$

The function $\phi \in X^0$ can be arbitrarily chosen in $C_0^\infty(\Omega)$ (as this is a subspace of X^0). The *Fundamental Theorem of the Calculus of Variations* [52, 60] then shows that the equality holds in each point, if the integrated function is continuous (which is given, if $f \in C^2(\Omega)$):

$$- f_{xx}\left(1 + f_y^2\right) - f_{yy}\left(1 + f_x^2\right) + 2 f_{xy} f_x f_y = 0 \text{ in } \Omega. \tag{3.25}$$

On the boundary, $f = g$ holds by construction. $\qquad\qquad\square$

Remark 3.78 (*Partial Differential Equations*) Equations of type (3.25) are called *partial differential equations*. The sought solution is a function $f : \Omega \to \mathbb{R}$ and a functional relationship between the various partial derivatives is described. Partial differential equations are often *boundary value problems*. In this excursus, we are dealing with the *Dirichlet problem*, where the function values of the sought function f are given on the boundary of the domain. The analysis of partial differential equations, i.e., the investigation of questions such as existence or uniqueness of solutions, is a large area of mathematics [31, 98]. The numerical approximation of partial differential equations is one main area of numerical mathematics [47]. $\qquad\qquad\blacklozenge$

The differential equation (3.25), which describes the solution of the minimal surface problem, is challenging due to its nonlinearity. Hence, we will simplify the equation and assume that the slope of the function f is small, i.e., $|\nabla f| \ll 1$ such that $|\nabla f|^2 \ll 1$ holds, i.e., we assume $f_x^2 = f_y^2 \approx 0$ and $f_{xy} \approx 0$. Strongly simplified, the solution of the minimal surface problem is therefore described by the linear Laplace problem.

Definition 3.79 (*Laplace Equation*) Let $\Omega = (0, 1)^2 \subset \mathbb{R}^2$ and g be a continuous function given on the boundary $\Gamma = \partial\Omega$. We are looking for a twice continuously differentiable function $f \in C^2(\bar\Omega)$ with

$$-\Delta f(x, y) := -f_{xx}(x, y) - f_{yy}(x, y) = 0 \text{ in } \Omega$$

with $f(x, y) = g(x, y)$ on the boundary Γ. The *Laplace operator*

$$\Delta := \frac{d^2}{dx^2} + \frac{d^2}{dy^2}$$

is the sum of the second partial derivatives.

The Laplace equation is the most important *elliptic partial differential equation*. Solutions of the Laplace equation can (usually simplified) describe many processes such as temperature propagation, diffusion processes or the minimal surface problem.

Discretization

The main problem in solving or approximating partial differential equations is the number of unknowns: We are looking for a function $f : \Omega \to \mathbb{R}$, i.e., function values in the infinitely many points of the domain Ω are sought. To make the problem manageable for an algorithmic approximation, it is first discretized, i.e., replaced by a finite-dimensional problem. Instead of a function $f : \Omega \to \mathbb{R}$, we will look for the solution only in some discrete points. We define:

Definition 3.80 *(Grid)* Let $\Omega \subset \mathbb{R}^2$ be a two-dimensional domain (an open, connected subset). By a *grid*

$$\Omega_h := \{\mathbf{x}_i = (x_i^1, x_i^2) \in \bar{\Omega}, \ i = 1, \ldots, N\},$$

we denote the set of pairwise different points in the domain. The set of *boundary points* is denoted by

$$\Gamma_h := \{\mathbf{x}_i \in \Omega_h, \ \mathbf{x}_i \in \Gamma := \partial\Omega\}$$

By the *grid size*

$$h_i := \min_{j \neq i} \|\mathbf{x}_i - \mathbf{x}_j\|,$$

we understand the minimal distance to the nearest point. The *maximum grid size* is given as

$$h := \max_{1 \leq i \leq N} h_i.$$

Since we are considering the simple domain $\Omega = (0, 1)^2$, we introduce a *uniform grid* with $N = (M + 1)^2$ points using the uniform grid size $h = 1/M$:

$$\Omega_h = \{\mathbf{x}_{ij} := (i \cdot h, j \cdot h), \ 0 \leq i, j \leq M\} \tag{3.26}$$

Instead of a continuous function $f : \Omega \to \mathbb{R}$, we are now looking for a grid function $f_h : \Omega_h \to \mathbb{R}$.

Definition 3.81 (*Grid Function*) Let Ω_h be a grid with $N \in \mathbb{N}$ points. By a grid function $\mathbf{f}_h : \Omega_h \to \mathbb{R}$ we understand the vector

$$(\mathbf{f}_h)_i = f_i \in \mathbb{R}, \quad 1 \le i \le N.$$

Next, we consider values of the vector $\mathbf{f}_h$, such that $f_i \approx f(\mathbf{x}_i)$ is a good approximate solution of the simplified minimal surface problem.

Finite Difference Approximation

We aim for an approximation of the differential equation

$$-\Delta f(x, y) = 0 \text{ in } \Omega, \quad f \equiv g \text{ on } \Gamma. \tag{3.27}$$

We must approximate the derivatives of the unknown solution $f : \Omega \to \mathbb{R}$ by the discrete values $\mathbf{f}_h$. In Sect. 9.3, we will investigate techniques for approximating derivatives. On a regular grid, derivatives can simply be expressed by difference quotients.

Theorem 3.82 (Difference Quotients) *Let $f \in C^4$. It holds*

$$-f_{xx}(x, y) = \frac{2f(x, y) - f(x + h, y) - f(x - h, y)}{h^2} + O(h^2),$$

$$-f_{yy}(x, y) = \frac{2f(x, y) - f(x, y + h) - f(x, y - h)}{h^2} + O(h^2).$$

From this, it immediately follows

$$-\Delta f(x, y) =$$
$$\frac{4f(x, y) - f(x + h, y) - f(x - h, y) - f(x, y + h) - f(x, y - h)}{h^2}$$
$$+ O(h^2).$$

Proof The proof follows using the Taylor expansion, see also Sect. 9.3.

$$f(x \pm h) = f(x) \pm hf'(x) + \frac{h^2}{2}f''(x) \pm \frac{h^3}{6}f'''(x) + \frac{h^4}{24}f^{(iv)}(\xi)$$

with an intermediate value $\xi \in [x - h, x + h]$. Thus it follows

$$\frac{-2f(x) + f(x+h) + f(x-h)}{h^2} = f''(x) + O(h^2).$$

The same applies for the y-direction, and the approximation of the Laplace operator is composed of the two directional derivatives.   □

On the regular grid (3.26), the derivatives of the partial differential equation (3.27) can be approximated using the difference quotients and the grid function $\mathbf{f}_h$. In each of the inner grid points (i.e., for $1 \leq i, j \leq M - 1$), it must hold

$$\frac{1}{h^2}\left(4f_{ij} - f_{i+1,j} - f_{i-1,j} - f_{i,j+1} - f_{i,j-1}\right) = 0. \tag{3.28}$$

On the boundary, we get

$$f_{ij} = g(\mathbf{x}_{ij}), \quad i \in \{0, M\} \text{ or } j \in \{0, M\}. \tag{3.29}$$

The Eqs. (3.28) and (3.29) form a linear system with $N = (M+1)^2$ equations and also $N = (M+1)^2$ variables. The outer points, i.e., at $i = 0$ or $i = M$, $j = 0$ or $j = M$ on the boundary, are already known so that we will remove these equations and unknowns from the linear system. With appropriate sorting of the unknowns f_{ij}, we can write the linear system in the compact form

$$\mathbf{A}_h\mathbf{f}_h = 0 \tag{3.30}$$

where $\mathbf{A}_h \in \mathbb{R}^{(M-1)^2 \times (M-1)^2}$ (since we only consider the inner points). In Fig. 3.10, we show the so-called *lexicographic* sorting of the inner degrees of freedom.

Fig. 3.10 Lexicographic arrangement of degrees of freedom on a uniform point grid. The arrangement of degrees of freedom does not change the solution of the linear system, but it has a large impact on the matrix structure. Different sorting of degrees of freedom corresponds to a simultaneous row and column permutation of the resulting matrix

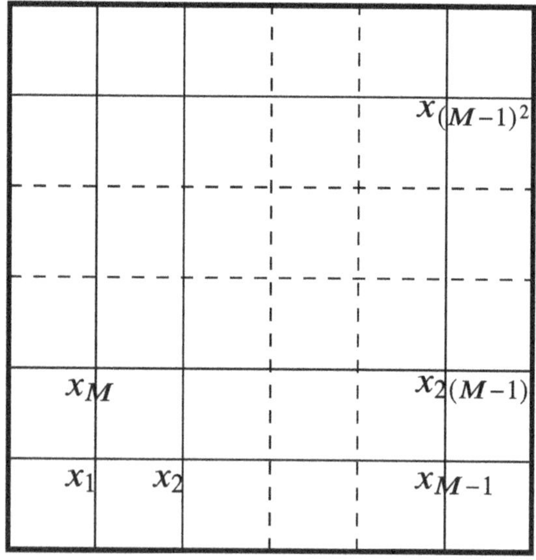

With this choice of arrangement of degrees of freedom, the following block shape of the matrix results, which we have already seen in the Sects. 3.4 and 3.7.1, where we introduced the *model matrix* :

$$
\mathbf{A} = \begin{pmatrix} \mathbf{B} & -\mathbf{I} & & \\ -\mathbf{I} & \mathbf{B} & \ddots & \\ & \ddots & \ddots & -\mathbf{I} \\ & & -\mathbf{I} & \mathbf{B} \end{pmatrix}, \quad
\mathbf{B} = \begin{pmatrix} 4 & -1 & & \\ -1 & 4 & \ddots & \\ & \ddots & \ddots & -1 \\ & & -1 & 4 \end{pmatrix}, \quad
\mathbf{I} = \begin{pmatrix} 1 & & & \\ & \ddots & & \\ & & \ddots & \\ & & & 1 \end{pmatrix}
$$

with $\mathbf{B}, \mathbf{I} \in \mathbb{R}^{(M-1)\times(M-1)}$ and thus $\mathbf{A} \in \mathbb{R}^{N\times N}$ where $N := (M-1)^2$.

Theorem 3.83 (Model Matrix) *The model matrix* $\mathbf{A} \in \mathbb{R}^{N\times N}$ *with* $N = (M-1)^2$ *is a band matrix with bandwidth* $M - 1$. *It is irreducible, diagonally dominant, satisfies the strong row sum criterion (see Theorem 3.66), and is symmetric positive definite. For its eigenvalues and eigenvectors, for* $k = (r, s)$ *with* $r, s \in \{1, \ldots, M - 1\}$

$$
(\omega^{r,s})_{i,j} = \sin\left(\frac{\pi r i}{M}\right) \sin\left(\frac{\pi s j}{M}\right), \quad i, j = 1, \ldots, M - 1
$$

$$
\lambda^{r,s} = \frac{1}{h^2}\left(4 - 2\cos\left(\frac{\pi r}{M}\right) - 2\cos\left(\frac{\pi s}{M}\right)\right). \tag{3.31}
$$

For the condition number of the matrix, it holds

$$
\mathrm{cond}_2(\mathbf{A}_h) = O(N) = O(M^2) = O\left(\frac{1}{h^2}\right).
$$

Remark 3.84 (*Condition of the Model Matrix*) The condition number of the model matrix behaves like

$$
\mathrm{cond}_2(\mathbf{A}_h) = O\left(\frac{1}{h^2}\right) = O(M^2),
$$

i.e., it grows quadratically with decreasing grid size. In Example 3.63, we examined the corresponding one-dimensional model matrix. This arises, for example, by discretizing the one-dimensional Laplace problem

$$
-\partial_{xx} f(x) = 0
$$

using the finite difference method. Here, very similar eigenvalues and eigenvectors are obtained—see Example 3.72. Also, in the one-dimensional case, the model matrix has a condition of order $O(h^{-2})$, i.e., quadratic in the grid size. In general, it turns out that the quadratic order is not related to the quadratic approximation quality of the central difference quotients or the dimension of the domain Ω, but is given by the *order of the differential operator*: The second derivative ∂_{xx} and the Laplace operator Δ are second-order differential operators.   ◆

Numerical Solution

We now investigate methods for solving linear systems of the form

$$\mathbf{A}_h \mathbf{x}_h = \mathbf{b}_h,$$

where $\mathbf{A}_h \in \mathbb{R}^{N \times N}$ with $N = (M - 1)^2$ is the model matrix, $\mathbf{x}_h \in \mathbb{R}^N$ is the vector of internal degrees of freedom, i.e., the unknowns, and $\mathbf{b}_h \in \mathbb{R}^N$ is the right-hand side. This vector can be given in a block form:

$$\mathbf{b}_h = \begin{pmatrix} \mathbf{b}_1 \\ \mathbf{b}_2 \\ \vdots \\ \mathbf{b}_{M-2} \\ \mathbf{b}_{M-1} \end{pmatrix}, \quad \mathbf{b}_i \in \mathbb{R}^{M-1},$$

where for the first and last block the form

$$\mathbf{b}_1 = \frac{1}{h^2} \begin{pmatrix} g_{0,1} + g_{1,0} \\ g_{2,0} \\ \vdots \\ g_{M-2,0} \\ g_{M,1} + g_{M-1,0} \end{pmatrix}, \quad \mathbf{b}_{M-1} = \frac{1}{h^2} \begin{pmatrix} g_{0,M-1} + g_{1,M} \\ g_{2,M} \\ \vdots \\ g_{M-2,M} \\ g_{M,M-1} + g_{M-1,M} \end{pmatrix},$$

and for the middle blocks, the form

$$\mathbf{b}_i = \frac{1}{h^2} \begin{pmatrix} g_{0,i} \\ 0 \\ \vdots \\ 0 \\ g_{M,i} \end{pmatrix}, \quad i = 1, \ldots, M - 1,$$

is assumed.

Solving the linear system can be done either with direct methods or with iterative methods from the previous sections.

We first consider direct methods. Due to the band structure and the symmetry of the matrix $\mathbf{A}_h$, the Cholesky factorization from Sect. 3.3 is particularly suitable. The linear system can be solved in $O(N^2)$ operations, where $N = O(M^2) = (h^{-2})$.

We have seen that simple iterative methods like Gauss-Seidel or Jacobi are not competitive. On the other hand, the matrix is very sparse and symmetric positive definite, such that the CG method from Sect. 6.2 is a good candidate. The CG method does not *solve* the system exactly but generates a sequence of approximations $\mathbf{x}_h^{(k)} \approx \mathbf{x}_h$. According to Theorem 11.15, the following estimate applies

$$\|\mathbf{x}^{(k)} - \mathbf{x}_h\|_{\mathbf{A}_h} \leq 2 \left(\frac{1 - 1/\sqrt{\kappa}}{1 + 1/\sqrt{\kappa}} \right)^k \|\mathbf{x}_h^{(0)} - \mathbf{x}_h\|_{\mathbf{A}_h}$$

and with the condition number $\kappa = O(h^{-2})$ we approximately get

$$\|\mathbf{x}^{(k)} - \mathbf{x}_h\|_{\mathbf{A}_h} \leq |1 - h|^k \|\mathbf{x}^{(0)} - \mathbf{x}_h\|_{\mathbf{A}_h}. \tag{3.32}$$

Each step reduces the error by $|1 - h|$ and for $k \to \infty$ we reach convergence to zero. However, the finite differences discussed in Lemma 3.82 also introduce a numerical error, which will come on top of the iteration error of the linear solver, such that the overall error can be split into two contributions

$$\|f - \mathbf{x}_h^{(k)}\| \leq \|f - \mathbf{x}_h\| + \|\mathbf{x}_h - \mathbf{x}_h^{(k)}\|. \tag{3.33}$$

In terms of the overall efficiency, it is optimal to balance all error contributions: spending a high effort on the finite difference approximation with very fine meshes is no benefit if the linear systems are not solved accurately; likewise, solving linear systems exactly is a waste of resources if the discretization error is too large. Unfortunately, we cannot directly use the finite difference error formula in one grid point to estimate the error of the complete finite difference solution. The more complex analysis is given in the following paragraph.

First, we study the effort of each CG step. It requires one matrix-vector product, two scalar products, and three vector additions. The effort per step for the model matrix (with a maximum of five entries per line) is therefore

$$E_{CG} = 1 \cdot 5N + 2 \cdot N + 3 \cdot N = 10 \, N. \tag{3.34}$$

The total effort depends on the required number of steps.

Finite Difference Error Analysis

We analyze the error between the solution of the simplified minimal surface problem

$$- \Delta f(x, y) = 0 \text{ in } \Omega, \quad f(x, y) = g(x, y) \text{ on } \Gamma,$$

and the finite difference approximation $\mathbf{x}_h$ with $\mathbf{x}_{i,j} \approx f(x_{i,j})$. The first question is how the *error* between a function $f(x, y)$ and some discrete values $\mathbf{x}_h$ at the grid points can be measured at all. For this, we define the discrete grid function of the exact values

$$\mathbf{f}_h \in \mathbb{R}^N, \quad \mathbf{f}_{ij} = f(x_{i,j}),$$

and the error $\mathbf{e}_h \in \mathbb{R}^N$

$$\mathbf{e}_h = \mathbf{f}_h - \mathbf{x}_h.$$

For this, we immediately get

$$\mathbf{e}_h = \mathbf{A}_h^{-1}\mathbf{A}_h\mathbf{e}_h = \mathbf{A}_h^{-1}(\mathbf{A}_h\mathbf{f}_h - \mathbf{b}_h),$$

since $\mathbf{A}_h\mathbf{x}_h = \mathbf{b}_h$. We define:

Definition 3.85 (*Truncation Error*)
The *truncation error* $\tau_h \in \mathbb{R}^N$, also called *consistency error*, is defined as

$$\tau_h = \frac{1}{h^2}\mathbf{A}_h(\mathbf{f}_h - \mathbf{b}_h).$$

In the case

$$\|\tau_h\| \to 0 \quad (h \to 0)$$

the finite difference method is called *consistent* with the differential equation.

The truncation error is a residual. It measures whether the exact solution of the differential equation also satisfies the discrete equation.

Theorem 3.86 (Truncation Error) *Let $f \in C^4(\Omega)$ be the solution of the equation*

$$-\Delta f(x, y) = b(x, y) \text{ in } \Omega, \quad f(x, y) = 0 \text{ on } \Gamma.$$

Then, for the truncation error, it holds

$$\max \|\tau_h\| = O(h^2).$$

Proof The result follows directly from Theorem 3.82. For the grid function $\mathbf{f}$ with $\mathbf{f}_{i,j} = f(x_{i,j})$ and $\mathbf{b}_{i,j} = b(x_{i,j})$ it holds

$$\frac{1}{h^2}(\mathbf{A}_h\mathbf{f}_h)_{ij} = \frac{4\mathbf{f}_{ij} - \mathbf{f}_{i+1,j} - \mathbf{f}_{i-1,j} - \mathbf{f}_{i,j+1} - \mathbf{f}_{i,j-1}}{h^2}$$
$$= -\Delta f(x_{i,j}) + O(h^2) = \mathbf{b}_{i,j} + O(h^2).$$

Hence,

$$\tau_{i,j} = O(h^2).$$

$\square$

The consistency error estimates how well the exact solution satisfies the discrete problem. Translated to linear systems, the consistency error takes the role of the defect $\mathbf{b} - \mathbf{A}\tilde{\mathbf{x}}$ for an approximation $\tilde{\mathbf{x}} \approx \mathbf{x} = \mathbf{A}^{-1}\mathbf{b}$. On the other hand, the discretization error measures the real error $\mathbf{x} - \tilde{\mathbf{x}}$. For linear systems, the connection between defect and error is by means of the matrix itself $\mathbf{b} - \mathbf{A}\tilde{\mathbf{x}} = \mathbf{A}(\mathbf{x} - \tilde{\mathbf{x}})$. We will derive a similar relation for the finite difference method.

Considering the maximum norm, it holds for $\mathbf{e}_h := \mathbf{f}_h - \mathbf{x}_h$

$$\|\mathbf{f}_h - \mathbf{x}_h\|_\infty = \|\mathbf{e}_h\|_\infty \leq \|\mathbf{A}_h^{-1}\tau_h\|_\infty \leq \|\mathbf{A}_h^{-1}\|_\infty\|\tau_h\|_\infty \leq C\|\mathbf{A}_h^{-1}\|_\infty h^2. \tag{3.35}$$

To obtain an error estimate, we need an estimate for the inverse of the matrix $\mathbf{A}_h$. For our model matrix, we know the eigenvalues, compare (3.31), and from $\lambda_{min} \geq c > 0$ (independent of $h > 0$), it follows that the eigenvalues of the inverse are bounded. Thus, we have

$$\|\mathbf{A}_h^{-1}\| \leq C \quad (h \to 0). \tag{3.36}$$

We continue with estimating the total error (3.33) and using (3.35) and (3.36), it holds

$$\|\mathbf{f} - \mathbf{x}_h^{(k)}\|_\infty \leq Ch^2 + \|\mathbf{x}_h - \mathbf{x}_h^{(k)}\|_\infty \tag{3.37}$$

with an unknown constant $C > 0$. An efficient use of the CG method for approximating the system must now ensure that the CG error does not dominate the total error. On the other hand, it is not worth solving the system too precisely, as the discretization error would then dominate. A possible criterion is to a balance the two error components

$$\|\mathbf{x}_h - \mathbf{x}_h^{(k)}\| \leq Ch^2,$$

where the constant $C > 0$ is not known. For simplicity, we choose $C = 1$. According to (3.32), we must determine the iteration number $k \in \mathbb{N}$ such that

$$(1 - h)^k \approx h^2 \quad \Leftrightarrow \quad k = \frac{\log(h^2)}{\log(1 - h)} = \frac{1}{h}\log\left(\frac{1}{h^2}\right) + O(\log(h)).$$

For $N = 1/h^2$, the required number of iterations is

$$k = \log(N)N^{\frac{1}{2}} + O(\log(h)).$$

Together with the effort per step of the CG method (see (3.34)), the total effort for approximating the problem is

$$E_{CG}(total) = 10\, N^{\frac{3}{2}} \log\left(N^{-\frac{1}{2}}\right),$$

compared to the direct solution of the Cholesky method

$$E_{Cholesky}(total) = N^2.$$

This difference does not seem large at first glance, but for a problem size of $M = 1\,000$, i.e., $N = 10^6$, it already makes a factor of 150, which leads to a reduction of the runtime from about one day to 10 min.

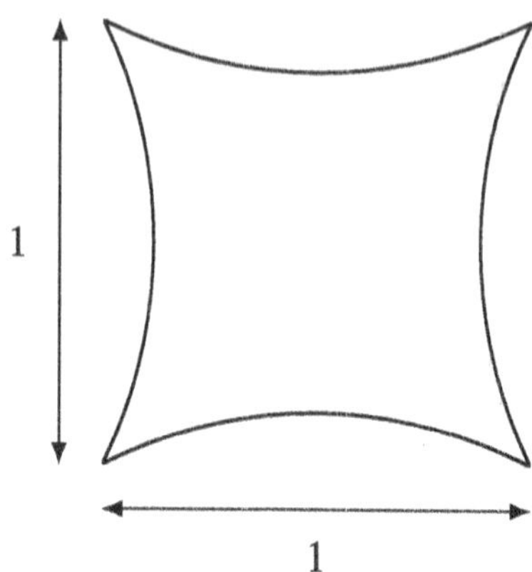

Fig. 3.11 The base area for the minimal surface problem is a square with rounded sides. The sides are circular arcs of circles with radius $r = \sqrt{5/4}$ and the centers $x_m^1 = (-1, 1/2)$, $x_m^2 = (2, 1/2)$, $x_m^3 = (1/2, -1)$, $x_m^4 = (1/2, 2)$

3.8.1 Examples

Finally, we consider two examples of numerical realizations of minimal surfaces. We will investigate how the error of the finite difference discretization behaves, i.e., whether we can verify Theorem 3.85 and the error estimate (3.37). Furthermore, we check whether the linear approximation, i.e., the simplification of the minimal surface equation (3.77) to the Laplace equation (Definition 3.79), is adequate.

Example 1: Music Pavilion

The first example is modeled after the *Music Pavilion*, which was erected for the Federal Horticultural Show in Kassel in 1955,[5] see also [6]. This is one of the first examples of the use of a minimal surface in architecture. Here, we consider, for simplicity, a square as the base area with rounded sides cut out (see Fig. 3.11)

$$\Omega = (0, 1) \times (0, 1) \setminus \left(\overline{K_1(2, 1/2)} \cup \overline{K_1(1/2, -1)} \cup \overline{K_1(-1, 1/2)} \cup \overline{K_1(1/2, 2)} \right),$$

and prescribe the desired deflection $g(x, y)$ on the boundary as

$$g(x, y) = \frac{1}{2}|1 - x - y|.$$

In Fig. 3.12, we show the solution. On the left side of the figure, the full nonlinear model is considered, on the right side the Laplace equation. In this example, no differences can be detected; the resulting minimal surfaces look the same in each case.

---

[5] The Music Pavilion in Kassel was designed by Frei Otto for the Federal Horticultural Show in 1955 and is one of the first *minimal surfaces* in architecture. Later, Frei Otto designed the roof of the Olympic Park in Munich.

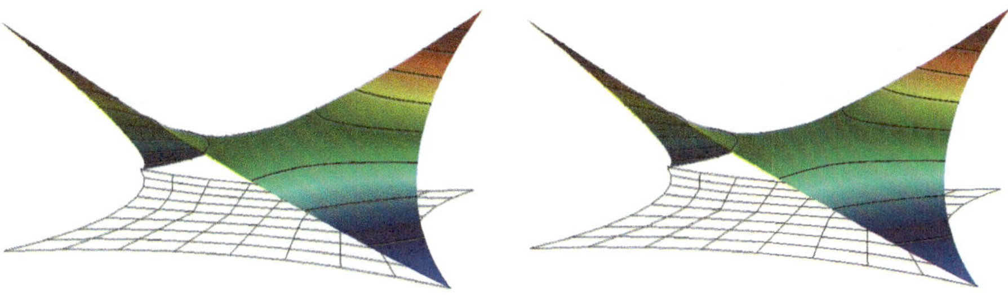

Fig. 3.12 Minimal surface of the music pavilion. Left: full, nonlinear model. Right: reduced Laplace equation. The lines indicate the level lines to the heights $h_i = i/20$ for $i = 0, \ldots, 10$. At first glance, no difference between the two models can be detected

However, a look at the resulting *minimal surface* $A(\mathcal{F}(f))$ in Table 3.3 shows a small difference. If we consider the nonlinear model, the resulting surface with $A \approx 0.73421$ is slightly smaller than $A \approx 0.73424$, the deviation is just 0.05%. The simplified model is a very good approximation.

The table lists the solution of the minimal surface problem on a sequence of grids with decreasing mesh spacing $h \to 0$. This allows us to validate the convergence of the method, in particular a validation of the estimate (3.37). As we do not know the exact solution to the problem, i.e., the function $f(x, y)$, we first create a *reference value*. This is done based on the numerical results themselves using numerical extrapolation. Extrapolation is a technique to estimate the limit of a sequence of discrete values, and we will discuss the details in Sect. 9.4.

We expect that the area of the discrete approximation $A(\mathcal{F}(f_h))$ converges to the real area, and make the following approach

$$A(\mathcal{F}(f)) = A(\mathcal{F}(f_h)) + Ch^\alpha + \mathcal{R} \tag{3.38}$$

with an unknown constant C, the unknown (we expect $\alpha = 2$) order of convergence and the unknown area $A(\mathcal{F}(f))$ of the exact parameterization of the surface $f(x, y)$. With $\mathcal{R}$, we denote higher order remainders. The three unknowns C, $A(\mathcal{F}(f))$ and α can be determined from three consecutive given in Table 3.3, by neglecting the remainder $\mathcal{R}$. Starting with a mesh size h and considering $h/2$ and $h/4$ fitting to (3.38) gives

$$
\begin{aligned}
\alpha &= \log_2 \left(\frac{A(\mathcal{F}(f_h)) - A(\mathcal{F}(f_{h/2}))}{A(\mathcal{F}(f_{h/2})) - A(\mathcal{F}(f_{h/4}))} \right), \\
\tilde{A}(\mathcal{F}(f)) &= \frac{A(\mathcal{F}(f_h))A(\mathcal{F}(f_{h/4})) - A(\mathcal{F}(f_{h/2}))^2}{A(\mathcal{F}(f_h)) - 2A(\mathcal{F}(f_{h/2})) + A(\mathcal{F}(f_{h/4}))}.
\end{aligned}
\tag{3.39}
$$

These formulas can be easily derived from (3.38). This first example shows almost perfect quadratic convergence of the numerical values.

Table 3.3 Calculation of the minimal surface of the music pavilion. Left: Calculation with the full, nonlinear model. Right: Reduced model based on the Laplace equation. Even for a very small problem size, very good approximations can be achieved. In both cases, the theoretically expected quadratic convergence in the grid width $O(h^2)$ is observed

N	$A(\mathcal{F}(f_h))$	Error
9	0.82393	0.08972
25	0.75703	0.02282
289	0.73564	0.00143
1089	0.73457	0.00036
4225	0.73430	0.00009
Exact	0.73421 (2.00)	

N	$A(\mathcal{F}(f_h))$	Error
9	0.82393	0.08969
25	0.75705	0.02281
81	0.73996	0.00572
289	0.73567	0.00143
1089	0.73460	0.00036
4225	0.73433	0.00009
Exact	0.73424 (1.98)	

Example 2: Minimal surface with reduced regularity

As a second example, we consider the configuration from Fig. 3.13. The "minimal surface" is suspended in the middle and attached to the ground at the four corners of a region. In between, it finds its position freely.

The result is shown in Fig. 3.14. Here, there is a large difference between the full nonlinear model (left) and the simplified Laplace equation (right). This time, the value of the area functional itself also shows a larger discrepancy. Using the nonlinear full model, the area is $A \approx 1.2260$, compared to $A \approx 1.2517$ for the Laplace equation—a deviation of almost 2%.

Extrapolating the results from Table 3.4, we obtain—contrary to the theoretical prediction—only linear convergence in the grid width $O(h)$. The expected quadratic convergence order is no longer present. The errors are much larger than in the first example.

What is the reason for this poor performance? The estimate shown in Theorem 3.85 does not appear to be correct for this second test case. The reason for the poorer performance in this example lies in specific assumptions of Theorem 3.85. It turns out that the solution for

Fig. 3.13 The minimal surface is suspended in the middle at a height $H = 0.5$. It is attached to the ground at $H = 0$ at the four corners. The surface is free along the remaining edges

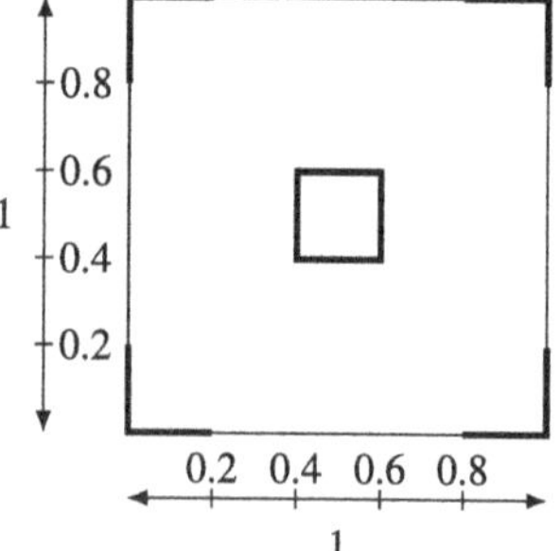

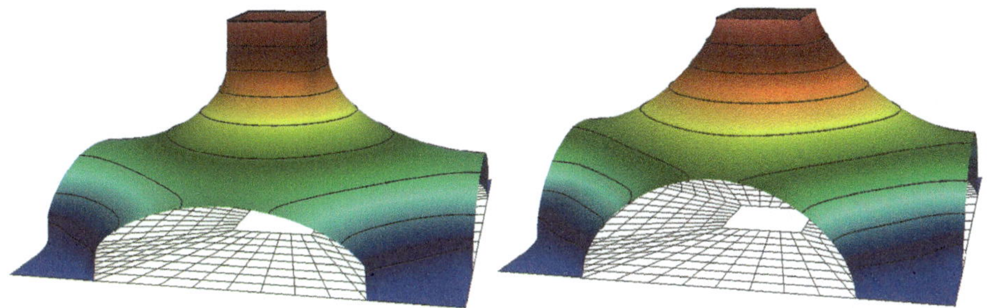

Fig. 3.14 Minimal surface for Example 2. On the left, the full nonlinear model was considered; on the right, the Laplace equation was considered. Especially near the *reentrant corners* in the middle of the region, there are very large differences

Table 3.4 Convergence behavior in Example 2. On the left is the full model, on the right is the reduced model. The area content in the full model is almost 2% lower. Here, the theoretically expected quadratic convergence is no longer present. Both methods only converge approximately linearly in the grid width $O(h)$. The "exact" values were determined as reference values by extrapolation

N	$A(\mathcal{F}(f_h))$	Error	N	$A(\mathcal{F}(f_h))$	Error
48	1.3211	0.0951	48	1.3244	0.0734
160	1.2755	0.0495	160	1.2823	0.0312
576	1.2532	0.0272	576	1.2650	0.0140
2 176	1.2410	0.0150	2 176	1.2574	0.0063
8 448	1.2342	0.0082	8 448	1.2539	0.0029
33 280	1.2304	0.0044	33 280	1.2524	0.0013
132 096	1.2284	0.0024	132 096	1.2517	0.0006
Exact	1.2260 (0.89)		Exact	1.2511 (1.13)	

this example has limited regularity. It does not hold $f \in C^4(\Omega)$ such that the theorem may not be applied. A close analysis shows that the derivatives of the solution at the four inner corners are not bounded. Corners that protrude into the domain are called *reentrant corners*. And at such reentrant corners, solutions of the Laplace equation have only a very limited regularity. In this region, no solution exists that is even one time continuously differentiable. Theorem 3.85 is only correct for $f \in C^4(\Omega)$ and cannot be applied here.

Orthogonalization Methods and QR Factorization 4

The LU factorization of a regular matrix $A \in \mathbb{R}^{n \times n}$ into the two triangular matrices L and U has been derived using Frobenius matrices, i.e., $U = FA$, with $L = F^{-1}$. It holds:

$$\operatorname{cond}(U) = \operatorname{cond}(FA) \leq \operatorname{cond}(F)\operatorname{cond}(A) = \|F\|\,\|L\|\,\operatorname{cond}(A)$$

With appropriate pivoting, all entries of the matrices F and L satisfy the estimate $|f_{ij}| \leq 1$ and $|l_{ij}| \leq 1$ (see Theorem 3.18). Nevertheless, the norms of L and F can generally not be estimated more favorably than $\|F\|_\infty \leq n$ and $\|L\|_\infty \leq n$. Thus, the following pessimistic estimate applies

$$\operatorname{cond}_\infty(U) \leq n^2 \operatorname{cond}_\infty(A).$$

The matrix U, which must be inverted for backward elimination, may have a much worse conditioning than the matrix A itself (which can already be very poorly conditioned).

In this chapter, we will analyze the QR factorization of a matrix A as a stable factorization of a matrix $A \in \mathbb{R}^{n \times n}$ into an upper right triangular matrix $R \in \mathbb{R}^{n \times n}$ and an orthogonal matrix Q. The QR factorization is more stable than the LU factorization and can be used to solve linear systems. In addition, we will extend the QR factorization to non-square matrices $A \in \mathbb{R}^{n \times m}$ with $m > n$ and use it in Sect. 4.5 to solve overdetermined linear systems. Finally, the QR factorization is also the basis for efficient algorithms for calculating eigenvalues. This is covered in Sect. 5.4.2.

Beyond the various applications of the QR factorization, we will consider the orthogonalization of vectors as an independent numerical problem. The simplest algorithm is the Gram-Schmidt orthogonalization method, see Theorem 2.10 in Chap. 2. We will also discuss more stable and robust alternatives.

© The Author(s), under exclusive license to Springer-Verlag GmbH, DE, part of Springer 133
Nature 2026
T. Richter et al., *Introduction to Numerical Mathematics*, Mathematics Study
Resources 25, https://doi.org/10.1007/978-3-662-72546-7_4

4.1 The QR Factorization

We are looking for a factorization $A = QR$ for a matrix $A \in \mathbb{R}^{n \times n}$ into a triangular matrix $R \in \mathbb{R}^{n \times n}$, which is numerically stable, by performing the factorization using matrices that are themselves well conditioned.

Definition 4.1 (*Orthogonal Matrix*) A matrix $Q \in \mathbb{R}^{n \times n}$ is called *orthogonal*, if its row and column vectors form an orthonormal basis of $\mathbb{R}^n$. It holds $Q^T Q = I$, thus $Q^{-1} = Q^T$.

Orthogonal matrices have the spectral condition $\mathrm{cond}_2(Q) = 1$, because for all eigenvalues λ of an orthogonal matrix Q with eigenvector ω it holds

$$\|\omega\|_2^2 = (Q^T Q \omega, \omega)_2 = (Q\omega, Q\omega)_2 = (\lambda\omega, \lambda\omega)_2 = |\lambda|^2 \|\omega\|_2^2 \quad \Rightarrow \quad |\lambda| = 1.$$

For a factorization of the form $A = QR$ with an orthogonal matrix Q, it thus holds $R = Q^T A$ and

$$\mathrm{cond}_2(R) = \mathrm{cond}_2(Q^T A) \leq \mathrm{cond}_2(Q^T)\,\mathrm{cond}_2(A) = \mathrm{cond}_2(A).$$

The triangular matrix R has at most the condition of the original matrix A. We first summarize some properties of orthogonal matrices.

Theorem 4.2 (Orthogonal Matrix) *Let $Q \in \mathbb{R}^{n \times n}$ be an orthogonal matrix. Then Q is regular, and it holds*

$$Q^{-1} = Q^T, \quad Q^T Q = I, \quad \|Q\|_2 = 1, \quad \mathrm{cond}_2(Q) = 1.$$

Further:

1. *It holds $\det(Q) = 1$ or $\det(Q) = -1$.*
2. *For two orthogonal matrices $Q_1, Q_2 \in \mathbb{R}^{n \times n}$, the product $Q_1 Q_2$ is also an orthogonal matrix. For any matrix $A \in \mathbb{R}^{n \times n}$ it holds $\|QA\|_2 = \|A\|_2$.*
3. *For arbitrary vectors $x, y \in \mathbb{R}^n$ it holds*

$$\|Qx\|_2 = \|x\|_2, \quad (Qx, Qy)_2 = (x, y)_2.$$

Proof We only prove those properties that are essential in the context of numerical stability. Mostly, this is the invariance of the norm considering multiplication with orthogonal matrices. It holds

$$\|QAx\|_2^2 = (QAx, QAx)_2 = (Q^T Q Ax, Ax)_2 = (Ax, Ax)_2 = \|Ax\|_2^2$$

and hence,

$$\|QA\|_2 = \sup_{x \neq 0} \frac{\|QAx\|_2}{\|x\|_2} = \sup_{x \neq 0} \frac{\|Ax\|_2}{\|x\|_2} = \|A\|_2.$$

$\square$

Definition 4.3 (*QR Factorization*) The factorization of a matrix $A \in \mathbb{R}^{n \times n}$ into an orthogonal matrix $Q \in \mathbb{R}^{n \times n}$ and an upper right triangular matrix $R \in \mathbb{R}^{n \times n}$ according to

$$A = QR$$

is called *QR Factorization*.

We define the QR factorization for general rectangular matrices. Once created, the QR factorization can be used to solve linear systems with the matrix A efficiently. The matrix Q is—unlike the matrix L of the LU factorization—generally dense and not a triangular matrix. Since its inverse is given by transposing it, the linear system can nevertheless be solved efficiently by

$$Ax = b \quad \Leftrightarrow \quad Q^T Ax = Q^T b \quad \Leftrightarrow \quad Rx = Q^T b.$$

Theorem 4.4 (QR Factorization) *Let $A \in \mathbb{R}^{n \times n}$ be a regular matrix. Then there exists a factorization $A = QR$ into an orthogonal matrix $Q \in \mathbb{R}^{n \times n}$ and an upper right triangular matrix $R \in \mathbb{R}^{n \times n}$.*

Proof Let $A = (a_1, \ldots, a_n)$ be the column vectors of the matrix A. Since A is regular, the vectors a_i are linearly independent. Now let $q_1, \ldots, q_n \in \mathbb{R}^n$ be an orthonormal system with the properties

$$(q_i, q_j)_2 = \delta_{ij} \text{ for } 1 \leq i, j \leq n, \text{ and } (q_i, a_j)_2 = 0 \text{ for } 1 \leq j < i \leq n.$$

Such a system of vectors can be generated with Gram-Schmidt orthogonalisation, see Theorem 2.10.

The matrix $Q = (q_1, \ldots, q_n) \in \mathbb{R}^{n \times n}$ is orthogonal and for the entries of the matrix $R = Q^T A \in \mathbb{R}^{n \times n}$ with $R = (r_{ij})_{ij}$ for $i, j = 1, \ldots, n$, it holds

$$r_{ij} = (q_i, a_j)_2 = 0 \text{ for } i > j. \tag{4.1}$$

Hereby, R is an upper right triangular matrix and, noting $(Q^T)^{-1} = (Q^{-1})^T = (Q^T)^T = Q$, it follows

$$A = QR.$$

$\square$

The QR factorization is not unique. Suppose, through $q_1, \ldots, q_n$ the orthonormal system of eigenvectors is given, then by changing the sign $q_1, \ldots, q_{i-1}, -q_i, q_{i+1}, \ldots, q_n$ another orthonormal system of eigenvectors is created and by (4.1), $\tilde{r}_{ij} = -r_{ij}$ results.

This consideration can be used for a simple normalization of the QR factorization. In each step, the sign of the vector q_i is chosen to produce positive diagonal entries $r_{ii} = (q_i, a_i) > 0$.

Theorem 4.5 (Uniqueness of the QR Factorization) *The QR factorization $A = QR$ of a regular matrix $A \in \mathbb{R}^{n \times n}$ is unique when choosing $r_{ii} > 0$.*

Proof Let $A = Q_1 R_1 = Q_2 R_2$ be two QR factorizations of A. Then it holds:

$$Q := Q_2^T Q_1 = R_2 R_1^{-1} \in \mathbb{R}^{m \times m}, \quad Q^T := Q_1^T Q_2 = R_1 R_2^{-1} \in \mathbb{R}^{n \times n}. \qquad (4.2)$$

The matrices Q and Q^T are both upper right triangular matrices, so Q must be a diagonal matrix. Further, it holds

$$Q^T Q = Q_1^T Q_2 Q_2^T Q_1 = R_1 R_2^{-1} R_2 R_1^{-1} = I,$$

i.e., Q is orthogonal. From (4.2) it follows

$$QR_1 = R_2$$

and for every unit vector e_i it holds

$$q_{ii} r_{ii}^{(1)} = e_i^T (QR_1) e_i = e_i^T (R_2) e_i = r_{ii}^{(2)} > 0. \qquad (4.3)$$

The diagonal elements $q_{ii} > 0$ are positive. Since the diagonal elements of a diagonal matrix are just the eigenvalues of the matrix, it follows from $|\lambda| = 1$ (since Q is orthogonal), that $\lambda = 1$ and thus $Q = I$. Finally $Q_1 = Q_2$ and therefore $R_1 = R_2$. $\qquad\square$

The crucial step in creating a QR factorization is the orthogonalization of a system of n independent vectors $a_1, \ldots, a_n \in \mathbb{R}^n$. This task will be examined in detail in the following sections. In particular, we develop orthogonalization procedures that are far more stable than the simple Gram-Schmidt orthogonalization method.

Remark 4.6 (*QR Factorization for Rectangular Matrices*) We have introduced the QR factorization for regular matrices $A \in \mathbb{R}^{n \times n}$. The algorithm can immediately be applied to rectangular matrices $A \in \mathbb{R}^{n \times m}$ with $n > m$ and full rank$(A) = m$. The column vectors $a_1, \ldots, a_m \in \mathbb{R}^n$ are still linearly independent and can, for example, be transformed into an orthonormal system $q_1, \ldots, q_m \in \mathbb{R}^n$ with the property

$$(q_i, q_j)_2 = \delta_{ij}, \quad (q_i, a_j)_2 = 0, \quad i, j = 1, \ldots, m. \qquad (4.4)$$

The matrix $Q = (q_1, \ldots, q_m) \in \mathbb{R}^{n \times m}$, as a non-square matrix, is not orthogonal, but it holds

$$Q^T Q = I \in \mathbb{R}^{m \times m}.$$

Similarly, with the orthonormality (4.4), it follows that $R = Q^T A \in \mathbb{R}^{m \times m}$ is an upper right triangular matrix.

If we supplement the vectors $q_1, \ldots, q_m$ with $\tilde{q}_{m+1}, \ldots, \tilde{q}_n$ to form an orthonormal basis of $\mathbb{R}^n$, and extend the matrix R with $n - m$ zero rows to the matrix $\tilde{R} \in \mathbb{R}^{n \times m}$, the usual notation applies

$$\tilde{Q} \tilde{R} = A \quad \Leftrightarrow \quad \left(\begin{array}{c|c} Q & \tilde{Q} \end{array} \right) \left(\begin{array}{c} R \\ \hline 0 \end{array} \right) = \left(\begin{array}{c} A \end{array} \right)$$

The enrichment of the vectors $q_1, \ldots, q_m$ by $\tilde{q}_{m+1}, \ldots, \tilde{q}_n$ is only for the purpose of aligning the notation. These $n - m$ vectors must not be practically determined in order to create the matrix $R = Q^T A$, as the lower rows of the extended matrix R are zero, anyway. $\blacklozenge$

4.2 Gram-Schmidt Orthogonalization

The Gram-Schmidt orthogonalization method, see Theorem 2.10, is simple and is based on projections of a vector onto already orthogonal components. The basic iteration formula for orthonormalizing vectors $a_1, a_2, \ldots, a_m$ is

$$q_1 := \frac{a_1}{\|a_1\|}, \quad \tilde{q}_i = a_i - \sum_{j=1}^{i-1} (a_i, q_j) q_j, \quad q_i := \frac{\tilde{q}_i}{\|\tilde{q}_i\|} \quad i = 2, 3, \ldots, m.$$

Here, (x, y) is a scalar product and $\|x\| = (x, x)^{\frac{1}{2}}$ is the associated norm. The effort of the process increases with the number of vectors in the basis. In step i of the process, the calculation of $i - 1$ scalar products, of $i - 1$ vector additions, as well as a norm are necessary. In the space $V = \mathbb{R}^n$, the number of operations in step n is thus $O(n^2)$. The total effort for orthogonalizing n vectors of $\mathbb{R}^n$ behaves like $O(n^3)$ (since each scalar product has the cost $O(n)$).

Example 4.7 (*Numerical Instability of the Gram-Schmidt Orthogonalization Method*) As an example, we consider the three vectors

$$a_1 = \begin{pmatrix} 1 \\ \frac{1}{2} \\ \frac{1}{3} \end{pmatrix}, \quad a_2 = \begin{pmatrix} \frac{1}{2} \\ \frac{1}{3} \\ \frac{1}{4} \end{pmatrix}, \quad a_3 = \begin{pmatrix} \frac{1}{3} \\ \frac{1}{4} \\ \frac{1}{5} \end{pmatrix},$$

and perform the calculation directly in Python. When using `half` as floating point numbers, even this small task shows a numerical instability. The Gram-Schmidt orthogonalization method is implemented as follows.

♣ Implementation 4.8: Gram-Schmidt Orthogonalization

```python
def gram_schmidt(A):
    n = A.shape[0]
    Q = np.zeros_like(A)

    for i in range(n):
        Q[:, i] = A[:, i]
        for j in range(i):
            Q[:, i] -= np.inner(A[:, i], Q[:, j]) * Q[:, j]
        Q[:, i] /= np.linalg.norm(Q[:, i])
    return Q
```

When we apply this to the 3×3 Hilbert matrix in `half` precision

```python
A = np.array([[1,     1 / 2, 1 / 3],
              [1 / 2, 1 / 3, 1 / 4],
              [1 / 3, 1 / 4, 1 / 5]], dtype=np.half)
Q = gram_schmidt(A)
print(Q)
```

we get

```
[[ 0.857  -0.4995  0.274 ]
 [ 0.4285  0.569  -0.7905]
 [ 0.2856  0.653   0.548 ]]
```

We test the orthogonality of this, i.e., we calculate $Q^T Q - I$ and get

```python
print(Q.T @ Q)
err = np.linalg.norm(Q.T @ Q - np.identity(Q.shape[0]), ord=2)
print(f'||Q * Q^T - I||_2 = {err}')
```
```
[[ 0.9995    0.002161  0.05252 ]
 [ 0.002161  0.9995   -0.2289  ]
 [ 0.05252  -0.2289    1.      ]]
||Q * Q^T - I||_2 = 0.33211513864862696
```

With an error of 33%, only a very inaccurate approximation of the identity matrix is achieved.   ◀

We consider an example.

Example 4.9 (*QR Factorization Using Gram-Schmidt Orthogonalization*) Let the system of linear equations $Ax = b$ be given with

$$A := \begin{pmatrix} 1 & 1 & 1 \\ 0.01 & 0 & 0.01 \\ 0 & 0.01 & 0.01 \end{pmatrix}, \quad b := \begin{pmatrix} 1 \\ 0 \\ 0.02 \end{pmatrix},$$

and the exact solution

$$x = \begin{pmatrix} -1 \\ 1 \\ 1 \end{pmatrix}.$$

We determine the corresponding orthonormal basis from the column vectors $A = (a_1, a_2, a_3)$ using the Gram-Schmidt orthogonalization method. With three-digit accuracy, we obtain

$$q_1 = \frac{q_1}{\|q_1\|} \approx \begin{pmatrix} 1 \\ 0.01 \\ 0 \end{pmatrix}.$$

Next,

$$\tilde{q}_2 = a_2 - (a_2, q_1)q_1 \approx a_2 - q_1 = \begin{pmatrix} 0 \\ -0.01 \\ 0.01 \end{pmatrix}, \quad q_2 = \frac{\tilde{q}_2}{\|\tilde{q}_2\|} \approx \begin{pmatrix} 0 \\ -0.709 \\ 0.709 \end{pmatrix},$$

and finally

$$\tilde{q}_3 = a_3 - (a_3, q_1)q_1 - (a_3, q_2)q_2 \approx a_3 - q_1 - 0 = \begin{pmatrix} 0 \\ 0 \\ 0.01 \end{pmatrix}, \quad q_3 = \begin{pmatrix} 0 \\ 0 \\ 1 \end{pmatrix}.$$

The QR factorization results in the "orthogonal matrix"

$$\tilde{Q} = \begin{pmatrix} 1 & 0 & 0 \\ 0.01 & -0.709 & 0 \\ 0 & 0.709 & 1 \end{pmatrix}$$

as well as the upper right triangular matrix $\tilde{R}$ with $\tilde{r}_{ij} = (q_i, a_j)$ for $j \geq i$

$$\tilde{R} := \begin{pmatrix} (q_1, a_1) & (q_1, a_2) & (q_1, a_3) \\ 0 & (q_2, a_2) & (q_2, a_3) \\ 0 & 0 & (q_3, a_3) \end{pmatrix} \approx \begin{pmatrix} 1 & 1 & 1 \\ 0 & 0.00709 & 0 \\ 0 & 0 & 0.01 \end{pmatrix}.$$

We solve using the QR factorization $\tilde{R}\tilde{x} = \tilde{Q}^T b$:

$$\begin{pmatrix} 1 & 1 & 1 \\ 0 & 0.00709 & 0 \\ 0 & 0 & 0.01 \end{pmatrix} \begin{pmatrix} \tilde{x}_1 \\ \tilde{x}_2 \\ \tilde{x}_3 \end{pmatrix} = \tilde{b} = \tilde{Q}^T b \approx \begin{pmatrix} 1 \\ 0.0142 \\ 0.02 \end{pmatrix}$$

It follows

$$\tilde{x} = \begin{pmatrix} -3 \\ 2 \\ 2 \end{pmatrix}$$

with the relative error of 140%, which results from

$$\frac{\|\tilde{x} - x\|_2}{\|x\|_2} \approx 1.41$$

When solving the system with this disturbed matrix, a huge error occurs. This is due to the fact that the matrix $\tilde{Q}$ only shows a very disturbed orthogonality

$$I \stackrel{!}{=} \tilde{Q}^T \tilde{Q} \approx \begin{pmatrix} 1 & -0.00709 & 0 \\ -0.00709 & 1.01 & 0.709 \\ 0 & 0.709 & 1 \end{pmatrix}, \quad \|\tilde{Q}^T \tilde{Q} - I\|_2 \approx 0.7.$$

We adapt the Python implementation of the Gram-Schmidt orthogonalization method to create the QR factorization:

♣ Implementation 4.10: QR Factorization Using the Gram-Schmidt Method

```python
def qr_gram_schmidt(A):
    n = A.shape[0]
    Q, R = np.zeros_like(A), np.zeros_like(A)

    for i in range(n):
        Q[:, i] = A[:, i]

        for j in range(i):
            Q[:, i] -= np.inner(A[:,i], Q[:, j]) * Q[:, j]
        Q[:, i] /= np.linalg.norm(Q[:, i])
        for j in range(i, n):
            R[i, j] = np.inner(Q[:, i], A[:, j])
    return Q, R
```

With the `backward` implemented in Chap. 3, we can solve the system. Using `half` precision floating point numbers, we get

```python
1 A = np.array([[1,     1,     1   ],
2                [0.01, 0,     0.01],
3                [0,     0.01, 0.01]], dtype=np.half)
4 b = np.array([1, 0, 0.02], dtype=np.half)
5 x_ex = np.array([-1, 1, 1])
6
7 Q, R = qr_gram_schmidt(A)
8 b2 = np.dot(Q.transpose(), b)
9 x = backward(R, b2)
10
11 print('x =', x)
12 print('x_ex = ', x_ex)
```

```
x = [-3. 2.  2.]
x_ex =  [-1  1  1]
```

```python
1 rel_err = np.linalg.norm(x - x_ex) / np.linalg.norm(x_ex)
2 print(f'||x - x_ex|| / ||x|| = {rel_err}')
```

```
||x - x_ex|| / ||x_ex|| = 1.414213562373095
```

Here, the orthogonality is similar to the QR factorization calculated by hand:

```python
1 err = np.linalg.norm(Q @ Q.transpose() - np.identity(Q.shape[0]), ord=2)
2 print(f'||Q * Q^T - I||_2 = {err}')
```

```
||Q * Q^T - I||_2 = 0.7072275245944134
```

◄

In principle, the QR factorization has better stability properties than the LU factorization, as the matrix R carries at most the condition number of the matrix A. However, this consideration only applies if Q and R can be generated without errors. The simple Gram-Schmidt method is not suitable for creating the orthogonal matrix Q. In the following sections, we will introduce alternative orthogonalization methods that rely on the application of well-conditioned operations. However, we first briefly discuss a variant of the Gram-Schmidt method. The orthogonalization step

$$\tilde{q}_i := a_i - \sum_{j=1}^{i-1}(a_i, q_j)q_j,$$

is replaced by an iterative rule:

$$\tilde{q}_i^0 := a_i, \quad k = 1,\ldots,i-1: \quad \tilde{q}_i^{(k)} = \tilde{q}_i^{(k-1)} - (\tilde{q}_i^{(k-1)}, q_k)q_k, \quad \tilde{q}_i := \tilde{q}_i^{(i-1)}.$$

In this way, newly arising errors are always taken into account during the projection.

Theorem 4.11 (Modified Gram-Schmidt Method) *Let $\{a_1,\ldots,a_n\}$ be a basis of a vector space V, and let $(\cdot,\cdot)$ be a scalar product with induced norm $\|\cdot\|$. The algorithm*

$$(i)q_1 := \frac{a_1}{\|a_1\|},$$

$$i = 2, \ldots, n: \quad (ii)\tilde{q}_i^{(0)} := a_i$$

$$\tilde{q}_i^{(k)} := \tilde{q}_i^{(k-1)} - (\tilde{q}_i^{(k-1)}, q_k)q_k, \quad k = 1, \ldots, i-1,$$

$$q_i := \frac{\tilde{q}_i^{(i-1)}}{\|\tilde{q}_i^{(i-1)}\|},$$

generates an orthonormal basis $\{q_1, \ldots, q_n\}$ of V. It also holds that

$$(q_i, a_j) = 0 \quad \forall 1 \le j < i \le n.$$

The proof can be carried out similarly to Theorem 2.10. We now repeat Example 4.7 with this modification.

Example 4.12 (*Modified Gram-Schmidt Method*) We again consider the vectors

$$a_1 = \begin{pmatrix} 1 \\ \frac{1}{2} \\ \frac{1}{3} \end{pmatrix}, \quad a_2 = \begin{pmatrix} \frac{1}{2} \\ \frac{1}{3} \\ \frac{1}{4} \end{pmatrix}, \quad a_3 = \begin{pmatrix} \frac{1}{3} \\ \frac{1}{4} \\ \frac{1}{5} \end{pmatrix}$$

and calculate the orthogonalization with the modified algorithm, which we implement in Python:

♣ **Implementation 4.13: Modified Gram-Schmidt Method**

```python
def gram_schmidt_mod(A):
    n = A.shape[0]
    Q = np.zeros_like(A)

    for i in range(n):
        Q[:, i] = A[:, i]
        for j in range(i):
            Q[:, i] -= np.inner(Q[:, i], Q[:, j]) * Q[:, j]
        Q[:, i] /= np.linalg.norm(Q[:, i])
    return Q
```

Since we only need to access $\tilde{q}_i^{(k-1)}$ in step k of part (ii) of the procedure, we can always overwrite this with $\tilde{q}_i^{(k)}$. We again apply the method to the 3×3 Hilbert matrix with `half` floating point numbers

```python
Q = gram_schmidt_mod(A)
err = np.linalg.norm(Q @ Q.transpose() - np.identity(Q.shape[0]), ord=2)
print(f'Q =\n{Q}')
print(f'||Q * Q^T - I||_2 = {err}')
```

and obtain

```
Q =
[[ 0.857  -0.4995  0.1514]
 [ 0.4285  0.569  -0.661 ]
 [ 0.2856  0.653   0.733 ]]
||Q * Q^T - I||_2 = 0.06270300839082639
```

Compared to Example 4.7, we achieve the (still significant) relative error of 6.3%, which is far smaller than the 33% we achieved with the original procedure.   ◄

Accordingly, we repeat the QR factorization of the matrix A from Example 4.9 now using the modified Gram-Schmidt method:

Example 4.14 (*QR Factorization With the Modified Gram-Schmidt Method*) Let's consider again the system of linear equations

$$A := \begin{pmatrix} 1 & 1 & 1 \\ 0.01 & 0 & 0.01 \\ 0 & 0.01 & 0.01 \end{pmatrix}, \quad b := \begin{pmatrix} 1 \\ 0 \\ 0.02 \end{pmatrix},$$

with the solution

$$x = \begin{pmatrix} -1 \\ 1 \\ 1 \end{pmatrix}$$

given. We will skip the detailed presentation of the three-digit calculation and jump straight to the Python implementation. To compute the QR factorization with the modified Gram-Schmidt method, we only need to extend our previous code by creating the matrix R in lines 10–11:

🍀 Implementation 4.15: QR Factorization With the Modified Gram-Schmidt Method

```python
 1  def qr_gram_schmidt_mod(A):
 2      n = A.shape[0]
 3      Q, R = np.zeros_like(A), np.zeros_like(A)
 4
 5      for i in range(n):
 6          Q[:, i] = A[:, i]
 7          for j in range(i):
 8              Q[:, i] -= np.inner(Q[:, i], Q[:, j]) * Q[:, j]
 9          Q[:, i] /= np.linalg.norm(Q[:, i])
10          for j in range(i, n):
11              R[i, j] = np.inner(Q[:, i], A[:, j])
12      return Q, R
```

Applied to the matrix

```
1  Q, R = qr_gram_schmidt_mod(A)
2  b3 = np.dot(Q.transpose(), b)
3  x2 = backward(R, b3)
4  print(f'x = {x}')
```

this yields

```
x = [-2. 2.  1.]
```

From this, we obtain the (relative) error in the solution and orthogonality:

```
1  rel_err = np.linalg.norm(x2 - x_ex) / np.linalg.norm(x_ex)
2  print(f'||x - x_ex|| / ||x_ex|| = {rel_err}')
3
4  err = np.linalg.norm(Q @ Q.transpose() - np.identity(Q.shape[0]), ord=2)
5  print(f'||Q * Q^T - I||_2 = {err}')
```

```
||x - x_ex|| / ||x_ex|| = 0.8164965809277261
||Q * Q^T - I||_2 = 0.010000213623046875
```

Although the orthogonality has improved by a factor of 70, the error of the solution has not even halved. The reason for the poor result lies in the structure of the method: the modified algorithm orthogonalizes in iteration i always with respect to the already calculated vectors $q_j, j < i$. This leads to improved orthogonality ($(q_i, q_j) \approx 0$, but not to better orthogonality with respect to the columns of the matrix A, i.e., $R = Q^T A$ generally does not have a triangular shape. ◄

Due to the orthogonality of the matrix Q, the QR factorization has the potential to be a very stable method. However, so far it has not been possible to carry out the factorization process in a sufficiently stable manner. The calculation of the entries of the matrix R turns out to be a problem.

4.3   Householder Transformations

The geometric principle behind the Gram-Schmidt method is the projection of the vectors a_i onto the already created orthogonal basis $q_1, \ldots, q_{i-1}$. This projection is poorly conditioned if a_i is almost parallel to the q_j with $j < i$, such as $a_i \approx q_j$. Then cancellation is imminent. We can write a step of the Gram-Schmidt method compactly using a matrix-vector product:

$$\tilde{q}_i = [I - G^{(i)}]a_i, \quad G^{(i)} = \sum_{l=1}^{i-1} q_l q_l^T .$$

Here, $q_e q_e^T \in \mathbb{R}^{n \times n}$ is the *dyadic product* of two vectors, i.e.,

$$a, b \in \mathbb{R}^n \quad \Rightarrow \quad (ab^T)_{ij} = a_i b_j, \ i, j = 1, \ldots, n.$$

The matrix $[I - G^{(i)}]$ is a projection onto $\mathbb{R}^n \to \mathbb{R}^n$, because

$$[I - G^{(i)}]^2 = I - 2 \sum_{l=1}^{i-1} q_l q_l^T + \sum_{k,l=1}^{i-1} q_l \underbrace{q_l^T q_k}_{=\delta_{lk}} q_k^T = [I - G^{(i)}].$$

Furthermore, it holds:

$$[I - G^{(i)}]q_k = q_k - \sum_{l=1}^{i-1} q_l \underbrace{q_l^T q_k}_{=\delta_{lk}} = 0, \quad k < i.$$

The matrix $I - G^{(i)}$ is therefore not regular. If in step i the vector a_i is almost parallel to the already orthogonal vectors, i.e., $\tilde{a}_i \in \delta a_i + \mathrm{span}\{q_1, \ldots, q_{i-1}\}$, then

$$\tilde{q}_i = [I - G^{(i)}]\tilde{a}_i = [I - G^{(i)}]\delta a_i \quad \Rightarrow \quad \frac{\|\delta q_i\|}{\|\tilde{q}_i\|} = \frac{\|\delta q_i\|}{\|[I - G^i]\delta a_i\|}.$$

For $\delta a_i \to 0$, an arbitrarily large error amplification is possible.

In the following, we are looking for a transformation of A to a triangular matrix R, which itself is based on orthogonal and thus stable operations. An orthogonal matrix $Q \in \mathbb{R}^{n \times n}$ with $\det(Q) = 1$ represents a rotation and for $\det(Q) = -1$ a reflection (or a rotation-reflection). The Householder transformations use a reflection matrix to transform a matrix $A \in \mathbb{R}^{n \times n}$ into an upper right triangular matrix.

Definition 4.16 (*Householder Transformation*)   For a vector $v \in \mathbb{R}^n$ with $\|v\|_2 = 1$, the *dyadic product* is defined by vv^T and the matrix

$$S := I - 2vv^T \in \mathbb{R}^{n \times n}$$

is called *Householder transformation*.

Theorem 4.17 (Householder Transformation) *Every Householder transformation $S = I - 2vv^T$ with $\|v\|_2 = 1$ is symmetric and orthogonal. The product of two Householder transformations $S_1 S_2$ is an orthogonal matrix itself.*

Proof *(i)* Symmetry holds

$$S^T = [I - 2vv^T]^T = I - 2(vv^T)^T = I - 2vv^T = S$$

and further

$$S^T S = I - 4vv^T + 4v \underbrace{v^T v}_{=1} v^T = I,$$

i.e., $S^{-1} = S^T$ and S are orthogonal to each other.

(ii) With two symmetric orthogonal matrices S_1 and S_2 it holds

$$(S_1 S_2)^T = S_2^T S_1^T = S_2^{-1} S_1^{-1} = (S_1 S_2)^{-1},$$

and the product $S_1 S_2$ is an orthogonal matrix. This relation only requires the orthogonality of S_1 and S_2. We have not used the fact that these matrices are Householder transformations.   □

We conclude the basic investigation of Householder transformations with a geometric characterization:

Remark 4.18 (*Householder Transformation as Reflection*) Let $v \in \mathbb{R}^n$ be an arbitrary normalized vector with $\|v\|_2 = 1$. Further, let $x \in \mathbb{R}^n$ be given with $x = \alpha v + w^\perp$, where $w^\perp \in \mathbb{R}^n$ is a vector with $v^T w^\perp = 0$ in the orthogonal complement to v. Then, for the Householder transformation $S = I - 2vv^T$, it holds:

$$S(\alpha v + w^\perp) = [I - 2vv^T](\alpha v + w^\perp) = \alpha(v - 2v \underbrace{v^T v}_{=1}) + w^\perp - 2v \underbrace{v^T w^\perp}_{=0} = -\alpha v + w^\perp.$$

That is, the Householder transformation describes a reflection on the plane perpendicular to v.   ◆

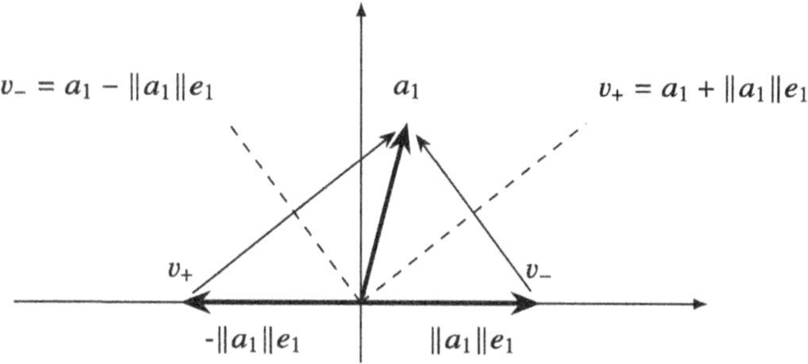

Fig. 4.1 Axes of reflection (dashed) and normals to the reflection v_+ and v_- for the reflection of a_1 onto span$\{e_1\}$

Using Householder transformations, a matrix $A \in \mathbb{R}^{n \times n}$ is to be transformed step by step into an upper right triangular matrix:

$$A^{(0)} = A, \quad A^{(i)} = S^{(i)} A^{(i-1)}, \quad S^{(i)} = I - 2v^{(i)}(v^{(i)})^T,$$

where $a_{kl}^{(i)} = 0$ for $l \leq i$ and $k > l$ has partial diagonal form. Finally, $R := A^{(n-1)}$.

We describe the first step of the procedure: Given is the regular matrix $A^{(0)} := A \in \mathbb{R}^{n \times n}$, written in its column vectors a_i for $i = 1, \ldots, n$. We are looking for an orthogonal Householder transformation $S^{(i)}$, such that its multiplication from the left, i.e., $A^{(1)} = S^{(1)} A^{(0)}$ reflects the first column of $A^{(0)}$ onto a multiple of the first unit vector e_1. The column vectors of $A^{(1)}$ we call $a_i^{(1)}$ for $i = 1, \ldots, n$, and expect

$$a_1^{(1)} = S^{(1)} a_1^{(0)} \in \text{span}\{e_1\}.$$

We have to reflect at the plane perpendicular to $a_1 \pm \|a_1\| e_1$, see Fig. 4.1. To determine the reflection vector, we have two options due to the choice of the sign. For stability reasons, we set

$$v^{(1)} := \frac{a_1 + \text{sign}(a_{11})\|a_1\| e_1}{\|a_1 + \text{sign}(a_{11})\|a_1\| e_1\|}, \tag{4.5}$$

to avoid the risk of cancellation by an optimal choice of the sign. With this choice, the following applies:

$$a_i^{(1)} = S^{(1)} a_i = a_i - 2(v^{(1)}, a_i)v^{(1)}, \quad i = 2, \ldots, n$$

$$\Rightarrow \quad a_1^{(1)} = -\text{sign}(a_{11})\|a_1\| e_1. \tag{4.6}$$

The resulting matrix $A^{(1)}$ is regular and $a_1^{(1)} \in \text{span}(e_1)$ holds

$$A := \begin{pmatrix} a_{11} & a_{12} & a_{13} & \cdots & a_{1n} \\ a_{21} & a_{22} & & & \vdots \\ a_{31} & a_{32} & & & \vdots \\ \vdots & & & & \vdots \\ a_{n1} & a_{n2} & \cdots & \cdots & a_{nn} \end{pmatrix} \Rightarrow A^{(1)} := S^{(1)} A = \begin{pmatrix} a_{11}^{(1)} & a_{12}^{(1)} & a_{13}^{(1)} & \cdots & a_{1n}^{(1)} \\ 0 & a_{22}^{(1)} & & & \vdots \\ 0 & a_{32}^{(1)} & & & \vdots \\ \vdots & & & & \vdots \\ 0 & a_{n2}^{(1)} & \cdots & \cdots & a_{nn}^{(1)} \end{pmatrix}$$

We continue the procedure with the submatrix $\tilde{A}^{(1)} := A_{kl>1}^{(1)} \in \mathbb{R}^{(n-1) \times (n-1)}$. For this, the shortened second column vector is defined as $\tilde{a}^{(1)} = (a_{22}^{(1)}, a_{32}^{(1)}, \ldots, a_{n2}^{(1)})^T \in \mathbb{R}^{n-1}$. Then we choose:

$$\tilde{v}^{(2)} := \frac{\tilde{a}^{(1)} + \text{sign}(a_{22}^{(1)}) \|\tilde{a}^{(1)}\| e_1}{\|\tilde{a}^{(1)} + \text{sign}(a_{22}^{(1)}) \|\tilde{a}^{(1)}\| e_1\|}, \quad S^{(2)} = \begin{pmatrix} 1 & 0 & \cdots\cdots & 0 \\ \hline 0 & & & \\ 0 & & & \\ \vdots & & I - 2\tilde{v}^{(2)}(v^{(2)})^T & \\ 0 & & & \end{pmatrix} \in \mathbb{R}^{n \times n}.$$

Multiplication with $S^{(2)}$ from the left, i.e., $A^{(2)} := S^{(2)} A^{(1)}$, leaves the first row unchanged. The block $A_{kl>1}^{(1)}$ is eliminated. The resulting matrix $A^{(2)}$ has zero subdiagonals for $a_{kl}^{(2)} = 0$ for $l = 1, 2$ at $k > l$. After $n - 1$ steps, it holds

$$R := \underbrace{S^{(n-1)} S^{(n-2)} \cdots S^{(1)}}_{=:Q^T} A.$$

All transformations are orthogonal, therefore $Q \in \mathbb{R}^{n \times n}$ is an orthogonal (and of course regular) matrix (see Theorem 4.2).

Theorem 4.19 (QR Factorization With Householder Transformations) *Let $A \in \mathbb{R}^{n \times n}$ be a regular matrix. Then the QR factorization of A according to Householder can be performed in*

$$\frac{2}{3}n^3 + O(n^2)$$

numerically stable elementary operations.

Proof *(i)* The feasibility of the QR factorization results from the algorithm. From the regularity of the matrices $A^{(i)}$, it follows that the subvector $\tilde{a}^{(i)} \in \mathbb{R}^{n-i}$ must be non-zero. Then $\tilde{v}^{(i)}$ is well-defined according to (4.5). The following elimination is performed column by column with (4.6):

$$\tilde{a}_k^{(i+1)} = \underbrace{[I - 2\tilde{v}^{(i)}(\tilde{v}^{(i)})^T]}_{=\tilde{S}^{(i)}} \tilde{a}_k^{(i)} = \tilde{a}_k^{(i)} - 2(\tilde{v}^{(i)}, \tilde{a}_k^{(i)})\tilde{v}^{(i)}$$

The numerical stability is guaranteed by $\text{cond}_2(S^{(i)}) = 1$, see Remark 3.2.

(ii) We now come to the estimation of the effort. In step $A^{(i)} \to A^{(i+1)}$, the vector $\tilde{v}^{(i)} \in \mathbb{R}^{n-i}$ must be calculated. This requires $2(n - i) + 2$ elementary operations. Subsequently, the column-wise elimination takes place. For each of the $n - i$ column vectors, a scalar product $(\tilde{v}^{(i)}, \tilde{a}^{(i)})$ (these are $n - i$ elementary operations) as well as a vector addition and the multiplication of $(\tilde{v}^{(i)}, \tilde{a}^{(i)})$ with $\tilde{v}^{(i)}$ (another $n - i$ elementary operations) are to be performed. This results in the total effort:

$$N_{QR} = 2 \sum_{i=1}^{n-1} (n-i) + (n-i)^2 = n(n-1) + 2 \frac{n(n-1)(2n-1)}{6} = \frac{2n^3}{3} + O(n^2).$$

$\square$

Remark 4.20 (*QR Factorization With Householder Transformations*) The Householder matrices $\tilde{S}^{(i)} = I - 2\tilde{v}^{(i)}(\tilde{v}^{(i)})^T$ are not explicitly constructed and stored. In an efficient implementation, only the vectors $\tilde{v}^{(i)} \in \mathbb{R}^{n-i+1}$ are memorized. The orthogonal matrix

$$Q := (S^{(1)})^T \cdots (S^{(n-1)})^T$$

is not explicitly calculated and stored. If the matrix is needed, for example, to calculate the right-hand side $\tilde{b} = Q^T b$, this is instead done step by step using the Householder transformations

$$\tilde{b} = Q^T b = S^{(n-1)} \cdots S^{(1)} b$$

and the product is determined based on a simple iteration directly working with the vectors $v^{(i)}$:

> **Input:** Vector b and Householder vectors $v_1, \ldots, v_{n-1}$
> 1 $b^{(0)} = b$
> 2   **for** $i = 1$ **to** $n - 1$ **do**
> 3 $b^{(i+1)} = b^{(i)} - 2(v^{(i)}, b^{(i)})v^{(i)}$
> **Result:** $\tilde{b} = b^{(n)}$

In addition to the upper triangular matrix R, the vectors $\tilde{v}^{(i)} \in \mathbb{R}^{n+1-i}$ must be stored. This requires storing $(n+1)n$ values in total. In contrast to the LU factorization, this cannot be done solely in the memory space of the matrix A since both R and the $\tilde{v}^{(i)}$ occupy the diagonal line. In practical implementations, an additional diagonal vector must be created. $\blacklozenge$

Finally, we calculate the QR factorization for Example 4.9 using Householder transformations:

Example 4.21 (*QR Factorization With Householder Transformations*) Let us again consider the system of linear equations $Ax = b$ with

$$A := \begin{pmatrix} 1 & 1 & 1 \\ 0.01 & 0 & 0.01 \\ 0 & 0.01 & 0.01 \end{pmatrix}, \quad b := \begin{pmatrix} 1 \\ 0 \\ 0.02 \end{pmatrix} \quad \text{and} \quad x = \begin{pmatrix} -1 \\ 1 \\ 1 \end{pmatrix}$$

be given. We skip the manual calculation at three-digit accuracy and immediately look at the Python implementation for creating the Householder vectors:

♣ Implementation 4.22: QR Factorization With Householder Transformations

```python
def qr_householder(A):
    n, m = A.shape
    V = np.zeros_like(A)

    for i in range(m):
        V[i:, i] = A[i:, i]
        ei = np.zeros(n - i, dtype=A.dtype)
        ei[0] = 1.0
        V[i:, i] += np.sign(A[i, i]) * np.linalg.norm(V[i:, i]) * ei
        V[i:, i] /= np.linalg.norm(V[i:, i])
        for k in range(i, m):
            A[i:, k] -= 2 * np.inner(V[i:, i], A[i:, k]) * V[i:, i]
    return V
```

To this end, we still need to implement the application of the transposed matrix Q^T to the right-hand side:

♣ Implementation 4.23: Applying the Householder Transformations

```python
def QT_apply(V, b):
    for i in range(b.shape[0]):
        b[i:] -= 2 * np.inner(V[i], b[i:]) * V[i]
    return None
```

We apply this to the matrix A using `half` floating point numbers:

```python
V = qr_householder(A)
```

If we calculate the matrix Q^T, which we get from these Householder vectors, then we have

$$Q^T = \begin{pmatrix} -1.0 & -0.01 & 0.0 \\ 0.007053 & -0.705 & 0.7065 \\ 0.00707 & -0.7065 & -0.707 \end{pmatrix}$$

and

$$Q^T Q \approx \begin{pmatrix} 1.0 & -1.073 \times 10^{-6} & -1.669 \times 10^{-6} \\ -1.073 \times 10^{-6} & 0.9966 & -1.330 \times 10^{-3} \\ -1.669 \times 10^{-6} & -1.33 \cdot 19^{-3} & 0.9990 \end{pmatrix}.$$

This corresponds to a relative error $\|Q^T Q - I\| \leq 4 \times 10^{-3}$. If we now use the orthogonal vectors to solve the system, we obtain the solution

```python
QT_apply(V, b)
x = backward(A, b)
print(f'x = {x}')
```
```
[-1.     0.999  1.001]
```

This has the relative error

```
1 rel_err = np.linalg.norm(x - x_ex) / np.linalg.norm(x_ex)
2 print(f'||x - x_ex|| / ||x_ex|| = {rel_err:.4e}')

  ||x - x_ex|| / ||x_ex|| = 7.974e-04
```

We compare these results with the previous ones from Examples 4.9 and 4.14. Compared to the QR factorization with the modified Gram-Schmidt method, we have improved the error of the solution by a factor of 17 at half (half) precision. In addition, we have determined the solution with a maximum component-wise error of 10^{-3}. When performed correctly, the QR factorization is a numerically very stable method.   ◄

4.4   Givens Rotations

The Gram-Schmidt method is based on projections, while Householder transformations are reflections. Another option is to use rotations as orthogonal transformations for the elimination.

Definition 4.24 (*Givens Rotation*) A *Givens rotation* in $\mathbb{R}^n$ is understood to be the rotation by the angle θ in the plane spanned by two unit vectors e_i and e_j. The transformation matrix is given by:

$$G(i, j, \theta) = \begin{pmatrix} 1 & & & & & & & & \\ & \ddots & & & & & & & \\ & & 1 & & & & & & \\ & & & c & & & -s & & \\ & & & & 1 & & & & \\ & & & & & \ddots & & & \\ & & & & & & 1 & & \\ & & & s & & & c & & \\ & & & & & & & 1 & \\ & & & & & & & & \ddots \\ & & & & & & & & & 1 \end{pmatrix}, \quad \begin{array}{l} c = \cos(\theta) \\[2mm] s = \sin(\theta) \end{array}$$

Theorem 4.25 (Givens Rotation) *The Givens rotation $G(i, j, \theta)$ is an orthogonal matrix with* $\det(G) = 1$. *It is* $G(i, j, \theta)^{-1} = G(i, j, -\theta)$.

Proof Check by calculation.   □

Like the Householder transformations, the Givens rotations are orthogonal matrices. Multiplying a matrix from the left, i.e., GA, or a vector, i.e., Gx, only changes the i-th and j-th row of A or x:

$$G(i, j, \theta)A = \begin{pmatrix} a_{11} & \cdots & a_{1n} \\ \vdots & \ddots & \vdots \\ a_{i-1,1} & \cdots & a_{i-1,n} \\ ca_{i1} - sa_{j1} & \cdots & ca_{in} - sa_{jn} \\ a_{i+1,1} & \cdots & a_{i+1,n} \\ \vdots & \ddots & \vdots \\ a_{j-1,1} & \cdots & a_{j-1,n} \\ sa_{i1} + ca_{j1} & \cdots & sa_{in} + ca_{jn} \\ a_{j+1,1} & \cdots & a_{j+1,n} \\ \vdots & \ddots & \vdots \\ a_{n1} & \cdots & a_{nn} \end{pmatrix}$$

The QR factorization based on Givens rotations again gradually transforms the matrix A into an upper right triangular matrix R. However, applying a Givens rotation can only eliminate a single subdiagonal element and not an entire column. Hence, $n - 1$ rotations are needed for the first column, $n - 2$ for the second column, and so forth. Although this sounds costly, we must consider that each Givens rotation consists of four non-zero or non-unit values only, and its application is much cheaper than that of a Householder step.

$$\begin{pmatrix} * & * & * & * \\ * & * & * & * \\ * & * & * & * \\ * & * & * & * \end{pmatrix} \rightarrow \begin{pmatrix} * & * & * & * \\ 0 & * & * & * \\ * & * & * & * \\ * & * & * & * \end{pmatrix} \rightarrow \begin{pmatrix} * & * & * & * \\ 0 & * & * & * \\ 0 & * & * & * \\ * & * & * & * \end{pmatrix} \rightarrow \begin{pmatrix} * & * & * & * \\ 0 & * & * & * \\ 0 & * & * & * \\ 0 & * & * & * \end{pmatrix}$$

$$\rightarrow \begin{pmatrix} * & * & * & * \\ 0 & * & * & * \\ 0 & 0 & * & * \\ 0 & * & * & * \end{pmatrix} \rightarrow \begin{pmatrix} * & * & * & * \\ 0 & * & * & * \\ 0 & 0 & * & * \\ 0 & 0 & * & * \end{pmatrix} \rightarrow \begin{pmatrix} * & * & * & * \\ 0 & * & * & * \\ 0 & 0 & * & * \\ 0 & 0 & 0 & * \end{pmatrix}$$

We consider one step of the procedure. For this, the matrix A is given in the form:

$$A = \begin{pmatrix} a_{11} & \cdots \cdots & \cdots & \cdots & \cdots & a_{1n} \\ 0 & \ddots & & \ddots & & \vdots \\ \vdots & \ddots & a_{ii} & & & \vdots \\ \vdots & & 0 & a_{i+1,i+1} & & \vdots \\ \vdots & & \vdots & \vdots & & \vdots \\ \vdots & & 0 & \vdots & & \vdots \\ \vdots & & a_{ji} & \vdots & \ddots & \vdots \\ \vdots & & \vdots & \vdots & & \vdots \\ 0 & \cdots & a_{ni} & a_{n,i+1} & \cdots & a_{nn} \end{pmatrix}$$

We are looking for the Givens rotation $G(i, j, \theta)$ for the elimination of a_{ji}. For GA we have

$$(G(i, j, \theta)A)_{ji} = sa_{ii} + ca_{ji}, \quad c := \cos(\theta), \quad s := \sin(\theta).$$

Instead of finding the angle θ, we directly determine the values c and s with the approach

$$sa_{ii} + ca_{ji} = 0, \quad c^2 + s^2 = 1 \quad \Rightarrow \quad c := \frac{a_{ii}}{\sqrt{a_{ii}^2 + a_{ji}^2}}, \quad s := -\frac{a_{ji}}{\sqrt{a_{ii}^2 + a_{ji}^2}}.$$

Application of $G(i, j, \theta)$ to A yields:

$$(G(i, j, \theta)A)_{ji} = \frac{-a_{ji}a_{ii} + a_{ii}a_{ji}}{\sqrt{a_{ii}^2 + a_{ji}^2}} = 0,$$

$$(G(i, j, \theta)A)_{ii} = \frac{a_{ii}^2 + a_{ji}^2}{\sqrt{a_{ii}^2 + a_{ji}^2}} = \sqrt{a_{ii}^2 + a_{ji}^2}.$$

The element a_{ji} is eliminated. To eliminate the i-th column (below the diagonal), $(n - i)$ Givens rotations are necessary. One of these rotations changes the $(n - i)$ matrix entries of two rows each. That is, to eliminate the i-th column, $4(n - i)^2$ elementary operations are necessary. In addition, there are $O(n - i)$ operations for calculating the coefficients, which we can neglect. In total, this results in

$$\sum_{i=1}^{n-1} 4(n - i)^2 = 4 \sum_{i=1}^{n-1} i^2 = \frac{2}{3}(n - 1)n(2n - 1) = \frac{4}{3}n^3 + O(n^2)$$

elementary operations. This is about twice the effort compared to the Householder method.

However, the QR factorization according to Givens becomes relevant when the matrix A is already sparsely populated. Only subdiagonal elements with $a_{ji} \neq 0$ need to be specifically eliminated. For so-called *Hessenberg matrices* (these are upper right triangular matrices, which additionally have a left subdiagonal), the QR factorization with Givens rotations can be performed in only $O(n^2)$ operations. The QR factorization of Hessenberg matrices plays a crucial role in the most critical method for calculating the eigenvalues of a matrix, see Chap. 5.

In Python, we can then implement the QR factorization with Givens rotations as follows. Here, we also allow the case that the matrix has more rows than columns.

♣ Implementation 4.26: QR Factorization using Givens Rotations

```python
def qr_givens(A):
    n, m = A.shape
    QT = np.identity(n, dtype=A.dtype)

    for i in range(m):
        for j in range(i + 1, n):
            c, s = A[i, i], -A[j, i]
            nrm = np.sqrt(c**2 + s**2)
            c, s = c / nrm, s / nrm
            for k in range(i, m):
                t1, t2 = A[i, k], A[j, k]
                A[i, k] = c * t1 - s * t2
                A[j, k] = s * t1 + c * t2
            for k in range(n):
                t1, t2 = QT[i, k], QT[j, k]
                QT[i, k] = c * t1 - s * t2
                QT[j, k] = s * t1 + c * t2
    return QT
```

Example 4.27 (*QR Factorization Using Givens Rotations*) As in the Examples 4.9 and 4.21 let us consider the following matrix and right side again:

$$A := \begin{pmatrix} 1 & 1 & 1 \\ 0.01 & 0 & 0.01 \\ 0 & 0.01 & 0.01 \end{pmatrix}, \quad b := \begin{pmatrix} 1 \\ 0 \\ 0.02 \end{pmatrix}.$$

If we apply our Python implementation with half precision (`half`),

```python
QT = qr_givens(A)
QTb = np.dot(QT, b)
x = backward(A, QTb)
```

we get the solution with a relative error of

```
||x - x_ex|| / ||x|| = 2.8191e-04
```

If we now look at the matrix Q^T in more detail, we see

$$Q^T = \begin{pmatrix} 1.0 & 0.01 & 0.0 \\ 0.007072 & -0.707 & 0.707 \\ 0.007072 & -0.707 & -0.707 \end{pmatrix}.$$

This result is almost identical to the matrix in Example 4.21. The approximation properties are, therefore, excellent, and we observe, that $QR \approx A$ and $QQ^T \approx I$ hold:

```
1 err = np.linalg.norm(A2 - QT.transpose() @ A, ord=2) / np.linalg.norm(A2, ord=2)
2 print(f'||A - QR||_2 / ||A||_2 = {err:.4e}')

  ||A - QR||_2 / ||A||_2 = 7.2264e-07

1 Id = np.identity(QT.shape[0], dtype=np.single)
2 err = np.linalg.norm(Id - QT @ QT.T, ord=2)
3 print(f'||I - Q^T*Q||_2 = {err:.4e}')

  ||I - Q*Q^T||_2 = 5.0022e-05
```

The matrix A2 is a copy of A in `single` floating point representation, so that we can determine the 2-norm of the error. We thus see that we have determined the QR factorization up to the machine precision of `half`, which is about $eps \approx 4 \times 10^{-4}$, and the matrix Q is orthogonal to the used machine precision.   ◀

4.5   Overdetermined Systems

In many applications, linear systems occur that have a different number of equations and unknowns:
$$Ax = b, \quad A \in \mathbb{R}^{n \times m}, \quad x \in \mathbb{R}^m, \quad b \in \mathbb{R}^n, \quad n \neq m.$$

In the case $n > m$ (i.e., more equations than unknowns), we speak of *overdetermined* systems; in the case $n < m$, we speak of *underdetermined* systems. An efficient method for determining solutions of overdetermined systems is based on the QR factorization.

The investigation of the unique solvability of general systems with a rectangular matrix A can no longer be determined by their regularity. Instead, we know that a system is solvable if and only if the rank of the matrix A is equal to the rank of the extended matrix $[A|b]$. If additionally $\text{rank}(A) = m$, then the solution is unique. An underdetermined linear system with $n < m$ can therefore never be uniquely solvable. We consider in this section exclusively overdetermined systems, i.e., the case $n > m$. In practice, such overdetermined systems occur very frequently in least squares calculations and in optimization tasks. An overdetermined system is generally not solvable.

Example 4.28 (*Overdetermined Systems*)   We are looking for the quadratic interpolation polynomial
$$p(x) = a_0 + a_1 x + a_2 x^2,$$

given by the rule:

$$p\left(-\frac{1}{4}\right) = 0, \quad p\left(\frac{1}{2}\right) = 1, \quad p(2) = 0, \quad p\left(\frac{5}{2}\right) = 1.$$

This results in four equations for only three unknowns:

$$a_0 - \tfrac{1}{4}a_1 + \tfrac{1}{16}a_2 = 0$$

$$a_0 + \tfrac{1}{2}a_1 + \tfrac{1}{4}a_2 = 1$$

$$a_0 + 2a_1 + 4a_2 = 0$$

$$a_0 + \tfrac{5}{2}a_1 + \tfrac{25}{4}a_2 = 1$$

We try to solve the linear system with Gaussian elimination:

$$
\left(\begin{array}{ccc|c}
1 & -\tfrac{1}{4} & \tfrac{1}{16} & 0 \\
1 & \tfrac{1}{2} & \tfrac{1}{4} & 1 \\
1 & 2 & 4 & 0 \\
1 & \tfrac{5}{2} & \tfrac{25}{4} & 1
\end{array}\right)
\rightarrow
\left(\begin{array}{ccc|c}
1 & -\tfrac{1}{4} & \tfrac{1}{16} & 0 \\
0 & 3 & \tfrac{3}{4} & 4 \\
0 & 9 & \tfrac{63}{4} & 0 \\
0 & 11 & \tfrac{99}{4} & 4
\end{array}\right)
\rightarrow
\left(\begin{array}{ccc|c}
1 & -\tfrac{1}{4} & \tfrac{1}{16} & 0 \\
0 & 3 & 3 & 4 \\
0 & 0 & \tfrac{27}{2} & -12 \\
0 & 0 & 22 & -\tfrac{32}{4}
\end{array}\right)
$$

This means, it should be $a_2 = -8/9 \approx -0.89$ as well as $a_2 = -4/11 \approx -0.36$.   ◀

In Chap. 9, we consider the interpolation problem in detail. We will see that this result was to be expected: For the unique interpolation of $n + 1$ points by a polynomial, this polynomial is to be chosen of degree n.

In the case of overdetermined systems, we must change the objective. We can no longer speak of *the* solution of the system. Instead, we will characterize a vector $x \in \mathbb{R}^m$ as the *best approximation* of the solution. How good an approximation is, we can determine by the residual:

$$r(x) = b - Ax \in \mathbb{R}^n$$

A best approximation $x \in \mathbb{R}^m$ is the one that minimizes the residual:

Definition 4.29 (*Best Approximation*) Let $A \in \mathbb{R}^{n \times m}$ with $n > m$ and $b \in \mathbb{R}^n$. Let $\| \cdot \|$ be a norm on $\mathbb{R}^n$. A vector $x \in \mathbb{R}^m$ is called *best approximation* to $Ax = b$, if

$$\|b - Ax\| \le \|b - Ay\| \quad \forall y \in \mathbb{R}^m.$$

Different norms will define different best approximations. For the following theorem, it is essential to consider the Euclidean norm, induced by the Euclidean scalar product.

Definition 4.30 (*Method of Least Squares*) Let $A \in \mathbb{R}^{n \times m}$ with $n > m$ and $b \in \mathbb{R}^n$. The *method of least squares* defines the best approximation $x \in \mathbb{R}^m$ as the approximation to $Ax = b$, whose residual assumes the smallest Euclidean norm:

$$\|b - Ax\|_2 = \min_{y \in \mathbb{R}^m} \|b - Ay\|_2 \tag{4.7}$$

The solution $x \in \mathbb{R}^m$ is also called *least squares solution*.

The search for the best approximation in the Euclidean norm is closely related to the best approximation of functions, a task that we will consider in Chap. 9.

Theorem 4.31 (Least Squares) *Assume that for the matrix $A \in \mathbb{R}^{n \times m}$ with $n > m$ it holds* rank$(A) = m$. *Then the matrix $A^T A \in \mathbb{R}^{m \times m}$ is positive definite and the best approximation* $x \in \mathbb{R}^m$ *is uniquely determined as the solution of the* system of normal equations

$$A^T A x = A^T b.$$

Proof *(i)* It holds rank$(A) = m$. This means, $Ax = 0$ only if $x = 0$. This implies the positive definiteness of the matrix $A^T A$

$$(A^T Ax, x)_2 = (Ax, Ax)_2 = \|Ax\|_2^2 > 0 \quad \forall x \neq 0,$$

and the equation system $A^T Ax = A^T b$ is uniquely solvable for every right side $b \in \mathbb{R}^n$.

(ii) Let $x \in \mathbb{R}^m$ be the solution of the system of normal equations. Then for arbitrary $y \in \mathbb{R}^m$

$$\|b - A(x + y)\|_2^2 = \|b - Ax\|_2^2 + \|Ay\|_2^2 - 2(b - Ax, Ay)_2$$
$$= \|b - Ax\|_2^2 + \underbrace{\|Ay\|_2^2}_{\geq 0} - 2(\underbrace{A^T b - A^T Ax}_{=0}, y)_2 \geq \|b - Ax\|_2^2.$$

(iii) Now let x be the minimum of (4.7). This means it holds for any vector y:

$$\|b - Ax\|_2^2 \leq \|b - A(x + y)\|_2^2$$
$$= \|b - Ax\|_2^2 + \|Ay\|_2^2 - 2(b - Ax, Ay)_2 \quad \forall y \in \mathbb{R}^m$$

From this follows:
$$2(A^T b - A^T Ax, y)_2 \leq \|Ay\|_2^2$$

We choose y as $y = se_i$ with the i-th unit vector and $s \in \mathbb{R}$. Then, it follows

$$2s(A^T b - A^T Ax)_i \leq |s|^2 \|A\|^2 \quad \Rightarrow \quad |(A^T b - A^T Ax)_i| \leq |s| \|A\|^2.$$

and $s \to 0$ shows $A^T Ax = A^T b$. $\qquad \square$

The best approximation (in the Euclidean norm) of an overdetermined system can be found by solving the system of normal equations. The naive approach, to determine the matrix $A^T A$ and then solve the normal equation with the Cholesky method, is numerically not advisable. First, the effort to calculate $A^T A$ is considerable, and the calculation of the matrix-matrix multiplication is poorly conditioned. Furthermore, the estimate

$$\text{cond}(A^T A) = \|A^T A\| \cdot \|(A^T A)^{-1}\|,$$

indicates a potentially large condition number of $A^T A$. For a regular matrix, the estimate $\text{cond}(A^T A) \leq \text{cond}(A)^2$ follows. In an example, we still examine this–presumably not stable–method. Once again, we consider the overdetermined system from Example 4.28.

Example 4.32 (*Solving the Normal Equation*) The exact solution of the system of normal equations $A^T A x = A^T b$ is given by

$$x \approx \begin{pmatrix} 0.3425 \\ 0.3840 \\ -0.1131 \end{pmatrix} \quad \Rightarrow \quad p(x) = 0.3425 + 0.3740x - 0.1131x^2. \tag{4.8}$$

It holds

$$\|b - Ax\|_2 \approx 0.947.$$

We set up the system of normal equations with three-digit calculation:

$$A^T A = \begin{pmatrix} 4 & 4.75 & 10.6 \\ 4.75 & 10.6 & 23.7 \\ 10.6 & 23.7 & 55.1 \end{pmatrix}, \quad A^T b = \begin{pmatrix} 2 \\ 3 \\ 6.5 \end{pmatrix}.$$

Using three-digit arithmetic, we determine the Cholesky factorization using the direct method from Algorithm 3.27.

$$
\begin{aligned}
l_{11} &= \sqrt{4} = 2 \\
l_{21} &= 4.75/2 \approx 2.38 \\
l_{31} &= 10.6/2 = 5.3 \\
l_{22} &= \sqrt{10.6 - 2.38^2} \approx 2.22 \\
l_{32} &= (23.7 - 2.38 \cdot 5.3)/2.22 \approx 5 \\
l_{33} &= \sqrt{55.1 - 5.3^2 - 5^2} \approx 1.42.
\end{aligned}
\qquad , \quad
L := \begin{pmatrix} 2 & 0 & 0 \\ 2.38 & 2.22 & 0 \\ 5.3 & 5 & 1.42 \end{pmatrix}.
$$

We solve

$$L \underbrace{L^T x}_{=y} = A^T b \quad \Rightarrow \quad y \approx \begin{pmatrix} 1 \\ 0.279 \\ -0.137 \end{pmatrix} \quad \Rightarrow \quad \tilde{x} \approx \begin{pmatrix} 0.348 \\ 0.345 \\ -0.0972 \end{pmatrix}$$

with the polynomial

$$p(x) = 0.348 + 0.345x - 0.0972x^2$$

and the residual $\|b - A\tilde{x}\| \approx 0.95$ as well as the relative error to the exact solution x from (4.8)

$$\frac{\|\tilde{x} - x\|}{\|x\|} \approx 0.08.$$

A deviation of 8% for a 4×3 matrix already indicates a significant error amplification. ◀

We develop an alternative method, which avoids the setup of the system of normal equations $A^T A x = A^T b$, and instead uses the QR factorization for non-square matrices, which has been introduced in Remark 4.6. For $A \in \mathbb{R}^{n \times m}$, let

$$A = QR$$

with $Q \in \mathbb{R}^{n \times m}$ and $R \in \mathbb{R}^{m \times m}$. Alternatively, R can be extended to $\tilde{R} \in \mathbb{R}^{n \times m}$ by $n - m$ zero rows and Q to $\tilde{Q} \in \mathbb{R}^{n \times n}$ by further $n - m$ orthogonal vectors.

With this generalized QR factorization, we get

$$A^T A x = A^T b \quad \Leftrightarrow \quad (QR)^T QRx = (QR)^T b \quad \Leftrightarrow \quad R^T Rx = R^T Q^T b.$$

The (small) matrix $R \in \mathbb{R}^{m \times m}$ is regular, and the solution of the system of normal equations is

$$Rx = Q^T b.$$

Example 4.33 (*Best Approximation of an Overdetermined System Using the QR Factorizations*) We start with the overdetermined linear system from Example 4.28

$$A = \begin{pmatrix} 1 & -\frac{1}{4} & \frac{1}{16} \\ 1 & \frac{1}{2} & \frac{1}{4} \\ 1 & 2 & 4 \\ 1 & \frac{5}{2} & \frac{25}{4} \end{pmatrix}, \quad b = \begin{pmatrix} 0 \\ 1 \\ 0 \\ 1 \end{pmatrix}.$$

In the first step of the Householder transformation (three-digit calculation) we choose

$$v^{(1)} = \frac{a_1 + \|a_1\| e_1}{\|a_1 + \|a_1\| e_1\|} \approx \begin{pmatrix} 0.866 \\ 0.289 \\ 0.289 \\ 0.289 \end{pmatrix}.$$

Then with $a_i^{(1)} = a_i - 2(a_i, v^{(1)}) v^{(1)}$ we have

$$\tilde{A}^{(1)} \approx \begin{pmatrix} -2 & -2.38 & -5.29 \\ 0 & -0.210 & -1.54 \\ 0 & 1.29 & 2.21 \\ 0 & 1.79 & 4.46 \end{pmatrix} =: (a_1^{(1)}, a_2^{(1)}, a_3^{(1)})$$

and the reduced vector $\tilde{a}_2^{(1)} = (-0.21, 1.29, 1.79)^T$ further leads to

$$\tilde{v}^{(2)} = \frac{\tilde{a}_2^{(1)} - \|\tilde{a}_2^{(1)}\| \tilde{e}_2}{\| \tilde{a}_2^{(1)} - \|\tilde{a}_2^{(1)}\| \tilde{e}_2 \|} \approx \begin{pmatrix} -0.740 \\ 0.393 \\ 0.546 \end{pmatrix}.$$

And hereby $\tilde{A}^{(2)}$ gets

$$\tilde{A}^{(2)} \approx \begin{pmatrix} -2 & -2.38 & -5.29 \\ 0 & 2.22 & 5.04 \\ 0 & 0 & -1.28 \\ 0 & 0 & -0.392 \end{pmatrix} =: (a_1^{(2)}, a_2^{(2)}, a_3^{(2)}).$$

We need to perform a third step and with $\tilde{a}_3^{(2)} = (-1.28, -0.392)$ we get by the same principle

$$\tilde{v}^{(3)} = \begin{pmatrix} -0.989 \\ -0.148 \end{pmatrix}.$$

Finally, we obtain

$$\tilde{R} = \tilde{A}^{(3)} \approx \begin{pmatrix} -2 & -2.38 & -5.29 \\ 0 & 2.22 & 5.04 \\ 0 & 0 & 1.34 \\ 0 & 0 & 0 \end{pmatrix}.$$

To solve the normal equation

$$A^T A x = A^T b \quad \Leftrightarrow \quad R x = \tilde{b}$$

we first determine the right-hand side according to $b^{(i)} = b^{(i-1)} - 2(b^{(i-1)}, v^{(i)}) v^{(i)}$:

$$\tilde{b} = \begin{pmatrix} -1 \\ 0.28 \\ -0.155 \end{pmatrix}$$

Finally, we solve by back substitution $Rx = \tilde{b}^{(3)}$

$$\tilde{x} \approx \begin{pmatrix} 0.343 \\ 0.389 \\ -0.116 \end{pmatrix}$$

and obtain the interpolation polynomial

$$p(x) = 0.343 + 0.389x - 0.116x^2$$

with the residual

$$\|b - A\tilde{x}\|_2 \approx 0.948$$

and the relative error to the exact least squares solution

$$\frac{\|\tilde{x} - x\|}{\|x\|} \approx 0.01,$$

i.e., an error of about 1% instead of almost 8% when directly solving the normal system. We summarize the results: The exact best approximation x, the Cholesky solution of the system of normal equations x_{LL} and the solution with extended QR factorization x_{QR} are given by

$$x \approx \begin{pmatrix} 0.343 \\ 0.384 \\ -0.113 \end{pmatrix}, \quad x_{LL} \approx \begin{pmatrix} 0.348 \\ 0.345 \\ -0.0972 \end{pmatrix}, \quad x_{QR} \approx \begin{pmatrix} 0.343 \\ 0.389 \\ -0.116 \end{pmatrix}.$$

The residual vector is given as

$$b - Ax \approx \begin{pmatrix} -0.240 \\ 0.493 \\ -0.659 \\ 0.403 \end{pmatrix}, \quad b - Ax_{LL} \approx \begin{pmatrix} -0.256 \\ 0.503 \\ -0.649 \\ 0.397 \end{pmatrix}, \quad b - Ax_{QR} \approx \begin{pmatrix} -0.238 \\ 0.492 \\ -0.657 \\ 0.410 \end{pmatrix}.$$

In Python, we can reuse our implementation of the QR factorization according to Householder. However, since we need a different right-hand side $\tilde{b}$, we also need to adjust the application of the matrix Q^T.

```python
def b_tilde(V, b):
    b = b.copy()
    for i in range(len(V)):
        b[i:] -= 2 * np.inner(V[i], b[i:]) * V[i]
    return np.array(b[:len(V)])
```

If we apply this to the above example, we get

```
1  V = qr_householder(A)
2  bt = b_tilde(V, b)
3  x_h = backward(A[:len(V),:], bt)
4  print(f'x = {x_h}')
```

```
   x = [ 0.3423  0.3804 -0.1113]
```

This results in the residual

```
1  print(f'||Ax - b|| = {np.linalg.norm(np.inner(A2, x_h) - b):.4e}')
```

```
   ||Ax - b|| = 9.4727e-01
```

Alternatively, we can use our QR factorization with Givens rotations. This results in the
solution

```
1  QT = qr_givens(A)
2  bt = np.dot(QT, b)[:A.shape[1]]
3  x_g = backward(A[:A.shape[1],:], bt)
4  print(f'x = {x_g}')
```

```
   x = [ 0.3425  0.384  -0.113 ]
```

with the residual

```
1  print(f'||Ax - b|| = {np.linalg.norm(np.inner(A2, x_g) - b):.4e}')
```

```
   ||Ax - b|| = 9.4727e-01
```

◀

In Sect. 9.5.1, we will consider the discrete Gauss approximation as a method for fitting
functions to discrete values. Linear regression is a special case of discrete Gauss approximation. Here, the task is to represent pairs of measured values (x_i, y_i) for $i = 1, \ldots, n$ as
well as possible through a linear relationship $y(x) = ax + b$ (see also Excursion 11.4). As
an application of the best approximation of overdetermined equation systems, this leads to
matrices of size $A \in \mathbb{R}^{n \times 2}$.

Another application of the QR factorization will be introduced in Chap. 5. One of the
most powerful methods for approximating the eigenvalues of a matrix $A \in \mathbb{R}^{n \times n}$ is based
on the repeated application of the QR factorization.

4.6   Excursus: Determining the Principal Axes of a Celestial Body

In this excursion, we present a historical application of overdetermined equation systems.
The original development of the method by Carl Friedrich Gauss was given exactly this
problem. The Italian Piazzi observed the planet Ceres for 40 days in 1801 and then lost
sight of it [37]. Gauss took the available data, used the method of least squares, and then
numerically (of course, without a computer) calculated the planet's orbit. As a result, the
astronomers were actually able to find the planet again. A celestial body moves on an
elliptical orbit around the origin

$$\frac{x^2}{a^2} + \frac{y^2}{b^2} = 1.$$

We can measure the position of the body at different times:

t	1	2	3	4	5	6
x	0.2	−0.7	−0.3	1.2	2.0	−2.4
y	1.8	1.7	−1.8	−1.6	1.1	0.5

From these data, we want to determine the lengths of the two major axes a and b. The unknown quantities a and b appear in a nonlinear context. Therefore, we first define the two auxiliary quantities

$$A = \frac{1}{a^2}, \quad B = \frac{1}{b^2}.$$

This results in the overdetermined system

$$\begin{pmatrix} 0.04 & 3.24 \\ 0.49 & 2.56 \\ 0.09 & 3.24 \\ 1.44 & 2.56 \\ 4.00 & 1.21 \\ 5.76 & 0.25 \end{pmatrix} \begin{pmatrix} A \\ B \end{pmatrix} = \begin{pmatrix} 1 \\ 1 \\ 1 \\ 1 \\ 1 \\ 1 \end{pmatrix}.$$

To solve this overdetermined problem, we could directly set up and then solve the normal equation. It holds

$$A^T A \approx \begin{pmatrix} 51.50 & 11.64 \\ 11.64 & 35.63 \end{pmatrix}, \quad A^T b \approx \begin{pmatrix} 11.82 \\ 13.06. \end{pmatrix}.$$

As a solution of $A^T A x = A^T b$ follows

$$x = \begin{pmatrix} A \\ B \end{pmatrix} \approx \begin{pmatrix} 0.158 \\ 0.315 \end{pmatrix}, \quad a = \frac{1}{\sqrt{A}} \approx 2.51, \quad b = \frac{1}{\sqrt{B}} \approx 1.78.$$

An alternative solution path consists of creating the QR factorization of A. For $A \in \mathbb{R}^{6 \times 2}$, we have to perform two steps. We get

$$A = QR, \quad R \approx \begin{pmatrix} -7.18 & -1.62 \\ 0 & -5.74 \\ \hline 0 & 0 \\ 0 & 0 \\ 0 & 0 \\ 0 & 0 \end{pmatrix}$$

and for the right side

Fig. 4.2 Positions of the celestial body on an ellipse at different times

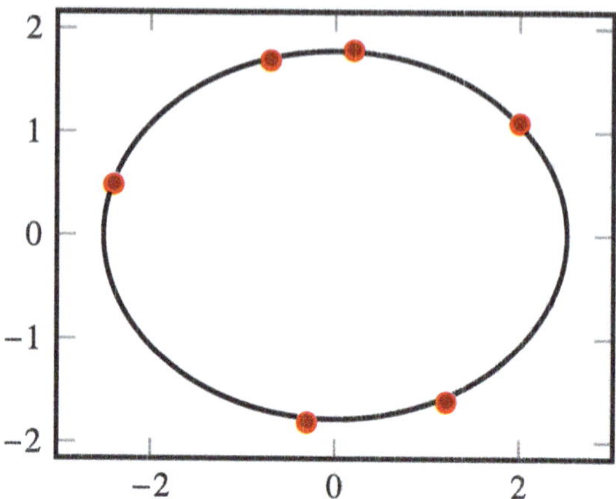

$$\tilde{b} = Q^T b \approx \begin{pmatrix} -1.65 \\ -1.81 \end{pmatrix}.$$

The solution of the system is obtained as above

$$\tilde{R}x = \tilde{b} \quad \Rightarrow \quad x \approx \begin{pmatrix} 0.158 \\ 0.315 \end{pmatrix}.$$

In Fig. 4.2, the result is graphically represented. For the residual of the least squares solution, it holds

$$\|b - Ax\|_2 \approx 0.13.$$

The solution $x \in \mathbb{R}^2$ is the best approximation, but not a solution of the overdetermined linear system. The residual of 0.13 says nothing about the quality of the numerical solution, but rather about the quality of the data. If the postulated elliptical law for the planet's orbit is exact, the residual gives a measure of the measurement error of the observations.

The Eigenvalue Problem 5

In this chapter, we deal with the eigenvalue problem, namely to find the eigenvalues (and possibly eigenvectors) of a matrix $A \in \mathbb{R}^{n \times n}$. We recall Definition 2.11 and formulate

Theorem 5.1 (Eigenvalues) *Let $A \in \mathbb{R}^{n \times n}$. It holds:*

1. *The matrix A has exactly n eigenvalues, counted with their multiplicity.*
2. *The eigenvalues are the roots of the* characteristic polynomial

$$\det(A - \lambda I) = 0.$$

3. *Real matrices can have complex eigenvalues. Complex eigenvalues of real matrices always occur as conjugate pairs $\lambda, \bar{\lambda}$.*
4. *Eigenvectors corresponding to different eigenvalues are linearly independent.*
5. *If n linearly independent eigenvectors exist, then there exists a regular matrix $S \in \mathbb{R}^{n \times n}$, such that $S^{-1} A S = D$ is a diagonal matrix. The column vectors of S are the eigenvectors, and the diagonal entries of D are the eigenvalues.*
6. *Real symmetric matrices $A \in \mathbb{R}^{n \times n}$ only have real eigenvalues. There exists an orthonormal basis of eigenvectors and a diagonalization $Q^T A Q = D$ with an orthogonal matrix consisting of the eigenvectors.*
7. *For triangular matrices, the eigenvalues are on the diagonal.*

Proofs for the individual statements can be found, for example, in [33]. We only consider the situation of real matrices $A \in \mathbb{R}^{n \times n}$ here.

For an eigenvalue $\lambda \in \mathbb{C}$, the eigenvectors are determined as the solution of the homogeneous linear system:

© The Author(s), under exclusive license to Springer-Verlag GmbH, DE, part of Springer Nature 2026

T. Richter et al., *Introduction to Numerical Mathematics*, Mathematics Study Resources 25, https://doi.org/10.1007/978-3-662-72546-7_5

$$(A - \lambda I)w = 0$$

Conversely, we immediately obtain

Theorem 5.2 (Rayleigh Quotient) *Let $A \in \mathbb{R}^{n \times n}$ and $w \in \mathbb{R}^n$ be an eigenvector. Then the corresponding eigenvalue is given by the* Rayleigh quotient:

$$\lambda = \frac{(Aw, w)_2}{\|w\|_2^2}$$

Using this definition, the Cauchy-Schwarz inequality provides a simple bound for the eigenvalues of a matrix, because

$$|\lambda| \leq \sup_{w \neq 0} \frac{(Aw, w)_2}{\|w\|_2^2} \leq \frac{\|A\|_2 \|w\|_2^2}{\|w\|_2^2} = \|A\|_2.$$

We have already seen in Sect. 2 that this bound for the absolute values of the eigenvalues holds in any matrix norm with a compatible vector norm.

5.1   Conditioning of the Eigenvalue Problem

Before we discuss specific methods for calculating eigenvalues, we analyze the conditioning of the problem, i.e., the dependence of the eigenvalues on the perturbation of the matrix. We start with a technical lemma.

Lemma 5.3 *Let $A, B \in \mathbb{R}^{n \times n}$ be arbitrary matrices. Then, for every eigenvalue λ of A that is not also an eigenvalue of B, the following estimate holds in any natural matrix norm*

$$\|(\lambda I - B)^{-1}(A - B)\| \geq 1.$$

Proof Let λ be an eigenvalue of A with corresponding eigenvector $w \neq 0$. Then

$$(A - B)w = (\lambda I - B)w.$$

If λ is not an eigenvalue of B, then the matrix $(\lambda I - B)$ is regular, i.e., it follows

$$(\lambda I - B)^{-1}(A - B)w = w.$$

Finally, by taking norms,

$$1 = \frac{\|(\lambda I - B)^{-1}(A - B)w\|}{\|w\|} \leq \sup_{x} \frac{\|(\lambda I - B)^{-1}(A - B)x\|}{\|x\|}$$

$$= \|(\lambda I - B)^{-1}(A - B)\|.$$

$\square$

On this basis, we obtain a simple criterion for bounding the eigenvalues of a matrix.

Theorem 5.4 (Gershgorin Circles) *Let $A \in \mathbb{R}^{n \times n}$. All eigenvalues λ of A lie in the union of the* Gershgorin circles:

$$K_i = \{z \in \mathbb{C} : |z - a_{ii}| \leq \sum_{k=1, k \neq i}^{n} |a_{ik}|\}, \quad i = 1, \ldots, n$$

Suppose that for the index sets $I_m = \{i_1, \ldots, i_m\}$ and $I'_m = \{1, \ldots, n\} \setminus I_m$, the unions $U_m = \bigcup_{i \in I_m} K_i$ and $U'_m = \bigcup_{i \in I'_m} K_i$ are disjoint. Then exactly m eigenvalues (counted with their algebraic multiplicity) lie in U_m and $n - m$ eigenvalues lie in U'_m.

Proof *(i)* Let $D \in \mathbb{R}^{n \times n}$ be the diagonal matrix $D = \text{diag}(a_{ii})$. Let λ be an eigenvalue of A with $\lambda \neq a_{ii}$. (In this case, the statement of the theorem would be trivially fulfilled.) Lemma 5.3 states for the choice of the maximum absolute row sum norm:

$$1 \leq \|(\lambda I - D)^{-1}(A - D)\|_\infty = \max_i \left\{ \sum_{j=1, j \neq i}^{n} |(\lambda - a_{ii})^{-1} a_{ij}| \right\}$$

$$= \max_i \left\{ |\lambda - a_{ii}|^{-1} \sum_{j=1, j \neq i}^{n} |a_{ij}| \right\}$$

This means, for every λ there exists an index $i \in \{1, \ldots, n\}$, such that:

$$|\lambda - a_{ii}| \leq \sum_{j=1, j \neq i}^{n} |a_{ij}|$$

Every eigenvalue λ is, therefore, in at least one of the Gershgorin circles.

(ii) Let I_m and I'_m be disjoint index sets with $I_m \cup I'_m = \{1, \ldots, n\}$ such that the unions U_m and U'_m of their circles are also disjoint. For $s \in \mathbb{R}$, we define the matrix

$$A_s := D + s(A - D)$$

with $A_0 = D$ and $A_1 = A$. We denote the eigenvalues of A_s by $\lambda_{s,i}$. Accordingly, we define the unions:

$$U_{m,s} := \bigcup_{i \in I_m} K_i(A_s), \quad U'_{m,s} := \bigcup_{i \in I'_m} K_i(A_s),$$

$$K_i(A_s) = \{z \in \mathbb{C}, \ |z - a_{ii}| \leq s \sum_{j \neq i} |a_{ij}|\}$$

We have $U_{m,1} = U_m$ and $U'_{m,1} = U'_m$ and these two sets are disjoint. For $s = 0$, all circles consist only of one point and each contains exactly one eigenvalue $\lambda_{i,0} = a_{ii}$. The circle radii depend linearly on s and, therefore, the circle unions remain disjoint for $s \in [0, 1]$. Finally, since the eigenvalues of the matrix depend continuously on the matrix entries, the statement of the theorem follows. $\qquad\qquad\square$

By choosing other matrix norms, similar criteria can be derived.

Remark 5.5 (*Eigenvalues of A^T*) When choosing the 1-norm, i.e., $\| \cdot \|_1$, the maximum column sum, it follows that

$$|\lambda - a_{ii}| \leq \sum_{i=1, \ j \neq i}^{n} |a_{ji}|,$$

i.e., the column sums form the radius of the corresponding circles. Accordingly, one could apply the Gershgorin criterion to the transposed matrix A^T, because from $\det(A) = \det(A^T)$ it follows that A and A^T have the same eigenvalues. We can therefore determine the circle radii either based on the row or the column sum. $\qquad\qquad\blacklozenge$

Remark 5.6 (*Continuous Dependence of the Eigenvalues on the Matrix Entries*) The eigenvalues are roots of the characteristic polynomial

$$p_A(\lambda) = \det(A - \lambda I) = \alpha_0 + \alpha_1 \lambda + \cdots + \alpha_n \lambda^n.$$

The coefficients $\alpha_0, \ldots, \alpha_n$ depend continuously on the entries a_{ij} of the matrix A. It is more difficult to show that the (complex) roots of a polynomial also depend continuously on the coefficients. We refer to the literature, for example, Sect. 3.9 in [93]. $\qquad\qquad\blacklozenge$

Example 5.7 (*Gershgorin Circles*) We consider the matrix

$$A = \begin{pmatrix} 2 & 0.1 & -0.5 \\ 0.2 & 3 & 0.5 \\ -0.4 & 0.1 & 5 \end{pmatrix}$$

with the eigenvalues $\Lambda(A) \approx \{1.91, 3.01, 5.08\}$. A first bound is provided by various matrix norms of A that are compatible with vector norms:

$$\|A\|_\infty = 5.5, \quad \|A\|_1 = 6, \quad \|A\|_2 \approx 5.1.$$

The Gershgorin circles of A are

$$K_1 = \{z \in \mathbb{C}, \ |z - 2| \leq 0.6\},$$
$$K_2 = \{z \in \mathbb{C}, \ |z - 3| \leq 0.7\},$$
$$K_3 = \{z \in \mathbb{C}, \ |z - 5| \leq 0.5\}.$$

This estimate can be refined, since A and A^T have the same eigenvalues. The radii of the Gershgorin circles can also be calculated as the sum of the column absolute values. Together, we get

$$K_1 = \{z \in \mathbb{C}, \ |z - 2| \leq 0.6\},$$
$$K_2 = \{z \in \mathbb{C}, \ |z - 3| \leq 0.2\},$$
$$K_3 = \{z \in \mathbb{C}, \ |z - 5| \leq 0.5\}.$$

All three circles are disjoint, i.e., exactly one eigenvalue lies in each circle.   ◀

Remark 5.8 (*Gershgorin Circles*) From the above example, one could easily infer the following conclusion:

All eigenvalues of a matrix A lie within the Gershgorin circles. These are defined as circles around the diagonal elements a_{ii} with radius $\min\{\sum_{j \neq i} |a_{ij}|, \sum_{j \neq i} |a_{ji}|\}$, i.e., the minimum of the row and column sum.

However, this tightening of the theorem is generally incorrect. Similarly, it would be wrong to conclude that at least one eigenvalue must lie within each of the circles, even if they are not disjoint. Counterexamples can easily be provided for both cases.   ◆

We can now prove the general stability theorem for the eigenvalue problem.

Theorem 5.9 (Stability of the Eigenvalue Problem) *Let $A \in \mathbb{R}^{n \times n}$ be a matrix with n linearly independent eigenvectors $w_1, \ldots, w_n$. Let $\tilde{A} = A + \delta A$ be an arbitrarily perturbed matrix. Then for each eigenvalue $\lambda(\tilde{A})$ of $\tilde{A} = A + \delta A$ there exists an eigenvalue $\lambda(A)$ of A such that with the matrix $W := (w_1, \ldots, w_n)$ it holds:*

$$|\lambda(A) - \lambda(\tilde{A})| \leq \text{cond}_2(W)\|\delta A\|$$

Proof For $i = 1, \ldots, n$, the relation $Aw_i = \lambda_i(A)w_i$ holds, or in matrix notation $AW = W \, \text{diag}(\lambda_i(A))$, thus

$$W^{-1}AW = \text{diag}(\lambda_i(A)).$$

Since the w_i are linearly independent, W is regular. We now consider an eigenvalue $\tilde{\lambda} = \lambda(\tilde{A})$. If $\tilde{\lambda}$ is also an eigenvalue of A, the claim holds. So let $\tilde{\lambda}$ not be an eigenvalue of A. Then it follows

$$\|(\tilde{\lambda} I - A)^{-1}\|_2 = \|W^{-1}[\tilde{\lambda} I - \text{diag}(\tilde{\lambda}_i(A))]^{-1} W\|_2$$

$$\leq \text{cond}_2(W)\|[\tilde{\lambda} I - \text{diag}(\lambda_i(A))]^{-1}\|_2.$$

For the (symmetric) diagonal matrix $\tilde{\lambda} I - \text{diag}(\lambda_i(A))$ it holds

$$\|[\tilde{\lambda} I - \text{diag}(\lambda_i(A))]^{-1}\|_2 = \max_{i=1,\dots,n} |\tilde{\lambda} - \lambda_i(A)|^{-1}.$$

Since $\delta A = \tilde{A} - A$, with Lemma 5.3 the desired result follows:

$$1 \leq \|[\tilde{\lambda} I - A]^{-1}\delta A\|_2 \leq \|[\tilde{\lambda} I - A]^{-1}\|_2 \|\delta A\|_2$$

$$\leq \text{cond}_2(W) \max_{i=1,\dots,n} |\tilde{\lambda} - \lambda_i(A)|^{-1} \|\delta A\|_2$$

$$\square$$

The conditioning of the eigenvalue problem of a matrix A depends on the condition number of the matrix of eigenvectors w_i. For symmetric (Hermitian) matrices, there exists an orthonormal basis of eigenvectors with $\text{cond}_2(W) = 1$. For such matrices, the eigenvalue problem is well-conditioned. For general matrices, the eigenvalue problem can be arbitrarily ill-conditioned.

5.2 Direct Method

The eigenvalues of a matrix A can, in principle, be calculated as the roots of the characteristic polynomial $\chi_A(z) = \det(zI - A)$. The calculation of the roots can be done, for example, with the Newton method or complex variants of it (Sect. 8.3), and the required initial values can be determined using the Gershgorin circles.

In Chap. 8, however, we will find that the calculation of the roots of a polynomial is a very ill-conditioned problem. We consider an example for this.

Example 5.10 (*Direct Calculation of Eigenvalues*) Let $A \in \mathbb{R}^{5 \times 5}$ be a matrix with the eigenvalues $\lambda_i = i, i = 1, \dots, 5$. Then for the characteristic polynomial $\chi_A(z) := \det(A - zI)$ it holds

$$\chi_A(z) = \prod_{i=1}^{5} (z - i) = z^5 - 15z^4 + 85z^3 - 225z^2 + 274z - 120.$$

Assume that the coefficient -15 in front of z^4 is perturbed with a relative error of 0.1%:

$$\tilde{\chi}_A(z) = z^5 - 0.999 \cdot 15z^4 + 85z^3 - 225z^2 + 274z - 120$$

This perturbed polynomial has the roots (i.e., eigenvalues)

$$\lambda_1 \approx 0.999, \quad \lambda_2 \approx 2.05, \quad \lambda_3 \approx 2.76, \quad \lambda_{4/5} \approx 4.59 \pm 0.430i.$$

The eigenvalues can therefore only be determined with a (from λ_3 on) substantial error. It holds, for example,

$$\frac{|4.59 + 0.43i - 5|}{5} \approx 0.1,$$

i.e., the error in the eigenvalues is 10%, which means an error amplification by a factor of 100. ◀

The setup of the characteristic polynomial requires a multitude of ill-conditioned additions and multiplications. For general matrices, this direct method is not advisable. Only for special matrices such as tridiagonal systems, the eigenvalues can be determined without having to calculate the coefficients of the characteristic polynomial first explicitly.

5.3 Iterative Methods

The previous example seemingly contradicts the (for Hermitian matrices) good conditioning of the eigenvalue problem. However, we must distinguish between the conditioning of the task and the stability of a particular method. Even if the task may be well-conditioned, the search for roots of the characteristic polynomial is not a stable method (see in particular Sects. 1.3 and 1.5).

As the first iterative method for approximating eigenvalues, we consider the *power iteration*. Suppose, for a given matrix $A \in \mathbb{R}^{n \times n}$, there exists a basis of eigenvectors $\{w_1, \ldots, w_n\}$, i.e., the matrix is diagonalizable. Furthermore, let $|\lambda_n| > |\lambda_{n-1}| \geq \cdots \geq |\lambda_1| > 0$. Then, let $x^{(0)} \in \mathbb{R}^n$ be given in basis representation

$$x^{(0)} = \sum_{j=1}^{n} \alpha_j w_j, \quad \alpha_j \in \mathbb{R} \text{ for } j = 1, \ldots, n.$$

We assume that $\alpha_n \neq 0$, i.e., the component of the vector $\alpha = (\alpha_1, \ldots, \alpha_n)$ corresponding to the eigenvector of the largest absolute eigenvalue is nontrivial. We define the iteration

$$x^{(i+1)} = \frac{Ax^{(i)}}{\|Ax^{(i)}\|}, \quad i = 1, 2, \ldots$$

For the iterates, it holds

$$x^{(i+1)} = \frac{A\frac{Ax^{(i-1)}}{\|Ax^{(i-1)}\|}}{\|A\frac{Ax^{(i-1)}}{\|Ax^{(i-1)}\|}\|} = \frac{A^2 x^{(i-1)}}{\|A^2 x^{(i-1)}\|} \frac{\|Ax^{(i-1)}\|}{\|Ax^{(i-1)}\|} = \cdots = \frac{A^{i+1} x^{(0)}}{\|A^{i+1} x^{(0)}\|}. \tag{5.1}$$

Further, with the basis representation of $x^{(0)}$

$$A^i x^{(0)} = \sum_{j=1}^{n} \alpha_j \lambda_j^i w_j = \alpha_n \lambda_n^i \left(w_n + \sum_{j=1}^{n-1} \frac{\alpha_j}{\alpha_n} \frac{\lambda_j^i}{\lambda_n^i} w_j \right). \tag{5.2}$$

Since $|\lambda_j/\lambda_n| < 1$ for $j < n$, it follows that

$$A^i x^{(0)} = \sum_{j=1}^{n} \alpha_j \lambda_j^i w_j = \alpha_n \lambda_n^i \left(w_n + o(1) \right).$$

From this, by normalization, we get

$$\lim_{i \to \infty} \left(\frac{A^i x^{(0)}}{\|A^i x^{(0)}\|} \right) = \lim_{i \to \infty} \left(\left(\frac{\alpha_n \lambda_n^i}{|\alpha_n \lambda_n^i|} \right) \frac{w_n}{\|w_n\|} + o(1) \right) \in \operatorname{span}\{w_n\})$$

The iteration converges to the space spanned by w_n. For a vector w, which is a multiple of an eigenvector, $w = s w_n$, it holds

$$Aw = s A w_n = s \lambda_n w_n = \lambda_n w.$$

This vector-valued equation holds in each component, and can therefore be solved for the eigenvalue:

$$\lambda_n = \frac{[Aw]_k}{w_k}$$

We summarize.

Theorem 5.11 (Power Assume) *Let $A \in \mathbb{R}^{n \times n}$ be a matrix with n linearly independent eigenvectors $\{w_1, \ldots, w_n\}$. The largest absolute eigenvalue is separated $|\lambda_n| > |\lambda_{n-1}| \geq \cdots \geq |\lambda_1|$. Let $x^{(0)} \in \mathbb{R}^n$ be a starting value with a nontrivial component with respect to w_n. For any index $k \in \{1, \ldots, n\}$, the iteration*

$$\tilde{x}^{(i)} = A x^{(i-1)}, \quad x^{(i)} := \frac{\tilde{x}^{(i)}}{\|\tilde{x}^{(i)}\|}, \quad \lambda^{(i)} := \frac{\tilde{x}_k^{(i)}}{x_k^{(i-1)}},$$

converges to the largest absolute eigenvalue:

$$|\lambda^{(i)} - \lambda_n| = O\left(\left| \frac{\lambda_{n-1}}{\lambda_n} \right|^i \right), \quad i \to \infty.$$

Proof We continue with the preliminary derivation, in particular Eq. (5.1):

$$\lambda^{(i)} = \frac{\tilde{x}_k^{(i)}}{x_k^{(i-1)}} = \frac{[A x^{(i-1)}]_k}{x_k^{(i-1)}} = \frac{[A^i x^{(0)}]_k}{[A^{i-1} x^{(0)}]_k}.$$

Further, with (5.2) it holds

$$
\lambda^{(i)} = \frac{a_n \lambda_n^i \left([w_n]_k + \sum_{j=1}^{n-1} \frac{\alpha_j}{\alpha_n} \frac{\lambda_j^i}{\lambda_n^i} [w_j]_k \right)}{a_n \lambda_n^{i-1} \left([w_n]_k + \sum_{j=1}^{n-1} \frac{\alpha_j}{\alpha_n} \frac{\lambda_j^{i-1}}{\lambda_n^{i-1}} [w_j]_k \right)}
$$

$$
= \lambda_n \frac{[w_n]_k + \sum_{j=1}^{n-1} \frac{\alpha_j}{\alpha_n} \frac{\lambda_j^i}{\lambda_n^i} [w_j]_k}{[w_n]_k + \sum_{j=1}^{n-1} \frac{\alpha_j}{\alpha_n} \frac{\lambda_j^{i-1}}{\lambda_n^{i-1}} [w_j]_k}
$$

$$
= \lambda_n \frac{[w_n]_k + \left(\frac{\alpha_{n-1}}{\alpha_n} + o(1) \right) \left| \frac{\lambda_{n-1}}{\lambda_n} \right|^i [w_{n-1}]_k}{[w_n]_k + \left(\frac{\alpha_{n-1}}{\alpha_n} + o(1) \right) \left| \frac{\lambda_{n-1}}{\lambda_n} \right|^{i-1} [w_{n-1}]_k}
$$

$$
= \lambda_n \left(1 + O\left(\left| \frac{\lambda_{n-1}}{\lambda_n} \right|^{i-1} \right) \right).
$$

The last estimate uses (for $0 < x < 1$) the relation

$$
\frac{1 + x^{i+1}}{1 + x^i} = 1 + \frac{x^{i+1} - x^i}{1 + x^i} = 1 + x^i \frac{x - 1}{1 + x^i} = 1 + O(|x|^i).
$$

$\square$

In Python, this iteration can be implemented as follows

♣ Implementation 5.12: Power Iteration

```python
def power_iteration(A, x, n, k=0):
    x0, x1 = x.copy(), np.zeros_like(x)
    lams = np.zeros(n)
    for i in range(n):
        x1[:] = np.inner(A, x0)
        lams[i] = x1[k] / x0[k]
        x0[:] = x1 / np.linalg.norm(x1)
    return lams
```

The power iteration is a straightforward and numerically stable method. However, it can only determine the largest absolute eigenvalue. The proof of convergence can be generalized to matrices whose r largest absolute eigenvalues are repeated. Convergence only holds, if $|\lambda_n| = |\lambda_{n-1}| = \cdots = |\lambda_{n-r}| > |\lambda_{n-r-1}|$ also induces $\lambda_n = \lambda_{n-1} = \cdots = \lambda_{n-r}$. There must not be two different eigenvalues, both having the same largest absolute value but representing different points in the complex plane. This excludes, for example, the case of two complex conjugate eigenvalues $\lambda \neq \bar{\lambda}$ as the largest in absolute value.

Remark 5.13 (*Power Iteration for Symmetric Matrices*) The power iteration can be improved in the case of symmetric matrices by using the Rayleigh quotient. The iteration

$$\lambda^{(i)} = \frac{(Ax^{(i-1)}, x^{(i-1)})_2}{(x^{(i-1)}, x^{(i-1)})_2} = (\tilde{x}^{(i)}, x^{(i-1)})_2$$

provides the error estimate

$$\lambda^{(i)} = \lambda_n + O\left(\left|\frac{\lambda_{n-1}}{\lambda_n}\right|^{2i}\right).$$

◆

Example 5.14 (*Power Iteration*) Let

$$A = \begin{pmatrix} 2 & 1 & 2 \\ -1 & 2 & 1 \\ 1 & 2 & 4 \end{pmatrix}$$

with the eigenvalues

$$\lambda_1 \approx 0.80131, \quad \lambda_2 = 2.2865, \quad \lambda_3 = 4.9122.$$

We start the power iteration with $x^{(0)} = (1, 1, 1)^T$ and choose the maximum norm for normalization. Further, we choose the index $k = 1$.

$$i = 1: \tilde{x}^{(1)} = Ax^{(0)} = \begin{pmatrix} 5 \\ 2 \\ 7 \end{pmatrix}, \qquad x^{(1)} \approx \begin{pmatrix} 0.714 \\ 0.286 \\ 1 \end{pmatrix}, \qquad \lambda^{(1)} \approx 5,$$

$$i = 2: \tilde{x}^{(2)} = Ax^{(1)} = \begin{pmatrix} 3.71 \\ 0.857 \\ 5.29 \end{pmatrix}, \qquad x^{(2)} \approx \begin{pmatrix} 0.702 \\ 0.162 \\ 1 \end{pmatrix}, \qquad \lambda^{(2)} \approx 5.19,$$

$$i = 3: \tilde{x}^{(3)} = Ax^{(2)} = \begin{pmatrix} 3.57 \\ 0.621 \\ 5.03 \end{pmatrix}, \qquad x^{(3)} \approx \begin{pmatrix} 0.710 \\ 0.124 \\ 1 \end{pmatrix}, \qquad \lambda^{(3)} \approx 5.077,$$

$$i = 4: \tilde{x}^{(4)} = Ax^{(3)} = \begin{pmatrix} 3.54 \\ 0.538 \\ 4.96 \end{pmatrix}, \qquad x^{(4)} \approx \begin{pmatrix} 0.715 \\ 0.108 \\ 1 \end{pmatrix}, \qquad \lambda^{(4)} \approx 4.999,$$

$$i = 5: \tilde{x}^{(5)} = Ax^{(4)} = \begin{pmatrix} 3.54 \\ 0.502 \\ 4.93 \end{pmatrix}, \qquad x^{(5)} \approx \begin{pmatrix} 0.717 \\ 0.102 \\ 1 \end{pmatrix}, \qquad \lambda^{(5)} \approx 4.950,$$

$$i = 6: \tilde{x}^{(6)} = Ax^{(5)} = \begin{pmatrix} 3.54 \\ 0.486 \\ 4.92 \end{pmatrix}, \qquad x^{(6)} \approx \begin{pmatrix} 0.719 \\ 0.0988 \\ 1 \end{pmatrix}, \qquad \lambda^{(6)} \approx 4.930.$$

If we take our Python implementation, we obtain

```
1 A = np.array([[2, 1, 2],
2              [-1, 2, 1],
3              [1, 2, 4]], dtype=np.double)
4 x = np.array([1, 1, 1], dtype=np.double)
5
6 l = power_iteration(A, x, 6, k=0)
7 print(l)

  [5.0, 5.2, 5.076923076923077, 4.99242424242424, 4.94992412746585, 4.92979767014101]
```

After six steps, we have determined an eigenvalue to the accuracy of two significant digits. ◀

In addition to focusing on the largest absolute eigenvalue, the power iteration has the disadvantage of very slow convergence if the eigenvalues are not sufficiently separated. A simple extension of the power iteration is the *inverse iteration* for calculating the smallest eigenvalue of a matrix. For the derivation, we use the fact that for an eigenvalue λ of a regular matrix $A \in \mathbb{R}^{n \times n}$, $\mu := \lambda^{-1}$ is an eigenvalue of the inverse matrix $A^{-1} \in \mathbb{R}^{n \times n}$:

$$Aw = \lambda(A)w \quad \Leftrightarrow \quad A^{-1}w = \lambda^{-1}w =: \mu w$$

The power iteration, applied to the inverse matrix, yields the largest absolute eigenvalue $\mu_{\max}$ of A^{-1}. The reciprocal of this eigenvalue is the smallest absolute eigenvalue of the matrix A, i.e, $\lambda_{\min} = \mu_{\max}^{-1}$. This principle can be further refined. Let λ be an eigenvalue of A with corresponding eigenvector $w \in \mathbb{R}^n$, $\sigma \in \mathbb{C}$ be any complex number (but not an eigenvalue of A). Then

$$(A - \sigma I)w = (\lambda - \sigma)w =: \lambda_\sigma w \quad \Leftrightarrow \quad [A - \sigma I]^{-1}w = \lambda_\sigma^{-1}w =: \mu_\sigma w.$$

The application of the power iteration to the matrix $[A - \sigma I]^{-1}$ yields, according to the preceding consideration, the smallest absolute eigenvalue of the matrix $[A - \sigma I]$, i.e., the eigenvalue of A that is closest to σ. If estimates for the eigenvalues of the matrix A are available, the exact values can be determined with the inverse iteration with shift.

Theorem 5.15 (Inverse Iteration with Shift) *Let $A \in \mathbb{R}^{n \times n}$ be a regular matrix with eigenvalues $\lambda_1, \dots, \lambda_n \in \mathbb{C}$. With $\sigma \approx \lambda_l \in \mathbb{C}$ for a $l \in \{1, \dots, n\}$, it holds*

$$\alpha_l := \min_{j \neq l} |\lambda_j - \sigma| > |\lambda_l - \sigma| > 0.$$

Let $x^{(0)} \in \mathbb{R}^n$ be a suitable normalized initial value. For $i = 1, 2, \dots$ the iteration

$$[A - \sigma I]\tilde{x}^{(i)} = x^{(i-1)}, \quad x^{(i)} := \frac{\tilde{x}^{(i)}}{\|\tilde{x}^{(i)}\|}, \quad \mu^{(i)} := \frac{\tilde{x}_k^{(i)}}{x_k^{(i-1)}}, \quad \lambda^{(i)} := \sigma + (\mu^{(i)})^{-1},$$

converges for every index $k \in \{1, \ldots, n\}$ to the eigenvalue λ_l of A:

$$|\lambda_l - \lambda^{(i)}| = O\left(\left|\frac{\lambda_l - \sigma}{\alpha_l}\right|^i\right)$$

Proof The proof is a simple consequence of Theorem 5.11. If σ is not an eigenvalue of A, then

$$B := A - \sigma I$$

is invertible. The matrix B^{-1} has the eigenvalues $\mu_1, \ldots, \mu_n$ with

$$\mu_i = (\lambda_i - \sigma)^{-1} \quad \Leftrightarrow \quad \lambda_i = \mu_i^{-1} + \sigma. \tag{5.3}$$

For these, it holds by assumption

$$|\mu_l| > |\mu_j| > 0 \quad j \neq i$$

The power iteration, Theorem 5.11, applied to the matrix B^{-1} provides an approximation for μ_l:

$$|\mu^{(i)} - \mu_l| = O\left(\left|\frac{\max_{j \neq l} |\mu_j|}{\mu_l}\right|^i\right)$$

The desired statement follows with (5.3). $\square$

In each step of the iteration, a linear system

$$(A - \sigma I)\tilde{x} = x$$

must be solved. This is best done with a factorization method, such as the LU factorization of the matrix. The factorization can be created once in $O(n^3)$ operations. Subsequently, in each step of the iteration, another $O(n^2)$ operations are necessary for the forward and backward substitution. The convergence speed of the inverse iteration with shift can be increased by a good estimate σ. A further improvement of the convergence speed can be achieved by constantly adjusting the estimate σ. If in each step the best approximation $\lambda^{(i)} = \sigma + 1/\mu^{(i)}$ is used as the new estimate, at least superlinear convergence can be achieved. For this, we choose $\sigma^{(0)} = \sigma$ and iterate

$$\sigma^{(i)} = \sigma^{(i-1)} + \frac{1}{\mu^{(i)}}.$$

However, each modification of the shift changes the linear system and requires the recreation of the LU factorization, which is expensive.

Example 5.16 (*Inverse Iteration With Shift*) Let

$$A = \begin{pmatrix} 2 & -0.1 & 0.4 \\ 0.3 & -1 & 0.4 \\ 0.2 & -0.1 & 4 \end{pmatrix}$$

with the eigenvalues

$$\lambda_1 = 1.954, \quad \lambda_2 \approx -0.983, \quad \lambda_3 = 4.029.$$

We aim to determine all eigenvalues using the inverse iteration with a shift. We obtain initial values by analyzing the Gershgorin circles:

$$K_1 = K_{0.5}(2), \quad K_2 = K_{0.2}(-1), \quad K_3 := K_{0.3}(4).$$

The three circles are disjoint, and we choose $\sigma_1 = 2$, $\sigma_2 = -1$, and $\sigma_3 = 4$ as shifts in the inverse iteration. The iteration is always started with $v = (1, 1, 1)^T$ and normalization is done with respect to the maximum norm. By choosing the first component for the calculation of the μ, we obtain the following approximations:

σ_i	$\mu_i^{(1)}$	$\mu_i^{(2)}$	$\mu_i^{(3)}$	$\mu_i^{(4)}$
2	-16.571	-21.744	-21.619	-21.622
-1	1.840	49.743	59.360	59.422
4	6.533	36.004	33.962	33.990

These approximations yield the following eigenvalue approximations:

$$\lambda_1 = \sigma_1 + 1/\mu_1^{(4)} \approx 1.954$$

$$\lambda_2 = \sigma_2 + 1/\mu_2^{(4)} \approx -0.983$$

$$\lambda_3 = \sigma_3 + 1/\mu_3^{(4)} \approx 4.029$$

All three approximations are exact to three decimal places.

In Python, we can implement this, for example, with our LU factorization with pivoting. Here, we need to be careful and store the number $x_k^{(i-1)}$ before we apply the pivoting.

♣ Implementation 5.17: Inverse Iteration With Shift

```python
def inverse_iteration(A, x, n, sigma=0, k=0):
    n1 = A.shape[0]
    B = np.array(A - sigma * np.identity(n1), dtype=A.dtype)
    pivot = lu_pivot(B)
    x0, x1, y = x.copy(), np.zeros_like(x), np.zeros_like(x)
    lams = np.zeros(n)
    for i in range(n):
        xk = x0[k]
        for p in pivot:
            x0[p] = x0[[p[1], p[0]]]
        y[:] = forward(B, x0)
        x1[:] = backward(B, y)
        mu = x1[k] / xk
        lams[i] = sigma + 1 / mu
        x0[:] = np.array(x1 / np.linalg.norm(x1, ord=np.inf))
    return lams
```

If we apply this to the above example, we get

```python
A = np.array([[2, -0.1, 0.4], [0.3, -1, 0.4], [0.2, -0.1, 4]])
v = np.array([1, 1, 1], dtype=np.double)

lam1 = inverse_iteration(A, v, 4, 2, k=0)
lam2 = inverse_iteration(A, v, 4, -1, k=0)
lam3 = inverse_iteration(A, v, 4, 4, k=0)

print(f'lam_1^({len(lam1)}) = {lam1[-1]}')
print(f'lam_2^({len(lam2)}) = {lam2[-1]}')
print(f'lam_3^({len(lam3)}) = {lam3[-1]}')

lam_1^(4) = 1.9537505736728318
lam_2^(4) = -0.9831712938368212
lam_3^(4) = 4.02942062252305
```

The eigenvalues are thus exact to the first six decimal places.   ◀

The example demonstrates that the inverse iteration with shift leads to a significant acceleration of the convergence if reasonable estimates of the eigenvalues are available.

Remark 5.18 (*Complex Eigenvalues*) Real matrices $A \in \mathbb{R}^{n \times n}$ can have complex eigenvalues. Simple iteration methods, however, are completely based on real arithmetic and will never be able to find complex eigenvalues. This is not a contradiction, because according to Theorem 5.1, complex eigenvalues of real matrices always appear as complex conjugate pairs $\lambda, \bar{\lambda} \in \mathbb{C}$. For eigenvalues of maximum magnitude with $\lambda \neq \bar{\lambda} \in \mathbb{C} \setminus \mathbb{R}$, the method does not converge. This can be easily tested with the matrix

$$A = \begin{pmatrix} 0 & 1 \\ -1 & 0 \end{pmatrix}.$$

By choosing a complex shift $\sigma \in \mathbb{C}$, the inverse iteration can also find complex eigenvalues. The method then requires complex arithmetic. $\blacklozenge$

5.4 Factorization Methods

A completely different approach to finding the eigenvalues of a matrix $A \in \mathbb{R}^{n \times n}$ is provided by so-called *factorization methods*. We make the following observation.

Theorem 5.19 (Eigenvalues in Multiplicative Factorizations) *Let $A \in \mathbb{R}^{n \times n}$ be a regular matrix. Assume the factorization*

$$A = BC$$

exists with two regular matrices $B, C \in \mathbb{R}^{n \times n}$. Then the matrices $A = BC$ and $A' = CB$ have the same eigenvalues. With the n eigenvalues (counted according to their multiplicity) $\lambda_1, \ldots, \lambda_n \in \mathbb{C}$ of A or of A', it holds

$$\mathrm{trace}(A) = \mathrm{trace}(A') = \sum_{i=1}^{n} \lambda_i.$$

Proof *(i)* Let $\lambda \in \mathbb{C}$ be an eigenvalue of A with corresponding eigenvector $w \in \mathbb{C}^n$. Then it holds

$$\lambda w = Aw = BCw = B(CB)B^{-1}w \quad \Leftrightarrow \quad \lambda(B^{-1}w) = CB(B^{-1}w) = A'(B^{-1}w)$$

so BC and CB have the same eigenvalue λ. Each eigenvector w of A is associated with a corresponding eigenvector $w' = B^{-1}w$ of A'.

(ii) The trace of a matrix is an *invariant* under similarity transformations, see e.g. [33]. With the eigenvalues λ_i as roots of the characteristic polynomial, it holds

$$\det(\lambda I - A) = \det(\lambda I - A') = \prod_{i=1}^{n}(\lambda - \lambda_i)$$

$$= \lambda^n - \lambda^{n-1}\sum_{i=1}^{n}\lambda_i + \cdots + \prod_{i=1}^{n}(-\lambda_i). \tag{5.4}$$

The sum of the eigenvalues appears in the coefficient before the second-highest monomial λ^{n-1}. Conversely, with the Leibniz formula

$$\det(\lambda I - A) = \sum_{\sigma \in S_n} \mathrm{sgn}(\sigma) \prod_{i=1}^{n} [\lambda I - A]_{i,\sigma(i)},$$

where the sum is taken over all permutations σ of the symmetric group S_n. With $\text{sgn}(\sigma)$, the *sign* of the permutation is denoted, i.e., $+1$ for an even, -1 for an odd permutation. To compare with (5.4), only the terms of interest are those where the exponent λ^{n-1} appears. This can be the case if σ is the identity, because otherwise at least two diagonal terms are missing. With this, we can write the determinant as

$$\det(\lambda I - A) = \prod_{i=1}^{n}(\lambda - a_{ii}) + p_{n-2}(\lambda),$$

where $p_{n-2}(\lambda)$ is a polynomial of degree $n - 2$. It follows

$$\det(\lambda I - A) = \lambda^{n} - \lambda^{n-1}\sum_{i=1}^{n} a_{ii} + q_{n-2}(\lambda) + p_{n-2}(\lambda)$$

with another polynomial $q_{n-2} \in P_{n-2}$. The comparison with (5.4) shows

$$\text{trace}(A) = \sum_{i=1}^{n} a_{ii} = \sum_{i=1}^{n} \lambda_i = \sum_{i=1}^{n} a'_{ii} = \text{trace}(A').$$

$\square$

The iteration

$$A_k = B_k C_k \mapsto C_k B_k =: A_{k+1}$$

thus defines a similarity transformation. All matrices A_k have the same eigenvalues and also the same trace. For many known factorizations (LU factorization, Cholesky factorization, QR factorization), a very surprising relationship is observed: The sequence of matrices $(A_k)_k \in \mathbb{R}^{n \times n}$ converges to a triangular matrix. Since the matrices are similar, the eigenvalues of the initial matrix $A_0 = A$ can be read from the diagonal.

5.4.1 Cholesky Iteration

If several or even all eigenvalues are sought, then factorization methods are the most efficient methods for determining eigenvalues. The mathematical analysis is far more complex than the analysis of the power iteration and the inverse iteration. We first consider a method based on the Cholesky factorization.

Algorithm 5.20: Cholesky Iteration

> **Input:** $A \in \mathbb{R}^{n \times n}$ symmetric positive definite matrix.
> 1 Set $A_0 := A$
> 2 **for** $i = 0, 1, 2, \ldots$ **do**
> 3   Compute the Cholesky factorization $A_i =: L_i L_i{}^T$
> 4   Compute $A_{i+1} := L_i{}^T L_i$

By using Implementation 3.28 of the Cholesky factorization, we can implement the Cholesky iteration as follows. We take care to avoid performing the full matrix multiplication due to the lower triangular structure of L.

♣ **Implementation 5.21: Cholesky Iteration**

```python
def cholesky_iter(A, k):
    A = A.copy()
    n, m = A.shape
    for i in range(k):
        L = cholesky(A)
        for i in range(n):
            for j in range(n):
                a = 0
                for k in range(max(i, j), n):
                    a += L[k, i] * L[k, j]
                A[i, j] = a
    return A
```

The special matrix-matrix product that we have implemented in the lines 6–11 can also be replaced with the full matrix-matrix product from `numpy`

```python
6 A = L.transpose() @ L
```

Since the matrix multiplication is efficiently implemented internally here, and no relatively slow Python loops are used or accesses to individual matrix entries occur, this may also be faster, even if many zero entries are calculated.

We first show that the Cholesky iteration is feasible, i.e., all matrices $A_{(i)}$ are again symmetric positive definite, so a Cholesky factorization exists.

Theorem 5.22 (Feasibility of the Cholesky Iteration) *Let $A \in \mathbb{R}^{n \times n}$ be symmetric positive definite and $A = LL^T$ the Cholesky factorization. The matrix $A' = L^T L$ is symmetric positive definite.*

Proof Let $A \in \mathbb{R}^{n \times n}$ be symmetric positive definite. Then the Cholesky factorization exists

$$A = LL^T$$

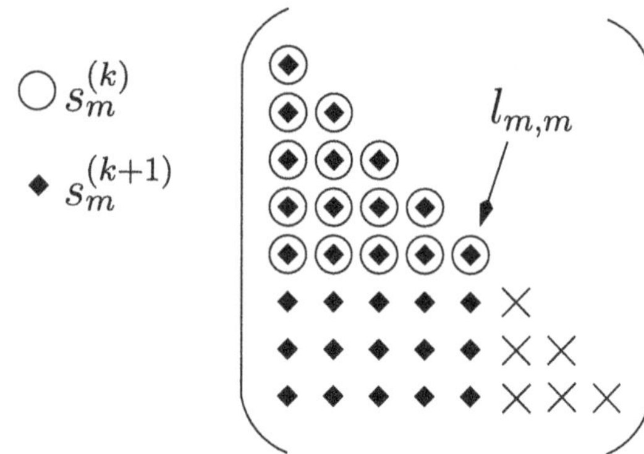

Fig. 5.1 Elements of $L = (l_{ij})_{ij}$, which appear in the partial sums $s_m^{(k)}$ (circles) and $s_m^{(k+1)}$ (diamonds). The difference is given by all indices, which are marked with diamonds but do not have circles

into a regular triangular matrix $L \in \mathbb{R}^{n \times n}$ and the matrix $A' = L^T L$ is again symmetric, because

$$A'^T = (L^T L)^T = L^T L.$$

Furthermore, the matrix A' is positive definite, because

$$(A'x, x) = (L^T L x, x) = (Lx, Lx) > 0 \quad \forall x \neq 0,$$

since L is regular. $\square$

The convergence of the Cholesky iteration can be proven relatively easily. It always converges, i.e., for every symmetric positive definite matrix.

Theorem 5.23 (Convergence of the Cholesky Iteration) *Let $A \in \mathbb{R}^{n \times n}$ be a symmetric positive definite matrix. The sequence of matrices A_k from the Cholesky iteration in Algorithm 5.20 converges to a diagonal matrix, whose entries are exactly the eigenvalues of A.*

Proof (i) To prove convergence, we consider two consecutive matrices A_k and A_{k+1}. Let

$$A_k = L_k L_k^T, \quad A_{k+1} = L_k^T L_k,$$

where L_k is the lower left triangular matrix of the Cholesky factorization. For the entries of the matrices $A_k = (a_{ij}^{(k)})_{ij}$ and $A_{k+1} = (a_{ij}^{(k+1)})_{ij}$ we have

$$a_{ij}^{(k)} = \sum_{k=1}^{n} l_{ik}^{(k)} l_{jk}^{(k)} = \sum_{k=1}^{\min\{i,j\}} l_{ik}^{(k)} l_{jk}^{(k)},$$

$$a_{ij}^{(k+1)} = \sum_{k=1}^{n} l_{ki}^{(k)} l_{kj}^{(k)} = \sum_{k=\max\{i,j\}}^{n} l_{ki}^{(k)} l_{kj}^{(k)}. \tag{5.5}$$

With $s_m^{(k)}$, we denote the partial trace, i.e.,

$$s_m^{(k)} := \sum_{i=1}^{m} a_{ii}^{(k)}, \quad s_m^{(k+1)} := \sum_{i=1}^{m} a_{ii}^{(k+1)}.$$

For these, with (5.5) we have

$$s_m^{(k)} = \sum_{i=1}^{m} \sum_{k=1}^{i} (l_{ik}^{(k)})^2, \quad s_m^{(k+1)} = \sum_{i=1}^{m} \sum_{k=i}^{n} (l_{ki}^{(k)})^2.$$

In Fig. 5.1, we graphically represent which entries of L appear in the two partial traces $s_m^{(k)}$ and $s_m^{(k+1)}$: For $s_m^{(k)}$ it is the submatrix to the left above $l_{mm}^{(k)}$, for $s_m^{(k+1)}$ it is the whole submatrix to the left of $l_{mm}^{(k)}$. The difference between the partial traces is thus given as

$$s_m^{(k+1)} - s_m^{(k)} = \sum_{i=m+1}^{n} \sum_{j=1}^{m} (l_{ij}^{(k)})^2 \geq 0. \tag{5.6}$$

The partial traces $s_m^{(k)}$ are each (for each m) a monotonically increasing sequence. However, these sequences are bounded, because positive definite matrices only have positive diagonal entries, see Theorem 2.15, and it holds

$$s_m^{(k)} \leq s_n^{(k)} = \text{trace}(A_k) = \text{trace}(A) = \sum_{i=1}^{n} \lambda_i.$$

As a monotonically increasing and bounded sequence, it converges

$$s_m^{(k)} \to s_m \leq \sum_{i=1}^{n} \lambda_i.$$

With (5.6) it then follows

$$\sum_{i=m+1}^{n} \sum_{j=1}^{m} (l_{ij}^{(k)})^2 = s_m^{(k+1)} - s_m^{(k)} \to 0 \quad (k \to \infty),$$

so for all $m \in \{1, \ldots, n\}$ and all $i > m$ and $j \leq m$:

$$l_{ij}^{(k)} \to 0 \text{ for } k \to \infty, \quad \forall i > m, j \leq m,$$

All off-diagonal elements converge to zero, such that

$$L_k \to \text{diag}(l_{11}, \ldots, l_{nn}) \quad \text{for } (k \to \infty).$$

(ii) We have already shown that the matrices A and A_k for $k \geq 0$ have the same eigenvalues. If $A_k \to \Lambda = \mathrm{diag}(l_1, \ldots, l_n)$, then the diagonal elements $l_1, \ldots, l_n$ must be the eigenvalues of A. We did not get any information about the order of the eigenvalues. $\square$

Example 5.24 *(Cholesky Iteration)* We consider the matrix

$$
A = \begin{pmatrix} 3 & -1 & 0 & 1 \\ -1 & 3 & 1 & 1 \\ 0 & 1 & 3 & 0 \\ 1 & 1 & 0 & 3 \end{pmatrix}.
$$

With $A_0 := A$ we perform some steps of the Cholesky iteration with our Python code. We also output some of the iteration matrices.

```python
A = np.array([[3, -1, 0, 1],
              [-1, 3, 1, 1],
              [0, 1, 3, 0],
              [1, 1, 0, 3]], dtype=np.double)
A_chol = cholesky_iter(A, 50)
```

```
A_0 =
[[ 3.66666667 -0.47140452 -0.17817416  0.79681907]
 [-0.47140452  3.70833333  0.74018043  1.12687234]
 [-0.17817416  0.74018043  2.7202381  -0.42591771]
 [ 0.79681907  1.12687234 -0.42591771  1.9047619 ]]

A_1 =
[[ 3.90909091 -0.23728987 -0.31241772  0.44884361]
 [-0.23728987  4.20305862  0.34963659  0.6942616 ]

 [-0.31241772  0.34963659  2.72441556 -0.42311459]
 [ 0.44884361  0.6942616  -0.42311459  1.1634349 ]]

A_2 =
[[ 4.         -0.19112739 -0.32003501  0.21701249]
 [-0.19112739  4.3390411   0.16839657  0.33700088]
 [-0.32003501  0.16839657  2.74715053 -0.2597052 ]
 [ 0.21701249  0.33700088 -0.2597052   0.91380838]]

...

A_49 =
[[ 4.47235427e+00 -6.46190915e-02 -4.96860043e-06  3.08058421e-18]
 [-6.46190915e-02  4.00884003e+00 -2.35753313e-05  3.17640652e-17]
 [-4.96860043e-06 -2.35753313e-05  2.68889218e+00 -2.25698099e-13]
 [ 3.08058421e-18  3.17640652e-17 -2.25698099e-13  8.29913513e-01]]
```

The eigenvalues can be determined using the `numpy.linalg` function `eig` as

```python
print(np.linalg.eig(A)[0])
```

```
[0.82991351 2.68889218 4.         4.4811943 ]
```

So the Cholesky iteration converges, and all eigenvalues can be determined. However, it converges very slowly. After 50 steps, we have determined the eigenvalues to a relative error of 0.2%.   ◄

The Cholesky iteration can, of course, only be applied to symmetric positive definite matrices. A generalization is a corresponding method, based on the LU factorization, i.e., an iteration of the kind

$$A_0 := A, \quad A_k =: L_k R_k, \quad A_{k+1} := R_k L_k.$$

This method also converges to a matrix whose diagonal elements are the eigenvalues of A. The method can, however, only be applied if the LU factorization can be carried out in each step without pivoting (compare Theorem 3.18). Assuming it is

$$PA = LU$$

with a pivot matrix $P \neq I$, then the matrices A and $A' := RL$ are *not* similar anymore.

5.4.2   The QR Iteration

We consider the QR factorization as an alternative to the LU factorization. The latter is not applicable to general regular matrices. This factorization, see Sect. 4.1, exists for any regular matrix and we know stable methods for its calculation, for example, with Householder transformations in Sect. 4.3 or the Givens rotations in Sect. 4.4. The disadvantage of the QR factorization is that it requires twice as much effort compared to the LU factorization.

Algorithm 5.25: Basic QR Iteration

 Input: $A \in \mathbb{R}^{n \times n}$.
1   Set $A_0 := A$
2   **for** $k = 0, 1, 2, \ldots$ **do**
3   Create the QR factorization $A_k =: Q_k R_k$
4   Calculate $A_{k+1} := R_k Q_k$

With our implementation of the QR factorization using Givens rotations, we can then implement this in Python as follows:

♣ Implementation 5.26: QR Iteration with Givens Rotations

```python
1 def qr_iter(A, k):
2     A = A.copy()
3     for i in range(k):
4         QT = qr_givens(A)
5         A = A @ QT.transpose()
6     return A
```

The QR iteration generates a sequence A_k, $k \geq 0$ of matrices, which, according to Theorem 5.19, are all similar to the matrix A. We will see that the diagonal entries of the sequence members A_k converge to the eigenvalues of the matrix A. The analysis of the method is more complex, so we only provide a brief sketch of the proof.

Theorem 5.27 (QR Iteration for Eigenvalue Calculation) *Let $A \in \mathbb{R}^{n \times n}$ be a matrix with separated eigenvalues*

$$|\lambda_1| > |\lambda_2| > \cdots > |\lambda_n|.$$

Then for the diagonal elements a_{ii}^k of the matrices A_k generated by the QR iteration, it holds that

$$\{\alpha_{11}^k, \ldots, \alpha_{nn}^k\} \to \{\lambda_1, \ldots, \lambda_n\} \quad for \quad (k \to \infty).$$

Proof The detailed proof can be found in [22, 71]. An essential part of the proof is the relation

$$A_{k+1} = \underbrace{Q_0 Q_1 \cdots Q_k}_{=\tilde{Q}_k} \underbrace{R_k R_{k-1} \cdots R_0}_{=\tilde{R}_k}$$

which we prove by induction. First, as claimed, $A_1 = Q_0 R_0 = \tilde{Q}_0 \tilde{R}_0$. The statement is thus correct for A^k. Then we have

$$A_{k+1} = A_1 A_k = Q_0 R_0 \underbrace{Q_0 Q_1 \cdots Q_{k-1}}_{=\tilde{Q}^{k-1}} \underbrace{R_{k-1} R_{k-2} \cdots R_0}_{=\tilde{R}^{k-1}}$$

$$= Q_0 \underbrace{R_0 Q_0}_{=A_1 = Q_1 R_1} Q_1 \cdots Q_{k-1} R_{k-1} \cdots R_0$$

$$= Q_0 Q_1 \underbrace{R_1 Q_1}_{=A_2 = Q_2 R_2} Q_2 \cdots Q_{k-1} R_{k-1} \cdots R_0$$

$$= \ldots$$

$$= Q_0 Q_1 \ldots Q_{k-2} Q_{k-1} \underbrace{R_{k-1} Q_{k-1}}_{=A_k = Q_k R_k} R_{k-1} \cdots R_0$$

$$= Q_0 Q_1 \ldots Q_{k-2} Q_{k-1} \cdots Q_k R_k R_{k-1} \cdots R_0 = \tilde{Q}_k \tilde{R}_k.$$

The product $\tilde{Q}_k$ is again an orthogonal matrix, and $\tilde{R}_k$, as a product of upper right triangular matrices, is itself one. In the course of the method, a factorization of the power A^k is thus created. This can be used to establish a connection to the power iteration. $\square$

In general, the diagonal elements of the sequence members A_k converge at least linearly to the eigenvalues. In special cases (e.g., symmetric matrices), however, with a good shift strategy, even quadratic or cubic convergence is achieved. The analysis is rather complex [97]. Compared to the power iteration and the inverse iteration, the QR iteration has better convergence properties and, at the same time, determines all eigenvalues of the matrix A.

Example 5.28 (*QR Iteration*) We consider the matrix from Example 5.24

$$A = \begin{pmatrix} 3 & -1 & 0 & 1 \\ -1 & 3 & 1 & 1 \\ 0 & 1 & 3 & 0 \\ 1 & 1 & 0 & 3 \end{pmatrix}.$$

We apply our implementation and consider for the first iterations $A_{k+1} = R_k Q_k$

```
1 A_new = qr_iter(A, 25)

  A_0 =
  [[ 3.90909091 -0.23728987 -0.31241772  0.44884361]
   [-0.23728987  4.20305862  0.34963659  0.6942616 ]
   [-0.31241772  0.34963659  2.72441556 -0.42311459]
   [ 0.44884361  0.6942616  -0.42311459  1.1634349 ]]

  A_1 =
  [[ 4.04651163 -0.19251315 -0.280484    0.10005528]
   [-0.19251315  4.36319045  0.09547029  0.15393536]
   [-0.280484    0.09547029  2.74000116 -0.14246604]
   [ 0.10005528  0.15393536 -0.14246604  0.85029676]]

  A_2 =
  [[ 4.10147992 -0.20988706 -0.19194259  0.02052083]

   [-0.20988706  4.35310663  0.0354996   0.03135503]
   [-0.19194259  0.0354996   2.71419158 -0.04148131]
   [ 0.02052083  0.03135503 -0.04148131  0.83122187]]

  ...

  A_24 =
  [[ 4.47235427e+00 -6.46190915e-02 -4.96860043e-06  4.84794998e-16]
   [-6.46190915e-02  4.00884003e+00 -2.35753313e-05 -2.25833466e-16]
   [-4.96860043e-06 -2.35753313e-05  2.68889218e+00 -2.25615109e-13]
   [ 3.08058421e-18  3.17640652e-17 -2.25698099e-13  8.29913513e-01]]
```

All four eigenvalues are determined again. After only 25 steps, all eigenvalues are again available with a relative error of less than 0.2%. ◄

The QR iteration can be significantly accelerated. The convergence speed again depends on the separation of the eigenvalues $|\lambda_i| > |\lambda_{i+1}|$. By choosing a *shift*, the spectrum can be distributed more favorably. The basis for this is the following theorem.

Theorem 5.29 (Similarity Transformation and Shift) *Let $A \in \mathbb{R}^{n \times n}$ be a regular matrix and $\mu \in \mathbb{R}$ such that $A - \mu I$ is regular. Let*

$$QR = A - \mu I,$$

be the QR factorization. Then the matrices A and

$$A' = RQ + \mu I$$

are similar.

Proof It holds

$$A' = RQ + \mu I = Q^T QRQ + \mu I = Q^T (A - \mu I) Q + \mu I = Q^T A Q.$$

$\square$

With the correct choice of shift, the QR iteration is extremely fast. We refer to the literature [22].

Example 5.30 (*QR Iteration With Shift*) We again consider the matrix

$$A = \begin{pmatrix} 3 & -1 & 0 & 1 \\ -1 & 3 & 1 & 1 \\ 0 & 1 & 3 & 0 \\ 1 & 1 & 0 & 3 \end{pmatrix}.$$

To apply the QR iteration with shift, we adapt our existing implementation of the QR iteration. As a shift, we always use the largest diagonal element $\mu_k = \max\{[A_k]_{ii}\}$.

Implementation 5.31: QR Iteration with Shift

```python
def qr_iter_shift(A, k):
    A = A.copy()
    Id = np.identity(A.shape[0], dtype=A.dtype)
    for l in range(k):
        mu = np.amax(np.diag(A))
        A[:, :] -= mu * Id
        QT = qr_givens(A)
        A[:, :] = A @ QT.transpose() + mu * Id
    return A
```

The first steps yield

```
mu_0 = 3.0
A_0 =
[[ 2.          -1.34164079  0.67082039  0.5        ]
 [-1.34164079  3.6          0.7         0.2236068 ]
 [ 0.67082039  0.7          3.4         0.4472136 ]
 [ 0.5         0.2236068    0.4472136   3.         ]]

mu_0 = 3.5999999999999996
A_1 =
[[ 0.91304348 -0.50102009  0.09837421  0.14239908]
 [-0.50102009  4.35563604  0.26225201 -0.03157971]
 [ 0.09837421  0.26225201  3.18483703  0.62151247]
 [ 0.14239908 -0.03157971  0.62151247  3.54648345]]

mu_0 = 4.355636041564521
A_2 =
[[ 8.30705670e-01 -4.05668796e-02 -3.96742622e-03 -1.68422780e-02]
 [-4.05668796e-02  3.19594681e+00  7.66669931e-01  8.13801166e-02]
 [-3.96742622e-03  7.66669931e-01  3.94422136e+00 -2.37077169e-01]
 [-1.68422780e-02  8.13801166e-02 -2.37077169e-01  4.02912616e+00]]

mu_0 = 4.029126163790677
A_3 =
[[ 8.29995403e-01 -1.24432457e-02 -3.77520654e-06  1.65271069e-04]
 [-1.24432457e-02  2.75040626e+00  3.26440965e-01 -2.25823280e-03]
 [-3.77520654e-06  3.26440965e-01  4.41933913e+00  1.21169434e-02]
 [ 1.65271069e-04 -2.25823280e-03  1.21169434e-02  4.00025921e+00]]

mu_0 = 4.419339127724223
A_4 =
[[ 8.29932204e-01 -5.89457686e-03 -3.08739043e-06 -1.90655467e-05]
 [-5.89457686e-03  2.68895970e+00  1.23244007e-02  5.50074407e-04]
 [-3.08739043e-06  1.23244007e-02  4.46881158e+00 -7.59412090e-02]
 [-1.90655467e-05  5.50074407e-04 -7.59412090e-02  4.01229652e+00]]

mu_0 = 4.468811575985431
A_5 =
[[ 8.29917985e-01 -2.88318998e-03 -2.45584035e-06  4.00544086e-07]
 [-2.88318998e-03  2.68888805e+00  6.66883821e-04 -2.36284725e-05]
 [-2.45584035e-06  6.66883821e-04  4.01246830e+00  7.64485048e-02]
 [ 4.00544087e-07 -2.36284725e-05  7.64485048e-02  4.46872566e+00]]
```

After only six steps, we have determined all eigenvalues to a relative error of less than 1%.   ◄

In each step of the iteration, a QR factorization of the iteration matrix A_k must be created. For a general matrix $A \in \mathbb{R}^{n \times n}$, this requires $O(n^3)$ arithmetic operations. To approximate the eigenvalues with sufficient accuracy, many steps are often necessary. In practical applications, the QR iteration is therefore always used in conjunction with a *reduction method* (see the following section), where the matrix A is first transformed into a "simple" similar form.

5.5   The QR Iteration for Matrices in Hessenberg Form

Creating the QR factorization of a matrix $A \in \mathbb{R}^{n \times n}$ using Householder matrices requires $O(n^3)$ arithmetic operations, see Theorem 4.19. Since in each step of the QR iteration, this factorization must be recreated, the method is very costly. However, if the matrix A has a special structure, for example, if it is a band matrix, then the QR factorization can be created with much less effort.

Theorem 5.32 (Similar Matrices) *Two matrices $A, B \in \mathbb{C}^{n \times n}$ are called* similar, *if there is a regular matrix $S \in \mathbb{C}^{n \times n}$ such that:*

$$A = S^{-1} B S.$$

Similar matrices have the same characteristic polynomial and the same eigenvalues. For an eigenvalue λ and eigenvector w of A, it holds:

$$B(Sw) = S(Aw) = \lambda Sw.$$

Our goal is to first transform the matrix A into a *similar matrix* with the same eigenvalues, for which the QR factorization can then be calculated more quickly or where the eigenvalues can even be read directly. Possible *normal forms*, where the eigenvalues can be read directly, are the *Jordan normal form*, the diagonalization of A, or the *Schur normal form*. The task of transforming a matrix A into one of these normal forms is usually only possible if the eigenvalues are known. A compromise, however, is the so-called *Hessenberg form*. This is a triangular matrix with an additional subdiagonal, and it can be constructed relatively easily.

Theorem 5.33 (Hessenberg Form) *For every matrix $A \in \mathbb{R}^{n \times n}$ there exists an orthogonal matrix $Q \in \mathbb{R}^{n \times n}$, such that*

$$H := Q^T A Q = \begin{pmatrix} * & \cdots & \cdots & \cdots & * \\ * & \ddots & \ddots & \ddots & \vdots \\ 0 & \ddots & \ddots & \ddots & \vdots \\ \vdots & \ddots & \ddots & \ddots & \vdots \\ 0 & \cdots & 0 & * & * \end{pmatrix}$$

is a Hessenberg *matrix, i.e., an upper right triangular matrix, with an additional lower subdiagonal, $H_{ij} = 0$ for $i > j + 1$. If A is symmetric (i.e., $A^T = A$), then $Q^T A Q$ is a tridiagonal matrix.*

Proof The construction of the Hessenberg matrix is similar to the QR factorization with Householder transformations. Here, we have created the upper right triangular matrix by successive multiplication with reflection matrices $S^{(k)}$, i.e.,

$$R = \underbrace{S^{(n-1)} \cdots S^{(2)} S^{(1)}}_{=Q^T} A.$$

However, the matrices R and A are not similar. To achieve this, we will perform the multiplications from the left and right, i.e., $A^{(i)} = S^{(i)} A^{(i-1)} S^{(i)}$ (the orthogonal reflection matrices $S^{(i)}$ are symmetric so that we can omit the transposition). $A^{(i)}$ and $A^{(i-1)}$ are then similar, so they have the same eigenvalues.

We describe the first step. In the QR factorization, the matrix $S^{(1)}$ is chosen so that the multiplication $S^{(1)} A^{(0)}$ maps the first column vector $a_1^{(0)}$ of $A^{(0)}$ to a multiple of the first unit vector e_1, i.e., the first subdiagonal of $S^{(1)} A^{(0)}$ is zero. The multiplication from the right with $S^{(1)}$ generally destroys this property again, and the entries of the first column are refilled. However, we achieve the desired result with a simple trick: $S^{(i)}$ is chosen so that the first column vector is mapped into the subspace $\mathrm{span}(e_1, e_2)$. For this, we determine $S^{(1)}$ according to

$$S^{(1)} = I - 2v^{(i)}(v^{(1)})^T, \quad v^{(i)} = \frac{\tilde{a}_1^{(0)} + \|\tilde{a}_1^{(0)}\| e_2}{\|\tilde{a}_1^{(0)} + \|\tilde{a}_1^{(0)}\| e_2\|},$$

where $\tilde{a}_1^{(0)}$ is the column vector shortened by the first element, i.e.,

$$\tilde{a}_1^{(0)} = (0, a_{21}^{(0)}, \ldots, a_{n1}^{(0)}).$$

Then $S^{(1)}$ has the form

$$S^{(1)} = \begin{pmatrix} 1 & 0 & \cdots & 0 \\ 0 & * & \cdots & * \\ \vdots & \vdots & \ddots & \vdots \\ 0 & * & \cdots & * \end{pmatrix}$$

and multiplying by $S^{(1)}$ from the left leaves the first row unchanged, and multiplying by $S^{(1)}$ from the right leaves the first column the same. We do not achieve a triangular shape with this, but the desired Hessenberg form:

$$A^{(1)} = S^{(1)} A^{(0)} S^{(1)} = \left(\begin{array}{c|ccc} a_{11} & a_{12} & \cdots & a_{1n} \\ \hline * & * & \cdots & * \\ 0 & \vdots & \ddots & \vdots \\ \vdots & \vdots & \ddots & \vdots \\ 0 & * & \cdots & * \end{array}\right) =: \left(\begin{array}{c|ccc} a_{11} & * & \cdots & * \\ \hline * & & & \\ 0 & & \tilde{A}_{(1)} & \\ \vdots & & & \\ 0 & & & \end{array}\right)$$

with the $(n-1) \times (n-1)$ matrix block $\tilde{A}^{(1)} \in \mathbb{R}^{n-1 \times n-1}$. In the second step, the corresponding procedure is applied to the matrix $\tilde{A}^{(1)}$. After $n-2$ steps, we obtain the matrix $A^{(n-2)}$, which has Hessenberg form:

$$Q^T A Q := \underbrace{S^{(n-2)} \cdots S^{(1)}}_{=:Q^T} A \underbrace{S^{(1)} \cdots S^{(n-2)}}_{=:Q}$$

In the case $A = A^T$, we have

$$(Q^T A Q)^T = Q^T A^T Q = Q^T A Q.$$

This means that $A^{(n-2)}$ is also symmetric. Symmetric Hessenberg matrices are tridiagonal matrices. $\qquad\Box$

The Python implementation of the transformation into Hessenberg form can then look as follows:

♣ Implementation 5.34: Transformation to Hessenberg Form

```python
def reduce_hessenberg(A):
    n, m = A.shape
    for i in range(m - 2):
        v = A[i + 1:, i].copy()
        ei = np.zeros(n - i - 1, dtype=A.dtype)
        ei[0] = 1
        v += np.sign(v[0]) * np.linalg.norm(v) * ei
        v /= np.linalg.norm(v)
        A[i + 1:, i:] -= 2 * np.outer(v, np.inner(v.T, A[i + 1:, i:].T))
        A[:, i + 1:] -= 2 * np.outer(np.inner(A[:, i + 1:], v), v)
```

Remark 5.35 (*Hessenberg Form*) The transformation of a matrix $A \in \mathbb{R}^{n \times n}$ into Hessenberg form requires $\frac{5}{3}n^3 + O(n^2)$ arithmetic operations when using the Householder transformations. In the case of symmetric matrices, the transformation into a tridiagonal matrix requires $\frac{2}{3}n^3 + O(n^2)$ operations. $\qquad\blacklozenge$

The transformation into Hessenberg form, therefore, requires slightly more arithmetic operations than *one* QR factorization. Subsequently, the QR factorizations of a Hessenberg matrix can be created with much less effort. Furthermore, for the QR factorization of a Hessenberg matrix A, the matrix RQ once again has Hessenberg form. Both of these facts are summarized in the following theorem.

Theorem 5.36 (QR Factorization of Hessenberg Matrices) *Let $A \in \mathbb{R}^{n \times n}$ be a Hessenberg matrix. Then the QR factorization $A = QR$ can be performed with Householder transformations in $2n^2 + O(n)$ arithmetic operations. The matrix RQ also has Hessenberg form.*

Proof *(i)* We have $a_{ij} = 0$ for $i > j + 1$. We show inductively that this property holds for all matrices $A^{(i)}$ that arise during the QR factorization. First, in the first step for $v^{(1)} = a_1 + \|a_1\| e_1$, we have $v_k^{(1)} = 0$ for all $k > 2$. The new column vectors are calculated as

$$a_i^{(1)} = a_i - (a_i, v^{(1)}) v^{(1)}. \tag{5.7}$$

Since $v_k^{(1)} = 0$ for $k > 2$, we have $(a_i^{(1)})_k = 0$ for $k > i + 1$, i.e., $A^{(1)}$ again has Hessenberg form. This property holds inductively for $i = 2, \ldots, n - 1$.

(ii) Since the (reduced) vector $\tilde{v}^{(i)}$ in each step only has two non-zero entries, it can be created in two arithmetic operations. The calculation of the $n - i$ new column vectors according to (5.7) requires four arithmetic operations. In total, the effort amounts to

$$\sum_{i=1}^{n-1} 2 + 4(n - i) = 2n^2 + O(n).$$

$\square$

The Python implementation can then take the following form. Here, we also explicitly calculate the matrix Q so that we can use it later in the calculation of the eigenvalues with the QR iteration.

Implementation 5.37: QR Factorization for Hessenberg Matrices

```python
def qr_hessenberg(A):
    n, m = A.shape
    dtype = A.dtype
    Q = np.identity(n, dtype=dtype)
    for i in range(m - 1):
        v = A[i: i + 2, i].copy()
        ei = np.zeros(2, dtype=dtype)
        ei[0] = 1.0
        v += np.sign(A[i,i]) * np.linalg.norm(v) * ei
        v /= np.linalg.norm(v)
        A[i:i + 2, i:] -= 2 * np.outer(v, np.inner(v.T, A[i:i + 2, i:].T))
        Q[:, i:i + 2] -= 2 * np.outer(np.inner(Q[:, i: i + 2], v), v)
    return Q
```

In conjunction with the reduction to Hessenberg form, or to tridiagonal form, the QR iteration is one of the most efficient methods for calculating the eigenvalues of a matrix $A \in \mathbb{R}^{n \times n}$. The QR factorization can be performed in $O(n^2)$ operations, and in the course of the QR iteration, only Hessenberg matrices are generated. We consider an example:

Example 5.38 *(Eigenvalue Calculation With Reduction and QR Iteration)* We consider the matrix

$$A = \begin{pmatrix} 338 & -20 & -90 & 32 \\ -20 & 17 & 117 & 70 \\ -90 & 117 & 324 & -252 \\ 32 & 70 & -252 & 131 \end{pmatrix}.$$

The matrix A is symmetric and has the eigenvalues

$$\lambda_1 \approx 547.407, \quad \lambda_2 \approx 297.255, \quad \lambda_3 \approx -142.407, \quad \lambda_4 \approx 107.745.$$

Step 1: Reduction to Hessenberg (Tridiagonal) Form.
In the first reduction step, we choose with $\tilde{a}_1 = (0, -20, -90, 32)$ the reflection vector $v^{(1)}$
as

$$v^{(1)} = \frac{\tilde{a}_1 - \|\tilde{a}_1\| e_2}{\| \tilde{a}_1 - \|\tilde{a}_1\| e_2 \|}.$$

In the first step of our code, we obtain with $S_{(1)} := I - 2v^{(1)}(v^{(1)})^T$ the matrix $A_{(1)} = S_{(1)} A S_{(1)}$:

```
[[ 3.38000000e+02  9.75909832e+01 -2.84217094e-14  7.10542736e-15]
 [ 9.75909832e+01  4.77579168e+02  2.77974214e+01  1.06979728e+02]
 [-2.84217094e-14  2.77974214e+01 -8.23446072e+01 -1.03493607e+02]
 [ 7.10542736e-15  1.06979728e+02 -1.03493607e+02  7.67654387e+01]]
```

With $\tilde{a}_2^{(1)} \approx (0, 0, 27.797, 106.980)^T$,

$$v^{(2)} = \frac{\tilde{a}_2 + \|\tilde{a}_2\| e_3}{\| \tilde{a}_2 + \|\tilde{a}_2\| e_3 \|} \quad \text{and} \quad S_{(2)} := I - 2v^{(2)}(v^{(2)})^T$$

the matrix $H := A_{(2)} = S_{(2)} A_{(1)} S_{(2)}$:

```
[[ 3.38000000e+02  9.75909832e+01  2.70632072e-16  2.92951775e-14]
 [ 9.75909832e+01  4.77579168e+02 -1.10532162e+02  1.42108547e-14]
 [-2.84217094e-14 -1.10532162e+02  1.63207694e+01 -1.29130642e+02]
 [ 7.10542736e-15  0.00000000e+00 -1.29130642e+02 -2.18999378e+01]]
```

The matrix H now has (up to rounding errors) tridiagonal form, and all further calculations
are based on

```
[[ 3.38000000e+02  9.75909832e+01  0.00000000e+00  0.00000000e+00]
 [ 9.75909832e+01  4.77579168e+02 -1.10532162e+02  0.00000000e+00]
 [ 0.00000000e+00 -1.10532162e+02  1.63207694e+01 -1.29130642e+02]
 [ 0.00000000e+00  0.00000000e+00 -1.29130642e+02 -2.18999378e+01]]
```

All transformations were similarity transformations. Therefore, H and A have the same
eigenvalues. If we numerically check the eigenvalues of A and H using the function
`np.linalg.eig`, we see that the difference is 10^{-13}, which is of the order of machine
precision.

Step 2: QR Method
We now perform a few steps of the QR iteration. For this, we use our QR factorization with
Householder transformations for the special case of Hessenberg matrices. The QR iteration
then takes the following form:

♣ Implementation 5.39: QR Iteration for Matrices in Hessenberg Form

```python
def qr_iter_hessenberg(A, k):
    reduce_hessenberg(A)
    for i in range(k):
        Q = qr_hessenberg(A)
        A[:, :] = A @ Q
    return np.diagonal(A)
```

We apply this to the above matrix and consider the results from the first iterations

```
A_0=
[[ 4.00759162e+02  1.23633693e+02  2.96068783e-14  6.09597216e-15]
 [ 1.23633693e+02  4.41337578e+02  3.21310279e+01 -3.62480544e-14]
 [ 3.12483731e-14  3.21310279e+01 -4.20816954e+01 -1.22497513e+02]
 [-6.82657267e-15 -1.21878868e-14 -1.22497513e+02  9.98495521e+00]]


A_1=
[[ 4.73938190e+02  1.13970724e+02  1.91482189e-15  1.80543844e-14]
 [ 1.13970724e+02  3.70410752e+02 -1.08311338e+01 -5.07842115e-15]
 [-2.98597627e-14 -1.08311338e+01 -7.46803080e+01 -1.11052737e+02]
 [ 1.01160781e-14  9.60055188e-15 -1.11052737e+02  4.03313663e+01]]


A_2=
[[ 5.20096493e+02  7.80160200e+01  1.23209238e-14 -1.25066927e-14]
 [ 7.80160200e+01  3.24515418e+02  4.34959013e+00 -3.78291085e-14]
 [ 2.94474464e-14  4.34959013e+00 -9.87169105e+01 -9.49422609e+01]
 [-1.20803921e-14 -7.42681841e-15 -9.49422609e+01  6.41050000e+01]]


...


A_9=
[[ 5.47401220e+02  1.21920894e+00 -1.20882609e-14  2.00836013e-14]
 [ 1.21920894e+00  2.97261149e+02 -2.16530512e-02  1.38847046e-14]
 [-2.91300457e-14 -2.16530512e-02 -1.41346782e+02 -1.62521231e+01]
 [ 1.38390076e-14  3.05621674e-15 -1.62521231e+01  1.06684413e+02]]
```

For the diagonal elements of the matrix A_9, i.e., after 10 iterations, we have

$$a_{11}^{(9)} \approx 547.401, \quad a_{22}^{(9)} \approx 297.261, \quad a_{33}^{(9)} \approx -141.346, \quad a_{44}^{(9)} \approx 106.684. \qquad (5.8)$$

These diagonal elements represent excellent approximations to *all* eigenvalues of the matrix A:

$$\frac{|a_{11}^{(10)} - \lambda_1|}{|\lambda_1|} \approx 1.085 \cdot 10^{-5}, \qquad \frac{|a_{22}^{(10)} - \lambda_2|}{|\lambda_2|} \approx 1.998 \cdot 10^{-5},$$

$$\frac{|a_{33}^{(10)} - \lambda_3|}{|\lambda_3|} \approx 0.0098, \qquad \frac{|a_{44}^{(10)} - \lambda_4|}{|\lambda_4|} \approx 0.0074.$$

The error for λ_3 and λ_4 is larger, as these eigenvalues are less separated. With a good shift strategy, the convergence can be further accelerated. ◀

5.6   Excursus: Natural Vibrations of a Multiple Spring Pendulum

In this section, we study the dynamics of spring pendulums, see Fig. 5.2. When describing mechanical systems like the spring pendulum, the eigenvalues of the system often play an important role. Here, they will explain the occurrence of the resonance effect, i.e., an amplification of acting forces.

A (mathematical) spring pendulum is a point-mass m (i.e., it has no extension), which hangs on a massless spring of length l. Due to the action of a force, e.g., the gravitational force, the mass is moved and thereby the spring is extended. The spring opposes this extension u with a force F, and we simplify by assuming that the force is proportional to the extension (and also to the compression of the spring), i.e.,

$$F(u) = \kappa \cdot u,$$

where $\kappa > 0$ is the so-called *spring constant*. This linear law is called *Hooke's law*.

Mathematical Modeling

To derive an equation for the spring, we apply the principle of *energy conservation*: The total energy of the system remains constant over time. We divide the energy into the *kinetic energy*

$$E_{kin}(t) = \frac{1}{2}mu'(t)^2,$$

where $v(t) := u'(t)$ is the velocity of the mass, and into the *potential energy*, which is given by the extension of the spring:

$$E_{pot}(t) = \frac{1}{2}\kappa u(t)^2$$

For the total energy $E(t) = E_{kin}(t) + E_{pot}(t)$, the conservation equation applies

$$0 \stackrel{!}{=} \frac{\mathrm{d}}{\mathrm{d}t}E(t) = mu''(t)u'(t) + \kappa u'(t)u(t) = \big(mu''(t) + \kappa u(t)\big) \cdot u'(t).$$

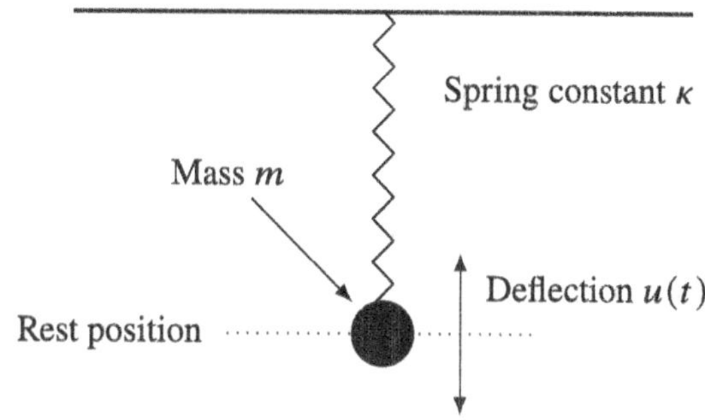

Fig. 5.2 Configuration of a mathematical spring pendulum. A mass m hangs on a massless spring. The force of the spring is proportional to the displacement $F = \kappa \cdot u$

For this condition to be fulfilled for all possible values of $u'(t)$, i.e., for all velocities, the *differential equation* must be fulfilled everywhere

$$u''(t) + \lambda^2 u(t) = 0, \quad \lambda^2 := \frac{\kappa}{m} > 0. \tag{5.9}$$

We are looking for a function $u(t)$, whose second derivative is exactly equal to $-\lambda^2$ times the function itself:

$$u''(t) = -\lambda^2 u(t)$$

The trigonometric functions are candidates for a solution, and we make the assumption:

$$u(t) = c_1 \sin(\omega_1 t) + c_2 \cos(\omega_2 t), \quad c_1, c_2 \in \mathbb{R}$$

with

$$u''(t) = -c_1 \omega_1^2 \sin(\omega_1 t) - c_2 \omega_2^2 \cos(\omega_2 t).$$

As we can easily see, a total of four unknown parameters must be specified. This is done using the material properties and the initial data. We can immediately see that

$$\omega_1 = \omega_2 = \lambda = \sqrt{\frac{\kappa}{m}}$$

must apply. This results in

$$u\left(t + \frac{2\pi}{\lambda} k\right) = u(t), \quad k \in \mathbb{Z}, \tag{5.10}$$

i.e., $2\pi/\lambda$ is the period of the oscillation and $\lambda/(2\pi)$ is the frequency of the spring pendulum.

We now must determine the two constants c_1 and c_2. To define these uniquely, conditions are missing. To obtain a well-defined system, we must define the state of the spring, i.e., its position and velocity, at the beginning of the observation. For instance, we assume that the mass is initially in position u^0 and at rest, i.e., $v(0) = u'(0) = 0$. From this, we get

$$u^0 = u(0) = c_2,$$
$$0 = v(0) = u'(0) = c_1.$$

The solution of the spring equation with these *initial conditions* is thus

$$u(t) = u^0 \cos\left(\sqrt{\frac{\kappa}{m}} t\right).$$

Model With Many Springs

Next, we assume that the system consists of N masses m_i, each experiencing a displacement $u_i(t)$. The masses m_{i-1} and m_i are each connected with a spring with spring constant κ

(for simplification, all springs are described with the same constant κ), see Fig. 5.3. We determine the kinetic energy as the sum of all partial energies

$$E_{kin}(t) = \sum_{i=1}^{N} \frac{1}{2} m_i u_i'(t)^2$$

and the potential energy as the sum of the potential energies of the individual springs, see Fig. 5.3,

$$E_{pot}(t) = \sum_{i=1}^{N} \frac{1}{2} \kappa (u_i(t) - u_{i-1}(t))^2,$$

where $u_0(t) \equiv 0$ is introduced just to simplify the notation. We apply the principle of energy conservation. For the derivative of the total energy $E(t) = E_{kin}(t) + E_{pot}(t)$, we have

$$E'(t) = \sum_{i=1}^{N} \left(m_i u_i''(t) u_i'(t) + \kappa (u_i'(t) - u_{i-1}'(t))(u_i(t) - u_{i-1}(t)) \right)$$

$$= \sum_{i=1}^{N-1} u_i'(t) \left(m_i u''(i) + \kappa (u_i(t) - u_{i-1}(t)) - \kappa (u_{i+1}(t) - u_i(t)) \right)$$

$$+ u_N'(t) \left(m_N u_N''(t) + \kappa (u_N(t) - u_{N-1}(t)) \right).$$

From the condition $E'(t) \equiv 0$ follows (for $N > 1$) the system of differential equations

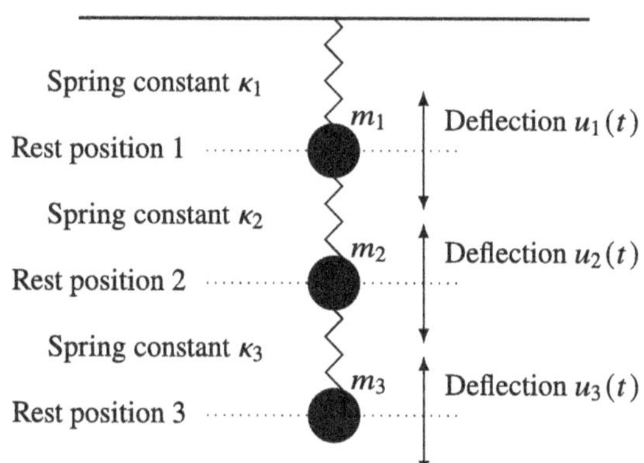

Fig. 5.3 Configuration of a mathematical spring pendulum with three masses m_1, m_2, m_3 and three springs with constants $\kappa_1, \kappa_2, \kappa_3$. We describe the displacement $u_i(t)$ of each mass relative to its rest position

$$u_1''(t) + \frac{2\kappa}{m_1}u_1(t) - \frac{\kappa}{m_1}u_2(t) = 0,$$

$$u_2''(t) - \frac{\kappa}{m_2}u_1(t) + \frac{2\kappa}{m_2}u_2(t) - \frac{\kappa}{m_2}u_3(t) = 0,$$

$$\vdots$$

$$u_i''(t) - \frac{\kappa}{m_i}u_{i-1}(t) + \frac{2\kappa}{m_i}u_i(t) - \frac{\kappa}{m_i}u_{i+1}(t) = 0,$$

$$\vdots$$

$$u_{N-1}''(t) - \frac{\kappa}{m_{N-1}}u_{N-2}(t) + \frac{2\kappa}{m_{N-1}}u_{N-1}(t) - \frac{\kappa}{m_{N-1}}u_N(t) = 0,$$

$$u_N''(t) - \frac{\kappa}{m_N}u_{N-1}(t) + \frac{\kappa}{m_N}u_N(t) = 0.$$

If only one spring exists, i.e., $N = 1$, we recover

$$u_1''(t) + \frac{\kappa}{m_1}u_1(t) = 0.$$

For the following we will always assume $N > 1$. By introducing the vector $u(t) \in \mathbb{R}^N$ we can compactly write the equation as

$$u''(t) + Au(t) = 0, \quad A = \kappa \begin{pmatrix} \frac{2}{m_1} & -\frac{1}{m_1} & 0 & \cdots & 0 \\ -\frac{1}{m_2} & \frac{2}{m_2} & -\frac{1}{m_2} & \ddots & \vdots \\ 0 & \ddots & \ddots & \ddots & 0 \\ \vdots & \ddots & -\frac{1}{m_{N-1}} & \frac{2}{m_{N-1}} & -\frac{1}{m_{N-1}} \\ 0 & \cdots & 0 & -\frac{1}{m_N} & \frac{1}{m_N} \end{pmatrix}. \tag{5.11}$$

Theorem 5.40 (Estimation of the Eigenvalues of the Spring Matrix) *The matrix A, given by (5.11), is always regular. All eigenvalues μ of A are real and positive and the estimate*

$$0 < \mu \le \max\left\{ \frac{\kappa}{m_N}, \max_{i<N} \frac{2\kappa}{m_i} \right\}$$

holds. In the case of uniform spring masses $m_i \equiv m$ for all i, the matrix A, given by (5.11), is symmetric positive definite.

Proof *(i)* We first show that the matrix A is regular. For this, let $M \in \mathbb{R}^{N \times N}$ be the regular diagonal matrix

$$M = \operatorname{diag}(m_1, \ldots, m_N).$$

Then, we have

$$
A = M^{-1}\kappa \underbrace{\begin{pmatrix} 2 & -1 & 0 & \cdots & 0 \\ -1 & 2 & -1 & \ddots & \vdots \\ 0 & \ddots & \ddots & \ddots & 0 \\ \vdots & \ddots & -1 & 2 & -1 \\ 0 & \cdots & 0 & -1 & 1 \end{pmatrix}}_{=:\bar{A}}.
$$

The matrix $\bar{A}$ is regular as the diagonal matrix M is regular and since $\bar{A}$ is symmetric positive definite, see Theorem 3.83.

(ii) Let $\mu \in \mathbb{C}$ be an eigenvalue of A with corresponding eigenvector $w \in \mathbb{C}^N$. Then

$$
\mu w = A w = \kappa M^{-1} \bar{A} w. \tag{5.12}
$$

We define the regular diagonal matrix

$$
M^{\frac{1}{2}} = \operatorname{diag}(\sqrt{m_1}, \ldots, \sqrt{m_N}).
$$

Multiplication of (5.12) with $M^{\frac{1}{2}}$ yields

$$
\mu(M^{\frac{1}{2}} w) = M^{-\frac{1}{2}} \kappa \bar{A} w = [M^{-\frac{1}{2}} \kappa \bar{A} M^{-\frac{1}{2}}](M^{\frac{1}{2}} w).
$$

Thus, μ is also an eigenvalue of the matrix

$$
\kappa M^{-\frac{1}{2}} \bar{A} M^{-\frac{1}{2}}.
$$

This matrix is symmetric, so μ must be real.

(iii) The estimation of the eigenvalues is done using the Gershgorin circle theorem, Theorem 5.4. In the case $m_i = m$, the matrix is symmetric. $\qquad\square$

Let $\lambda_1^2, \ldots, \lambda_N^2 \in \mathbb{R}_+$ be the positive real eigenvalues of the matrix A. We can easily show that the matrix is diagonalizable. Because there exists an orthogonal matrix $\bar{W}$, such that for the symmetric positive definite matrix $\kappa\, M^{-\frac{1}{2}} \bar{A} M^{-\frac{1}{2}}$ it holds:

$$
\bar{W}^T [M^{-\frac{1}{2}} \kappa \bar{A} M^{-\frac{1}{2}}] \bar{W} = \Lambda
$$

with a diagonal matrix

$$
\Lambda = \operatorname{diag}(\lambda_1^2, \ldots, \lambda_N^2).
$$

Since $\bar{W}^T = \bar{W}^{-1}$, it follows

$$
\Lambda = \bar{W}^{-1} [M^{-\frac{1}{2}} \kappa \bar{A} M^{-\frac{1}{2}}] \bar{W} = \underbrace{\bar{W}^{-1} M^{\frac{1}{2}}}_{=:W^{-1}} \underbrace{M^{-1} \kappa \bar{A}}_{=A} \underbrace{M^{-\frac{1}{2}} \bar{W}}_{=:W}.
$$

The matrix A is thus diagonalizable with the regular matrix $W = M^{-\frac{1}{2}} \bar{W}$.

With this matrix W, we can transform the differential equation. We define

$$\bar{u}(t) := W u(t)$$

and by multiplying the differential equation from the left with W^{-1}, we get the diagonalized system

$$\bar{u}''(t) + \Lambda \bar{u}(t) = 0$$

or elementwise as

$$\bar{u}_i''(t) + \lambda_i^2 \bar{u}_i(t) = 0, \quad i = 1, \ldots, N.$$

This system is solved using the ansatz

$$\bar{u}_i(t) = a_i \sin(\lambda_i t) + b_i \cos(\lambda_i t).$$

Assuming that the deflection u^0 is given at time $t = 0$ and that the velocity is zero, the solution is given as

$$\bar{u}_i(t) = \bar{u}_i^0 \cos(\lambda_i t), \quad u(t) = W^{-1} \bar{u}(t).$$

The roots of the eigenvalues determine the frequencies of the individual oscillation components.

Remark 5.41 (*Significance of Natural Vibrations*) The eigenvalues of mechanical systems are of great importance. We now assume that the spring (we consider a single one) is not only deflected at the initial time, but that a force is constantly acting. The differential equation (5.9) then writes as

$$u''(t) + \lambda^2 u(t) = f(t), \quad u(0) = u^0, \quad u'(0) = 0, \tag{5.13}$$

where $f(t)$ is the acting force. We assume that this force is arbitrarily small, $|f(t)| \le \epsilon$, with an $\epsilon > 0$, but oscillates at the same frequency:

$$f(t) = \epsilon \sin(\lambda t)$$

For the approach

$$u(t) = u^0 \cos(\lambda t) + \frac{\epsilon}{2\lambda} t \sin(\lambda t),$$

we can verify that a solution of Eq. (5.13) is given. For this solution, the following holds

$$|u(t)| \to \infty \quad (t \to \infty).$$

No matter how small the periodically acting force $|f(t)| \le \epsilon$, the oscillation is always amplified. This effect is called *resonance* and occurs when a periodic force exactly hits a natural frequency. It plays a role in many structures that are subject to regular loads.

For instance, strong loads can result from wind, football stadiums with jumping fans, or oil platforms in strong, regular wave motion. In vehicles, the engine vibration should not excite the exhaust system, as the vibrations would cause cracks in the long term, leading to breakage. When designing structures and parts, care must be taken that all relevant natural frequencies lie in a range that is not periodically excited by the expected load, in order to avoid natural vibrations. For this reason, the largest and smallest eigenvalues play a special role. ♦

Uniform Multiple Spring Pendulum

We start with the simple case

$$m_i = m, \quad \kappa_i = \kappa, \quad i = 1, \ldots, N.$$

Then the matrix is given as

$$A = \frac{\kappa}{m} \begin{pmatrix} 2 & -1 & 0 & \cdots & 0 \\ -1 & 2 & -1 & \ddots & \vdots \\ 0 & \ddots & \ddots & \ddots & 0 \\ \vdots & \ddots & -1 & 2 & -1 \\ 0 & \cdots & 0 & -1 & 1 \end{pmatrix}.$$

We choose $m = 1$ and $\kappa = 5$. In the following, we will calculate the largest and smallest eigenvalues λ^2_{max} and λ^2_{min} of this matrix. The Gershgorin Theorem 5.4 gives us the bound

$$0 \le \lambda_i^2 \le \frac{4\kappa}{m} \le 20.$$

We can thus estimate the highest eigenfrequency occurring in the system as

$$F_{max} = \frac{\lambda_{max}}{2\pi} \le \frac{\sqrt{20}}{2\pi} \approx 0.71.$$

To calculate the smallest eigenvalue, we use the inverse iteration.

N	1	2	4	8	16
λ^2_{min}	5	1.91	0.60	0.17	0.045
F_{min}	0.35	0.22	0.12	0.066	0.034
λ^2_{max}	5	13.01	17.58	18.53	18.55
F_{max}	0.35	0.58	0.67	0.69	0.69

For increasing numbers of springs, the smallest eigenvalue and thus the smallest frequency decrease. The largest eigenvalue slowly approaches the limit 20.

Numerical Approximation of the Differential Equation

Even though methods for the numerical approximation of differential equations fill entire books, we want to briefly describe a simple way to approximate the solution of

$$u''(t) + Au(t) = 0 \text{ for } t > 0 \text{ and } u(0) = u^0 \in \mathbb{R}^N \text{ and } v(0) := u'(0) = 0.$$

The basic idea of an approximation is that we are not looking for the entire function $u(t)$, but only for individual discrete function values $u(t_n)$ for

$$t_n = n \cdot h, \quad n = 0, 1, 2, \ldots$$

with a parameter $h > 0$, which we call the *grid size*. A Taylor expansion of the solution yields

$$u(t_{n+1}) = u(t_n + h) = u(t_n) + hu'(t_n) + \frac{h^2}{2}u''(t_n) + O(h^3).$$

We can express the first derivative by the velocity $v(t_n) = u'(t_n)$ and the second derivative by the differential equation $u''(t_n) = -Au(t_n)$. So we get

$$u(t_{n+1}) = u(t_n) + hv(t_n) - \frac{h^2}{2}Au(t_n) + O(h^3). \tag{5.14}$$

Applied to the numerical approximation $u_n \approx u(t_n)$ and $v_n \approx v(t_n)$ we get the iteration

$$u_{n+1} = \left(I - \frac{h^2}{2}A\right)u_n + hv_n.$$

If we know u_n and v_n, u_{n+1} can be easily calculated. Afterwards, we need an approximation of v_{n+1} which we determine as

$$v(t_{n+1}) = v(t_n) + \int_{t_n}^{t_{n+1}} v'(s)ds,$$

approximated with the *trapezoidal rule*, see Definition 10.4. It holds

$$v(t_{n+1}) = v(t_n) + \int_{t_n}^{t_{n+1}} v'(s)ds \approx v(t_n) + \frac{h}{2}v'(t_n) + \frac{h}{2}v'(t_{n+1}) + O(h^3)$$

With $v'(t) = u''(t) = -Au(t)$ we finally get

$$v(t_{n+1}) = v(t_n) - \frac{h}{2}Au(t_n) - \frac{h}{2}Au(t_{n+1}) + O(h^3). \tag{5.15}$$

Combined, the following simple iteration results:

$$\left.\begin{array}{ll}(i) & u_{n+1} = u_n - \dfrac{h^2}{2} A u_n + h v_n \\[2ex] (ii) & v_{n+1} = v_n - \dfrac{h}{2} A u_n - \dfrac{h}{2} A u_{n+1}\end{array}\right\} \quad n = 1, 2, \dots \qquad (5.16)$$

Theorem 5.42 (Convergence of the Approximation Rule) *For exact initial values $u_0 = u(0)$ and $v_0 = v(0)$, the error estimate for the approximation generated by (5.16) is*

$$\max_{1 \le k \le n} \{\|u(t_k) - u_k\|, \|v(t_k) - v_k\|\} \le c(t_n) h^2,$$

with a constant depending on the interval length t_n.

Proof Let $n \in \mathbb{N}$ and $t_n = hn$. If we insert the true solution $u(t)$ and $v(t)$ at time t_n into the numerical approximation rule, a remainder term remains:

$$\begin{aligned} u(t_{n+1}) &= u(t_n) + h v(t_n) - \frac{h^2}{2} A u(t_n) + O(h^3) \\[2ex] v(t_{n+1}) &= v(t_n) - \frac{h}{2} A u(t_n) - \frac{h}{2} A u(t_{n+1}) + O(h^3). \end{aligned} \qquad (5.17)$$

This gives an estimate for the error that is developing in each step, given that the initial value $u(t_n)$ and $v(t_n)$ were still exact. We call this local error the *truncation error*.

To get an estimate for the real discretization errors that also takes into account the accumulation of errors in all steps, we introduce the notation

$$e_n^u := u(t_n) - u_n, \quad e_n^v := v(t_n) - v_n$$

From (5.16) and (5.17), we get:

$$\begin{array}{ll}(i) & e_{n+1}^u = e_n^u - \dfrac{h^2}{2} A e_n^u + h e_n^v + O(h^3) \\[2ex] (ii) & e_{n+1}^v = e_n^v - \dfrac{h}{2} A e_n^u - \dfrac{h}{2} A e_{n+1}^u + O(h^3)\end{array}$$

If we substitute the first into the second line, we can eliminate e_{n+1}^u:

$$e_{n+1}^v = e_n^v - \frac{h}{2} A e_n^u - \frac{h}{2} A e_n^u - \frac{h^2}{2} A e_n^v + \frac{h^3}{4} A e_n^u + O(h^3)$$

From (i) and this error formula for e_n^v we get in the spectral norm

$$\|e_{n+1}^u\|_2 \le \left\|I - \frac{h^2}{2} A\right\|_2 \|e_n^u\|_2 + h \|e_n^v\|_2 + O(h^3),$$

$$\|e_{n+1}^v\|_2 \le \left\|I - \frac{h^2}{2} A\right\|_2 \|e_n^v\|_2 + \left(h + \frac{h^3}{4}\right) \|A\|_2 \|e_n^u\|_2 + O(h^3).$$

Now let $E_n := \max\{\|e_n^u\|_2, \|e_n^v\|_2\}$ to estimate

$$\|E_{n+1}\| \leq \left(\|I - \frac{h^2}{2}A\|_2 + h + \left(h + \frac{h^3}{4}\right)\|A\|_2\right)\|E_n\|_2 + O(h^3).$$

In Theorem 5.40, we showed that the eigenvalues λ^2 of the matrix A are all real and positive. For $0 < h < 2/\sqrt{spr(A)}$ it holds

$$\|I - \frac{h^2}{2}A\|_2 \leq 1.$$

With this, we get

$$\|E_{n+1}\|_2 \leq \alpha(h)\|E_n\|_2 + O(h^3), \quad \alpha(h) := 1 + h + \left(h + \frac{h^3}{4}\right)\|A\|_2. \tag{5.18}$$

Repeated application of this estimate gives:

$$\|E_n\|_2 \leq \alpha(h)^n\|E_0\| + O(h^3) \cdot \sum_{k=0}^{n-1}\alpha(h)^k.$$

It is $E_0 = 0$, since the initial values are known. Further,

$$\alpha(h)^k = \left(1 + h + \left(h + \frac{h^3}{4}\right)\|A\|_2\right)^k = e^{k\log\left(1+h+\left(h+\frac{h^3}{4}\right)\|A\|_2\right)}.$$

Because $\log(1 + x) \leq x$ for $x > 0$, with $k = t_k/h$ we get

$$\alpha(h)^k \leq e^{t_k\left(1+\left(1+\frac{h^2}{4}\right)\|A\|_2\right)} \leq e^{c(A)t_k} \leq e^{c(A)t_n}.$$

With (5.18), we finally obtain the result by summation

$$\|E_n\|_2 \leq O(h^3) \cdot ne^{c(A)t_n} \leq O(h^2)t_ne^{c(A)t_n}.$$

$\square$

Remark 5.43 The numerical approximation scheme examined here is a so-called *difference method for initial value problems*, see e.g. [26]. Our method is of second order (in the grid size h) and the analysis is already rather complex.

The differential equation $u'' + Au = 0$ is of second order. This means that the second derivative of $u(t)$ appears. By introducing the velocity $v = u'$ as an auxiliary variable, we have rewritten the equation into a *first-order system*. The method examined here belongs to the *Runge-Kutta methods*. There is extensive literature on the numerics of ordinary differential equations [80, 86]. $\blacklozenge$

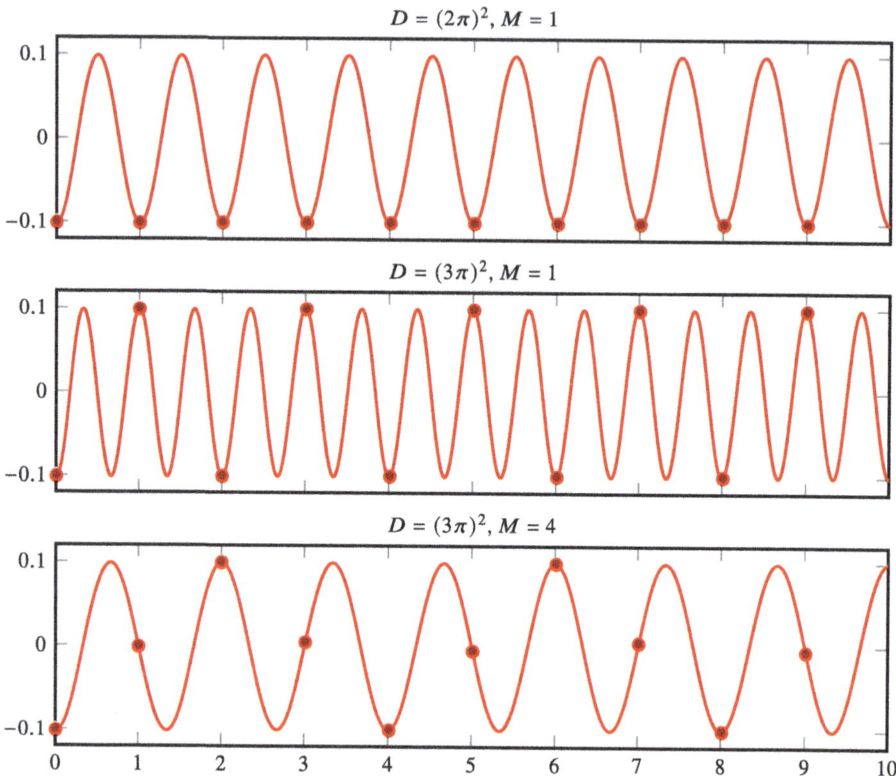

Fig. 5.4 Oscillation of a simple spring pendulum with different spring constants and different masses. In addition to the course of the spring position, the points indicate the position of the spring at each second. The theoretically expected frequencies can be seen in each case. Top: $F = 2\pi/\sqrt{(2\pi)^2/1} = 1$. Middle: $F = 2\pi/\sqrt{(3\pi)^2/1} = 2/3$. Bottom: $F = 2\pi/\sqrt{(3\pi)^2/4} = 4/3$

We now perform some simple simulations. First, we study the oscillation of a simple pendulum, which is initially deflected with $u(0) = -0.1$ and $v(0) = 0$. In Fig. 5.4, we show three different situations. At the very top, the spring has the spring constant $\kappa = (2\pi)^2$ and the mass $m = 1$. In the middle, $\kappa = (3\pi)^2$ and still $m = 1$, at the very bottom $\kappa = (3\pi)^2$ and $m = 4$. The frequencies expected according to (5.10) are obtained in each case. For the case $\kappa = (3\pi)^2$ and $m = 4$, we evaluate the numerically calculated deflection and velocity at $T = 1/2$ for different grid sizes $h > 0$ and show the error. The exact solution is given by $u(t) = u^0 \cos(\sqrt{\kappa/m}t)$ through

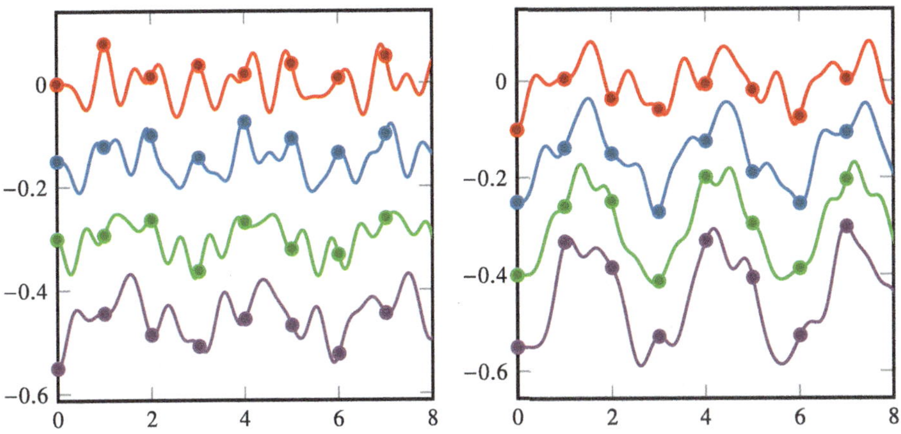

Fig. 5.5 Oscillation of a four-fold spring pendulum. In both figures, $\kappa = (2\pi)^2$ and $m = 1$. We examine two initial conditions: On the left, the lowest mass is displaced by -0.1, on the right, all masses are initially displaced by -0.1

$$u\left(\frac{1}{2}\right) \approx 0.707107, \quad v\left(\frac{1}{2}\right) \approx 0.333216.$$

h	0.1	0.05	0.025	0.0125
$|u(10) - u_N|$	$1.56 \cdot 10^{-3}$	$3.87 \cdot 10^{-4}$	$9.64 \cdot 10^{-5}$	$2.41 \cdot 10^{-5}$
$|v(10) - v_N|$	$1.67 \cdot 10^{-2}$	$4.14 \cdot 10^{-3}$	$1.03 \cdot 10^{-3}$	$2.58 \cdot 10^{-4}$

The quadratic order, i.e., a quartering of the error when halving the step size, is clearly visible.

As a second example, we consider a quadruple spring pendulum. We choose the uniform spring constant

$$\kappa = (2\pi)^2$$

and initially, also a uniform weight distribution

$$m_1 = m_2 = m_3 = m_4 = 1.$$

In Fig. 5.5, we show the oscillation behavior of the quadruple spring pendulum. We choose different initial values. On the left, we have

$$u_1(0) = u_2(0) = u_3(0) = 0, \quad u_4(0) = -0.1$$

and on the right

$$u_1(0) = u_2(0) = u_3(0) = u_4(0) = -1.$$

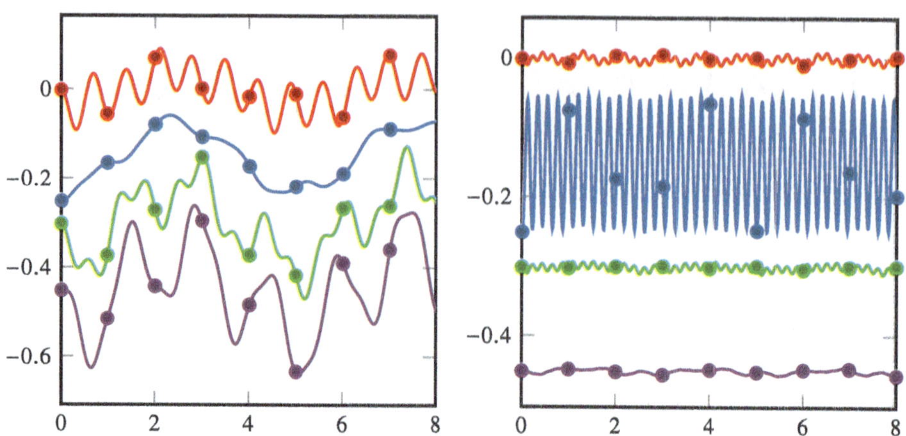

Fig. 5.6 Oscillation of a four-spring pendulum. In both figures, $\kappa = (2\pi)^2$. The second spring is initially deflected. On the left, $m_2 = 10$, and on the right, $m_2 = 0.1$

The oscillation behavior is very complex and hardly predictable.

In the following example, we keep $\kappa = (2\pi)^2$ for all springs. Of the four masses, we choose three to be equal, with the second one either larger or smaller:

$$m_1 = m_3 = m_4 = 1, \quad m_2 = 10 \text{ or } m_2 = 0.1.$$

Initially, only the second spring is displaced, so

$$u_1(0) = u_3(0) = u_4(0) = 0, \quad u_2(0) = -0.1.$$

In Fig. 5.6, we show the respective oscillation behavior. This example shows that the eigenvalues of the matrix depend significantly on the mass. The calculation of the eigenvalues using the power iteration or the inverse iteration gives

$$
\begin{aligned}
m = 0.1 \quad & \lambda^2 \in [5.63, 831.1], \quad && F \in [0.37, 5.59], \\
m = 10 \quad & \lambda^2 \in [1.58, 88.8], \quad && F \in [0.2, 1.5].
\end{aligned}
\tag{5.19}
$$

In the right figure, a very high frequency occurs.

Finally, we want to examine the excitation of natural oscillations briefly. For this, we continue to consider the example with $\kappa = (2\pi)^2$, $m_1 = m_3 = m_4 = 1$ and $m_2 = 0.1$. At the beginning, let the mass m_4 be deflected by $u_4(0) = -0.1$. For the other deflections, let $u_1(0) = u_2(0) = u_3(0) = 0$. We now drive the dynamics through a force and change the differential equation to

$$u''(t) + Au(t) = f(t), \quad u(0) = u^0, \quad u'(0) = 0.$$

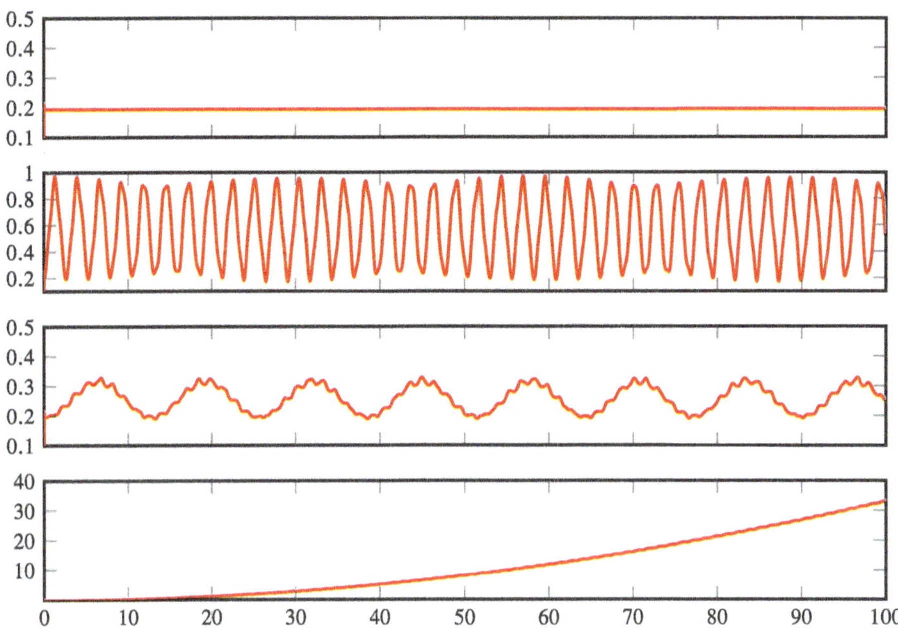

Fig. 5.7 Excitation of multiple springs by a force f. From top to bottom: $f_1 = 0$ and $f_2 = -0.1$ constant forces, $f_3(t) = 0.1 \cos(2\pi \cdot 0.3)$ and $f_4(t) = 0.1 \cos(2\pi \cdot 0.3777t)$. The oscillation is stable, unless—as with $f_4(t)$—exactly one eigenfrequency is excited. Here, resonance occurs

Instead of looking at the course of the oscillations, we now consider the total energy of the system, i.e., the expression

$$E_n = \sum_{i=1}^{N} \frac{1}{2} m_i |u_i'(t_n)|^2 + \sum_{i=1}^{N} \frac{1}{2} \kappa |u_i(t_n) - u_i(t_{n-1})|^2,$$

consisting of kinetic and potential energy. At the top of Fig. 5.7, we show the course of the energy for the constant force $f_1 = 0$. Here, no external force is acting, and the principle of energy conservation states that $E_n = E_0$ must hold for all $n \geq 0$. With $\kappa = (2\pi)^2$ and the initial deflections $u_1(0) = u_2(0) = u_3(0) = 0$ and $u_4(0) = -0.1$ we get

$$E_0 = \frac{(2\pi)^2}{2}(-0.1)^2 \approx 0.197,$$

as can be read from the top graph of Fig. 5.7. The second figure belongs to the constant force $f_2 = -1$. The energy is not constant, but it stays bounded. This case corresponds to the effect of gravity. This naturally deflects the balls, but it does not lead to a constant amplification of the oscillations. The third representation in Fig. 5.7 corresponds to the oscillatory force $f_3(t) = \sin(2\pi \cdot 0.3t)$. Again, the energy in the system changes, but remains bounded. The

0.3 describes a frequency that is outside the spectrum of this multiple spring pendulum. We found approximately 0.37 as the smallest and 5.59 as the largest frequency, see 5.19. There is no resonance effect. Finally, the oscillating force $f_4(t) = \sin(2\pi \cdot 0.3777t)$ generates the bottom graph in Fig. 5.7. This frequency corresponds to the smallest eigenfrequency of the system. The energy increases very quickly.

If the exciting force hits one of the system's eigenfrequencies, even the smallest force eventually leads to an arbitrarily large deflection of the springs.

5.7   Singular Value Decomposition

Eigenvalues are defined for square matrices, i.e., for mappings $f : V \to V$ from a vector space to itself. In this section, we will extend this concept and introduce *singular values* for general rectangular matrices $A \in \mathbb{R}^{n \times m}$.

In Sect. 4.5, we dealt with overdetermined linear systems $Ax = b$ with $A \in \mathbb{R}^{n \times m}$ and $n > m$. Access to a solution concept was the normal equation system $A^T A x = A^T b$. We consider the matrix $A^T A \in \mathbb{R}^{m \times m}$, regardless of whether n or m is larger, or whether the matrix is square. The symmetric positive semi-definite matrix $A^T A$ has a system of real and nonnegative eigenvalues as well as an orthonormal basis $\{v_1, \ldots, v_m\}$ with $v_i \in \mathbb{R}^m$ for $i = 1, \ldots, m$ with

$$A^T A v_i = \lambda_i v_i, \quad (v_i, v_j) = \delta_{ij}, \quad i, j = 1, \ldots, m. \tag{5.20}$$

Without loss of generality, we assume that the eigenvalues are sorted, i.e.,

$$\lambda_1 \geq \lambda_2 \geq \cdots \geq \lambda_r > 0, \tag{5.21}$$

where $1 \leq r \leq m$ is the rank of the matrix $A^T A$ and also of the matrix A, see Theorem 2.18. Further, $\lambda_{r+1} = \cdots = \lambda_m = 0$.

Definition 5.44 (*Singular Values*) Let $A \in \mathbb{R}^{n \times m}$. Furthermore, let

$$\lambda_1 \geq \lambda_2 \geq \cdots \geq \lambda_r > 0$$

be the non-zero eigenvalues of $A^T A$. Then, $\sigma_i = \sqrt{\lambda_i}$ for $1 \leq i \leq r$ are called the *singular values* of the matrix A.

Theorem 5.45 (Singular Values) *Let $A \in \mathbb{R}^{n \times m}$ with rank $r \leq \min\{n, m\}$. Then there exists a system of r singular values $\sigma_1 \geq \cdots \geq \sigma_r > 0$ and $\sigma_{r+1} = \cdots = \sigma_m = 0$ (in the case $r < m$), as well as two orthonormal systems $\{v_1, \ldots, v_m\}$ with $v_i \in \mathbb{R}^m$ and $\{u_1, \ldots, u_r\}$ with $u_i \in \mathbb{R}^n$, such that*

$$A^T A v_i = \sigma_i^2 v_i \quad i = 1, \ldots, m, \tag{5.22}$$

$$A v_i = \sigma_i u_i \quad i = 1, \ldots, r, \tag{5.23}$$

$$A^T u_i = \sigma_i v_i \quad i = 1, \ldots, r. \tag{5.24}$$

The vectors $u_1, \ldots, u_r$ span the image space of A.

Proof With the $r = \text{rank}(A)$ positive eigenvalues λ_i of $A^T A$, we define the singular values $\sigma_i = \sqrt{\lambda_i}$, see (5.20)-(5.21). Since $A^T A$ is symmetric positive semi-definite, there exists an orthonormal system of eigenvectors $v_1, \ldots, v_m$ for which (5.22) holds. For $i = 1, \ldots, r$, we define

$$u_i = \frac{1}{\sigma_i} A v_i, \tag{5.25}$$

which form an orthonormal system, because

$$(u_i, u_j) = \frac{1}{\sigma_i \sigma_j}(A v_i, A v_j) = \frac{\sigma_i^2}{\sigma_i \sigma_j}(v_i, v_j) = \delta_{ij}.$$

With (5.25), we directly obtain (5.23) and multiplication with A^T yields (5.24).

The rank r of $A^T A$ corresponds to the rank of A, such that $u_1, \ldots, u_r$ span the image space of A. $\qquad\square$

Remark 5.46 (*Singular Values of Symmetric Matrices*) For a square symmetric matrix $A \in \mathbb{R}^{n \times n}$ with $A^T = A$, the singular values are exactly the absolute values of the eigenvalues. $\qquad\blacklozenge$

The orthonormal system $\{u_1, \ldots, u_r\}$ introduced in Theorem 5.45 can be extended to an orthonormal basis of $\mathbb{R}^n$ with vectors $u_{r+1}, \ldots, u_n$. These vectors span the kernel of A^T, i.e., $A^T u_j = 0$ for $j = r + 1, \ldots, n$. Thus, we have

$$(u_j, A v_i) = \begin{cases} \sigma_i & 1 \leq i = j \leq r \\ 0 & 1 \leq i \leq m, \ 1 \leq j \neq i \leq n, \end{cases}$$

or, if we introduce the two orthogonal matrices $V \in \mathbb{R}^{m \times m}$ and $U \in \mathbb{R}^{n \times n}$

$$U^T A V = \Sigma := \begin{pmatrix} \tilde{\Sigma} & 0 \\ 0 & 0 \end{pmatrix} \in \mathbb{R}^{n \times m}, \tag{5.26}$$

where $\tilde{\Sigma} \in \mathbb{R}^{r \times r}$ is with $\tilde{\Sigma} = \text{diag}(\sigma_1, \ldots, \sigma_r)$.

Definition 5.47 (*Singular Value Decomposition*) Let $A \in \mathbb{R}^{n \times m}$. A decomposition of A of the form

$$A = U \Sigma V^T$$

into the orthogonal matrices $V \in \mathbb{R}^{m \times m}$ and $U \in \mathbb{R}^{n \times n}$ and $\Sigma \in \mathbb{R}^{n \times m}$ according to (5.26) is called singular value decomposition.

Theorem 5.48 (Singular Value Decomposition) *For every matrix $A \in \mathbb{R}^{n \times m}$ there exists a singular value decomposition with two orthonormal matrices $V \in \mathbb{R}^{m \times m}$ and $U \in \mathbb{R}^{n \times n}$ and the diagonal matrix $\Sigma \in \mathbb{R}^{n \times m}$ with $\Sigma_{ii} = \sigma_i > 0$ for $i = 1, \ldots, r$ and $\Sigma_{ij} = 0$ otherwise, such that*

$$U^T A V = \Sigma.$$

The singular values $\sigma_1 \geq \cdots \geq \sigma_r > 0$ are unique, but the decomposition itself is generally not, since the eigenspaces of A, but also the vectors $u_{r+1}, \ldots, u_n$ are not uniquely determined.

Since the matrices U and V are orthogonal to each other, it holds

$$A = U \Sigma V^T, \quad A^T = V^T \Sigma U$$

which, in sum form, turns to

$$A = \sum_{i=1}^{r} \sigma_i u_i v_i^T, \quad A^T = \sum_{i=1}^{r} \sigma_i v_i u_i^T.$$

The r singular values, together with the vectors $v_1, \ldots, v_r$ and $u_1, \ldots, u_r$ thus characterize the entire matrix A.

We conclude this section with three relatively simple examples, which are intended to provide an understanding of the singular value decomposition.

Example 5.49 The matrix for a rotation by the angle α is given by

$$A = \begin{pmatrix} \cos \alpha & -\sin \alpha \\ \sin \alpha & \cos \alpha \end{pmatrix}.$$

A has the singular value decomposition

$$A = \underbrace{\begin{pmatrix} \cos \alpha & -\sin \alpha \\ \sin \alpha & \cos \alpha \end{pmatrix}}_{=U} \underbrace{\begin{pmatrix} 1 & 0 \\ 0 & 1 \end{pmatrix}}_{=\Sigma} \underbrace{\begin{pmatrix} 1 & 0 \\ 0 & 1 \end{pmatrix}}_{=V^T}.$$

◀

Example 5.50 We determine the singular value decomposition of the singular matrix

$$A = \begin{pmatrix} 1 & 1 \\ 7 & 7 \end{pmatrix}.$$

Obviously, the matrix has rank 1, as the two columns are linearly dependent. Accordingly, there is one singular value. We first determine the matrix $A^T A$ as

$$A^T A = \begin{pmatrix} 50 & 50 \\ 50 & 50 \end{pmatrix}.$$

This matrix has the eigenvalues $\lambda_1 = 100$ and $\lambda_2 = 0$. The corresponding eigenvectors are

$$v_1 = \frac{1}{\sqrt{2}} \begin{pmatrix} 1 \\ 1 \end{pmatrix}, \quad v_2 = \frac{1}{\sqrt{2}} \begin{pmatrix} 1 \\ -1 \end{pmatrix}.$$

Thus, we can determine the matrix V:

$$V = [v_1, v_2] = \frac{1}{\sqrt{2}} \begin{pmatrix} 1 & -1 \\ 1 & 1 \end{pmatrix}$$

The singular value is $\sigma_1 = \lambda(A^T A) = \sqrt{100} = 10$. The values of the matrix U can now be determined as shown above:

$$u_1 = \frac{A v_1}{10} = \frac{1}{\sqrt{50}} \begin{pmatrix} 1 \\ 7 \end{pmatrix},$$

$$u_2 = \frac{A v_2}{10} = \frac{1}{\sqrt{50}} \begin{pmatrix} -7 \\ 1 \end{pmatrix}.$$

Thus, we have

$$U = [u_1, u_2] = \frac{1}{\sqrt{50}} \begin{pmatrix} 1 & -7 \\ 7 & 1 \end{pmatrix}.$$

Obviously, $U^T U = I$. The final decomposition is then given by

$$A = U \Sigma V^T = \frac{1}{\sqrt{50}} \begin{pmatrix} 1 & -7 \\ 7 & 1 \end{pmatrix} \begin{pmatrix} 10 & 0 \\ 0 & 0 \end{pmatrix} \frac{1}{\sqrt{2}} \begin{pmatrix} 1 & 1 \\ -1 & 1 \end{pmatrix}.$$

◀

Example 5.51 Finally, we present the singular value decomposition of the matrix from Example 4.28 from Sect. 4.5. Here, a rectangular matrix $A \in \mathbb{R}^{4 \times 3}$ is given, which will illustrate the special structure of the matrix Σ in the following. The matrix A is given by

$$A = \begin{pmatrix} 1 & -\frac{1}{4} & \frac{1}{16} \\ 1 & \frac{1}{2} & \frac{1}{4} \\ 1 & 2 & 4 \\ 1 & \frac{5}{2} & \frac{25}{4} \end{pmatrix}.$$

The singular value decomposition of A is given as:

$$A = U\Sigma V^T \approx \begin{pmatrix} -0.017 & 0.696 & 0.671 & -0.253 \\ -0.073 & 0.680 & -0.511 & 0.521 \\ -0.556 & 0.149 & -0.430 & -0.695 \\ -0.827 & -0.175 & 0.321 & 0.426 \end{pmatrix}$$

$$\cdot \begin{pmatrix} 8.217 & 0 & 0 \\ 0 & 1.380 & 0 \\ 0 & 0 & 0.523 \\ 0 & 0 & 0 \end{pmatrix} \begin{pmatrix} -0.179 & -0.391 & -0.903 \\ 0.979 & 0.0203 & -0.203 \\ 0.0978 & -0.920 & 0.379 \end{pmatrix}$$

◄

5.7.1 Numerical Approximation of the Singular Value Decomposition

The singular value decomposition could be created using the eigenvalues and eigenvectors of the matrix $A^T A$. However, this method is numerically unstable, as the condition number of $A^T A$ can increase quadratically with respect to the condition number of A. The singular value decomposition can also be approximated directly based on the matrix A. We sketch an algorithm that was presented in 1970 by Golub and Reinsch [42]. There are now more efficient variants, but the basic principle remains, see e.g. [43, 92]. A recent summary of [42] and extension was recently presented in [81].

For the following, we assume without loss of generality that $A \in \mathbb{R}^{n \times m}$ with $n \geq m$ is given. Otherwise, we consider the transposed matrix A^T, because we know that for

$$A = U\Sigma V^T,$$

the SVD of the transposed is given by

$$A^T = V\Sigma U^T.$$

The restriction serves to avoid case distinctions. The method proceeds in two steps:

1. First, the matrix $A \in \mathbb{R}^{n \times m}$ is brought to upper *bidiagonal form* by orthogonal transformations. That is, with two orthogonal matrices $H_L \in \mathbb{R}^{n \times n}$ and $H_R \in \mathbb{R}^{m \times m}$ with $H_L^T H_L = I \in \mathbb{R}^{n \times n}$ and $H_R^T H_R = I \in \mathbb{R}^{m \times m}$, we decompose A according to

$$H_L A H_R = B \in \mathbb{R}^{n \times m}, \tag{5.27}$$

   where $B_{ij} = 0$ for $j < i$ and $j > i + 1$.

2. In the second step, the singular value decomposition $B = R_L \Sigma R_R^T$ of the matrix B is approximated. Usually, this step is not performed exactly. Instead, we introduce an iteration that converges to the singular value decomposition.

Finally, the (approximate) singular value decomposition of A is given by

$$A = H_L^T R_L \Sigma R_R^T H_R^T.$$

5.7.1.1 Transformation to Bidiagonal Form

The procedure for transforming a matrix $A \in \mathbb{R}^{n \times m}$ with $n \geq m$ to bidiagonal form using Householder transformations, see Sect. 4.3, is similar to the procedure for QR factorization, or for creating the Hessenberg form. We multiply the matrix A from the left with reflection matrices $S_l^{(i)}$ with the aim of eliminating the subdiagonal of the i-th column. This requires m steps. After each step, we multiply the resulting matrix with another reflection matrix $S_r^{(l)}$ from the right, this time with the aim of eliminating the entries to the right of the first subdiagonal element. In this second step, as with the Hessenberg form, we must use a shortened reflection vector so that the first i columns of the matrix are not modified. We show the transformation $A^{(i)} \rightarrow A^{(i+1)}$, omitting the index (i), which indicates the iteration, for simplicity:

$$\begin{pmatrix} b_{11} & b_{12} & 0 & \cdots & \cdots & \cdots & 0 \\ 0 & \ddots & \ddots & \ddots & & & \vdots \\ \vdots & \ddots & b_{i-1,i-1} & b_{i-1,i} & 0 & \cdots & 0 \\ \vdots & & 0 & a_{i,i} & a_{i,i+1} & \cdots & a_{i,m} \\ \vdots & & 0 & a_{i+1,i} & a_{i+1,i+1} & \cdots & a_{i+1,m} \\ \vdots & & \vdots & \vdots & \vdots & \ddots & \vdots \\ 0 & \cdots & 0 & a_{n,i} & a_{n,i+1} & \cdots & a_{n,m} \end{pmatrix} \tag{5.28}$$

$$S_L^{(i)} \cdot \downarrow$$

$$\begin{pmatrix} b_{11} & b_{12} & 0 & \cdots & \cdots & \cdots & 0 \\ 0 & \ddots & \ddots & \ddots & & & \vdots \\ \vdots & \ddots & b_{i-1,i-1} & b_{i-1,i} & 0 & \cdots & 0 \\ \vdots & & 0 & b_{i,i} & \tilde{a}_{i,i+1} & \cdots & \tilde{a}_{i,m} \\ \vdots & & 0 & 0 & \tilde{a}_{i+1,i+1} & \cdots & \tilde{a}_{i+1,m} \\ \vdots & & \vdots & \vdots & \vdots & \ddots & \vdots \\ 0 & \cdots & 0 & 0 & \tilde{a}_{n,i+1} & \cdots & \tilde{a}_{n,m} \end{pmatrix}$$

$$\downarrow \cdot S_R^{(i)}$$

$$
\begin{pmatrix}
b_{11} & b_{12} & 0 & \cdots & \cdots & \cdots & 0 \\
0 & \ddots & \ddots & \ddots & & & \vdots \\
\vdots & 0 & b_{i-1,i-1} & b_{i-1,i} & 0 & \cdots & 0 \\
\vdots & & 0 & b_{i,i} & b_{i,i+1} & 0 & 0 \\
\vdots & & 0 & 0 & \hat{a}_{i+1,i+1} & \cdots & \hat{a}_{i+1,m} \\
\vdots & & \vdots & \vdots & \vdots & \ddots & \vdots \\
0 & \cdots & 0 & 0 & \hat{a}_{n,i+1} & \cdots & \hat{a}_{n,m}
\end{pmatrix}
$$

The Householder transformation from the left is given as

$$
S_L^{(i)} = I - 2v_L^{(i)}(v_L^{(i)})^T, \quad v_L^{(i)} = \frac{\tilde{a}_i + \operatorname{sign}(\tilde{a}_i)\|\tilde{a}_i\|e_i}{\|\tilde{a}_i + \operatorname{sign}(\tilde{a}_i)\|\tilde{a}_i\|e_i\|}
$$

with the shortened column vector

$$
\tilde{a}_i = (\underbrace{0, \ldots, 0}_{i-1}, \underbrace{a_{i,i}, \ldots, a_{n,i}}_{n+1-i})^T.
$$

Multiplication with $S_L^{(i)}$ leaves the first $i-1$ rows and columns unchanged. Subsequently, a Householder transformation from the right is multiplied

$$
S_R^{(i)} = I - 2v_R^{(i)}(v_R^{(i)})^T, \quad v_R^{(i)} = \frac{\hat{a}_i + \operatorname{sign}(\hat{a}_i)\|\hat{a}_i\|e_{i+1}}{\|\hat{a}_i + \operatorname{sign}(\hat{a}_i)\|\hat{a}_i\|e_{i+1}\|}.
$$

Here, $\hat{a}_i$ is a shortened column vector, which is formed from the row subdiagonal of the intermediate matrix $S_L^{(i)}A^{(i)}$, see (5.28)

$$
\hat{a}_i = (\underbrace{0, \ldots, 0}_{i}, \underbrace{\tilde{a}_{i,i+1}, \ldots, \tilde{a}_{i,m}}_{n-i})^T.
$$

with m multiplications from the left (or $m-1$, if the matrix A is square) and $m-2$ multiplications from the right are required. In total, we get

$$
B = \underbrace{S_L^{(m)} \cdots S_L^{(1)}}_{S_L} A \underbrace{S_R^{(1)} \cdots S_R^{(m-2)}}_{S_R}.
$$

We summarize:

Theorem 5.52 (Bidiagonalization) *Let $A \in \mathbb{R}^{n \times m}$ with $n \geq m$. Then there exist two orthogonal matrices $S_L \in \mathbb{R}^{n \times n}$ and $S_R \in \mathbb{R}^{m \times m}$ and an upper right bidiagonal matrix $B \in \mathbb{R}^{n \times m}$ such that*

$$
B = S_L A S_R.
$$

The factorization can be created with the effort of $4nm^2 - \frac{4}{3}m^3$ when using Householder transformations from the left and right.

5.7.1.2 Singular Value Decomposition of a Bidiagonal Matrix

We now assume that $B \in \mathbb{R}^{n \times m}$ is a bidiagonal matrix and we are looking for the singular value decomposition of B. Without loss of generality, we assume that $m = n$, because if a singular value decomposition $\Sigma = R_L B R_R$ is found for $B \in \mathbb{R}^{m \times m}$, then Σ and B can be extended by additional zero rows and R_L by additional zero rows and columns. So $B \in \mathbb{R}^{m \times m}$ is of the form

$$B = \begin{pmatrix} \alpha_1 & \beta_1 & 0 & \cdots & 0 \\ 0 & \alpha_2 & \beta_2 & \ddots & \vdots \\ \vdots & \ddots & \ddots & \ddots & 0 \\ \vdots & & \ddots & \ddots & \beta_{n-1} \\ 0 & \cdots & \cdots & 0 & \alpha_n \end{pmatrix}. \tag{5.29}$$

The algorithm by Golub and Kahan [41] approximates the singular value decomposition of B iteratively by multiplying with Givens rotations from the right and left

$$B^{(0)} = B, \quad l = 1, 2, \ldots, m : \quad B^{(l)} = G_l^{(l)} B^{(l-1)} G_r^{(l)}. \tag{5.30}$$

The matrices $G_l^{(l)}$ and $G_r^{(l)}$ themselves are products of $m - 1$ elementary Givens rotations

$$G_l^{(l)} = G_l^{(l),m-1} G_l^{(l),m-2} \cdots G_l^{(l),1}, \quad G_r^{(l)} = G_r^{(l),1} G_r^{(l),2} \cdots G_r^{(l),m-1}. \tag{5.31}$$

The goal of the iteration (5.30) is to successively transform the matrix B into a diagonal matrix with the singular values on the diagonal.

The first multiplication is from the right

$$B^{(l),1} = B^{(l-1)} G_r^{(l),1}$$

with a Givens rotation $G_r^{(l),1} = G(1, 2, \theta^{(l)})$. We refer to Definition 4.24 for the construction of the Givens matrix $G(i, j, \theta)$. The angle $\theta^{(l)}$ is initially free and will be chosen later to achieve convergence.

This multiplication creates a new non-zero entry $B_{2,1}^{(l),1} \neq 0$ below the diagonal, and the second multiplication (from the left)

$$B^{(l),1'} = G_l^{(l),1} B^{(l),1}$$

with a Givens rotation $G_l^{(l),1'} = G(1, 2, \theta^{(l),1'})$ now serves to eliminate this just created entry. In turn, a new non-zero entry $B_{1,3}^{(l),1'}$ is created. This continues until the matrix has the initial form again. The initially free parameter $\theta^{(l)}$ can be determined so that the last subdiagonal β_{n-1} converges to zero as quickly as possible; all other angles depend on this first one. As soon as a subdiagonal is small enough, e.g.

$$|B^{(l)}_{i-1,i}| \leq \epsilon_{\text{tol}}\big(|B^{(l)}_{i-1,i-1}| + |B^{(l)}_{i,i}|\big), \quad \epsilon_{\text{tol}} > 0$$

the value is set to zero $B^{(l)}_{i-1,i} = 0$ and the value α_n is the first identified singular value. Subsequently, the algorithm continues with the remaining reduced matrix $B^{(l)}_{1 \leq i,j < m}$.

After l steps, an approximation to the singular value decomposition of B is obtained

$$\Sigma^{(l)} = \underbrace{G^{(l)}_l G^{(l-1)}_l \cdots G^{(1)}_l}_{=G_l}\, B\, \underbrace{G^{(1)}_r G^{(2)}_r \cdots G^{(l)}_r}_{=G_r},$$

where $G^{(l)}_l$ and $G^{(l)}_r$ are constructed according to (5.31). We summarize the effort of the method.

Theorem 5.53 (Effort of a Golub-Kahan Iteration) *Let $B \in \mathbb{R}^{m \times m}$ be an upper right bidiagonal matrix. Each iteration of the Golub-Kahan algorithm has the effort of $24m + O(1)$. If the orthogonal matrices G_l and G_r are also calculated in addition to the singular values, the effort increases to $4mn + O(m^2)$.*

Proof Per iteration, $m - 1$ Givens rotations, each from the left and right, are applied to a matrix $B^{(l),i} \in \mathbb{R}^{m \times m}$. The matrix $B^{(l),i}$ is a matrix with a maximum of 3 entries per row, i.e., a Givens rotation requires 12 operations. In total, $24(m - 1)$ operations are therefore necessary for the entire sequence.

If the orthogonal matrices $G_l \in \mathbb{R}^{n \times n}$ and $G_r \in \mathbb{R}^{m \times m}$ are also calculated, then $(m - 1)4n$ and $(m - 1)4m$ operations are added. $\qquad\square$

5.7.2   Applications of Singular Value Decomposition

In the following, we will discuss some prototype applications of singular value decompositions.

5.7.2.1 Image and Kernel of Homogeneous Linear Mappings

Let $f : \mathbb{R}^m \to \mathbb{R}^n$ be a linear mapping with the representation matrix $A \in \mathbb{R}^{n \times m}$. We are looking for the image and the kernel of A, i.e., the subspaces

$$\text{Ker}(A) = \text{span}\{x \in \mathbb{R}^m \mid Ax = 0\} \subset \mathbb{R}^m,$$
$$\text{Img}(A) = \text{span}\{Ax \in \mathbb{R}^n \;\forall x \in \mathbb{R}^m\} \subset \mathbb{R}^n.$$

Let $A = U \Sigma V^T$ be the singular value decomposition of A with singular values $\sigma_1 \geq \sigma_2 \geq \cdots \geq \sigma_r > 0$ with $r \leq m$. The number r, i.e., the number of non-zero singular values, indicates the rank of the matrix A. The singular value decomposition can be created stably [23], i.e., the relative accuracy of the determined singular values depends only on the condition of the problem and the computational accuracy. Thus, for the numerical approximation $\tilde{\sigma}_i$ to σ_i it holds

$$\frac{|\tilde{\sigma}_i - \sigma_i|}{|\sigma_i|} \leq \kappa\epsilon \quad \Rightarrow \quad |\tilde{\sigma}_i - \sigma_i| \leq \kappa\epsilon|\sigma_i|.$$

Even singular values close to zero can be determined stably. However, the singular value decomposition cannot decide whether a singular value is exactly zero or merely close to zero. This question is essential in determining the rank, image, and kernel. Numerically, we therefore choose a tolerance, e.g., $\epsilon_{\mathrm{Ker}}\sigma_1$ and determine the rank of A by

$$r := \arg \max_{1 \leq i \leq n} \{\sigma_i > \epsilon_{\mathrm{Ker}}\sigma_1\}.$$

The dimension of the image is then r, and $n - \mathrm{rank}(A)$ is the dimension of the kernel. We modify Σ to $\tilde{\Sigma}$, so that for all diagonal elements $\tilde{\Sigma}_{ii} = 0$ holds for $i > \mathrm{rank}(A)$. The homogeneous linear system for determining the kernel is

$$U\tilde{\Sigma}V^T x = 0 \quad \Rightarrow \quad \tilde{\Sigma}y = 0, \quad x = Vy.$$

Hereby, $y_1 = \cdots = y_r = 0$ and the $y_{r+1}, \ldots, y_n$ can be chosen freely, e.g., as unit vectors. The kernel of A is then spanned by

$$\mathrm{Ker}(A) = \mathrm{span}\{v_{\mathrm{rank}(A)+1}, \ldots, v_n\}, \tag{5.32}$$

where v_i are the column vectors of the matrix A. The image of A can be determined just as easily. It is

$$y = Ax = U\tilde{\Sigma}V^T x = U\tilde{\Sigma}z.$$

Since V is regular, $x, z \in \mathbb{R}^m$ can be used interchangeably. Only the first r singular values are non-zero, hence

$$\mathrm{Img}(A) = \mathrm{span}\{u_1, \ldots, u_r\},$$

where u_i are the column vectors of U.

5.7.2.2 Pseudoinverse

The singular value decomposition of a matrix can be used to generalize the concept of the inverse of a matrix.

Definition 5.54 (*Moore-Penrose Inverse*) Let $A \in \mathbb{R}^{n \times m}$. The *pseudoinverse* $A^+ \in \mathbb{R}^{m \times m}$ of A is a matrix with the properties

$$
\begin{aligned}
(i) \quad & AA^+A = A \\
(ii) \quad & A^+AA^+ = A^+ \\
(iii) \quad & (AA^+)^T = AA^+ \\
(iv) \quad & (A^+A)^T = A^+A
\end{aligned}
\tag{5.33}
$$

The inverse A^{-1} of a square regular matrix fulfills all these properties.

Theorem 5.55 (Pseudoinverse) *Let $A \in \mathbb{R}^{n \times m}$ with singular value decomposition $A = U\Sigma V^T$ with $U \in \mathbb{R}^{n \times n}$, $V \in \mathbb{R}^{m \times m}$, $\Sigma \in \mathbb{R}^{n \times m}$ and singular values $\sigma_1 \geq \sigma_2 \geq \cdots \geq \sigma_r > 0$. We define the matrix $\Sigma^+ \in \mathbb{R}^{m \times n}$ with*

$$
\Sigma^+ = \mathrm{diag}(\sigma_1^{-1}, \ldots, \sigma_r^{-1}, 0, \ldots, 0).
$$

The matrices Σ^+ and $A^+ = V\Sigma^+U^T$ are Moore-Penrose inverses to Σ and to A.

Proof The properties from Definition 5.54 can all be calculated elementarily. $\square$

A possible application of the pseudoinverse is the solution of linear systems $Ax = b$ with a non-square matrix $A \in \mathbb{R}^{n \times m}$, i.e., in the case $n > m$ of overdetermined and in the case $n < m$ of underdetermined linear systems. We have studied the case $n > m$ with full rank$(A) = m$ in Sect. 4.5. Since in the general case an exact solution $Ax = b$ is not to be expected, we have introduced the best approximation

$$
x \in \mathbb{R}^m \quad \|Ax - b\|_2 \leq \|Ay - b\|_2 \quad \forall y \in \mathbb{R}^m.
$$

The solution is uniquely determined by the normal system

$$
A^T Ax = A^T b,
\tag{5.34}
$$

when $A^T A$ is regular, i.e., when A has full rank. Here, we see the connection to the singular value decomposition, because the singular values are precisely the roots of the eigenvalues of $A^T A$, and it holds with $A = U\Sigma V^T$

$$
A^T A = V\Sigma^2 V^T,
$$

so (5.34) can be written as

$$
V\Sigma^2 V^T x = V\Sigma U^T b \quad \Leftrightarrow \quad \Sigma V^T x = U^T b \quad \Leftrightarrow \quad x = V\tilde{\Sigma}^+ U^T b = A^+ b.
$$

With this preparation, the method can also be extended to systems with reduced rank. For these, the best approximation is not unique.

Definition 5.56 (*Pseudonormal Solution*) Let $A \in \mathbb{R}^{n \times m}$ and $b \in \mathbb{R}^n$. Further, let $\bar{x} \in \mathbb{R}^m$ be a best approximation in the Euclidean norm, i.e.,

$$\|A\bar{x} - b\|_2 \leq \|Ay - b\|_2 \quad \forall y \in \mathbb{R}^m.$$

Then the vector $x^+ \in \mathbb{R}^m$ according to

$$(i) \quad \|Ax^+ - b\|_2 \leq \|Ay - b\|_2 \quad \forall y \in \mathbb{R}^m$$
$$(ii) \quad \|x^+\|_2 \leq \|x\|_2 \quad \forall x \in \{\bar{x} + \mathrm{Ker}(A)\},$$

is the *pseudonormal solution* to the linear system $Ax = b$.

Only the second condition $\|x^+\|_2 \leq \|x\|_2$ leads to a unique solution, and thus among all possible best approximations, the pseudonormal solution is the one with the smallest Euclidean norm.

Theorem 5.57 (Pseudonormal Solution) *Let $A \in \mathbb{R}^{n \times m}$ and $b \in \mathbb{R}^n$. The pseudonormal solution to $Ax = b$ is uniquely determined by*

$$x^+ = A^+ b$$

Proof Let x^+ be a solution of the normal system and thus a best approximation:

$$A^T A x^+ = (V \Sigma^2 V^T)(V \Sigma^+ U^T) b = V \Sigma U^T b = A^T b$$

We write the kernel of the matrix A with (5.32) as

$$\mathrm{Ker}(A) = \mathrm{span}\{v_{r+1}, \ldots, v_n\},$$

where $p = \mathrm{rank}(A)$. For arbitrary $z \in \mathrm{Ker}(A)$ it holds

$$\|x^+ + z\|_2^2 = \|x^+\|_2^2 + 2(x^+, z) + \|z\|_2^2.$$

Further, with $z = \sum_{i=r+1}^{n} \alpha_i v_i$

$$(x^+, z) = \sum_{i=r+1}^{n} \alpha_i (V \Sigma^+ U^T b, v_i) = \sum_{i=r+1}^{n} \alpha_i (\Sigma^+ U^T b, V^T v_i)$$
$$= \sum_{i=r+1}^{n} \alpha_i (\Sigma^+ U^T b, e_i) = 0,$$

since all $\sigma_i = 0$ for $i > p$. Thus, the orthogonal factorization holds

$$\|x^+ + z\|_2^2 = \|x^+\|_2^2 + \|z\|_2^2 \quad \forall z \in \mathrm{Ker}(A)$$

which shows that x^+ is indeed the unique pseudonormal solution. $\qquad \square$

Remark 5.58 In the case of a square matrix $A \in \mathbb{R}^{n \times n}$ and in the case of a unique solution $\bar{x} = A^{-1}b$, the pseudonormal solution x^+ coincides with the classical solution $\bar{x}$.   ♦

In the definition of the pseudonormal solution as well as in Theorem 5.57, it did not matter whether $n = m$, $n < m$, or $n > m$. We have extended the concept of solutions of linear systems to the general case, which includes underdetermined and overdetermined equations.

Example 5.59 (*Calculation of a Pseudoinverse A^+*) From Example 5.51, we know A, U, Σ and V^T. We use

$$B := V \tilde{\Sigma} U^T, \quad \tilde{\Sigma} \in \mathbb{R}^{m \times n}, \quad \tilde{\Sigma}_{ij} := \tau_i \cdot \delta_{ik}, \tag{5.35}$$

with the vector $\tau \in \mathbb{R}^n$, given as

$$\tau_i := \begin{cases} \sigma_i^{-1} & \sigma_i \neq 0, \\ 0 & \sigma_i = 0, \end{cases} \tag{5.36}$$

to calculate the pseudoinverse to $A \in \mathbb{R}^{4 \times 3}$: $A^+ = V \tilde{\Sigma} U^T$. First, the matrix $\tilde{\Sigma} \in \mathbb{R}^{3 \times 4}$ is given by (5.35) and (5.36)

$$\tilde{\Sigma} = \begin{pmatrix} 0.121705 & 0 & 0 & 0 \\ 0 & 0.724587 & 0 & 0 \\ 0 & 0 & 1.91011 & 0 \end{pmatrix}.$$

Then, in particular, the property $\Sigma \cdot \tilde{\Sigma} \in \mathbb{R}^{4 \times 4}$ holds:

$$\Sigma \cdot \tilde{\Sigma} = \begin{pmatrix} 1 & 0 & 0 & 0 \\ 0 & 1 & 0 & 0 \\ 0 & 0 & 1 & 0 \\ 0 & 0 & 0 & 0 \end{pmatrix}$$

For the pseudoinverse $A^+ \in \mathbb{R}^{3 \times 4}$, we now obtain with the orthogonal matrices U and V^T, calculated in Example 5.51 and (5.35),

$$A^+ = V \tilde{\Sigma} U^T \approx \begin{pmatrix} 0.620 & 0.388 & 0.0378 & -0.0459 \\ -1.169 & 0.911 & 0.785 & -0.527 \\ 0.386 & -0.462 & -0.273 & 0.349 \end{pmatrix}.$$

The test $A^+ A \in \mathbb{R}^{3 \times 3}$ yields

$$A^+ A \approx \begin{pmatrix} 0.620 & 0.388 & 0.0378 & -0.0459 \\ -1.169 & 0.911 & 0.785 & -0.527 \\ 0.386 & -0.462 & -0.273 & 0.349 \end{pmatrix} \cdot \begin{pmatrix} 1 & -\frac{1}{4} & \frac{1}{16} \\ 1 & \frac{1}{2} & \frac{1}{4} \\ 1 & 2 & 4 \\ 1 & \frac{5}{2} & \frac{25}{4} \end{pmatrix} \approx \begin{pmatrix} 1 & 0 & 0 \\ 0 & 1 & 0 \\ 0 & 0 & 1 \end{pmatrix}.$$

◀

5.7.2.3 Low-Rank Approximation

Similar to the eigenvalues and eigenvectors, the singular values can be used to characterize the matrix and the linear mapping $f : x \mapsto Ax$, respectively.

Corollary 5.60 (Singular Value Decomposition and Matrix Norm) *Let $A \in \mathbb{R}^{n \times m}$ be of rank r and have singular the value decomposition V, U, Σ. Then*

$$\|A\|_2 = \sigma_1, \quad \|A\|_F = \sqrt{\sum_{i=1}^{r} \sigma_i^2}.$$

Proof The singular values σ_i (as well as possibly additional zeros) are precisely the squares of the eigenvalues of $A^T A$. Hence,

$$\|A\|_2 = \max\{\lambda^{\frac{1}{2}} \mid \lambda \text{ is eigenvalue of } A^T A\}.$$

For the Frobenius norm, we have

$$\|A\|_F^2 = \mathrm{tr}(A^T A) = \sum_{i=1}^{r} \sigma_i^2.$$

$\square$

Since the vectors u_i, v_i are normalized and σ_i are decreasing, the sequence

$$(\sigma_1, v_1, u_1), (\sigma_2, v_2, u_2), \dots$$

takes a role in characterizing the matrix A. We define

Definition 5.61 (*Low-Rank Approximation*) Let $A \in \mathbb{R}^{n \times m}$ be a matrix of rank r with singular value decomposition V, U, Σ. Then the matrix $A_l \in \mathbb{R}^{n \times m}$ for $l \leq r$ according to

$$A_l := \sum_{i=1}^{l} \sigma_i u_i v_i^T$$

is called the *rank-l approximation* of the matrix A. With the matrix

$$\Sigma_l = \begin{pmatrix} \tilde{\Sigma}_l & 0 \\ 0 & 0 \end{pmatrix} \in \mathbb{R}^{n \times m}, \quad \tilde{\Sigma}_l := \mathrm{diag}(\sigma_1, \ldots, \sigma_l, 0, \ldots, 0) \in \mathbb{R}^{r \times r}$$

we write

$$A_l = U \Sigma_l V^T.$$

The rank-l approximation is in a certain sense a best approximation to the matrix A in the vector space of all matrices with maximum rank l. The *Eckart-Young-Mirsky Theorem* shows this connection in general matrix norms and also in the infinite-dimensional case. We give a simple variant:

Theorem 5.62 (**Rank-l Approximation**) *Let $A \in \mathbb{R}^{n \times m}$ be of rank r. Let $l < r$. Then the rank-l approximation $A_l \in \mathbb{R}^{n \times m}$ according to Definition 5.61 is the best approximation to A in the vector space of all matrices up to rank l, i.e.,*

$$\|A - A_l\|_2 \leq \min_{B_l \in \mathbb{R}^{n \times m},\ \mathrm{rank}(B_l) \leq l} \|A - B_l\|_2.$$

It holds

$$\|A - A_l\|_2 = \sigma_{l+1}.$$

Proof *(i)* Let V, U, Σ be the singular value decomposition of the matrix A and $r = \mathrm{rank}(A)$. It holds

$$A - A_l = \sum_{i=l+1}^{r} \sigma_i u_i v_i^T \tag{5.37}$$

and the error representation

$$\|A - A_l\|_2 = \sigma_{l+1},$$

follows from Corollary 5.60.

(ii) Now let $B_l \in \mathbb{R}^{n \times m}$ be an arbitrary matrix with rank $l < r$. The kernel of B_l has the dimension $\mathrm{Ker}(B_l) = m - l$ and we denote by $\{x_1, \ldots, x_{m-l}\}$ an orthonormal basis of the kernel. This means that the intersection

$$\{x_1, \ldots, x_{m-l}\} \cap \{v_1, \ldots, v_{l+1}\} \neq \emptyset$$

is non-trivial. We now choose a vector $z \in \mathbb{R}^m$ with $\|z\|_2 = 1$ from this intersection. For it, $B_l z = 0$ and thus

$$(A - B_l)z = Az = \sum_{i=1}^{l+1} \sigma_i u_i v_i^T z$$

and consequently

$$\|A - B_l\|_2 = \max_{w \in \mathbb{R}^m} \frac{\|(A - B_l)w\|_2}{\|w\|_2} \geq \|(A - B_l)z\|_2 = \|Az\|_2.$$

We write z in the form

$$z = \sum_{k=1}^{l+1} \alpha_k v_k$$

with $\alpha_k = u_k v_k^T z$ using the orthonormality of the u_i and v_i. It holds

$$\|Az\|_2^2 = \sum_{i=1}^{l+1} \sigma_i^2 (u_i v_i^T z)^T (u_i v_i^T z) = \sum_{i=1}^{l+1} \sigma_i^2 \alpha_i^2 \geq \sigma_{l+1}^2 \sum_{i=1}^{l+1} \alpha_i^2 = \sigma_{l+1}^2.$$

$\square$

5.8   Excursus: Singular Value Decomposition for Image Compression

We want to use the singular value decomposition of a matrix as a simple method for image compression. The basis for this is the *low-rank representation* of a matrix based on the singular value decomposition. For this, we consider the photo from Fig. 5.8. The image has a resolution of 2000×1000 and we store the three color channels *red, green* and *blue* in three separate matrices

$$A_{\text{red}}, A_{\text{green}}, A_{\text{blue}} \in [0, 1]^{2000 \times 1000}.$$

For the color channels red, green, and blue, we each create a singular value decomposition of the matrix. On the right in Fig. 5.8, we show the logarithmic course of the singular values. It can be seen that the values drop very quickly, which offers potential for a reduced representation with the singular value decomposition.

Especially the first singular values drop quickly. The figure on the right shows that the first 50 singular values already contain 99% of the information, as the values drop from 10^3 to 10. To reduce the information, we shorten the representation and introduce the notation

$$A_q := U \Sigma_q V_q^T \tag{5.38}$$

where

$$\Sigma_q := \text{diag}(\sigma_1, \ldots, \sigma_q, 0, \ldots, 0)$$

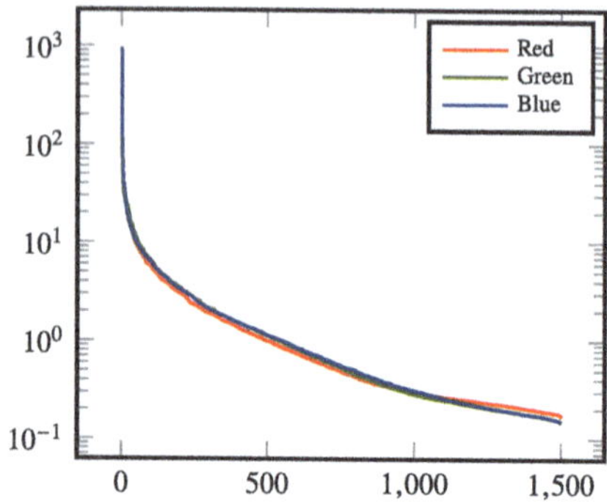

Fig. 5.8 Left: Photo of the research icebreaker Polarstern. Right: Singular values of the matrices A_{red}, A_{green}, A_{blue}, which represent the three color channels of the image. Each entry, e.g. $A_{\text{red},ij} \in [0, 1]$ corresponds to the color intensity in the pixel (i, j)

is, i.e., the diagonal matrix of the first q singular values. The reduced representation A_q according to (5.38) can be written as

$$A_q = \sum_{i=1}^{q} \sigma_i u_i v_i^T,$$

and instead of nm pixels it requires only the first q rows of V and columns of U, thus $q(m + n)$ pixels. In Fig. 5.9 we show the reduced representation for different values of q as well as a small section containing the vessel's name "Polarstern". The quality of the approximation can be measured using

$$\varepsilon(q) = \frac{\sum_{i=1}^{q} \sigma_i^2}{\sum_{i=1}^{r} \sigma_i^2}.$$

A significant feature of the singular value decomposition is that structures with sharp edges, e.g., the inscription "Polarstern", are already represented with the first singular value. Only $3(n + m) = 9\,000$ values are stored instead of $3nm = 6\,000\,000$ pixel information.

In practice, one would not use singular value decomposition for image compression in this way. The cubic scaling would lead to very long runtimes for large images. A solution is to rasterize the image. This can initially be divided into blocks, e.g., of size 16×16, and then a singular value decomposition is created from each block. In this way, the poor scaling does not come into play. Furthermore, image compression methods like JPEG, see the corresponding excursion in Sect. 9.8, take physiological factors into account. The information is not reduced on a purely mathematical basis. Instead, human perception is taken into account, and the data is weighted accordingly.

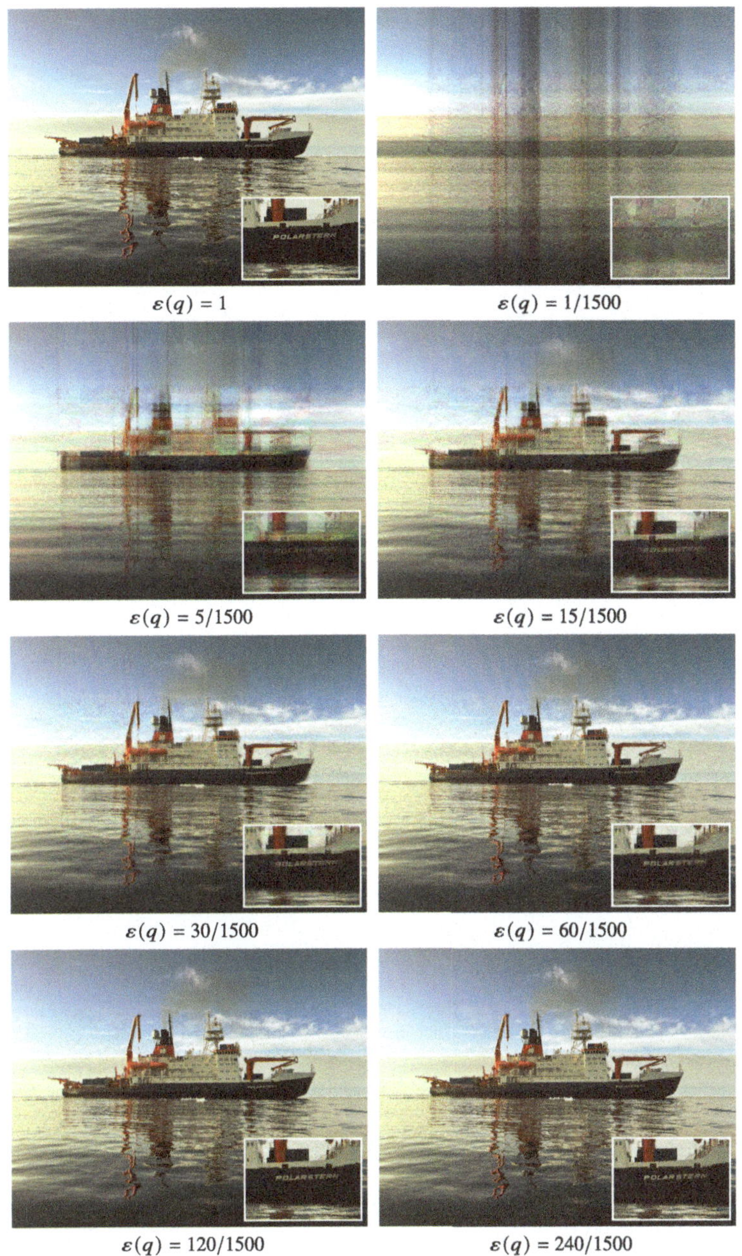

Fig. 5.9 Image compression with singular value decomposition. The original image (detail in the top right) is represented as a matrix, and singular value decompositions are created from the three channels, red, green, and blue. The reconstructions are each based only on a portion of the singular values and associated orthogonal vectors

Krylov Subspace Methods

6

In this chapter, we extend the methods already discussed for solving $Ax = b$. As a reminder, in Chap. 3 and Chap. 4 we studied direct methods (i.e., LU, Cholesky, and QR factorization) and iterative methods (i.e., Richardson, Jacobi, Gauss-Seidel). A motivation for iterative methods is the lower memory requirement and generally lower cost complexity. The factorization methods (LU, Cholesky, and QR factorization) from Chap. 3 and Chap. 4 all have a cubic runtime $O(n^3)$. For band matrices with bandwidth m, the effort is reduced to $O(nm^2)$, but for sparse matrices cannot be well utilized. Especially for very large problems, the effort grows so quickly that the solution cannot be achieved in a reasonable time, see Table 3.1. In addition to runtime, memory usage plays a central role. To store a dense matrix $A \in \mathbb{R}^{n \times n}$ with $n = 1\,000\,000$, about 7 terabytes of memory are needed at double precision. This is only possible on large parallel computers. The linear systems resulting from most applications (such as the discretization of differential equations) are very sparsely populated, i.e., only a few entries are in each row. A sparse matrix with $n = 1\,000\,000$, but only 100 entries per row, requires just 750 MB of storage and fits on any laptop. In Sect. 3.4, we learned sorting methods to make this sparse occupancy pattern usable for an LU factorization. In general, however, the matrices L and R are not sparse and thus by far exceed the available memory.

Another disadvantage of factorization methods is numerical stability. Due to rounding errors in the factorization process, the LU factorization usually only approximately satisfies $A \neq \tilde{L}\tilde{R}$. This means that although the LU factorization is a direct method, the system cannot be solved exactly.

The fixed-point methods of the form discussed in Sect. 3.7

$$x^{k+1} = x^k + C(b - Ax^k)$$

© The Author(s), under exclusive license to Springer-Verlag GmbH, DE, part of Springer Nature 2026

T. Richter et al., *Introduction to Numerical Mathematics*, Mathematics Study Resources 25, https://doi.org/10.1007/978-3-662-72546-7_6

with a matrix $C \approx A^{-1}$ solve this second problem. If the spectral radius of the matrix $I - CA$ is small, in the case $\rho = \rho(I - CA) < 1$, the methods converge robustly. However, convergence is usually very slow.

In this chapter, we will always assume that the matrix $A \in \mathbb{R}^{n \times n}$ is sparse, has only $O(1)$ entries per row, and that a matrix-vector multiplication can thus be performed in $O(n)$ operations. We will refrain from modifying matrix entries entirely, as this could lead to *fill-ins* that could drive up the effort again. Instead, the algorithms presented in the following will be based exclusively on matrix-vector multiplications.

The resulting methods are not direct methods, and the solution of a linear system is not calculated exactly but only approximated.

6.1   Galerkin Equation and Krylov Subspaces

We consider a linear system $Ax = b$ with a matrix $A \in \mathbb{R}^{n \times n}$ and the right-hand side $b \in \mathbb{R}^n$. If $x \in \mathbb{R}^n$ solves the linear system $Ax = b$, then the residual is zero

$$r(x) = b - Ax = 0.$$

In other words, for the solution $x \in \mathbb{R}^n$ in particular

$$(b - Ax, y) = 0 \quad \forall y \in \mathbb{R}^n. \tag{6.1}$$

This equation is called the *variational equation*. Conversely, it holds

Theorem 6.1   (Linear Systems as a Variational Problem) *Let $A \in \mathbb{R}^{n \times n}$ be regular, $b \in \mathbb{R}^n$. The solution of the linear system $Ax = b$ is uniquely determined as the solution of the variational problem*

$$\textit{Find } x \in \mathbb{R}^n : \quad (b - Ax, y) = 0 \quad \forall y \in \mathbb{R}^n.$$

Proof If $Ax = b$ is the (unique) solution, then $b - Ax = 0$ and also the variational equation is satisfied for all $y \in \mathbb{R}^n$. Conversely, if this equation is satisfied for all $y \in \mathbb{R}^n$, then also for $y = b - Ax$, so

$$0 = (b - Ax, b - Ax) = \|b - Ax\|^2 = 0.$$

Since $\| \cdot \|$ is a norm, $b - Ax = 0$, therefore $Ax = b$. $\square$

This is where the *Krylov subspace methods* come in. We create an approximation to the solution $x \in \mathbb{R}^n$ by restricting the variational problem to a subspace. We look for $x^k \in X_k \subset \mathbb{R}^n$, such that

$$(b - Ax^k, \tilde{y}) = 0 \quad \forall \tilde{y} \in Y_k, \tag{6.2}$$

with a second subspace $Y_k \subset \mathbb{R}^n$. The restriction of a variational form to a subspace is called a *Galerkin equation*. The space X_k in which the approximation is sought is called the *trial space* and the space Y_k is called the *test space*. The relationship (6.2) is also called *Galerkin orthogonality* because Eq. (6.2) states that the residual $b - Ax^k$ is orthogonal to the subspace $Y_k \subset \mathbb{R}^n$. This class of methods is also called *projection methods*.

Assume that $\{x_1, \ldots, x_k\} \subset X_k$ and $\{y_1, \ldots, y_k\} \subset Y_k$ are bases of X_k and Y_k with dimension $\dim(X_k) = \dim(Y_k) = k$. We then write the desired solution to (6.2) as $x^k = \sum_{i=1}^k \alpha_i x_i$ and the Galerkin equation (6.2) is equivalent to

$$\alpha_1, \ldots, \alpha_k \in \mathbb{R} : \quad \left(b - A \sum_{i=1}^k \alpha_i x_i, y_j\right) = 0, \quad j = 1, \ldots, k,$$

which we reformulate as follows

$$\alpha_1, \ldots, \alpha_k \in \mathbb{R} : \quad \sum_{i=1}^k (Ax_i, y_j)\alpha_i = (b, y_j), \quad j = 1, \ldots, k.$$

This is equivalent to the smaller linear system $\tilde{A}\alpha = \tilde{b}$, where $\alpha = (\alpha_1, \ldots, \alpha_k)$ and $\tilde{A} \in \mathbb{R}^{k \times k}$ with $\tilde{A}_{ji} = (Ax_i, y_j)$ and $\tilde{b} \in \mathbb{R}^k$ with $\tilde{b}_j = (b, y_j)$. Conditions for the solvability of the reduced system and approximation properties for general projection methods are discussed, for example, in [[82], Chap. 5].

We limit ourselves here to the *Krylov subspace methods*, which result from a special choice of the subspaces X_k and Y_k. We define:

Definition 6.2 (*Krylov Subspace*) Let $Ax = b$ be a linear system with matrix $A \in \mathbb{R}^{n \times n}$ and right-hand side $b \in \mathbb{R}^n$. Furthermore, let $x^0 \in \mathbb{R}^n$ be arbitrarily given and $d^0 := b - Ax^0$. Then the subspaces spanned by

$$K_k := K_k(d^0, A) := \mathrm{span}\underbrace{\{d^0, Ad^0, \ldots, A^{k-1}d^0\}}_{k \text{ vectors}} \subset \mathbb{R}^n$$

are called *Krylov subspaces*.

We can think of $x^0 \in \mathbb{R}^n$ as an initial guess for the solution. The approximation in the subspace is then sought as $x^k \in x^0 + \hat{x}$ with

$$\hat{x} \in K_k(d^0, A) : \quad \left(b - A(x^0 + \hat{x}), y\right) = 0 \quad \forall y \in Y_k.$$

If no good starting guess is known, we choose $x^0 = 0$.

The Krylov subspaces are uniquely determined by the initial value x^0 and the matrix A. It holds that $K_k(d^0, A) \subset K_{k+1}(d^0, A)$, as vectors are only added for larger k but never removed from the space.

6.2   The Conjugate Gradient Method

The *Conjugate Gradient method*, or CG method for short, is a simple and very efficient Krylov subspace method for solving linear systems $Ax = b$ with symmetric positive definite matrix $A \in \mathbb{R}^{n \times n}$. As trial space and as test space, we choose the Krylov subspace $K_k(d^0, A)$, the Galerkin equation in step k is

$$x^k \in x^0 + K_k(d^0, A) : \quad (b - Ax^k, y) = 0 \quad \forall y \in K_k(d^0, A). \tag{6.3}$$

We have discussed that the Krylov spaces form a sequence of subspaces $K_{k-1}(d^0, A) \subset K_k(d^0, A) \subset K_{k+1}(d^0, A)$ for increasing k. However, it is not said that $K_{k+1}(d^0, A)$ is really larger than $K_k(d^0, A)$. Suppose for instance that d^0 was an eigenvector of A, then $K_k(d^0, A) = \mathrm{span}\{d^0\}$ for all k. This raises the question of whether this construction is suitable at all for determining the solution of the linear system. The following theorem gives a positive answer:

Theorem 6.3 *Suppose $A^k d^0 \in K_k(d^0, A)$. Then the solution $x \in \mathbb{R}^n$ of $Ax = b$ is already contained in the k-th Krylov space $K_k(d^0, A)$.*

Proof Let $K_k := K_k(d^0, A)$ be given and $x^k \in x_0 + K_k$ be the best approximation, which satisfies the Galerkin equation (6.3). Let $r^k := b - Ax^k$. Since

$$r^k = b - Ax^k = \underbrace{b - Ax^0}_{=d^0} + A \underbrace{(x^0 - x^k)}_{\in K_k} \in d^0 + A K_k,$$

it holds that $r^k \in K_{k+1}$. Suppose now, $K_{k+1} \subset K_k$. Then it holds $r^k \in K_k$. The Galerkin equation states $r^k \perp K_k$, i.e., it necessarily holds $r^k = 0$ and $Ax^k = b$. $\qquad\square$

We can therefore abort the iteration when the next Krylov space has not become genuinely larger and stop with the solution.

We now assume that the Krylov subspace $K_k(d^0, A)$ is spanned by k independent vectors $d^0, \ldots, d^{k-1}$. The Galerkin equation for finding the solution coefficients $\alpha_0, \ldots, \alpha_{k-1}$ of

$$x^k = x^0 + \sum_{i=0}^{k-1} \alpha_i d^i$$

can be written as

$$\sum_{i=0}^{k-1} (Ad^i, d^j)\alpha_i = (b - Ax^0, d^j) \quad \forall j = 0, \ldots, k - 1. \tag{6.4}$$

In order to solve this reduced system quickly, we will determine the directions d^i as an orthogonal basis of $K_k(d^0, A)$. We have assumed the matrix A to be symmetric positive definite. Thus, (Ax, y) is a scalar product. Now suppose the vectors d^j are pairwise A-orthogonal, i.e.,

$$(Ad^j, d^i) = 0 \quad i \neq j, \tag{6.5}$$

then the solution to (6.4) can be easily read off as

$$\alpha_j = \frac{(b - Ax^0, d^j)}{(Ad^j, d^j)} \quad j = 0, \ldots, k - 1.$$

The A-orthogonal basis $\{d^0, \ldots, d^{k-1}\}$ of the Krylov space $K_k(d^0, A)$ could, for example, be calculated using the Gram-Schmidt orthogonalization method. The name *conjugate gradients* comes from the orthogonality of the new directions d^j, which step by step span the Krylov space.

The disadvantages of the Gram-Schmidt orthogonalization method are the high effort and the lack of numerical robustness. To orthogonalize a vector with respect to an already existing basis $\{d^0, \ldots, d^{k-1}\}$, k scalar products are required. For the CG method, we can exploit the special structure of the symmetric positive definite matrix A and the A-scalar product to formulate a more robust and much faster orthogonalization scheme.

Lemma 6.4 (Two-stage Recursion Formula for Orthogonalization) *Let $A \in \mathbb{R}^{n \times n}$ be symmetric positive definite and $x^0 \in \mathbb{R}^n$ and $d^0 := b - Ax^0$. The iteration*

$$r^k := b - Ax^k, \quad \beta_{k-1} := -\frac{(r^k, Ad^{k-1})}{(d^{k-1}, Ad^{k-1})}, \quad d^k := r^k + \beta_{k-1}d^{k-1}, \quad k = 1, 2 \ldots$$

generates an A-orthogonal basis with $(Ad^r, d^s) = 0$ for $r \neq s$. Here, x^k is defined in step k as the Galerkin solution $(b - Ax^k, d^j) = 0$ for $j = 0, \ldots, k - 1$.

Proof Let $\{d^0, \ldots, d^{k-1}\}$ be an A-orthogonal basis of $K_k(d^0, A)$. Further, let $x^k \in x^0 + K_k(d^0, A)$ be the Galerkin solution to (6.2). Let $r^k := b - Ax^k \in K_{k+1}$ and we demand that $r^k \notin K_k(d^0, A)$. Otherwise, the iteration stops according to Lemma 6.3. We determine d^k with the approach

$$d^k = r^k + \sum_{j=0}^{k-1} \beta_j^{k-1} d^j. \tag{6.6}$$

The orthogonality condition states:

$$0 \stackrel{!}{=} (d^k, Ad^i) = (r^k, Ad^i) + \sum_{j=0}^{k-1} \beta_j^{k-1}(d^j, Ad^i)$$

$$= (r^k, Ad^i) + \beta_i^{k-1}(d^i, Ad^i), \quad i = 0, \ldots, k - 1$$

It holds $(r^k, Ad^i) = (b - Ax^k, Ad^i) = 0$ for $i = 0, \ldots, k - 2$, since $r^k \perp AK_{k-1} \subset K_k$. From this follows $\beta_i^{k-1} = 0$, for $i = 0, 1, \ldots, k - 2$. For $i = k - 1$, it holds

$$\beta_{k-1} := \beta_{k-1}^{k-1} = -\frac{(r^k, Ad^{k-1})}{(d^{k-1}, Ad^{k-1})}.$$

Finally, with (6.6), it holds $d^k = r^k + \beta_{k-1}d^{k-1}$. $\square$

With these preparations, we can put together all the components of the CG method: Let x^0 be a starting solution and $r^0 := b - Ax^0$ the starting residual. Assuming, $K_k := \text{span}\{d^0, \ldots, d^{k-1}\}$ and $x^k \in x^0 + K_k$ and the residual $r^k = b - Ax^k$ are given. Then d^k is calculated according to Lemma 6.4 as

$$\beta_{k-1} = -\frac{(r^k, Ad^{k-1})}{(d^{k-1}, Ad^{k-1})}, \quad d^k = r^k + \beta_{k-1}d^{k-1}. \tag{6.7}$$

For the new coefficient α_k in $x^{k+1} = x^0 + \sum_{i=0}^{k} \alpha_i d^i$, we test the Galerkin equation (6.2) with d^k

$$0 \stackrel{!}{=} \left(\underbrace{b - Ax^0}_{=d^0} - \sum_{i=0}^{k} \alpha_i Ad^i, d^k \right) = (b - Ax^0, d^k) - \alpha_k(Ad^k, d^k)$$

$$= (b - Ax^k - A\underbrace{(x^0 - x^k)}_{\in K_k \perp Ad^k}, d^k) - \alpha_k(Ad^k, d^k).$$

Thus,

$$\alpha_k = \frac{(r^k, d^k)}{(Ad^k, d^k)}, \quad x^{k+1} = x^k + \alpha_k d^k. \tag{6.8}$$

From this, the new residual r^{k+1} can be directly determined as

$$r^{k+1} = b - Ax^{k+1} = b - Ax^k - \alpha_k Ad^k = r^k - \alpha_k Ad^k. \tag{6.9}$$

We summarize (6.7)–(6.9) and formulate the classical CG method:

Algorithm 6.5: Conjugate Gradient Method

Input: $A \in \mathbb{R}^{n \times n}$ symmetric positive definite, $x^0 \in \mathbb{R}^n$ and $r^0 = d^0 = b - Ax^0$.

1 **for** $k = 0, 1, \ldots$ **do**
2      $\alpha_k = \frac{(r^k, d^k)}{(Ad^k, d^k)}$
3      $x^{k+1} = x^k + \alpha_k d^k$
4      $r^{k+1} = r^k - \alpha_k Ad^k$
5   **if** $\|r^{k+1}\| = 0$ **then**
6   Stop
7      $\beta_k = -\frac{(r^{k+1}, Ad^k)}{(d^k, Ad^k)}$
8      $d^{k+1} = r^{k+1} + \beta_k d^k$

Result: Approximate solution $x = x^{k+1}$ of $Ax = b$.

For our Python implementation, we can make the method even more efficient. Due to the *Galerkin orthogonality* and since the Krylov spaces form a sequence of subspaces, it holds

$$(r^k, d^j) = 0, \quad j = 0, \ldots, k-1, \qquad (r^k, r^j) = 0, \quad j = 0, \ldots, k-1.$$

With this, we have with (6.7), that

$$(d^k, r^k) = (r^k - \beta_{k-1} d^{k-1}, r^k) = \|r^k\|^2,$$

and finally with (6.8), we have

$$\alpha_k = \frac{\|r^k\|^2}{(Ad^k, d^k)}. \tag{6.8b}$$

To simplify β_k, we first observe that from (6.8) it follows $Ax^{k+1} = Ax^k + \alpha_k Ad^k$. With the pairwise orthogonality of the residuals and (6.8b), we then have, with a zero addition,

$$(r^{k+1}, Ad^k) = \frac{1}{\alpha_k}(r^{k+1}, Ax^{k+1} - Ax^k) = \frac{1}{\alpha_k}(r^{k+1}, r^k - r^{k+1}) = -\frac{\|r^{k+1}\|}{\|r^k\|}(Ad^k, d^k).$$

Thus,

$$\beta_k = \frac{\|r^{k+1}\|}{\|r^k\|}.$$

This saves us a scalar product and the separate evaluation of the norm $\|r^{k+1}\|$.

🐍 Implementation 6.6: Conjugate Gradient Method

```python
def cg(A, b, x, k=100, tol=1e-5):
    x = x.copy()
    d, Ad, r = [np.zeros_like(r) for _ in range(3)]
    r[:] = b - A.dot(x)
    nrm_r2 = r.dot(r)
    tol = tol**2
    for i in range(k):
        if nrm_r2 < tol:
            break
        Ad[:] = A.dot(d)
        dAd = d.dot(Ad)
        alpha = nrm_r2 / dAd
        x[:] += alpha * d
        r[:] -= alpha * Ad
        nrm_r2_new = r.dot(r)
        beta = nrm_r2_new / nrm_r2
        d[:] = r + beta * d
        nrm_r2 = nrm_r2_new
    else:
        print(f'CG method did not converge after {k} iterations.')
    return x, i
```

In exact arithmetic, the CG method delivers a solution for an n-dimensional problem after (at most) n steps and can therefore, in principle, be considered as a direct method.

Theorem 6.7 (CG as a Direct Method) *The CG method terminates for any starting vector* $x^0 \in \mathbb{R}^n$ *in round-off free calculation after at most n steps with $x^n = x$.*

Proof We need to distinguish between two cases. Suppose the iteration breaks in the k-th step, with $k \leq n$. Then, according to Lemma 6.3, we have

$$b - Ax^k \in K_k(d^0, A)$$

and due to Galerkin orthogonality, $r^k = 0$, so $x^k = x$.

If the nth step of the CG method is reached, then again according to Lemma 6.3

$$K_0(d^0, A) \subsetneq K_1(d^0, A) \subsetneq \cdots \subsetneq K_{n-1}(d^0, A) = \mathbb{R}^n.$$

Thus, $x^n, r^n \in \mathbb{R}^n$. We can therefore test the Galerkin equation (6.3) with $y = r^n = b - Ax^n$. This implies $\|r^n\| = 0$, which is exactly the case when $x^n = x$. $\qquad\qquad\square$

Although the CG method is in principle a direct method, it is used in practice as an approximate, iterative method. Due to rounding errors, the search directions $\{d^0, \ldots, d^{k-1}\}$ will never be exactly orthogonal.

The analysis of the convergence speed of the CG method is quite complex and is based on aspects of approximation theory, which we will only introduce in Sect. 9.5.2 of this book. The proof for the following convergence result is therefore given as Sect. 11.3 in Chap. 11.

Theorem 6.8 (Convergence of the CG Method) *Let $A \in \mathbb{R}^{n\times n}$ be symmetric positive definite, $b \in \mathbb{R}^n$ be the right-hand side and $x^0 \in \mathbb{R}^n$ be any initial value. Then, for the iterates of the CG method, it holds*

$$\|x^k - x\|_A \leq 2 \left(\frac{1 - 1/\sqrt{\kappa}}{1 + 1/\sqrt{\kappa}} \right)^k \|x^0 - x\|_A, \quad k \geq 0,$$

with the spectral condition $\kappa = \mathrm{cond}_2(A)$ of the matrix A.

Example 6.9 We once more consider the model matrix $A \in \mathbb{R}^{n\times n}$ with $n = m^2$ and

$$A = \begin{pmatrix} B & -I & & \\ -I & B & \ddots & \\ & \ddots & \ddots & -I \\ & & -I & B \end{pmatrix}, \quad B = \begin{pmatrix} 4 & -1 & & \\ -1 & 4 & \ddots & \\ & \ddots & \ddots & -1 \\ & & -1 & 4 \end{pmatrix}, \quad I = \begin{pmatrix} 1 & & & \\ & \ddots & & \\ & & \ddots & \\ & & & 1 \end{pmatrix}$$

Table 6.1 Results for the CG method, applied to the model matrix

n	# Steps	Time [s]	Residual
100	16	0.0044	$3.58 \cdot 10^{-14}$
400	36	0.0090	$4.87 \cdot 10^{-7}$
900	55	0.0094	$6.42 \cdot 10^{-7}$
1600	74	0.0271	$6.92 \cdot 10^{-7}$
2500	93	0.1072	$6.86 \cdot 10^{-7}$
3600	112	0.2532	$7.18 \cdot 10^{-7}$
4900	131	1.0982	$7.47 \cdot 10^{-7}$
6400	150	2.5969	$7.76 \cdot 10^{-7}$
8100	169	4.6177	$8.36 \cdot 10^{-7}$
10000	188	7.9617	$8.60 \cdot 10^{-7}$

with $B, I \in \mathbb{R}^{m \times m}$ from Example 3.74 and apply our Python implementation to it. We take $m = 10, 20, \ldots, 100$ and set the error tolerance to $\mathtt{tol} = 10^{-6}$. Since we know that $\kappa \approx n$, we expect that the method will converge more slowly as n increases. We also observe this in the summary of the results in Table 6.1. ◀

6.3   Preconditioned Conjugate Gradient Method

The convergence rate of the CG method depends on the condition number of the system matrix. It holds

$$\rho_{CG} = \frac{1 - \frac{1}{\sqrt{\kappa}}}{1 + \frac{1}{\sqrt{\kappa}}} = 1 - \frac{2}{\sqrt{\kappa}} + O\left(\frac{1}{\kappa}\right).$$

Example 6.10 For discretizations of elliptic partial differential equations in two dimensions, we have to expect $\kappa = O(N)$, i.e.,

$$\rho_{CG} \sim 1 - \frac{1}{\sqrt{N}},$$

so that the convergence rate becomes worse with increasing problem size N. ◀

The idea of preconditioning is a reformulation of the linear system $Ax = b$ by multiplication with a (regular) matrix $P^{-1} \approx A^{-1}$, i.e.,

$$Ax = b \quad \Leftrightarrow \quad P^{-1}Ax = P^{-1}b \tag{6.10}$$

with the goal that the new matrix $P^{-1}A$ is better conditioned than the matrix A itself. The choice of P is again determined by the following goals: P^{-1} must be easy to calculate and should at the same time be a good approximation to A^{-1}, in any case improve the condition number, i.e., $\text{cond}_2(P^{-1}A) \ll \text{cond}_2(A)$. The application of the preconditioner from the left in (6.10) is called *left preconditioning*, correspondingly the application from the right is called *right preconditioning*

$$Ax = b \quad \Leftrightarrow \quad AP^{-1}y = b, \quad Px = y.$$

A preconditioner does not have to be represented and stored as a matrix. It can just as well be the application of a simple method, e.g., a step of the Jacobi iteration.

Example 6.11 (*Preconditioning*) The matrix $A \in \mathbb{R}^{2\times 2}$

$$A = \begin{pmatrix} 10000 & 0.9 \\ 0.9 & 1 \end{pmatrix}$$

has the spectral condition $\text{cond}_2(A) \approx 10000$. For preconditioning we choose the Gauss-Seidel method from Sect. 3.7 with $P_{GS}^{-1} = (L + D)^{-1}$, where $A = L + D + R$. The preconditioned matrix $P_{GS}^{-1}A$ then has the condition number $\text{cond}_2(P^{-1}A) \approx 2.4$. ◄

The CG method can only be applied to linear systems with a symmetric matrix. That is, the preconditioned matrix $P^{-1}A$ must also be symmetric. For the following let $P \in \mathbb{P}^{n\times n}$ be a matrix, which allows the notation

$$P = KK^T.$$

Then, it holds

$$Ax = b \quad \Leftrightarrow \quad \underbrace{K^{-1}AK^{-T}}_{=:\tilde{A}} \underbrace{K^T x}_{=:\tilde{x}} = \underbrace{K^{-1}b}_{=:\tilde{b}},$$

thus

$$\tilde{A}\tilde{x} = \tilde{b},$$

with a matrix $\tilde{A}$, which is again symmetric. In the case

$$\text{cond}_2(\tilde{A}) \ll \text{cond}_2(A)$$

and if the application of K^{-1} is "cheap", a significant acceleration can be achieved by considering the *preconditioned system* $\tilde{A}\tilde{x} = \tilde{b}$.

The CG method with preconditioning can be formulated as follows:

Algorithm 6.12: CG Method With Preconditioning

> **Input:** $A \in \mathbb{R}^{n \times n}$ symmetric positive definite, $P = KK^T$ symmetric preconditioner and
> $x^0 \in \mathbb{R}^n$.
>
> 1 $r^0 = b - Ax^0$
> 2 $Pp^0 = r^0$
> 3 **for** $k = 0, 1, \ldots$ **do**
> 4 $\quad \alpha_k = \frac{(r^k, d^k)}{(Ad^k, d^k)}$
> 5 $\quad x^{k+1} = x^k + \alpha_k d^k$
> 6 $\quad r^{k+1} = r^k - \alpha_k Ad^k$
> 7 $\quad Pp^{k+1} = r^{k+1}$
> 8 $\quad \beta_k = \frac{(r^{k+1}, p^{k+1})}{(r^k, p^k)}$
> 9 $\quad d^{k+1} = p^{k+1} + \beta_k d^k$
>
> **Result:** Approximate solution $x = x^{k+1}$ of $Ax = b$.

In each step of the procedure, the application of the preconditioner P is added as additional effort. When selecting the preconditioner, care must be taken that it allows the notation $P = KK^T$—even if the parts K and K^T are not used in the procedure. The usual approximation methods from Sect. 3.7 can be used as preconditioners:

- *Jacobi preconditioning*
  We choose $P \approx D^{-1}$, where D is the diagonal part of the matrix A. Here, it holds

$$D = D^{\frac{1}{2}}(D^{\frac{1}{2}})^T,$$

  that is, for $D_{ii} > 0$ this preconditioner is admissible. For the preconditioned matrix, it holds

$$\tilde{A} = D^{-\frac{1}{2}} A D^{-\frac{1}{2}} \quad \Rightarrow \quad \tilde{a}_{ii} = 1$$

  and the preconditioning causes a scaling of the matrix entries. This simple type of preconditioning can improve the condition.

- *SSOR preconditioning*
  The SSOR method is a symmetric variant of the SOR method and is based on the factorization

$$P = (D + \omega L)D^{-1}(D + \omega R) = \underbrace{(D^{\frac{1}{2}} + \omega L D^{-\frac{1}{2}})}_{K}\underbrace{(D^{\frac{1}{2}} + \omega D^{-\frac{1}{2}} R)}_{=K^T},$$

  which in the case of symmetric matrices $L = R^T$ fulfills the necessary condition for a CG preconditioner.

  For the model matrix of the discretization of the Laplace equation, with optimal choice of ω (which is a non-trivial task), the relation

$$\mathrm{cond}_2(\tilde{A}) = \sqrt{\mathrm{cond}_2(A)}$$

can be shown. Thus, the convergence rate improves significantly. The number of steps to achieve an error reduction by the fixed factor ϵ improves to

$$t_{CG}(\epsilon) = \frac{\log(\epsilon)}{\log(1 - \kappa^{-\frac{1}{2}})} \approx -\frac{\log(\epsilon)}{\sqrt{\kappa}}, \quad \tilde{t}_{CG}(\epsilon) = \frac{\log(\epsilon)}{\log(1 - \kappa^{-\frac{1}{4}})} \approx \frac{\log(\epsilon)}{\sqrt[4]{\kappa}}.$$

Instead of, for example, 100, only ten steps are needed. However, the determination of an optimal parameter ω is generally complicated.

- *Incomplete Cholesky Decomposition*
  Finally, we consider preconditioning by an incomplete Cholesky factorization. For the symmetric positive definite matrix A, we create the approximate factorization

$$A \approx C^T C,$$

where the factor C uses the same structure as the lower left part of the matrix A. That is, $C_{ij} \neq 0$ only when $A_{ij} \neq 0$. It holds

$$A = C^T C + N,$$

where the sparsity (i.e., the number of non-zeros) and significance of N depend heavily on the sorting of the matrix A (see Sect. 3.4). An analysis of this preconditioning is complex. In practice, however, this method proves to be far superior to the other methods. In particular, it does not require the determination of the parameter ω. The application is, of course, much "more expensive" than the application of Jacobi or SSOR preconditioning.

Example 6.13 We continue Example 6.9 and consider the preconditioning of the CG method using the model matrix. In particular, we consider Jacobi and SOR preconditioning, as we know the optimal relaxation parameter from Example 3.74. The results can be seen in Table 6.2. We observe that the Jacobi preconditioning hardly reduces the number of necessary steps. However, the application of this preconditioner is very cheap, so the computation times are very similar to the CG method in Example 6.9. In contrast, the number of necessary steps in the SOR-preconditioned CG method grows very slowly. However, the application of this preconditioner in this implementation (with `numpy.linalg.solve`) is so expensive that the computation times have worsened. ◀

6.4 The Arnoldi Method

The efficiency of the CG method lies in the fast two-stage orthogonalization of the Krylov space $K_k(d^0, A)$. We now consider the case of general, i.e., non-symmetric matrices $A \in \mathbb{R}^{n \times n}$. For $v \in \mathbb{R}^n$ arbitrary, let $K_k(v, A) = \text{span}\{v, Av, \ldots, A^{k-1}v\}$ be the Krylov subspace. The vector $v \in \mathbb{R}^n$ does not initially have to be related to a linear system. The

Table 6.2 Results for Example 6.13: Results for the CG method, applied to the model matrix, for two different preconditioners

n	Jacobi			SOR		
	# Steps	Time [s]	Residual	# Steps	Time [s]	Residual
100	16	0.0007	3.58×10^{-14}	12	0.0040	6.05×10^{-7}
400	35	0.0008	1.25×10^{-6}	17	0.0246	9.23×10^{-7}
900	54	0.0048	1.40×10^{-6}	22	0.1803	4.55×10^{-7}
1600	73	0.0344	1.21×10^{-6}	26	0.8807	4.67×10^{-7}
2500	91	0.1126	1.73×10^{-6}	29	3.3557	8.18×10^{-7}
3600	110	0.2500	1.60×10^{-6}	33	11.6160	6.75×10^{-7}
4900	128	1.1094	1.98×10^{-6}	36	29.7653	6.94×10^{-7}
6400	147	2.4710	1.78×10^{-6}	39	71.3470	5.25×10^{-7}
8100	166	4.6795	1.71×10^{-6}	–	–	–
10000	185	7.7390	1.63×10^{-6}	–	–	–

Arnoldi method is an iteration for building an orthogonal basis of the Krylov space $K_k(v, A)$. The simplest variant of the Arnoldi method can be written using the Gram-Schmidt orthogonalization method, see Sect. 4.2, as well as Theorem 2.10.

Algorithm 6.14: Arnoldi Method with Gram-Schmidt

> **Input:** Matrix $A \in \mathbb{R}^{n \times n}$ and vector $v \in \mathbb{R}^n$
> 1 $v^1 = v^1 / \|v^1\|$
> 2 **for** $i = 1$ **to** k **do**
> 3   **for** $j = 1$ **to** i **do**
> 4 $h_{ji} = (Av^i, v^j)$
> 5 $w^i = Av^i - \sum_{j=1}^{i} h_{ji} v^j$
> 6 $h_{i+1,i} = \|w^i\|$
> 7   **if** $h_{i+1,i} = 0$ **then**
> 8   Stop
> 9 $v^{i+1} = w^i / h_{i+1,i}$
> **Result:** Matrix $H_k = (h_{ji})_{ji} \in \mathbb{R}^{k \times k}$ and vectors $v^1, \ldots, v^k \in \mathbb{R}^n$. Possibly $h_{k+1,k}$ and vector v^{k+1}.

The Arnoldi algorithm is the classic Gram-Schmidt method for orthonormalizing $K_k(v, A)$. One difference from the basic algorithm from Theorem 2.10 is that the vectors to be orthogonalized appear in the course of the iteration, i.e., Av^i in line 4.

Theorem 6.15 *If the Arnoldi method does not stop at line 7, then the vectors $v^1, \ldots, v^k$ form an orthonormal system of $K_k(v, A)$ with respect to the Euclidean scalar product.*

Proof The orthonormality follows by construction. It is also easy to see that

$$v^i \in K_i(v, A)$$

holds. This can be shown by induction. For $i = 1$, the statement is clear. So let $v^i \in K_i(v, A)$. Then $Av^i \in K_{i+1}(v, A)$ and $\sum_{j=1}^{i} h_{ji} v^j \in K_i(v, A) \subset K_{i+1}(v, A)$, see line 5. This implies $w^i \in K_{i+1}(v, A)$ and with line 9 also $v^{i+1} \in K_{i+1}(v, A)$.

A system of k orthonormal vectors must have dimension k. Thus, the complete space $K_k(v, A)$ is generated. $\qquad\qquad\square$

If in line 7 $h_{i+1,i} = 0$ would hold, the vector Av^i already lies in the space spanned by $\{v^1, \ldots, v^i\}$, see Lemma 6.3.

Remark 6.16 (*Orthogonality in Arnoldi and CG*) The CG method generates an orthonormal basis with respect to the scalar product $(A\cdot, \cdot)$. In the Arnoldi method, orthonormality is achieved with respect to the Euclidean scalar product $(\cdot, \cdot)$. The choice of the weighted scalar product in the CG method is motivated by the fact that the Galerkin equation (6.3) is formulated in such a way that the test space and the trial space coincide:

$$x^k \in x^0 + K_k(d^0, A) : \quad (b - Ax^k, y) = 0 \quad \forall y \in K_k(d^0, A).$$

If we insert the representation

$$x^k = x^0 + \sum_{i=1}^{k} \alpha_i d^i,$$

we get

$$\sum_{i=1}^{k} \alpha_i (Ad^i, d^j) = (d^0, d^j) \quad \forall j = 1, \ldots, k \quad \Rightarrow \quad \alpha_i = (d^0, d^i),$$

and the solution can be directly determined if the orthonormal basis is known. This approach works in the CG method, of course, only if $(A\cdot, \cdot)$ is really a scalar product, i.e., if the matrix A is symmetric positive definite. This is not possible in the general case. $\qquad\blacklozenge$

Theorem 6.17 (Properties of the Arnoldi Method) *Let $A \in \mathbb{R}^{n \times n}$, $v^1 \in \mathbb{R}^n$ with $\|v^1\| = 1$ and h_{ij} for $i, j = 1, \ldots, k$ and $h_{k,k+1}$ be the coefficients created by the Arnoldi method. Further, let $v^1, \ldots, v^k$ be the computed orthonormal vectors and w^k be the last orthogonalized vector. With the matrices $V_k = (v^1, \ldots, v^k) \in \mathbb{R}^{n \times k}$ and*

$$
H_k = \begin{pmatrix} h_{1,1} & h_{1,2} & \cdots & & \cdots & h_{1,k} \\ h_{2,1} & h_{2,2} & \cdots & & \cdots & h_{2,k} \\ 0 & \ddots & \ddots & & & \vdots \\ \vdots & \ddots & \ddots & \ddots & & \vdots \\ 0 & \cdots & 0 & h_{k,k-1} & h_{k,k} \end{pmatrix} \in \mathbb{R}^{k \times k}, \quad \tilde{H}_k = \left(\begin{array}{c} H_k \\ \hline 0 \cdots \ 0 \ \ h_{k+1,k} \end{array} \right) \in \mathbb{R}^{k+1 \times k},
$$

it holds

$$
AV_k = V_k H_k + w^k e_k^T = V_{k+1} \tilde{H}_k, \tag{6.11}
$$

and

$$
V_k^T A V_k = H_k. \tag{6.12}
$$

Proof Lines 5, 4 and 9 yield in the case $h_{i+1,i} \neq 0$

$$
Av^i = w^i + \sum_{j=1}^{i} h_{ji} v^j.
$$

For $i < k$, $w^i = h_{i+1,i} v^{i+1}$, i.e., it holds

$$
Av^i = \sum_{j=1}^{i+1} h_{ji} v^j \quad i = 1, \ldots, k-1.
$$

Together, this yields exactly (6.11). Then, (6.12) follows when multiplying with V_k^T and exploiting the orthonormality, i.e., that $V_k^T V_k$ is the identity matrix in $\mathbb{R}^{n \times n}$. Finally, $V_k^T v^k = 0$, since either w^k is zero, or w^k is orthogonal to all $v^1, \ldots, v^k$. $\qquad\square$

Remark 6.18 (*Arnoldi Method and Hessenberg Form*) The Arnoldi method is an iterative method for transforming a matrix A into Hessenberg form, which fundamentally differs from the algorithms examined in Sect. 5.5.

We recall: the construction of the Hessenberg form based on Householder transformations is a similarity transformation that multiplies the matrix from the left and right with Householder matrices

$$
Q^T A Q = H,
$$

where $Q \in \mathbb{R}^{n \times n}$ is an orthogonal matrix. The Arnoldi iteration delivers a similar result

$$
V_k^T A V_k = H_k.
$$

However, $V_k \in \mathbb{R}^{n \times k}$ is not square and of course not regular for $k < n$. So it is not a similarity transformation. If the Arnoldi method does not terminate prematurely at n iterations but runs until $V_n \in \mathbb{R}^{n \times n}$, it also generates the Hessenberg form and a similarity transformation. $\blacklozenge$

The effort to create the Hessenberg form with Householder transformations is of $\frac{2}{3}n^3 + O(n^2)$, compare Remark 5.35. The essential difference of the Arnoldi method, however, is that it can be performed for $k \ll n$ and that it does not rely on matrix transformations but only on matrix-vector products and scalar products.

Theorem 6.19 (Effort of the Arnoldi Method) *Let $A \in \mathbb{R}^{n \times n}$. Assume that the Arnoldi method does not terminate prematurely. Then, to carry out the method using Algorithm 6.14, essentially k matrix-vector products as well as $\frac{k^2}{2} + O(k)$ scalar products in $\mathbb{R}^n$ need to be calculated.*

Proof The product $\tilde{v}^i = A v^i$ only needs to be calculated once per iteration and can then be stored. In step i, a total of i scalar products need to be calculated in line 4, so in total $\frac{k(k+1)}{2}$. In addition, there are a total of k vector additions, which are included in the effort of the scalar products.

For a sparse matrix with only $O(n)$ non-zero entries, the effort is $O(k^2 n + kn)$, and for a dense one, the effort is $O(k^2 n + kn^2)$ operations. If the iteration is complete, the effort is cubic.

Remark 6.20 (Robustness of the Arnoldi Method) In Sect. 4.2, we discussed the stability problems of the Gram-Schmidt method. These also apply to the Arnoldi method. The algorithm presented above is simple, but suffers from stability problems. In practical applications, the method should therefore be created based on the modified Gram-Schmidt method or with the help of Householder transformations.   ♦

6.4.1 The Eigenvalue Problem

The similarity transformation to Hessenberg form is performed because the eigenvalues of the Hessenberg matrix can be calculated very efficiently, e.g., with the QR iteration in Theorem 5.36. The same also applies to the Arnoldi method and the much smaller Hessenberg matrix $H_k \in \mathbb{R}^{k \times k}$. However, the transformation $V_k^T A V_k = H_k$ is not a similarity transformation, as the generally non-square matrix V_k is not invertible. From Theorem 6.17 the following relation follows:

Corollary 6.21 *Let $A \in \mathbb{R}^{n \times n}$, $V_k \in \mathbb{R}^{n \times k}$, $H_k \in \mathbb{R}^{k \times k}$ and $h_{k+1,k} \in \mathbb{R}$ be the result of the Arnoldi method. In the case $h_{k+1,k} = 0$, the eigenvalues of H_k are also eigenvalues of A. For $\lambda \in \mathbb{C}$ and $x \in \mathbb{C}^k$ with*

$$H_k x = \lambda x$$

it holds

$$A(V_k x) = \lambda(V_k x).$$

Proof In the case $h_{k+1,k} = 0$, (6.11) simplifies to

$$A V_k = V_k H_k.$$

For a pair of eigenvalue and eigenvector, it then immediately follows that

$$A(V_k x) = V_k H_k x = \lambda(V_k x).$$

$\square$

If the Arnoldi iteration terminates prematurely with $h_{i+1,i} = 0$, it allows a very efficient calculation of some of the eigenvalues of the matrix A. Of course, the case $h_{k+1,k} \neq 0$, which is usually to be assumed, is more interesting. With $v^k = h_{k+1,k} v^{k+1}$ and $\|v^{k+1}\| = 1$, compare Algorithm 6.14, it then holds for an eigenvalue $\lambda \in \mathbb{C}$ and eigenvector $x \in \mathbb{C}^k$ of H_k that

$$A(V_k x) = \lambda(V_k x) + h_{k+1,k} v^{k+1} e_k^T x \quad \Leftrightarrow \quad (A - \lambda I)(V_k x) = h_{k+1,k} v^{k+1} e_k^T x.$$

The pair $\lambda \in \mathbb{C}$ and $V_k x \in \mathbb{C}^n$ do not satisfy the eigenvalue equation. However, multiplying the equation by V_k^T shows

$$V_k^T (A - \lambda I)(V_k x) = 0,$$

since $V_k = (v_1, \ldots, v_k)$ is orthonormal to v^{k+1}. In other words, the following Galerkin equation is obtained

$$\big((A - \lambda I)(V_k x), v^j\big) = 0 \quad j = 1, \ldots, m.$$

The residual of the eigenvalue equation is orthogonal to the Krylov space. For $k \to n$, a better approximation can therefore be assumed. The eigenvalues and eigenvectors of H_k are called the *Ritz values* and *Ritz vectors* of A with respect to the Krylov space $K_k(v, A)$.

The analysis of the Arnoldi method is not yet complete. It turns out that especially separated eigenvalues converge very quickly, i.e., eigenvalues that are isolated in magnitude. This is similar to the power method and the inverse iteration, where the separation of the eigenvalues also influences the convergence speed. However, the Arnoldi method converges not only towards the largest eigenvalues but throughout the spectrum. For further analyses, we refer to the literature [83, 87].

When applied to symmetric positive definite matrices $A \in \mathbb{R}^{n \times n}$, the Arnoldi iteration can be significantly simplified. As with the CG method from Sect. 6.2, the Gram-Schmidt method can be replaced by a two-stage iteration. The so-called *Lanczos method* generates a symmetric tridiagonal matrix instead of a Hessenberg matrix. The eigenvalues of this can again be calculated very quickly. For details, we again refer to the two books mentioned above.

6.5   The GMRES Method

The idea of the CG method can be transferred to linear systems $Ax = b$ with general, non-symmetric matrices $A \in \mathbb{R}^{n \times n}$. We choose a starting vector $x^0 \in \mathbb{R}^n$ and with the residual $d^0 = (b - Ax^0)$ we form the Krylov space $K_k(d^0, A)$.

With the Arnoldi method, we can create an orthonormal basis $\{v^1, \ldots, v^k\}$ of $K_k(d^0, A)$. We consider the Galerkin equation

$$x^k \in x^0 + K_k(d^0, A): \quad (b - Ax^k, y) = 0 \quad \forall y \in Y_k \tag{6.13}$$

without specifying the test space Y_k for the moment. For the unknown solution x^k, we make the approach

$$x^k = x^0 + \sum_{i=1}^{k} \alpha_i d^i$$

with free coefficients α_i and we get

$$x^k \in x^0 + K_k(d^0, A): \quad \sum_{i=1}^{k} \alpha_i (Ad^i, y) = (d^0, y) \quad \forall y \in Y_k. \tag{6.14}$$

In contrast to the CG method (Remark 6.16), this system cannot be directly resolved for the coefficients α_i. Even when choosing the test space $Y_k = K_k(d^0, A)$, it is not true that $(Ad^i, d^j) = \delta_{ij}$, because the orthonormality is with respect to the Euclidean product. For a general, i.e., non-symmetric and non-positive definite matrix A, $(A\cdot, \cdot)$ does not give a scalar product at all.

The GMRES method, instead, is based on the representation of the solution as a minimization problem. We determine the approximation x^k as

$$x^k \in x^0 + K_k(d^0, A) \quad \|b - Ax^k\|_2 \leq \|b - Ay^k\|_2 \quad \forall y^k \in x^0 + K_k(d^0, A).$$

The name GMRES comes from this notation, standing for *Generalized Minimal Residual method*. We minimize the residual, which is a generalization that can also be applied to non-symmetric matrices. Similar minimization problems also occur in the least squares method of Sect. 4.5.

Assuming x^k is a minimum, then it must necessarily be a stationary point. It must therefore hold for arbitrary $y \in K_k(d^0, A)$ that

$$0 \stackrel{!}{=} \frac{d}{ds} \frac{1}{2} \|b - A(x + sy)\|^2 \Big|_{s=0} = \left(b - A(x + sy), -Ay\right)\Big|_{s=0} = -(b - Ax, Ay).$$

The condition for this corresponds exactly to the Galerkin equation

$$x^k \in x^0 + K_k(d^0, A) : \quad (b - Ax^k, y) = 0 \quad \forall y \in AK_k(d^0, A). \tag{6.15}$$

In the GMRES method, the trial space $x^0 + K_k(d^0, A)$ and the test space $AK_k(d^0, A)$ are chosen. The minimization of $\|b - Ax^k\|$ in the subspace $x^k \in x^0 + K_k(d^0, A)$ is (except for the initial value x^0) precisely the solution of an overdetermined linear system, as discussed in Sect. 4.5. Instead of directly calculating the solution of this overdetermined system with the QR factorization as in Chap. 4, we can use the orthonormal representation obtained with the Arnoldi method and greatly simplify the problem:

Theorem 6.22 (GMRES—Solving the Minimization Problem) *Let $A \in \mathbb{R}^{n \times n}$, $b \in \mathbb{R}^n$ and $x^0 \in \mathbb{R}^n$. For $d^0 = b - Ax^0$ and $k \in \mathbb{N}$, let $V_k \in \mathbb{R}^{n \times k}$, $\tilde{H}_k \in \mathbb{R}^{k+1,k}$ be the matrices created by the Arnoldi method for orthonormalizing $K_k(d^0, A)$. Then*

$$\min_{x \in x^0 + K_k(d^0, A)} \|b - Ax\|_2 = \min_{y \in \mathbb{R}^k} \left\| \|d^0\| e_1 - \tilde{H}_k y \right\|_2,$$

where $e_1 \in \mathbb{R}^{k+1}$ is the first unit vector in $\mathbb{R}^{k+1}$.

Proof With the ansatz

$$x = x^0 + \sum_{i=1}^{k} \alpha_i v^i$$

and with Theorem 6.17, we have

$$b - Ax = b - Ax^0 - \sum_{i=1}^{k} \alpha_i Av^i = d^0 - AV_k y = d^0 - V_{k+1}\tilde{H}_k y.$$

The first column vector in V_{k+1} is $v^1 = d^0/\|d^0\|$, so

$$b - Ax = \|d^0\| v^1 - V_{k+1}\tilde{H}_k y = V_{k+1}\left(\|d^0\| e_1 - \tilde{H}_k y\right).$$

Now, due to the orthonormality of the vectors v^i, for arbitrary vectors $z \in \mathbb{R}^{k+1}$ in the Euclidean scalar product, we have

$$\|V_{k+1} y\|_2^2 = (V_{k+1}y, V_{k+1}y)_2 = (V_{k+1}^T V_{k+1}y, y)_2 = (y, y)_2 = \|y\|_2^2.$$

It follows that

$$\|b - Ax\|_2 = \left\| \|d^0\| e_1 - \tilde{H}_k y \right\|_2.$$

$\square$

To solve the minimization problem in the Krylov space $K_k(d^0, A)$, it is sufficient to solve an overdetermined system with the matrix $\tilde{H}_k \in \mathbb{R}^{k+1,k}$. This system has the form

$$
\begin{pmatrix}
h_{11} & h_{12} & \cdots & \cdots & h_{1k} \\
h_{21} & h_{22} & \ddots & & \vdots \\
0 & h_{32} & \ddots & & \vdots \\
0 & \ddots & \ddots & \ddots & \vdots \\
\vdots & & \ddots & h_{k,k-1} & h_{kk} \\
0 & \cdots & \cdots & 0 & h_{k+1,k}
\end{pmatrix}
\begin{pmatrix}
y_1 \\ y_2 \\ \vdots \\ \vdots \\ \vdots \\ y_k
\end{pmatrix}
=
\begin{pmatrix}
\|d^0\| \\ 0 \\ \vdots \\ \vdots \\ \vdots \\ 0
\end{pmatrix}.
\tag{6.16}
$$

The matrix $\tilde{H}_k$ is almost upper right triangular. To create a QR factorization $\tilde{H}_k = Q_k R_k$, only the elements below the diagonal, i.e., $h_{21}, h_{32}, \ldots, h_{k+1,k}$, need to be eliminated. This can be achieved, for example, with k steps of Givens rotations, see Sect. 4.4. The effort for this is $O(k^2)$. With the help of the QR factorization, the normal equation system can be solved. It is

$$
\tilde{H}_k^T \tilde{H}_k y = \|d^0\| \tilde{H}_k^T e_1 \quad \Leftrightarrow \quad R_k^T R_k y = \|d^0\| R_k^T Q_k^T e_1,
$$

and if the Arnoldi iteration has not been prematurely terminated, then the matrix R_k is regular and the solution can be determined with back substitution in $O(k^2)$ operations

$$
R_k y = \|d^0\| Q_k^T e_1.
$$

It holds

Theorem 6.23 (GMRES Method) *Let $A \in \mathbb{R}^{n \times n}$ be a regular matrix and $x^0 \in \mathbb{R}^n$ be an arbitrary starting vector and let $k \leq n$. Assume that the matrix A is sparse and the effort for a matrix-vector product is $O(n)$ operations. Assuming the Arnoldi method for orthonormalizing $K_k(b - Ax^0, A)$ does not terminate prematurely, then the GMRES iteration is well defined, it provides a unique solution, and it requires*

$$
O(k^2 n) + O(k^2)
$$

operations.

Proof If the method does not terminate prematurely, then V_k is a matrix of rank k, thus forming an orthonormal basis of $K_k(b - Ax^0, A)$. Since A itself is regular, it follows that $H_k = V_k^T A V_k$ is also regular. For $y \in \mathbb{R}^k$ we have

$$
y \neq 0 \quad \Rightarrow \quad V_k y \neq 0 \quad \Rightarrow \quad A V_k y \neq 0 \quad \Rightarrow \quad H_k = V_k^T A V_k y \neq 0.
$$

The overdetermined linear system is therefore uniquely solvable.

The number of operations results from Theorem 6.19 as well as from the creation of the QR factorization of $\tilde{H}_k$ and the subsequent backward elimination. $\square$

Remark 6.24 (*GMRES and Orthonormalization*) In the GMRES method, the orthonormal basis is not used to calculate the solution coefficients directly. However, the special factorization is still essential, as it allows us to replace the optimization problem $\|b - Ax\|$ for $x \in x^0 + K_k(b - Ax^0, A)$ with an overdetermined system with the matrix $\tilde{H}_k \in \mathbb{R}^{k+1,k}$. This significantly reduces the effort, as we only have to work with "short" vectors of length $O(k)$ instead of vectors of length $O(n)$. $\blacklozenge$

The convergence analysis of the GMRES method, like the Arnoldi method, is not complete. However, in exact arithmetic, as in the case of the CG method, the GMRES method terminates after at most n steps with the exact solution. Furthermore, the GMRES method is monotonically convergent, which is clear since it is based on minimizing $\|b - Ax\|$ in ever larger subspaces. Finally, it can be shown that the GMRES method, when applied to symmetric positive definite matrices, converges as well as the CG method.

6.5.1   Practical Aspects and Implementation

The simple variant of the GMRES method discussed so far still has some weaknesses, which we will examine in more detail in this section:

1. We know that the Gram-Schmidt method for orthonormalization is not stable. For a robust algorithm, this part must be improved and replaced, for example, by Householder transformations or Givens rotations.
2. The underlying Arnoldi Algorithm 6.14 is formulated in such a way that we have to decide at the beginning how many steps k we want to perform. Then, it follows the orthonormalization, and only at the very end, the solution of the linear system. It would be more natural if we proceed step by step as in the CG method and orthonormalize a new vector in each iteration and then immediately improve the approximation of the linear system $x^k \mapsto x^{k+1}$. This would allow the method to be terminated when the desired tolerance is reached.
3. The cost of the GMRES method increases quadratically with the number of steps k and is $O(k^2 n)$. If a massive number of steps is necessary, the method becomes too slow.

6.5.1.1 Robust Variants

A particularly robust variant of GMRES can be achieved based on Householder transformations. However, the efficient implementation is quite complex if care is to be taken to use only the really necessary memory. We therefore refer to the literature [82]. A simpler, but not so stable alternative is the modified Gram-Schmidt algorithm, which we will also use for our implementation. It differs mainly in the order of operations from the original Gram-Schmidt method and is mathematically equivalent if no rounding errors are made, compare Theorem 4.11.

Algorithm 6.25: Arnoldi Method With Modified Gram-Schmidt

> **Input:** Matrix $A \in \mathbb{R}^{n \times n}$ and vector $v \in \mathbb{R}^n$
> 1 $v^1 = v^1 / \|v^1\|$
> 2 **for** $i = 1$ **to** k **do**
> 3 $w^i = Av^i$
> 4   **for** $j = 1$ **to** i **do**
> 5 $h_{ji} = (w^i, v^j)$
> 6 $w^i = w^i - h_{ji}v^j$
> 7 $h_{i+1,i} = \|w^i\|$
> 8   **if** $h_{i+1,i} = 0$ **then**
> 9   Stop
> 10 $v^{i+1} = w^i / h_{i+1,i}$
> **Result:** Matrix $H_k = (h_{ji})_{ji} \in \mathbb{R}^{k \times k}$ and vectors $v^1, \ldots, v^k \in \mathbb{R}^n$. Possibly $h_{k+1,k}$ and vector v^{k+1}.

This variant has better stability properties, but does not reach the robustness of the Householder reflections.

6.5.1.2 Stepwise Approximation

We now consider the second problem mentioned and derive a variant that allows a current approximation x^i to be determined in each step $i = 1, 2, 3, \ldots$. For this purpose, we first repeat the procedure for transforming the Householder matrix $\tilde{H}_k$ with Givens rotations. The system, compare (6.16), has the form

$$
\begin{pmatrix}
h_{11} & h_{12} & \cdots & & \cdots & h_{1k} \\
h_{21} & h_{22} & \ddots & & & \vdots \\
0 & h_{32} & \ddots & & & \vdots \\
0 & \ddots & \ddots & \ddots & & \vdots \\
\vdots & & \ddots & h_{k,k-1} & h_{kk} & \\
0 & \cdots & \cdots & 0 & h_{k+1,k}
\end{pmatrix}
\begin{pmatrix}
y_1 \\ y_2 \\ \vdots \\ \vdots \\ \vdots \\ y_k
\end{pmatrix}
=
\begin{pmatrix}
\|d^0\| \\ 0 \\ \vdots \\ \vdots \\ \vdots \\ 0
\end{pmatrix} .
$$

We multiply from the left with a Givens rotation of the form

$$
G_1 =
\begin{pmatrix}
c_1 & s_1 & & & \\
-s_1 & c_1 & & & \\
& & 1 & & \\
& & & \ddots & \\
& & & & 1
\end{pmatrix}
\in \mathbb{R}^{k+1 \times k+1},
$$

and choose

$$s_1 = \frac{h_{21}}{\sqrt{h_{11}^2 + h_{21}^2}}, \quad c_1 = \frac{h_{11}}{\sqrt{h_{11}^2 + h_{21}^2}}.$$

Multiplying the entire system from the left results in

$$\begin{pmatrix} h_{11}^{(1)} & h_{12}^{(1)} & \cdots & \cdots & h_{1k}^{(1)} \\ 0 & h_{22}^{(1)} & \ddots & & \vdots \\ 0 & h_{32} & \ddots & & \vdots \\ 0 & \ddots & \ddots & \ddots & \vdots \\ \vdots & & \ddots & h_{k,k-1} & h_{kk} \\ 0 & \cdots & \cdots & 0 & h_{k+1,k} \end{pmatrix} \begin{pmatrix} y_1 \\ y_2 \\ \vdots \\ \vdots \\ \vdots \\ y_k \end{pmatrix} = \begin{pmatrix} c_1 \|d^0\| \\ -s_1 \|d^0\| \\ 0 \\ \vdots \\ \vdots \\ 0 \end{pmatrix}.$$

We proceed accordingly with G_2 and

$$s_2 = \frac{h_{32}}{\sqrt{(h_{22}^{(1)})^2 + h_{32}^2}}, \quad c_2 = \frac{h_{22}^{(1)}}{\sqrt{(h_{22}^{(1)})^2 + h_{32}^2}}.$$

After k steps, we obtain

$$\begin{pmatrix} h_{11}^{(1)} & h_{12}^{(1)} & \cdots & \cdots & h_{1k}^{(1)} \\ 0 & h_{22}^{(2)} & \ddots & & \vdots \\ 0 & & \ddots & \ddots & \vdots \\ 0 & & \ddots & \ddots & \ddots & \vdots \\ \vdots & & & 0 & h_{kk}^{(k)} \\ 0 & \cdots & \cdots & 0 & 0 \end{pmatrix} \begin{pmatrix} y_1 \\ y_2 \\ \vdots \\ \vdots \\ \vdots \\ y_k \end{pmatrix} = \begin{pmatrix} \gamma_1^{(k)} \\ \gamma_2^{(k)} \\ \vdots \\ \vdots \\ \vdots \\ \gamma_{k+1}^{(k)} \end{pmatrix}.$$

With the orthogonal matrix

$$G^{(k)} = G_k G_{k-1} \cdots G_1$$

and with

$$\tilde{R}_k = \tilde{H}_k^{(k)} = G^{(k)} \tilde{H}_k$$

as well as the right-hand side

$$\tilde{b}_k = G^{(k)} \|d^0\| e_1 = \begin{pmatrix} \gamma_1^{(k)} \\ \gamma_2^{(k)} \\ \vdots \\ \vdots \\ \vdots \\ \gamma_{k+1}^{(k)} \end{pmatrix},$$

it holds

$$\min_{x \in x^0 + K_k(b - Ax^0, A)} \|b - Ax\|_2 = \min_{y \in \mathbb{R}^k} \left\| \|d^0\| e_1 - \tilde{H}_k y \right\|_2 = \min_{y \in \mathbb{R}^k} \|\tilde{b}_k - \tilde{R}_k y\|_2.$$

This resulting overdetermined linear system is easy to solve, because the normal equation
is

$$R_k^T R_k y = R_k^T \tilde{b}_k \quad \Leftrightarrow \quad R_k y = \begin{pmatrix} \gamma_k^{(k)} \\ \vdots \\ \gamma_k^{(k)} \end{pmatrix},$$

since the last row of $\tilde{R}_k$, thus the last column of $\tilde{R}_k^T$ is zero. The normal equation can be
solved by backward substitution from

$$R_k y = b_k$$

where $R_k \in \mathbb{R}^{k \times k}$ and $b_k \in \mathbb{R}^k$ are obtained from $\tilde{R}_k$ and $\tilde{b}_k$ respectively by omitting the
last row. This also makes it clear that for the solution y^k, and correspondingly for $x^k = x^0 + V_k y^k$, it holds

$$\|b - Ax^k\|_2 = \|\tilde{b}_k - \tilde{R}_k y^k\|_2 = |\gamma^{(k+1)}|.$$

The residual of the current approximation x^k is thus obtained in the course of solving the
small (we always assume $k \ll n$) overdetermined system without having to determine y^k
and thus x^k at all.

We now assume that for iteration k the factorization $\tilde{R}_k = G^{(k)} \tilde{H}_k$ and $\tilde{b}_k = G^{(k)} \|d^0\| e_1$
has been created. If we then jump to $i = k + 1$ in Algorithm 6.25, the entries $h_{j,k+1}$ for
$j = 1, \ldots, k + 1$ are created in lines 4–6 and the entry $h_{k+2,k+1}$ is created in line 7, i.e.,

$$\tilde{H}_{k+1} = \left(\begin{array}{c|c} \tilde{H}_k & \begin{matrix} h_{1,k+1} \\ \vdots \\ h_{k,k+1} \\ h_{k+1,k+1} \end{matrix} \\ \hline 0 & h_{k+2,k+1} \end{array} \right).$$

Multiplication from the left and the matrix

$$
\tilde{G}^{(k)} = \left(
\begin{array}{c|c}
 & 0 \\
G^{(k)} & \vdots \\
 & 0 \\
\hline
0 \quad \cdots \quad 0 & 1
\end{array}
\right)
$$

yields

$$
\tilde{G}^{(k)}\tilde{H}_{k+1} = \left(
\begin{array}{c|c}
 & \tilde{h}_{1,k+1} \\
\tilde{R}_k & \vdots \\
 & \tilde{h}_{k,k+1} \\
 & \tilde{h}_{k+1,k+1} \\
\hline
0 & h_{k+2,k+1}
\end{array}
\right)
= \left(
\begin{array}{ccccc}
h_{11}^{(k)} & \cdots & h_{1k}^{(k)} & \tilde{h}_{1,k+1} \\
0 & \ddots & \vdots & \vdots \\
0 & \ddots & h_{kk}^{(k)} & \tilde{h}_{k,k+1} \\
\vdots & \ddots & 0 & \tilde{h}_{k+1,k+1} \\
0 & \cdots & 0 & \tilde{h}_{k+2,k\cdot}
\end{array}
\right).
$$

The right-hand side of the overdetermined linear system is only extended by a zero entry

$$
\tilde{G}^{(k)}\|d^0\|\tilde{e}_1 = \left(
\begin{array}{c}
\gamma_1^{(k)} \\
\vdots \\
\gamma_{k+1}^{(k)} \\
0
\end{array}
\right) \in \mathbb{R}^{k+2},
$$

where we denote by $\tilde{G}^{(k)} \in \mathbb{R}^{k+2,k+1}$ the matrix extended by a zero row and by $\tilde{e}_1 \in \mathbb{R}^{k+2}$ the first unit vector.

With another Givens rotation G_{k+1} and the coefficients

$$
s_{k+1} = \frac{h_{k+2,k+1}}{\sqrt{h_{k+1,k+1}^2 + h_{k+2,k+1}^2}}, \qquad
c_{k+1} = \frac{h_{k+1,k+1}}{\sqrt{h_{k+1,k+1}^2 + h_{k+2,k+1}^2}},
$$

the linear system becomes

$$
\tilde{R}_{k+1} = G_{k+1}\tilde{H}_{k+1} = \left(
\begin{array}{cccc}
h_{11}^{(k)} & \cdots & h_{1k}^{(k)} & h_{1,k+1}^{(k+1)} \\
0 & \ddots & \vdots & \vdots \\
0 & \ddots & h_{kk}^{(k)} & h_{k,k+1}^{(k+1)} \\
\vdots & \ddots & 0 & h_{k+1,k+1}^{(k+1)} \\
0 & \cdots & 0 & 0
\end{array}
\right)
\left(
\begin{array}{c}
y_1 \\
\vdots \\
\vdots \\
y_{k+2}
\end{array}
\right)
= \left(
\begin{array}{c}
\gamma_1^{(k)} \\
\vdots \\
\gamma_k^{(k)} \\
c_6\gamma^{(k)} \\
-s_6\gamma^{(k)}
\end{array}
\right)
=: \underbrace{\left(
\begin{array}{c}
\gamma_1^{(k+1)} \\
\vdots \\
\gamma_k^{(k+1)} \\
\gamma_{k+1}^{(k+1)} \\
\gamma_{k+2}^{(k+1)}
\end{array}
\right)}_{=:\tilde{b}_{k+1}},
$$

and the residual $\min_{x \in x^0 + K_{k+1}(b-Ax^0,A)} \|b - Ax\|_2 = |\gamma_{k+2}^{(k+1)}|$ can be read off. Assuming the desired tolerance is reached, the system can be solved and $y^{k+1} \in \mathbb{R}^{k+1}$ and $x^{k+1} = x^0 + V_{k+1}y^{k+1} \in \mathbb{R}^n$ are determined.

Algorithm 6.26: GMRES With Modified Gram-Schmidt

Input: Matrix $A \in \mathbb{R}^{n \times n}$, the right-hand side $b \in \mathbb{R}^n$ and an initial value $x^0 \in \mathbb{R}^n$. Desired tolerance $\epsilon > 0$ and maximum number of steps $m \in \mathbb{N}$

1. $d^0 = b - Ax^0$ `// calculate initial residual`
2. $\beta = \|d^0\|$
3. $v^1 = d^0 / \beta$ `// first orthonormal vector`
4. $b^1 = \beta$ `// first right-hand side`
5.   **for** $k = 1$ **to** m **do**
6. $w^k = Av^k$
7.   **for** $j = 1$ **to** k **do**
8. $h_{jk} = (w^k, v^j)$
9. $w^k = w^k - h_{jk} v^j$
10. $h_{k+1,k} = \|w^k\|$
11.   **if** $h_{k+1,k} = 0$ **then**
12.   Jump to line 25
13. $v^{k+1} = w^k / h_{k+1,k}$
14.   **for** $j = 1$ **to** $k - 1$ **do**
15. $\begin{pmatrix} h_{jk} \\ h_{j+1,k} \end{pmatrix} = \begin{pmatrix} c_j & s_j \\ -s_j & c_j \end{pmatrix} \begin{pmatrix} h_{jk} \\ h_{j+1,k} \end{pmatrix}$ `// old Givens rotations`
16. $\alpha = \sqrt{h_{k,k}^2 + h_{k+1,k}^2}$ `// Calculation of Givens rotation`
17. $c_k = h_{k,k} / \alpha$
18. $s_k = h_{k+1,k} / \alpha$
19. $h_{k,k} = \alpha$ `// Apply to $\tilde{H}_k$`
20. $h_{k+1,k} = 0$
21. $z_{k+1} = -s_k z_k$ `// Apply to right-hand side`
22. $z_k = c_k z_k$
23.   **if** $|z_{k+1}| < \epsilon$ **then**
24.   Jump to line 25

    `// Termination of the orthonormalization, as tolerance or number of steps is reached. The matrix` h_{ij} `for` $i, j = 1, \ldots, k$ `is a triangular matrix.`

25. $y_k = z_k / h_{kk}$ `// Calculation of` $y \in \mathbb{R}^k$ `with back substitution`
26.   **for** $i = k - 1$ **to** 1 **do**
27. $y_i = \left(z_i - \sum_{j=i+1}^{k} h_{ij} y_j \right) / h_{ii}$
28. $x^k = x^0 + \sum_{i=1}^{k} y_i v^i$ `// Calculation of the solution`

Result: Approximation $x^k \in \mathbb{R}^n$ of $x \in \mathbb{R}^n$ in $Ax = b$

In Python, we can implement this as follows:

♣ Implementation 6.27: GMRES With Modified Gram-Schmidt

```python
def gmres(A, b, x, m, eps):
    n = A.shape[0]
    x = x.copy()
    d = b.copy() - A.dot(x)
    b = np.zeros(m + 1)
    b[0] = np.linalg.norm(d)
    v = np.zeros((n, m + 1))
    h = np.zeros((m + 1, m))
    R = np.zeros((m, 2, 2))
    v[:, 0] = d / b[0]
    for k in range(m):
        w = A.dot(v[:, k])
        for j in range(k + 1):
            h[j, k] = np.dot(w, v[:, j])
            w -= h[j, k] * v[:, j]
        h[k + 1, k] = np.linalg.norm(w)
        if np.abs(h[k + 1, k]) > 1e-12:
            v[:, k + 1] = w / h[k + 1, k]
            for j in range(k):
                h[j:j + 2, k] = R[j].dot(h[j:j + 2, k])
            alpha = np.linalg.norm(h[k:k + 2, k])
            R[k, 0, 0] = h[k, k] / alpha
            R[k, 0, 1] = h[k + 1, k] / alpha
            R[k, 1, 1], R[k, 1, 0] = R[k, 0, 0], -R[k, 0, 1]
            h[k:k + 2, k] = alpha, 0
            b[k + 1] = R[k, 1, 0] * b[k]
            b[k] = R[k, 0, 0] * b[k]
            if np.abs(b[k + 1]) < eps:
                break
        else:
            break
    y = np.zeros(k + 1)
    y[-1] = b[-1] / h[k, k]
    for i in range(k - 2, -1, -1):
        y[i] = (b[i] - h[i, i + 1:k + 1].dot(y[i + 1:])) / h[i, i]
    for i in range(k):
        x += y[i] * v[:, i]
    return x
```

This variant of the GMRES method stops after m steps, or when the residual is reached. The actual approximation x^k is only calculated when it is clear that the residual is small enough. The main effort of the algorithm still lies in $O(k)$ matrix-vector products with the matrix A, as well as $O(k^2)$ scalar products in $\mathbb{R}^n$. The application of the Givens rotations and the calculation of the minimization solution needs $O(k^2)$ operations and is negligible for $k \ll n$. For sparse matrices, the effort behaves like $O(k^2 n)$.   ◄

Example 6.28 We again consider the model matrix $A \in \mathbb{R}^{n \times n}$ from Examples 3.74, 6.9 and 6.13, and apply our implementation of the GMRES method. We take $m = 20, 30, \ldots, 100$ and the tolerance $\epsilon = 10^{-6}$. The number of GMRES steps, the computational time, the final residual, and the error of the solution can be found in Table 6.3. We see that the required number of steps is comparable to the CG method. The computational time is about twice as large, but the GMRES method can be applied to general matrices.

Table 6.3 Results for the GMRES method, applied to the model matrix

n	# Steps	Time [s]	Residual	Error
400	34	0.0069	2.95×10^{-6}	1.54×10^{-6}
900	53	0.0248	2.18×10^{-6}	2.50×10^{-6}
1600	71	0.0913	2.43×10^{-6}	5.57×10^{-6}
2500	90	0.2866	1.77×10^{-6}	6.10×10^{-6}
3600	109	0.6605	1.41×10^{-6}	6.94×10^{-6}
4900	127	1.9946	1.47×10^{-6}	1.11×10^{-5}
6400	146	4.0009	1.25×10^{-6}	1.24×10^{-5}
8100	164	7.1323	1.26×10^{-6}	1.69×10^{-5}
10000	182	11.8184	1.25×10^{-6}	2.15×10^{-5}

6.5.1.3 Restarted GMRES and Truncated GMRES

For large k, the effort can increase too quickly, which addresses the third of the initially mentioned problems of the method. In addition, the error in the orthonormalization will also increase for larger k. Therefore, truncated versions of GMRES are often used, in which the iteration is either completely restarted after a fixed number of steps $m \in \mathbb{N}$. The previously calculated approximation is chosen as the starting value $x^{(0)}$. Alternatively, the orthonormalization in lines 7–9 of Algorithm 6.26 is only carried out over the last m vectors, i.e., these lines are replaced by

7 **for** $j = \max(1, k - m)$ **to** k **do**
8     $h_{jk} = (w^k, v^j)$;
9     $w^k = w^k - h_{jk} v^j$;

The first variant is called *restarted GMRES* and the second variant *truncated GMRES*. We refer to the literature [82].

6.5.2 Preconditioned GMRES Method

Just like with the CG method, the convergence of GMRES depends on the condition number of the matrix A, and preconditioning can sometimes significantly accelerate convergence. Since GMRES does not require symmetric positive definite matrices, there is more freedom in the choice of preconditioners.

Let $P \in \mathbb{R}^{n \times n}$ be the preconditioner with $P^{-1} \approx A^{-1}$. We transform the linear system to be solved into the form,

$$P^{-1} A x = P^{-1} b.$$

This corresponds to left preconditioning, which we also considered in the CG method. To apply the GMRES iteration to this system, only a few lines of Algorithm 6.26 need to be modified. The initial residual in line 1 is determined as

$$Pd^0 = b - Ax^0$$

and the vector to be orthogonalized in line 6 must be calculated as

$$Pw^k = Av^k$$

per iteration.

Since the preconditioned GMRES iteration corresponds to the minimization of $\| P^{-1}(b - Ax) \|$, the preconditioned residual $|z_{k+1}| = \| P^{-1}(b - Ax^k) \|$ is determined in line 21. This can differ significantly from the actual residual $\| b - Ax^k \|$. To ensure that the residual is really minimized sufficiently, a calculation of $b - Ax^k$ should be performed after solving the system in lines 26–28.

Like the CG method, the GMRES method can be preconditioned "from the right", i.e., the linear system

$$AP^{-1}\hat{x} = b, \quad Px = \hat{x}.$$

is considered. This variant has the advantage that the residual $|z_{k+1}|$ calculated during the process really corresponds to the residual of the actual linear system. Secondly, right preconditioning allows the so-called FGMRES (flexible GMRES), where the preconditioner P can be adjusted in each iteration step. For example, this is the case when no fixed matrix P is chosen. Instead, another numerical method, such as another iterative method, is used, so that a nested overall method is created. We refer to the literature [82].

Part II
Numerical Methods of Analysis

Real analysis studies sequences, series, and real-valued functions, with a focus on foundational concepts such as convergence, approximation, continuity, differentiability, and integration. Numerical aspects of analysis concern the explicit computation and practical evaluation of these quantities. The intermediate value theorem, for example, states that the function

$$f(x) = x(1 + \exp(x)) + 10 \sin(3 + \log(x^2 + 1))$$

has at least one root in the interval $[-10, 2]$ as it holds

$$x > 2 \Rightarrow f(x) > 0, \; x < -10 \Rightarrow f(x) < 0.$$

The actual calculation of this or possibly several of these roots, however, can be extremely complicated even for very simple functions. This is where numerical mathematics comes into play. Another example is the calculation of integrals. From the continuity of the function f we can conclude that the integral

$$I_{ab}(f) = \int_a^b f(x)\mathrm{d}x$$

exists for all $a, b \in \mathbb{R}$ but its explicit calculation "by hand" is challenging in many cases. Analytical tools often fail.

This part of the book will deal with solution methods for problems such as *finding roots, interpolation and approximation*, num int ANN.

As already mentioned in Part I (Numerical Methods of Linear Algebra), numerics will generally not be able to provide exact solutions, but will usually provide approximations to the solution. This means that not only the approximate solutions are explicitly calculated, but also their "distance", in a suitable error measure with respect to the (unknown) exact solution. This ultimately leads to the question of how quickly the numerical solution provides satisfactory results, which leads to the analysis of the convergence speed. Analogous to Part I, we will classify our results based on the *concepts of numerical mathematics*. In addition, all topics will again be illustrated with numerous examples and excursions.

Fundamentals of Calculus

7

In the second part of the book, we deal with numerical algorithms to solve problems from real analysis. To this end, we use methods from analysis to analyze these numerical schemes. Central is the concept of convergence of a sequence $(x_k)_{k\in\mathbb{N}} \subset \mathbb{R}^n$, for $n \in \mathbb{N}$, towards a limit $z \in \mathbb{R}^n$. In analysis, this is usually a qualitative term meaning that for every $\epsilon > 0$ there is a $k_0 \in \mathbb{N}$, such that

$$\|x_k - z\| \le \epsilon \quad \forall k \ge k_0.$$

Sequences appear, for example, as intermediate solutions in iterative schemes. To compare different numerical methods, we will introduce the terms *convergence speed* and *convergence order* in Chap. 8.

One of the major tasks is the search for zeros of a function. That is, for a given $f : D \subset \mathbb{R}^n \to \mathbb{R}^n$, we are looking for a point $z \in D$ in the domain D with $f(z) = 0$. For scalar functions, the *intermediate value theorem* gives some information on the existence of a root:

Theorem 7.1 (Intermediate Value Theorem) *Let $f : [a, b] \to \mathbb{R}$ be a continuous function with a sign change in the interval, i.e., $f(a)f(b) < 0$. Then there exists at least one root $z \in (a, b)$ with $f(z) = 0$.*

Proof In the proof, we construct sequences of intervals $[a_k, b_k]$, which always enclose a root $a_k \le z \le b_k$. They are built such that a_k is monotonically increasing, b_k is monotonically decreasing, and for the interval length we have $b_k - a_k \to 0$.

Without loss of generality, let $a_0 := a < b =: b_0$. It holds by assumption $f(a_0)f(b_0) < 0$. For $k = 1, 2, \ldots$, we construct the sequence of midpoints

$$c_k := \frac{a_{k-1} + b_{k-1}}{2} \in (a_{k-1}, b_{k-1}).$$

If $f(c_k) = 0$, then $z = c_k$ is the desired root and the scheme terminates. Otherwise, we choose

© The Author(s), under exclusive license to Springer-Verlag GmbH, DE, part of Springer Nature 2026

T. Richter et al., *Introduction to Numerical Mathematics*, Mathematics Study Resources 25, https://doi.org/10.1007/978-3-662-72546-7_7

261

$$a_k := \begin{cases} a_{k-1} & f(a_{k-1})f(c_k) < 0 \\ c_k & f(a_{k-1})f(c_k) > 0 \end{cases}, \quad b_k := \begin{cases} c_k & f(a_{k-1})f(c_k) < 0 \\ b_{k-1} & f(a_{k-1})f(c_k) > 0. \end{cases}$$

In both cases, $f(a_k)f(b_k) < 0$, i.e., there is again a change of sign in the interval.

Furthermore, due to $c_k \in (a_{k-1}, b_{k-1})$ by construction, $a_k \geq a_{k-1}$ and $b_k \leq b_{k-1}$. Finally, it holds

$$b_k - a_k = \frac{b_{k-1} - a_{k-1}}{2} = \frac{(b-a)}{2^k} \to 0 \quad \text{for } k \to \infty.$$

Therefore, the sequence c_k is bounded from above and below, and it holds

$$\lim_{k \to \infty} a_k = \lim_{k \to \infty} b_k = \lim_{k \to \infty} c_k = z.$$

Furthermore,

$$f(a_k)f(b_k) \leq 0 \quad \Rightarrow \quad \lim_{k \to \infty} f(a_k)f(b_k) = f(z)^2 \leq 0 \quad \Rightarrow \quad f(z) = 0.$$

$\square$

The proof can be directly implemented as a numerical method to find the root of a scalar function. This is called the bisection method, which we will examine in Sect. 8.2. The theorem cannot be extended to the higher-dimensional case. Here, we will explore additional tools for characterizing and finding zeros.

Another central aspect of analysis is the differentiation and integration of functions. We will cover these topics in Sect. 9.3 and Chap. 10, respectively. A first method to numerically approximate first-order derivatives is the difference quotient

$$D_h f(x) := \frac{f(x+h) - f(h)}{h} \approx f'(x). \tag{7.1}$$

The mean value theorem of differential calculus states that the difference quotient coincides with the derivative at an intermediate point:

Theorem 7.2 (Mean Value Theorem of Differential Calculus) *Let $f : [a, b] \to \mathbb{R}$ be a continuously differentiable function. Then there exists $\xi \in (a, b)$, such that*

$$f'(\xi) = \frac{f(b) - f(a)}{b - a}.$$

A special case of the mean value theorem is Rolle's theorem

Theorem 7.3 (Rolle) *If $f(a) = f(b)$ holds additionally in Theorem 7.2, then there exists $\xi \in (a; b)$, such that $f'(\xi) = 0$.*

The mean value theorem also has a corresponding counterpart in integral calculus.

Theorem 7.4 (Mean Value Theorem of Integral Calculus) *Let $f : [a, b] \to \mathbb{R}$ be a continuous function. Then there exists $\xi \in (a, b)$, such that*

$$\int_a^b f(x)\mathrm{d}x = f(\xi) \cdot (b - a).$$

The error in the approximation of the derivative with the difference quotient $D_h f(x) \approx f'(x)$ in (7.1) can be investigated using the Taylor expansion of a function.

Theorem 7.5 (Taylor) *Let $f \in C^{r+1}[a, b]$ with $r \geq 0$. Then for all $x, x_0 \in [a, b]$, it holds*

$$f(x) = \sum_{l=0}^{r} \frac{f^{(l)}(x_0)}{l!}(x - x_0)^l + \frac{f^{(r+1)}(\xi_{x,x_0})}{(r + 1)!}(x - x_0)^{r+1},$$

where $\xi_{x,x_0} \in [\min(x, x_0), \max(x, x_0)]$ is an intermediate value. This is known as the Taylor approximation with differential remainder. Similarly, it holds

$$f(x) = \sum_{l=0}^{r} \frac{f^{(l)}(x_0)}{l!}(x - x_0)^l + \frac{1}{r!} \int_{x_0}^{x} (x - x_0)^r f^{(r+1)}(t)\mathrm{d}t.$$

This is the Taylor approximation with integral remainder. For $r + 1$ times continuously differentiable functions $f : X \subset \mathbb{R}^n \to \mathbb{R}$, and for all $x, x_0 \in X$, it holds

$$f(x) = \sum_{|\alpha|=0}^{r} \frac{1}{|\alpha|!} D^\alpha f(x_0)(x - x_0)^\alpha + \sum_{|\alpha|=r+1} \frac{1}{|\alpha|!} D^\alpha f(\xi)(x - x_0)^\alpha$$

with an intermediate value $\xi \in X$. Here, we use the multi-index notation

$$\alpha \in \mathbb{N}^n, \quad |\alpha| = \alpha_1 + \cdots + \alpha_n, \quad x^\alpha = x_1^{\alpha_1} \cdots x_n^{\alpha_n}, \quad D^\alpha = \frac{\partial^{\alpha_1}}{\partial x_1} \cdots \frac{\partial^{\alpha_n}}{\partial x_n}.$$

Proofs can be found, for example, in [5, Chap. IV] and [91, Chap. 4]. This directly leads us to an error bound for the difference quotient (7.1), if the function f is twice continuously differentiable, i.e., $D_h f(x) = f'(x) + \frac{h}{2} f'(\xi)$.

Theorem 7.5 describes a local approximation of general (differentiable) functions with Taylor polynomials. Approximation theory is another focus of numerical analysis: Given a function f, find a simple function p, e.g., a polynomial or a piecewise linear function, that approximates f as well as possible. A theoretical basis is provided by the *Weierstrass approximation theorem*.

Theorem 7.6 (Weierstrass Approximation Theorem) *Let $f \in C[a, b]$. For every $\varepsilon > 0$, there exists a polynomial p defined on $[a, b]$, such that*

$$|f(x) - p(x)| < \varepsilon \quad \forall x \in [a, b].$$

For a proof, we refer to the literature [5] and [91]. The Weierstrass approximation theorem initially states that it is possible to approximate any continuous function arbitrarily well by a polynomial. A typical proof is constructed with a sequence of *Bernstein polynomials* that converge to the function f. However, this convergence is very slow, the realization is complex, and it does not lead to efficient numerical approximation methods. The *Stone-Weierstrass theorem* generalizes the result in such a way that the space of functions used for the approximation is much broader. *Fourier analysis* chooses trigonometric polynomials for the approximation of a function f. We give a simple version:

Theorem 7.7 (Fourier Analysis) *Let $f \in C(\mathbb{R})$ be a 2π-periodic function, i.e., $f(t) = f(t + 2\pi)$ for all $t \in \mathbb{R}$. Then the Fourier sum*

$$F_n^f(x) = \frac{1}{2}a_0 + \sum_{k=1}^{n} \left(a_k \cos(kx) + b_k \sin(kx) \right)$$

with the coefficients

$$a_0 = \frac{1}{\pi} \int_0^{2\pi} f(x)\mathrm{d}x, \quad a_k = \frac{1}{\pi} \int_0^{2\pi} f(x)\cos(kx)\mathrm{d}x, \quad b_k = \frac{1}{\pi} \int_0^{2\pi} f(x)\sin(kx)\mathrm{d}x$$

converges in the mean square to the function f, i.e.,

$$\|f - F_n^f\|_{L^2[0,2\pi]}^2 = \int_0^{2\pi} |f(x) - F_n^f(x)|^2\mathrm{d}x \to 0 \quad \text{for } n \to \infty.$$

For a proof, we again refer to the literature [52]. Unlike the Taylor approximation, Fourier analysis does not require the function f to be differentiable. The *discrete Fourier analysis* is of great practical importance. Discrete numerical measurements $a_0, \ldots, a_n$ of a periodic signal are represented in the form of trigonometric polynomials. We will discuss this in Sect. 9.7.

Fixed-point iterations play a role in many iterative algorithms as we already have seen in the linear algebra chapters.

Definition 7.8 (Fixed Point) Let $g : \mathbb{R}^n \to \mathbb{R}^n$. A point $x \in \mathbb{R}^n$ is called a fixed point of g, if $g(x) = x$.

Many algorithms are formulated in the form of fixed point iterations, i.e., starting from an initial value $x_0 \in \mathbb{R}^n$, a sequence

$$x_k = g(x_{k-1}), \quad k = 1, 2, \ldots$$

is constructed and examined for convergence.

Definition 7.9 (Lipschitz Continuity, Contraction) Let $G \subset \mathbb{R}^n$ be a non-empty set. A mapping $g : G \to \mathbb{R}^n$ is called Lipschitz continuous, if

$$\|g(x) - g(y)\| \le L\|x - y\|, \quad x, y \in G,$$

with $L > 0$. If $0 < L < 1$, then g is called a contraction on G.

Every function that is differentiable on a bounded set is Lipschitz continuous, and every Lipschitz continuous function is (uniformly) continuous. An example of a uniformly continuous but not Lipschitz continuous function on $[0, 1]$ is the square root function $f(x) = \sqrt{x}$.

The Banach fixed point theorem now states that every self-map $g : G \to G$, which is a contraction, has a fixed point:

Theorem 7.10 (Banach's Fixed Point Theorem) *Let $G \subset \mathbb{R}^n$ be a non-empty and closed set and $g : G \to G$ a Lipschitz continuous self-map with Lipschitz constant $q < 1$ (a contraction). Then there exists exactly one fixed point $z \in G$ of g and the iteration sequence (x_k) converges for every starting point $x_0 \in G$ to this fixed point.*

There holds the a priori estimate

$$\|x_k - z\| \le \frac{q^k}{1 - q}\|x_1 - x_0\|$$

and the a posteriori estimate

$$\|x_k - z\| \le \frac{q}{1 - q}\|x_k - x_{k-1}\|.$$

Proof Although the Banach fixed point theorem is covered in every lecture on real analysis [5] and [91], we provide the proof here, as it contains numerically relevant error estimates.

(i) Existence of a limit. Since g is a self-map in G, all iterations $x_k = g(x_{k-1}) = \ldots = g_k(x_0)$ are defined when choosing any starting point $x_0 \in G$. Due to the contraction property, we have

$$\|x_{k+1} - x_k\| = \|g(x_k) - g(x_{k-1})\|$$
$$\le q\|x_k - x_{k-1}\| \le \ldots \le q^k\|x_1 - x_0\|.$$

Next, we show that $(x_k)_k$ is a Cauchy sequence. For every $l \ge m$ we obtain

$$
\begin{aligned}
\|x_l - x_m\| &\leq \|x_l - x_{l-1}\| + \ldots + \|x_{m+1} - x_m\| \\
&\leq (q_{l-1} + q_{l-2} + \ldots + q_m)\|x_1 - x_0\| \\
&= q_m \frac{1 - q_{l-m}}{1 - q}\|x_1 - x_0\| \\
&\leq q_m \frac{1}{1 - q}\|x_1 - x_0\| \to 0 \quad \text{for } l \geq m \to 0.
\end{aligned}
\tag{7.2}
$$

That is, $(x_l)_{l \in \mathbb{N}}$ is a Cauchy sequence. Since all sequence members lie in G and G is a closed subset of $\mathbb{R}^n$, the limit $x_l \to z \in G$ exists.

(ii) Fixed point property. We prove that z is indeed a fixed point of g. From the continuity of g it follows that $g(x_k) \to g(z)$. Then, taking the limit leads to

$$
z \leftarrow x_{k+1} = g(x_k) \to g(z) \quad \text{for } k \to \infty.
$$

(iii) Uniqueness. The uniqueness follows from the contraction property. Let z and $\hat{z}$ be two fixed points of g. Then

$$
\|z - \hat{z}\| = \|g(z) - g(\hat{z})\| \leq q\|z - \hat{z}\|.
$$

For $q < 1$, it must be $\|z - \hat{z}\| = 0$ and therefore $z = \hat{z}$ is the unique fixed point.

(iv) A priori error estimate. With (7.2) it holds

$$
\|z - x_m\| \underset{l \to \infty}{\longleftarrow} \|x_l - x_m\| \leq q_m \frac{1}{1 - q}\|x_1 - x_0\|.
$$

(v) A posteriori error estimate. Again with (7.2) it holds

$$
\|x_m - z\| \leq q \frac{1}{1 - q}\|x_m - x_{m-1}\|.
$$

Every continuously differentiable function is also Lipschitz continuous, and the derivative can be used as a bound for the Lipschitz constant.

Theorem 7.11 *Let the mapping $g : G \subset \mathbb{R}^n \to \mathbb{R}^n$ be continuously differentiable on a convex and open set G. Let*

$$
L := \sup_{\xi \in G} |Dg(\xi)| < \infty.
$$

Then

$$
\|g(x) - g(y)\| \leq L\,\|x - y\|, \quad x, y \in G,
$$

with the Jacobian

$$
Dg(x) = \left(\frac{\partial g_i}{\partial x_j}\right)_{i,j=1,\ldots,n} \in \mathbb{R}^{n \times n}.
$$

If $\sup_{\xi \in G} \|Dg(\xi)\| < 1$, then g is a contraction on G. In particular, in the scalar case, it holds

$$q := \max_{\xi \in [a,b]} |Dg(\xi)|.$$

Another classical problem of real analysis is the search for minima and maxima of continuous functions $f : G \subset \mathbb{R}^n \to \mathbb{R}$.

Definition 7.12 (Minimization) Let $f : X \to \mathbb{R}$ with $X \subset \mathbb{R}^n$ be a continuous function. Any $x \in \mathbb{R}^n$ with $x \in X$ is called *feasible point*. The point $x \in X$ is called *local minimum*, if there exists an $\epsilon > 0$, such that

$$f(x) \leq f(y) \quad \forall y \in B_\epsilon(x) := \{z \in \mathbb{R}^n \cap X \mid \|z - x\| < \epsilon\}.$$

It is called *strict or isolated minimum*, if there exists an $\epsilon > 0$, such that

$$f(x) < f(y) \quad \forall y \in B_\epsilon(x) := \{z \in \mathbb{R}^n \cap X \mid \|z - x\| < \epsilon\}.$$

It is called *global minimum*, if

$$f(x) \leq f(y) \quad \forall y \in X,$$

and *global strict minimum*, if

$$f(x) < f(y) \quad \forall y \in X.$$

In the scalar case $f : \mathbb{R} \to \mathbb{R}$, the following theorem gives the condition for the existence of minima and maxima.

Theorem 7.13 (Minimum and Maximum (in $\mathbb{R}$)) *Let $[a, b]$ be an interval and $f : [a, b] \to \mathbb{R}$ a continuous function. Then f is bounded and attains the minimum and maximum on $[a, b]$. That is, there exist two values $x_{min}, x_{max} \in [a, b]$ with*

$$f(x_{min}) \leq f(x) \leq f(x_{max}) \quad \forall x \in [a, b].$$

Proof *(i)* Let Y be the range of f on $[a, b]$. Then for every sequence $y_k \in Y$, there exists a sequence $x_k \in [a, b]$ with $y_k = f(x_k)$. Since $x_k \in [a, b]$ is bounded, there exists a subsequence that converges, $x_{k'} \to x^*$. Since $[a, b]$ is closed, $x^* \in [a, b]$ holds and due to the continuity of f we have $y^* := f(x^*) \in M$.

(ii) We now show that f (i.e., Y) must be bounded. Suppose f is not bounded from below. Then there exists a divergent sequence $y_k \in Y$. Let $x_k \in [a, b]$ be the corresponding sequence with $y_k = f(x_k)$. Since y_k diverges, every subsequence $y_{k'}$ must also diverge. However, this is excluded by *(i)* since $[a, b]$ is bounded. Thus, f is bounded from below by

$$-\infty < f_{min} := \inf_{x\in[a,b]} f(x).$$

With similar argumentation, we can derive the upper bound $f_{max} < \infty$.

(iii) Now, let $y_k \in Y$ be a sequence converging to f_{min} and define $x_k \in [a, b]$ as well as $y_k = f(x_k)$. This sequence has a convergent subsequence $x_{k'} \to x_{min}$ and for the limit $f(x_{k'}) \to f_{min} = f(x_{min})$. Thus, the minimum is attained at x_{min}. $\qquad\Box$

In contrast to the proof of the intermediate value theorem, Theorem 7.1, this proof is not constructive, as we select multiple subsequences. The proof cannot be realized as an algorithm. Also, the theorem does not predict the uniqueness of the minimum. If there exist two different points x, x', each with $f(x) = f(x') = f_{min}$, the proof does not specify which of these two points is reached with the sequence.

The problem of finding the minimum or maximum of a function is much more complex than finding roots, and there are generally no algorithms that are both efficient and reliably identify the minimum. This is especially true for global minima and maxima. It becomes easier if only local minima and maxima are sought and if additional properties of the function are known. In the following, we present various *criteria for minima and maxima*. These conditions go beyond continuity, serving to characterize and search for extreme points. The simplest criterion can be applied when the function f is differentiable. We immediately deal with the general case of higher-dimensional functions.

Theorem 7.14 (Necessary Criterion for Minima and Maxima) *Let $f : X \subset \mathbb{R}^n \to \mathbb{R}$ be differentiable on the open set X and have a (local) minimum or a (local) maximum at the point $x_0 \in (a; b)$. Then $\nabla f(x_0) = 0$ holds.*

Definition 7.15 (Stationary points) A point $x \in X \subset \mathbb{R}^n$ with $\nabla f(x) = 0$ is called a *stationary point*. A stationary point that is neither a minimum nor a maximum is called a *saddle point*.

A stationary point does not necessarily have to be a minimum or maximum.

Theorem 7.16 (Sufficient Criterion for Minima and Maxima) *Let $f : X \subset \mathbb{R}^n \to \mathbb{R}$ be twice continuously differentiable and the necessary criterion $\nabla f(x_0) = 0$ hold at the point $x_0 \in X$. Then f has a minimum at x_0 if and only if the Hessian matrix $H_f(x_0)$ is positive definite, and a maximum if and only if $H_f(x_0)$ is negative definite.*

Further details and proofs can be found, for example, in [70].

Root Finding Problems

8

Solving non-linear equations plays a fundamental role in mathematics. Often, such tasks are formulated as root-finding problems. Determining the roots of linear or quadratic polynomials is the simplest example and is already known from school. In particular, closed formulas, such as the quadratic formula, can be given here. However, for polynomials of degree five and higher, there are no such closed formulas for the solutions.

In the following, we consider a general function

$$f : D(f) \to \mathbb{R}$$

and search for the (one or several) roots $z \in D(f)$ in the domain. Any non-linear equation $N(x) = b$ can be transformed into a root-finding problem using $f(x) = N(x) - b$.

Definition 8.1 (*Root*) Let $f : D \subset \mathbb{R}^n \to \mathbb{R}^m$ be a continuous function. Then $z \in D$ is called a *real root of f*, if

$$f(z) = 0.$$

Suppose that $f : \mathbb{R} \to \mathbb{R}$ is at least k-times continuously differentiable for a $k \in \mathbb{N}$, and it holds

$$f(z) = 0, \ f'(z) = 0, \ \ldots, f^{(k-1)}(z) = 0, \quad f^{(k)}(z) \neq 0,$$

then $z \in D$ is called a *k-fold root of f*.

Roots usually have to be approximated with an *iterative method*. Starting from an initial value x_0, we obtain a sequence of approximations $(x_k)_{k \in \mathbb{N}}$, whose limit converges to the root $z \in \mathbb{R}$:

$$f(x_0) \to f(x_1) \to f(x_2) \cdots \to f(z) = 0$$

© The Author(s), under exclusive license to Springer-Verlag GmbH, DE, part of Springer
Nature 2026

T. Richter et al., *Introduction to Numerical Mathematics*, Mathematics Study
Resources 25, https://doi.org/10.1007/978-3-662-72546-7_8

The iteration is stopped at a finite $N \in \mathbb{N}$ and the sequence member x_N is considered as an approximation of the root $x_N \approx z$.

Most of the *concepts of numerical mathematics* introduced in the introduction can be discussed for the root-finding problem.

- How is the root-finding problem conditioned, and what stability properties do different methods have?
- How do we construct *efficient* and *stable* numerical methods?
- How quickly does an approximation $(x_k)_{k\in\mathbb{N}}$ converge to the root z? How can we quantify the terms *order of convergence* and *rate of convergence* as measures for the speed of convergence? In analysis, it is investigated whether sequences converge: $x_k \to z$ for $k \to \infty$? In numerics, this statement is not sufficient. The *concept of efficiency* calls for sufficiently fast convergence such that a good approximation is found quickly.
- Does the selected numerical method converge on arbitrarily large intervals and with arbitrarily chosen initial values x_0, or only if the initial value x_0 is already sufficiently close to the solution? In other words: Is the method *locally* or *globally* convergent?
- Can we provide an estimate for the error $|x_k - z|$ for the iteration x_k? Can this estimate be evaluated numerically to assess the error of the already calculated approximation $x_k, x_{k-1}, \ldots$? In this case, call the estimate an *a posteriori estimate*.

In the rest of this chapter, we will discuss the above questions for various methods and specify the key terms.

8.1   Stability and Conditioning

We first consider the condition of the general root-finding problem. Let $f : D \subset \mathbb{R} \to \mathbb{R}$ be a differentiable and invertible function on D with root $f(z) = 0$ i.e., $z = f^{-1}(0)$. We consider the problem

$$N : f \mapsto f^{-1}(0)$$

and want to investigate how a disturbance in the function f can affect the disturbance in the root. To be able to calculate the condition number with Definition 1.16, we first need to clarify what a disturbance in the input means here. The input is the function f, a disturbed input is therefore a disturbed function $\tilde{f} = f + \delta f$. To calculate the condition number, we need to differentiate $N(\cdot)$ in the direction of the input f. This type of derivative is more complicated than the usual one. We simplify the situation by choosing an arbitrary differentiable fixed disturbance function δf and considering the disturbed function

$$\tilde{f}_s = f + s\delta f$$

with a parameter $s \in \mathbb{R}$. The "direction" of the disturbance is fixed (but arbitrary), but we let the strength of the disturbance s vary. Then we can calculate the derivative with respect to the scalar parameter s in an elementary way. It holds at $N(f) = f^{-1}(0) = z$ in the form of the directional derivative by utilizing the derivative rules of the inverse function

$$\frac{d}{ds} N(f + s\delta f) = \frac{1}{(f' + s\delta f')(z)} \frac{d}{ds}\left(f(z) + s\delta f(z)\right) = \frac{\delta f(z)}{f'(z) + s\delta f'(z)}.$$

With this, we can approximate the amplification of the relative error and thus the conditioning of the task for small disturbances

$$\frac{N(f + s\delta f) - N(f)}{N(f)} = \frac{N(f + s\delta f) - N(f)}{s} \cdot \frac{s}{N(f)}$$

$$\approx \frac{d}{ds} N(f + s\delta f) \cdot \frac{s}{N(f)} = \frac{1}{f'(z) + s\delta f'(z)} \frac{s\delta f(z)}{N(f)}. \tag{8.1}$$

For small disturbances with $s \approx 0$, $s\delta f'(z)$ is negligible compared to $f'(z)$, and we approximate the expression as

$$\frac{N(f + s\delta f) - N(f)}{N(f)} \approx \frac{1}{f'(z)} \frac{s\delta f(z)}{z}. \tag{8.2}$$

The condition of the root-finding problem thus behaves like $\kappa = 1/f'(z)$. If the derivative at the root is small, arbitrarily large error amplifications can occur.

In the following, we examine a specific situation for f and consider the quadratic polynomial

$$f(x) = x^2 - px + q, \quad p, q \in \mathbb{R},$$

and investigate the conditioning of root-finding as well as the stability of simple algorithms. The quadratic formula is a direct solution method:

$$x_{1/2} = \frac{p}{2} \pm \sqrt{\frac{p^2}{4} - q}$$

In Fig. 8.1, we show four different situations:

  (I)   There are two real roots that are widely separated.
 (II)   There is only one real root. This root is a *double root* with $f(z) = f'(z) = 0$.
(III) There are two real roots that are very close together.
(IV) There is no real root. The function then has two complex roots.

Example 8.2 (Conditioning of Root-Finding of Quadratic Polynomials) We calculate the conditioning of root determination depending on the coefficients. For this, let

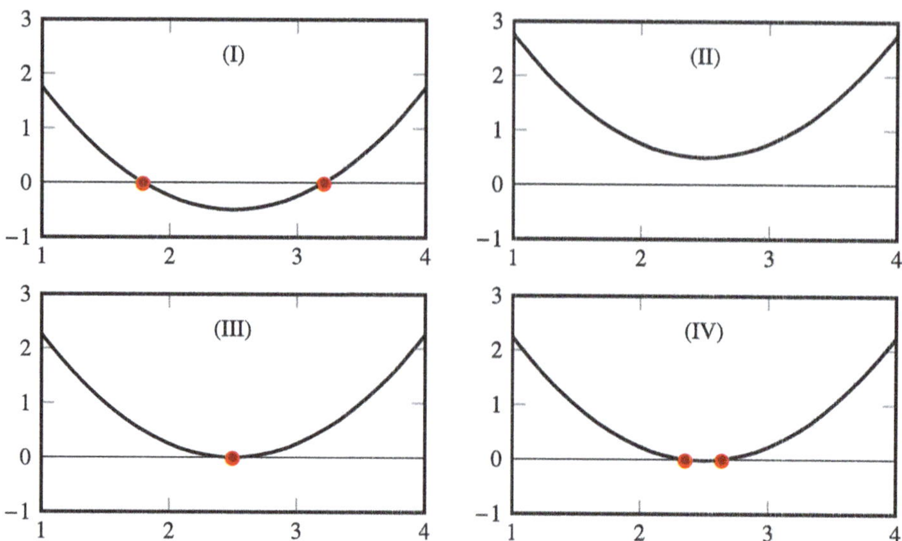

Fig. 8.1 Four possible situations in the root-finding of quadratic functions. From top left to bottom right: (I) two clearly separated roots, (II) no real root, (III) a double root, and (IV) two real roots close to each other

$$f(x) = x^2 - px + q = 0, \quad x \in \mathbb{R}.$$

For the roots $x_{1,2} := x_{1,2}(p, q)$, we have

$$x_{1,2} = \frac{p}{2} \pm \sqrt{\frac{p^2}{4} - q}.$$

Due to

$$f(x) = (x - x_1)(x - x_2) = x^2 - (x_1 + x_2)x + x_1 x_2,$$

it holds

$$p = x_1 + x_2, \quad q = x_1 x_2.$$

This is the statement of the *Vieta's theorem*. We now calculate the condition numbers of the root-finding problem, where in particular $x_{1,2}(p, q)$ represents the function, and p and q are the variables:

$$k_{1p} = \frac{\partial x_1}{\partial p} \frac{p}{x_1} = \frac{1 + x_2/x_1}{1 - x_2/x_1}, \qquad k_{1q} = \frac{\partial x_1}{\partial q} \frac{q}{x_1} = \frac{1}{1 - x_1/x_2},$$

$$k_{2p} = \frac{\partial x_2}{\partial p} \frac{p}{x_2} = -\frac{1 + x_2/x_1}{1 - x_2/x_1}, \qquad k_{2q} = \frac{\partial x_2}{\partial q} \frac{q}{x_2} = \frac{1}{1 - x_2/x_1}.$$

The calculation of the roots x_1, x_2 is therefore poorly conditioned if $x_1/x_2 \sim 1$, i.e., if both roots are close together. In this case, the condition numbers can become arbitrarily large. This is illustrated in Fig. 8.1 (III).

The example is in good agreement with the conditioning of the general root problem (8.2), because if both roots almost coincide, the derivative of the quadratic polynomial nearly vanishes there.

Example 8.3 (Stability of the Quadratic Formula) We consider specific values as an example. Specifically, let $p = 4$ and $q = 3.9999$:

$$f(x) = x^2 - 4x + 3.9999 = 0$$

with roots $x_{1,2} = 2 \pm 0.01$. Then the condition number is

$$|k_{1q}| = \left| \frac{1}{1 - x_1/x_2} \right| = 99.5.$$

This corresponds to a possible error amplification by a factor of 100. With a disturbance of p by 1% to $\tilde{p} = 4.04$, we get the disturbed roots

$$\tilde{x}_1 \approx 2.304, \quad \tilde{x}_2 \approx 1.736,$$

with relative errors of about 15%.

With a disturbance of p by minus 1% to $\tilde{p} = 3.96$ we no longer get real but two complex roots

$$\tilde{x}_1 \approx 1.98 \pm 0.28196i.$$

◀

Determining the roots of quadratic polynomials can be very poorly conditioned if the two roots are close together. We now examine the stability of the quadratic formula for calculating the roots in the well-conditioned case, i.e., in the case of widely separated roots. Let us consider a polynomial

$$f(x) = x^2 - px + q$$

with $|q| \ll p^2/4$. The two solutions are

$$x_{1,2} = \frac{p}{2} \pm \sqrt{\frac{p^2}{4} - q},$$

i.e., $x_1 \approx p$ and $x_2 \approx 0$. From the previous examples, we have learned that the task is well conditioned for $|x_1/x_2| \gg 1$. We calculate both roots with the quadratic formula:

Algorithm 8.4: Root Calculation With the Quadratic Formula

> **Input:** Coefficients $p, q \in \mathbb{R}$ of the function $f(x) = x^2 - px + q$.
> 1 $a_1 = p^2/4$
> 2 $a_2 = a_1 - q$
> 3 $a_3 = \sqrt{a_2} \geq 0$
> 4 $x_1 = p/2 - a_3$
> 5 $x_2 = p/2 + a_3$
> **Result:** Roots $x_{1,2}$ of the polynomial $f(x)$.

The stability of this algorithm can be examined in detail using the tools of Sect. 1.5. We only consider the last two steps, i.e., the calculation of x_1 and x_2. We assume that a_3 is given with a relative error ϵ_a and $p/2$ with a relative error ϵ_p. In each case, let $\epsilon_p = \epsilon_a = O(\epsilon)$. Then

$$\tilde{x}_{1/2} = \left(\frac{p}{2}(1 + \epsilon_p) \pm a_3(1 + \epsilon_a) \right)(1 + \epsilon)$$

$$= x_{1/2}(1 + \epsilon) + \frac{p}{2}\epsilon_p \pm a_3\epsilon_a + O(\epsilon^2).$$

For the relative error, we get

$$\frac{|\tilde{x}_{1/2} - x_{1/2}|}{|x_{1/2}|} \leq \epsilon + \frac{|p/2| + |a_3|}{|p/2 \pm a_3|}\epsilon.$$

For $p/2 \pm a_3 \approx 0$, the error can be arbitrarily amplified, and it depends on the sign of $p/2$. Suppose $p/2 \approx a_3$ with the same sign. Then the calculation of the root

$$x_1 = \frac{p}{2} - a_3$$

might be highly unstable, whereas the calculation of the root

$$x_2 = \frac{p}{2} + a_3$$

is robust. To avoid cancellation for $p < 0$, we must compute the second root using the quadratic formula $\tilde{x}_2 = p/2 - a_3$. For $p < 0$, the first one $\tilde{x}_1 = p/2 + a_3$ is computed with this technique. The other zero must be determined using Vieta's theorem as

$$q = x_1 x_2 \quad \Leftrightarrow \quad x_1 = \frac{q}{x_2}, \quad x_2 = \frac{q}{x_1}.$$

Because of the good conditioning of the division, this procedure is always stable. We summarize:

Algorithm 8.5: Stable Calculation of Roots of Quadratic Functions

Input: Coefficients $p, q \in \mathbb{R}$ of the function $f(x) = x^2 - px + q$.
1   $a_1 = p^2/4$
2   $a_2 = a_1 - q$
3   **if** $a_2 < 0$ **then**
4   Abort, no real root!
5   $a_3 = \sqrt{a_2}$
6   **if** $p < 0$ **then**
7        $x_1 = p/2 - a_3$
8        $x_2 = q/x_1$
9   **else**
10        $x_2 = p/2 + a_3$
11        $x_1 = q/x_2$
     Result: Approximate roots $x_{1,2}$ of f.

Remark 8.6 (Conditioning and Stability) This analysis emphasizes that conditioning is a property of the problem, and stability is a property of the algorithm. If the real roots are separated, the task itself is well-conditioned. However, the calculation of both roots with the quadratic formula can still be arbitrarily unstable. With proper use of Vieta's theorem, both roots can always be calculated stably. ◆

Example 8.7   We calculate the root of

$$x^2 - 4x + 0.01 = 0.$$

It holds

$$x_1 \approx 0.002502, \quad x_2 \approx 3.997498.$$

Four-digit calculation yields the following for the stable computation

$$a_1 = p^2/4 = 4.000$$
$$a_2 = a_1 - q = 3.990$$
$$a_3 = \sqrt{a_2} = 1.997$$
$$x_2 = p/2 + a_3 = 3.997$$
$$x_1 = q/x_2 = 0.0025.$$

The first four significant digits are correct. If we were to calculate both roots using the quadratic formula, the result would be

$$x_1 = p/2 - a_3 = 0.003$$

with a relative error of 20%.

If a numerical method provides the approximation $\tilde{z}$ to the root z, the error $|z - \tilde{z}|$ can be easily estimated by the following *a posteriori error estimate* if additionally, the function f is differentiable:

Theorem 8.8 (A Posteriori Error Estimate) *Let $f \in C^1[a, b]$ with*

$$0 < m \le |f'(x)| \le M < \infty.$$

Furthermore, let $z \in [a, b]$ with $f(z) = 0$. Then for each $\tilde{z} \in [a, b]$

$$\frac{|f(\tilde{z})|}{M_{z,\tilde{z}}} \le |\tilde{z} - z| \le \frac{|f(\tilde{z})|}{m_{z,\tilde{z}}},$$

where

$$I_{z,\tilde{z}} := [\min(z, \tilde{z}), \max(z, \tilde{x})], \quad M_{z,\tilde{z}} := \max_{x \in I_{z,\tilde{z}}} |f'(x)|, \quad m_{z,\tilde{z}} := \min_{x \in I_{z,\tilde{z}}} |f'(x)|.$$

Proof With Taylor expansion, we have

$$f(\tilde{z}) = f(z) + f'(\xi)(\tilde{z} - z) = f'(\xi)(\tilde{z} - z) \tag{8.3}$$

with an intermediate point $\xi \in M_{z;\tilde{z}}$. The error estimate follows after dividing by $f'(\xi)$ and then taking the minimum and maximum. $\qquad\square$

This simple error estimator can be used, for example, as a termination criterion in a numerical method.

8.2 The Bisection Method

Let us consider the function

$$f(x) = x(1 + \exp(x)) + 10 \sin(3 + \log(x^2 + 1)) \tag{8.4}$$

from the introduction. With the intermediate value theorem, Theorem 7.1, we can conclude that there is at least one root in the interval $[-10, 2]$ since $f(-10) \cdot f(2) < 0$ holds. The proof of the intermediate value theorem constructs a sequence of intervals whose boundaries converge to the root. This method can be directly implemented as a numerical method, the so-called *bisection method*.

We start with $a_0 = a$ and $b_0 = b$, calculate the mean $x_k = \frac{1}{2}(a_{k-1} + b_{k-1})$ in each step and check whether there is a sign change in the subintervals $[a_{n-1}, x_n]$ or $[x_n, b_{n-1}]$. The iteration then continues with this interval. In Fig. 8.2, we show the first eight steps applied to the function $f(x)$. In addition to the endpoints $a = -10$ and $b = 2$, we indicate the first

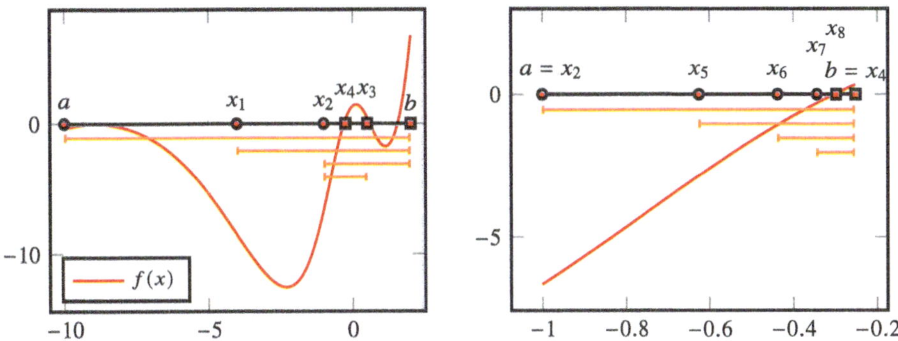

Fig. 8.2 Application of bisection method to the function $f(x)$ from (8.4). The orange segments each show the current interval in which a root must lie

4 interval midpoints x_k on the left. On the right, we illustrate the smaller intervals starting from x_5. We highlight the intervals in orange.

After 8 steps, we obtain the interval $(x_7, x_8) = (-0.34375, -0.296875)$ and the root is confined here. If the root is found exactly in one step, i.e., $f((a + b)/2) = 0$, the iteration can be terminated.

Algorithm 8.9: Bisection Method

Input: Interval $I = [a, b]$, function $f \in C[a, b]$ and tolerance $\epsilon > 0$
1 Set $a_0 = a$ and $b_0 = b$
2 **for** $n = 1, 2, \ldots$ **do**
3      $x_n = (a_{n-1} + b_{n-1})/2$
4   **if** $f(x_n) = 0$ **then**
5   Terminate with root x_n
6   **else if** $f(a_{n-1})f(x_n) < 0$ **then**
7          $a_n = a_{n-1}$
8          $b_n = x_n$
9   **else**
10          $a_n = x_n$
11          $b_n = b_{n-1}$
12   **if** $|f(x_n)| < \epsilon$ **then**
13   Terminate with root x_n
Result: (Approximate) root x_n

In Python, we can implement this algorithm as follows:

♣ Implementation 8.10: Bisection Method

```python
def bisection(f, a, b, n, tol=1e-10):
    for i in range(n):
        x = (a + b) / 2
        if abs(f(x)) < tol:
            return x
        elif f(a) * f(x) < 0:
            b = x
        else:
            a = x

    print(f'The Bisection Method did not converge after {i + 1} iterations')
    print(f'x = {x}, abs(f(x)) = {abs(f(x))} > {tol}')
    return x
```

We summarize.

Theorem 8.11 (Bisection Method) *Let $f \in C[a, b]$ with $f(a)f(b) < 0$. Let a_n, b_n, x_n be the sequence of intervals and midpoints determined by Algorithm 8.9. It holds*

$$x_n \to z, \quad f(z) = 0,$$

as well as the error estimate

$$|x_n - z| \le \frac{b - a}{2^{n+1}}.$$

Proof *(i)* For $a < b$ we always preserve

$$a \le a_1 \le \cdots \le a_n \le a_{n+1} \le x_n \le b_{n+1} \le b_n \le \cdots \le b_1 \le b.$$

Therefore, the method is feasible. The sequence x_n is bounded by two monotonic (increasing and decreasing) sequences a_n and b_n. The sequence is, therefore, convergent. The method only stops if $f(x_n) = 0$, i.e., when $x_n = z$ is a root.

(ii) It holds

$$b_{n+1} - a_{n+1} = \begin{cases} x_n - a_n = \frac{b_n - a_n}{2} & f(a_n)f(x_n) < 0 \\ b_n - x_n = \frac{b_n - a_n}{2} & f(a_n)f(x_n) > 0. \end{cases}$$

In both cases, it follows

$$|b_n - a_n| = \frac{|b_0 - a_0|}{2^n}. \tag{8.5}$$

Since

$$a_n \le x_n \le b_n,$$

the convergence of $x_n \to z$ to a $z \in [a_n, b_n] \subset [a, b]$ follows.

(iii) With the continuity of $f(\cdot)$ it holds

$$0 \le f(z)^2 = \lim_{n \to \infty} f(a_n)f(b_n) \le 0 \quad \Rightarrow \quad f(z) = 0.$$

The limit is thus the sought root. Finally, we can derive the error estimator from (8.5):

$$|x_n - z| \le \frac{|b_n - a_n|}{2} \le \frac{|b - a|}{2^{n+1}}.$$

Remark 8.12 (A Priori Error Estimate) The error estimate given in Theorem 8.11

$$|x_n - z| \le \frac{b - a}{2^{n+1}} \tag{8.6}$$

is an *a priori error estimate*. It can already be evaluated *before* the calculation to determine, for example, the required number of steps n. The estimate from Theorem 8.8 is an a posteriori estimation. To evaluate it, the approximation $\tilde{z} = x_n$ must already be known.

The advantage of the a priori estimate (8.6) is that no further information is necessary for evaluation. For the estimate from Theorem 8.8, we need bounds for the minimum and maximum of the first derivative. For example, let us look for a root of the function $f(x) = \cos(x)$ on $[0, 2]$. We start the bisection method with $a_0 = 0$ and $b_0 = 2$. After 10 steps, it holds

$$b_{10} - a_{10} = 2^{-10}(b - a) \approx 0.002.$$

We can conclude that the error exists at least with an accuracy of $|x_{10} - z| < 0.01$ (x_{10} lies in the center, i.e., we gain another factor $\frac{1}{2}$). We next run 10 iterations of the bisection method:

$$a_{10} = 1.5703125, \quad b_{10} = 1.572265625, \quad x_{11} = 1.5712890625$$

and $f(x_{11}) \approx -0.00049$. It is $z = \frac{\pi}{2}$, so for the exact error it holds

$$|z - x_{11}| \approx 0.000493.$$

For the cosine function it holds $|\cos'(x)| = |\sin(x)| \le 1$ and close to the root z approximately $|\cos'(\tilde{z})| \approx 1$. The a posteriori estimate provides

$$|x_{11} - z| \approx |f(x_{11})| \approx 0.00049,$$

which is very close to the real error.

The bisection method is very stable. From its construction, it directly follows that the root is always within the iteration interval. However, the method converges slowly. Due to its stability, the bisection method can be used as a starting method for other methods. Unlike other methods we will see, the bisection method is limited to real scalar functions. Furthermore, the bisection method is not suitable for finding double roots, i.e., when $f(z) = f'(z) = 0$.

8.3 Newton's Method

The bisection method only requires the function f to be continuous. The idea of Newton's method is to use more information about the function to improve the scheme. Let $f \in C^1[a, b]$ be continuously differentiable. Close to a point x_0, we approximate $f(x)$ by the linear Taylor approximation, i.e., the tangent

$$f(x) \approx t_{x_0}(x) = f(x_0) + f'(x_0)(x - x_0) \approx f(x).$$

Subsequently, we approximate the root z of $f(x)$ by the root of the tangent

$$f(z) \approx t_{x_0}(z) = f(x_0) + f'(x_0)(z - x_0) \stackrel{!}{=} 0 \quad \Rightarrow \quad z = x_0 - \frac{f(x_0)}{f'(x_0)}.$$

We choose the root of the tangent as a better approximation to the root of the function f.
    Starting from the initial value x_0, we define the iteration

$$x_{k+1} := x_k - \frac{f(x_k)}{f'(x_k)}.$$

This is the classical Newton's method. A geometric illustration is provided in Fig. 8.3.

Algorithm 8.13: Newton's Method

> **Input:** Function $f \in C^1[a, b]$, an initial value $x_0 \in [a, b]$ and a tolerance ϵ.
> 1 $k = 0$
> 2 **while** $|f(x_k)| \geq \epsilon$ **do**
> 3 $x_{k+1} = x_k - f(x_k)/f'(x_k)$
> 4 $k = k + 1$
> **Result:** Approximate root $x \approx x_k$.

In Python, we can implement this as follows

♣ **Implementation 8.14: Newton's Method**

```python
def newton(f, df, x, n=10, tol=1e-10):
    for i in range(n):
        x -= f(x) / df(x)
        if abs(f(x)) < tol:
            return x
    print(f"Newton's method did not converge after {n} iterations")
    return x
```

Remark 8.15 We refer to Sect. 1.4 for examples of the interaction of machine precision and tolerances (termination criteria) of iterative methods. The latter should be chosen sufficiently large to avoid infinite loops due to rounding errors, but at the same time small enough to guarantee sufficient accuracy of the numerical solution. Of course, this applies to all iterative methods, such as fixed-point methods, the bisection method, etc.

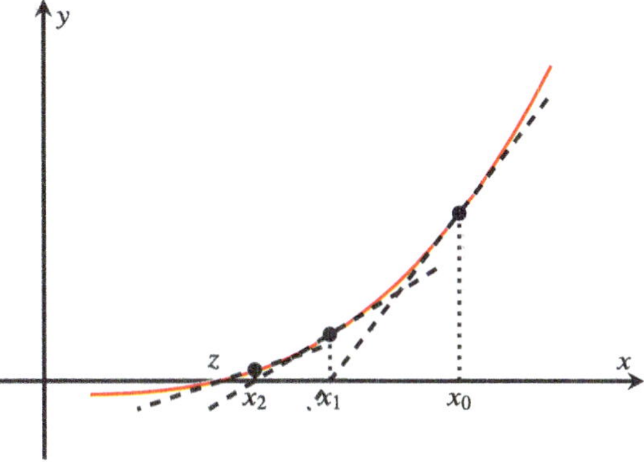

Fig. 8.3 Geometric interpretation of Newton's method. The new iterate x_{k+1} is always the root of the tangent to f at the point $(x_k, f(x_k))$

Algorithm 8.13 belongs to the class of fixed-point iterations with the iteration function

$$g(x) := x - \frac{f(x)}{f'(x)}, \quad x_{k+1} = g(x_k). \tag{8.7}$$

For a fixed point $z = g(z)$, it is obviously $f(z) = 0$.

Theorem 8.16 (Newton's Method) *Let $f \in C^2[a, b]$ have a root z inside the interval $[a, b]$ and let*

$$m := \min_{a \leq x \leq b} |f'(x)| > 0, \quad M := \max_{a \leq x \leq b} |f''(x)| \tag{8.8}$$

be finite bounds for the first derivative from below and the second derivative from above. Let $\rho > 0$ be chosen, such that

$$q := \frac{M}{2m}\rho < 1, \quad B_\rho(z) := \{x \in \mathbb{R} : |x - z| \leq \rho\} \subset [a, b]. \tag{8.9}$$

Then, for every starting point $x_0 \in B_\rho(z)$, the Newton iterates $x_k \in B_\rho(z)$ are well-defined and converge to the root z. Furthermore, the a priori error estimate

$$|x_k - z| \leq \frac{2m}{M}q^{2^k}, \quad k \in \mathbb{N}$$

and the a posteriori error estimate

$$|x_k - z| \leq \frac{1}{m}|f(x_k)|, \quad k \in \mathbb{N}$$

hold.

Proof *(i)* The proof is a direct application of the Taylor approximation. Let $z \in [a, b]$ be the root and $x \in [a, b]$. Then

$$f(z) = f(x) + f'(x)(z - x) + \frac{f''(\xi)}{2}(z - x)^2$$

with an intermediate value $\xi \in [x, z]$ or $\xi \in [z, x]$. For $f'(x) \neq 0$, which is always given in $[a, b]$ because $|f'(x)| \geq m > 0$, we obtain

$$\frac{f(x)}{f'(x)} = (x - z) - \frac{f''(\xi)}{2f'(x)}(x - z)^2. \tag{8.10}$$

(ii) We first show that the iteration is feasible under the given conditions. Suppose $x_k \in B_\rho(z)$. Then, with (8.10) and

$$x_{k+1} - z = x_k - \frac{f(x_k)}{f'(x_k)} - z = (x_k - z) - \left((x_k - z) - \frac{f''(\xi)}{2f'(x_k)}(x_k - z)^2 \right),$$

we have

$$x_{k+1} - z = \frac{f''(\xi)}{2f'(x_k)}(x_k - z)^2$$

or, in terms of an inequality,

$$|x_{k+1} - z| = \frac{|f''(\xi)|}{2|f'(x_k)|}|x_k - z|^2 \leq \frac{M}{2m}|x_k - z|^2. \tag{8.11}$$

Given $x_k \in B_\rho(z)$, $|x_k - z| \leq \rho$, it follows that

$$|x_{k+1} - z| \leq \underbrace{\frac{M}{2m}\rho}_{=q<1} \cdot \rho \leq \rho,$$

and $x_{k+1} \in B_\rho(z) \subset [a, b]$ remains in the vicinity of the root and the domain of $f(\cdot)$, such that the method is feasible.

(iii) We can now prove the error estimate. With

$$\rho_k := \frac{M}{2m}|x_k - z|$$

and (8.11), we have

$$\rho_k = \frac{M}{2m}|x_k - z| \leq \frac{M}{2m}\frac{M}{2m}|x_{k-1} - z|^2 = \rho_{k-1}^2,$$

and it follows

$$\rho_k \leq \rho_0^{2^k}.$$

Thereby

$$|x_k - z| \le \frac{2\,m}{M} \left(\frac{M}{2\,m} |x_0 - z| \right)^{2^k} \le \frac{2\,m}{M} q^{2^k}.$$

(iv) It remains to prove the *a posteriori* estimate. The first Taylor expansion is

$$f(x_k) = f(z) + f'(\xi)(x_k - z)$$

and due to $|f'(\xi)| \ge m$, the error estimate follows

$$|x_k - z| \le \frac{|f(x_k)|}{m}.$$

$\square$

This theorem shows the quadratic convergence of Newton's method if the function f is twice continuously differentiable, has a root in the interval, and if the first derivative is bounded away from zero. The starting value must already be sufficiently close to the root.

Remark 8.17 (A Posteriori Error Estimate) The a posteriori error estimate can be further refined by the distance between two consecutive terms of the solution sequence. With the Taylor expansion and the definition of the Newton iteration, we get

$$f(x_k) = f\left(x_{k-1} - \frac{f(x_{k-1})}{f'(x_{k-1})} \right)$$

$$= f(x_{k-1}) + f'(x_{k-1}) \left(-\frac{f(x_{k-1})}{f'(x_{k-1})} \right) + \frac{f''(\xi)}{2} \left(\frac{f(x_{k-1})}{f'(x_{k-1})} \right)^2.$$

We then obtain the estimate

$$|x_k - z| \le \frac{1}{m}|f(x_k)| \le \frac{M}{2\,m}|x_k - x_{k-1}|^2, \quad k \in \mathbb{N}.$$

In general, we do not know the constants in the second inequality. Moreover, it is a weaker estimate of the root. Therefore, this is usually not suitable for checking how well the method has converged.

Remark 8.18 (Termination Criteria of Newton's Method) Along with the a posteriori estimate, the theorem provides a criterion for termination. If for a given tolerance $\epsilon > 0$ it holds

$$\frac{|f(x_k)|}{m} < \epsilon$$

or

$$\frac{M}{2}|x_k - x_{k-1}|^2 < \epsilon$$

we know that $|x_k - z| < \epsilon$. These stopping criteria require knowledge of the constants m and M, i.e., the bounds of the derivatives of $f(x)$ are introduced. Generally, they are not available. Therefore, we provide a simpler heuristic, but verifiable criterion.

The iteration is terminated (i.e., we have found an acceptable solution x_{k+1} in the k-th iteration), if

$$\frac{|x_{k+1} - x_k|}{|x_k|} < \epsilon. \tag{8.12}$$

The reasoning behind this approach is based on the fast quadratic convergence, which gives $|x_{k+1} - z| \ll |x_k - z|$

$$\frac{|x_k - z|}{|z|} \leq \frac{|x_{k+1} - x_k| + |x_{k+1} - z|}{|z|} = \frac{|x_{k+1} - x_k|}{|z|} + O(|x_k - z|^2),$$

if we are within the regime of quadratic convergence. Finally, $|z|$ is replaced by $|x_k|$ in the denominator to obtain a computable quantity.

Example 8.19 (Newton's Method and the Bisection Method) We consider the function f from the introduction

$$f(x) = x(1 + \exp(x)) + 10\sin(3 + \log(x^2 + 1)).$$

We have argued that $f(x) < 0$ for $x < -10$ and $f(x) > 0$ for $x > 2$. Thus, there is at least one root in the interval $[-10, 2]$. In fact, the function has five roots in this interval. It holds (to six significant digits)

$$z_1 = -9.16359, \quad z_2 = -8.65847, \quad z_3 = -0.304487,$$
$$z_4 = 0.608758, \quad z_5 = 1.54776.$$

We first perform the bisection method:

```
k  a           b           x_k           f(a)       f(b)       b - a
00 -1.000e+01  1.000e+01   0.000000e+00 -2.844e-01  2.203e+05 2.000e+01
01 -1.000e+01  0.000e+00  -5.000000e+00 -2.844e-01  1.411e+00 1.000e+01
02 -5.000e+00  0.000e+00  -2.500000e+00 -5.285e+00  1.411e+00 5.000e+00
03 -2.500e+00  0.000e+00  -1.250000e+00 -1.235e+01  1.411e+00 2.500e+00
...
15 -3.046e-01 -3.040e-01  -3.042603e-01 -5.523e-04  3.766e-03 6.104e-04
```

After 16 steps we obtain an approximation $\tilde{x} \approx -0.3042603$, which has at most the error $|\tilde{x} - z| \leq 0.0006104$. Compared to the exact root, $|\tilde{x} - z_3| \approx 0.00023$.

Next, we study Newton's method and must choose a starting value satisfying the assumptions of Theorem 8.16. This, however, is difficult for the function f (and in the general case). Instead, for lack of a better choice, we take $x^{(0)} = 0$ and get

```
k x_k           f(x_k)
0  0.00000000   1.4112e+00
1 -0.70560004  -3.6485e+00
2 -0.34944175  -3.3204e-01
3 -0.30623519  -1.2393e-02
4 -0.30449044  -2.1400e-05
5 -0.30448741  -6.4398e-11
6 -0.30448741  -2.8866e-15
```

After just four steps, the first five significant digits of the root z_3 are determined. If we choose the end of the left interval $x^{(0)} = a = -10$ as the starting value we get:

```
k x_k           f(x_k)
0 -10.0000000  -2.8437e-01
1 -9.46454160  -6.3943e-02
2 -9.24183447  -1.2229e-02
3 -9.17249436  -1.2325e-03
4 -9.16373585  -1.9882e-05
5 -9.16358985  -5.5322e-09
6 -9.16358981   0.0000e+00
```

Newton's method now converges to the root z_1. We again have very fast convergence, but to a different root than z_3. If Newton's method is in the area of quadratic convergence, it is extremely fast. However, if all five roots are to be found, the problem of good starting values must be solved. These could, for example, be found by rasterizing the interval, i.e., by calculating all $f(x)$ for $x \in \{-10, -9, \ldots, 10\}$.

Remark 8.20 For a twice continuously differentiable function f, there always exists a (possibly tiny) neighborhood $B_\rho(z)$ for each *simple* root z, for which the conditions of Theorem 8.16 are fulfilled. The problem with Newton's method is the determination of a starting point x_0 located in the *domain of quadratic convergence* of the root z (local convergence). This starting value can, for example, be calculated using the bisection method (see the corresponding remark above). Then Newton's method converges very quickly (quadratic convergence) towards the root z.

If the starting point x_0 is in the area of quadratic convergence, then Newton's method converges very quickly towards the desired root. Let us say $q \leq \frac{1}{2}$, then ten iterations give

$$|x_{10} - z| \leq \frac{2m}{M} q^{2^{10}} \sim \frac{2m}{M} 10^{-300}.$$

Example 8.21 (Calculating the n-th Root) The n-th root of a number $a > 0$ is the root of the function $f(x) = x^n - a$. Newton's method to calculate $z = \sqrt[n]{a} > 0$ is

$$x_{k+1} = x_k - \frac{x_k^n - a}{n x_k^{n-1}} = \frac{1}{n}\left\{(n-1)x_k + \frac{a}{x_k^{n-1}}\right\}.$$

The following special cases can be derived from this:

$$x_{k+1} = \frac{1}{2}\left\{ x_k + \frac{a}{x_k} \right\} \quad \text{(Square root)}$$

$$x_{k+1} = \frac{1}{3}\left\{ 2x_k + \frac{a}{x_k^2} \right\} \quad \text{(Cube root)}$$

$$x_{k+1} = (2 - ax_k)x_k \quad \text{(Reciprocal)}$$

Due to Theorem 8.16, $x_k \to z$ converges for $k \to \infty$, if x_0 is chosen close enough to z.

However, in the case of the square root, global convergence can be shown for all starting values $x_0 > 0$. Regardless of x_0, the first iterate satisfies

$$x_1 \geq \sqrt{a}.$$

Subsequently, monotone convergence can be demonstrated:

$$x_{k+1} < x_k \quad k = 2, 3, \ldots$$

This results in global convergence (for all positive initial values). Quadratic convergence is generally only present after a few iterations.

Remark 8.22 (Practical Calculation of the Root) Newton's method is used on computers for the calculation of many quantities that cannot be determined elementarily. Since the iteration only requires multiplications, divisions, and additions, it can be implemented very efficiently. In Excursus 8.6, we discuss a particularly efficient algorithm for approximating the reciprocal of the square root.

8.3.1   Variants of Newton's Method

If the function $f(x)$ has a double root in $[a, b]$ with $f(z) = f'(z) = 0$, then $m = \min|f'| = 0$ then the assumptions of Theorem 8.16 cannot be met. There is no convergence radius $\rho > 0$, such that

$$\frac{M}{2m}\rho < 1$$

holds.

Example 8.23 (Newton's Method for Double Roots) We consider the function $f(x) = x^2$ with the double root $f(0) = f'(0) = 0$. We perform some iterations

$$x_{k+1} = x_k - \frac{f(x_k)}{f'(x_k)},$$

and choose $x_0 = 1$ as the starting value. It applies

$$x_0 = 1, \quad x_1 = 0.5, \quad x_2 = 0.25, \quad x_3 = 0.125, \ldots$$

The values approach the root $z = 0$, but the error $x_k - z$ seems to only halve in each step. We look at the example in more detail and get

$$x_{k+1} = x_k - \frac{f(x_k)}{f'(x_k)} = x_k - \frac{x_k^2}{2x_k} = \frac{1}{2}x_k.$$

This shows that Newton's method for $f(x) = x^2$ does indeed converge towards the root. There is even global convergence for any starting values; however, convergence is just linear.

The question arises whether the observed behavior is just a special case or whether we can generally expect convergence even with multiple roots. The problem is, that as $x_k \to z$ both $f(x_k) \to 0$ and $f'(x_k) \to 0$ apply. For the quotient, the *rule of de l'Hospital* [62] applies to a double root with $f''(z) \neq 0$

$$\lim_{x_k \to z} \frac{f(x_k)}{f'(x_k)} = \lim_{x_k \to z} \frac{f'(x_k)}{f''(x_k)} = \frac{f'(z)}{f''(z)} = 0.$$

From this, we can conclude that the iteration is indeed well-defined, even if $f'(z) = 0$ but $f''(z) \neq 0$. Given that f'' also has a zero for z, another step of l'Hospital might be necessary.

Theorem 8.24 (Newton's Method for Multiple Roots) *Let $z \in [a, b]$ be a p-fold root of the function $f \in C^{p+1}([a, b])$, i.e.,*

$$f(z) = f'(z) = \cdots = f^{(p-1)}(z) = 0, \quad f^{(p)}(z) \neq 0.$$

Furthermore, let

$$m := \min_{x \in [a,b]} |f^{(p)}(x)| > 0, \quad M := \max_{x \in [a,b]} |f^{(p+1)}(x)| < \infty.$$

Then Newton's method

$$x_{k+1} = x_k - \frac{f(x_k)}{f'(x_k)}$$

converges locally to this root. The convergence is only linear. The modified Newton's method

$$x_{k+1} = x_k - p\frac{f(x_k)}{f'(x_k)} \tag{8.13}$$

converges locally to the root quadratically.

Proof We refrain from giving a quantitative estimate for the radius of convergence here. With a $\omega > 0$ let

$$x_{k+1} = x_k - \omega \frac{f(x_k)}{f'(x_k)}.$$

For the error $x_{k+1} - z$ it holds as in the proof of Theorem 8.16

$$x_{k+1} - z = \left(1 - \omega \frac{f(x_k)}{(x_k - z)f'(x_k)}\right)(x_k - z). \tag{8.14}$$

We expand $f(x_k)$ around the root:

$$f(x_k) = \underbrace{\sum_{i=0}^{p-1} \frac{f^{(i)}(z)}{i!}(x_k - z)^i}_{=0} + \frac{f^{(p)}(\xi_{x_k,z})}{p!}(x_k - z)^p = \frac{f^{(p)}(\xi_{x_k,z})}{p!}(x_k - z)^p$$

with an intermediate point $\xi_{x_k,z}$ from $[x_k, z]$ or $[z, x_k]$. Differentiating this formula with respect to x_k yields

$$f'(x_k) = \frac{f^{(p)}(\xi_{x_k,z})}{p!}p(x_k - z)^{p-1} + \frac{(x_k - z)^p}{p!}\underbrace{\frac{\mathrm{d}}{\mathrm{d}x_k}f^{(p)}(\xi_{x_k,z})}_{=:R(x_k,z,p)}.$$

The remainder $R(x_k, z, p)$ contains the $p + 1$-th derivative of f. This is bounded by assumption. We insert these expansions into (8.14):

$$x_{k+1} - z = \left(1 - \omega \frac{\frac{f^{(p)}(\xi_{x_k,z})}{p!}(x_k - z)^p}{\frac{f^{(p)}(\xi_{x_k,z})}{(p-1)!}(x_k - z)^p + \frac{R(x_k,z,p)}{p!}(x_k - z)^{p+1}}\right)(x_k - z)$$

$$= \left(1 - \frac{\omega}{p}\frac{1}{1 + \frac{R(x_k,z,p)}{pf^{(p)}(\xi_{x_k,z})}(x_k - z)}\right)(x_k - z)$$

By assumption, it holds

$$\frac{|R(x_k, z, p)|}{p|f^{(p)}|} \leq \frac{M}{mp},$$

therefore, for $|x_k - z|$ small enough (in particular, smaller than 1)

$$\frac{|R(x_k, z, p)|}{p|f^{(p)}|}|x_k - z| \leq 1.$$

With $(1 + h)^{-1} = 1 + O(h)$ it finally follows

$$x_{k+1} - z = \left(1 - \frac{\omega}{p} + O(|x_k - z|)\right)(x_k - z).$$

For $\omega = 1$, i.e., the usual Newton's method, linear convergence follows, for $\omega = p$ local quadratic convergence follows. $\qquad\square$

Remark 8.25 This theorem shows that Newton's method is also suitable for multiple roots. If the multiplicity p of the root is known, even quadratic convergence can be achieved. In general, however, it will be challenging to determine p optimally.

Example 8.26 (Double-Well Potential) The double-well potential is a polynomial of degree four

$$f(x) = (1 - x^2)^2.$$

At first glance, this looks rather innocent and simple. We plot the function in Fig. 8.4.

The equation appears in various contexts, e.g., in quantum mechanics, phase field equations, and multiphase flows [28]. We are looking for one of the two roots $x^* = \pm 1$ with Newton's method. Here, we find that the multiplicity is $p = 2$, and the residual step should be modified according to (8.13). If we start with $x_0 = 10^{-8}$ (i.e., close to zero), then we find (without modification) the root $x = 1$ in 77 iterations.

```
1 newton(f, df, 1e-8, x_ex=1, n=100, tol=1e-10)

  k  x_k          f(x_k)     |x_k-x|/|x|
  --------------------------------------------
  0  0.00000001 1.000e+00 1.000e+00
  1  25000000.0 3.906e+29 2.500e+07
  2  18750000.0 1.236e+29 1.875e+07

     ...
  75  1.00001837 1.350e-09 1.837e-05
  76  1.00000919 3.375e-10 9.185e-06
  77  1.00000459 8.437e-11 4.593e-06
```

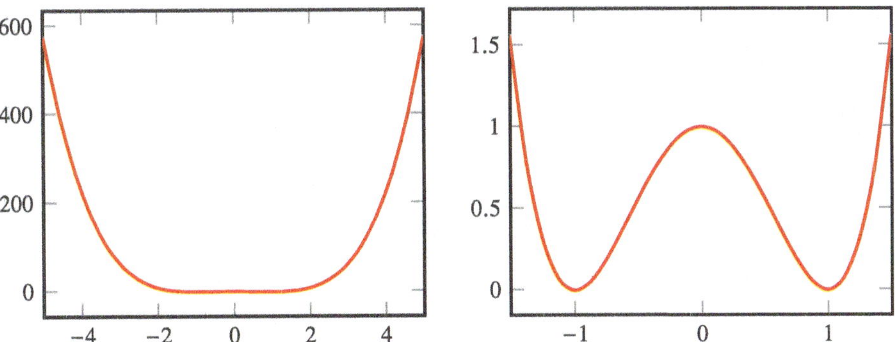

Fig. 8.4 Visualization of the double-well function with focus on the two roots in the right image

At $x_0 \approx 0$, the slope of the tangent is close to zero. This means that the first intersection with the x-axis (see geometric idea of Newton's method) is then far away at $x_1 = 2.5 \cdot 10^7$. From this value, the method then approaches the point $x = 1$ again. By utilizing (8.13), we obtain the following results:

```
1 newton_mod(f, df, 1e-8, x_ex=1, p=2, n=100, tol=1e-10)

  k  x_k          f(x_k)      |x_k-x|/|x|
  -----------------------------------
  0  1.000e-08 1.000e+00 1.000e+00
  1  5.000e+07 6.250e+30 5.000e+07

        ...
      27  1.147e+00 9.894e-02 1.465e-01
      28  1.009e+00 3.540e-04 9.364e-03
      29  1.000e+00 7.547e-09 4.343e-05
      30  1.000e+00 3.559e-18 9.433e-10
```

Still, many steps are required, but once we are close to zero, the modified Newton method shows quadratic convergence.

Newton's method converges once we get close to the zero. A disadvantage especially the high dimensional case is the need to recompute the derivative $f'(x_k)$ in every step. This can be time-consuming. In many cases, one is interested in variants of the method that avoid the calculation of the derivative. One possibility is to reuse the derivative for several steps:

Algorithm 8.27: Simplified Newton Method

Input: Function $f \in C[a, b]$, tolerance ϵ, initial value $x_0 \in [a, b]$ and $c \in [a, b]$.
1 Calculate $f'(c)$
2 $k = 0$
3 **while** $|f(x_k)| \geq \epsilon$ **do**
4 $\quad x_{k+1} = x_k - f(x_k)/f'(c)$
5 $\quad k = k + 1$
Result: Approximate root $x_k \approx x$.

We can adapt our Python code for this as follows:

✤ Implementation 8.28: Simplified Newton Method

```python
1 def newton_simpl(f, df, x, c, n=10, tol=1e-10):
2     df_inv =  1 / df(c)
3     for i in range(n):
4         x -= df_inv * f(x)
5         if abs(f(x)) < tol:
6             return x
7     print(f'The simplified Newton method did not converge after {n} iterations')
8     return x
```

A reasonable choice is, for example, $c = x_0$.

Theorem 8.29 (Convergence of the Simplified Newton Method) *Let $f \in C^2([a, b])$ and $z \in (a, b)$ be a root. Further, let*

$$m = \min_{x \in [a,b]} |f'(x)| > 0, \quad M = \max_{x \in [a,b]} |f''(x)| < \infty,$$

and $\rho \in \mathbb{R}$ with

$$q := \frac{2M}{m}\rho < 1, \quad B_\rho(z) \subset [a, b].$$

Then for all $x_0, c \in B_\rho(z)$, the simplified Newton method converges linearly towards the root. It holds

$$|x_k - z| \le q^k |x_0 - z|.$$

Proof The proof is a simple modification of Theorem 8.16. $\qquad\square$

The simplified Newton method only converges linearly. Convergence can still be fast if $f'(c)$ is a good approximation to $f'(x_k)$. The simplified Newton method then plays out its advantages if the first derivative is time-consuming to calculate. This is not the case in the examples considered here, but can occur when searching for roots in $\mathbb{R}^n$ (see Sect. 8.4) or with problems involving differential equations. A compromise between fast convergence and an efficient algorithm can be achieved if the derivative is not rebuilt in every step, but, e.g., in every n-th step. Then, superlinear convergence can be inferred. A simple way to balance effort and efficiency is to recompute the derivative whenever the convergence factor $|f(x_k)/f(x_{k-1})| > \gamma$ exceeds a certain threshold $\gamma > 0$. The simplified Newton method is a special kind of *Quasi-Newton method*, which is a general framework for all approaches that use approximations to the derivative.

Example 8.30 (Simplified Newton Method) We again consider

$$f(x) = x(1 + \exp(x)) + 10\sin(3 + \log(x^2 + 1)),$$

and apply the simplified Newton method to it. We already know that the function has a root at

$$z_1 \approx -9.1635898.$$

Therefore, we choose the initial value $x_0 = -10$ and use this point also for the evaluation of the derivative, i.e., $c = -10$. With a residual tolerance of 10^{-7}, we get

```
k  x_k            f(x_k)
00 -10.0000000 -2.8437e-01
01 -9.46454160 -6.3943e-02
02 -9.34414034 -3.2920e-02
03 -9.28215371 -1.9753e-02
...
40 -9.16359138 -2.1427e-07
41 -9.16359098 -1.5936e-07
42 -9.16359068 -1.1852e-07
43 -9.16359045 -8.8146e-08
```

Here, 43 steps are needed to get the residual below 10^{-7}. In comparison, the full Newton's method in Example 8.19 only needed six steps to determine the solution up to a residual of machine precision. We also observe the linear convergence proven in Theorem 8.29. After the first iteration steps, the residual is reduced by a factor of 0.743 in each step.

However, the convergence strongly depends on the choice of c. If we instead take $c = -9.344$, we get the solution after 20 steps with the same accuracy.

```
k  x_k            f(x_k))
00 -10.0000000 -2.8437e-01
01 -8.75202698  1.0689e-02
02 -8.79893618  1.4171e-02
03 -8.86112739  1.6883e-02
...
18 -9.16358732  3.3841e-07
19 -9.16358880  1.3629e-07
20 -9.16358940  5.4886e-08
```

Another variant of Newton's method is based on the approximation of the derivative using a difference quotient. We present a simple algorithm:

Algorithm 8.31: Approximated Newton Method

Input: Function $f \in C[a, b]$, tolerance ϵ, initial value $x_0 \in [a, b]$ and $\delta > 0$.
1   $k = 0$
2   **while** $|f(x_k)| \geq \epsilon$ **do**
3      $y_k = f(x_k)$
4      $z_k = f(x_k + \delta)$
5      $x_{k+1} = x_k - \delta y_k/(z_k - y_k)$
6      $k = k + 1$
   Result: Approximate root $x_k \approx x$.

In Python, this can look like this:

♣ Implementation 8.32: Approximated Newton Method

```python
def newton_approx(f, x, delta, n=10, tol=1e-10):
    for i in range(n):
        y = f(x)
        z = f(x + delta)
        x -= delta * y / (z - y)
        if abs(f(x)) < tol:
            return x
    print(f'The approximated Newton method did not converge after {n} iterations')
    return x
```

Each iteration step requires the function f to be evaluated twice. Here, we have chosen the one-sided difference quotient. The central difference quotient

$$f'(x) = \frac{f(x+\delta) - f(x-\delta)}{2\delta} + O(\delta^2)$$

can also be chosen. Then three evaluations of the function f must be performed in each step. For the simple difference quotient, it holds

$$f'(x) = \frac{f(x+\delta) - f(x)}{\delta} + O(\delta).$$

This leads to the iteration

$$x_{k+1} = x_k - \frac{f(x_k)}{f'(x_k) + \frac{f''(\xi)}{2}\delta} = x_k - \frac{f(x_k)}{f'(x_k)}\left(\frac{1}{1 + \frac{f''(\xi)}{f'(x_k)}\delta}\right).$$

Since δ is small, with $(1+\delta)^{-1} = 1 + O(\delta)$ it further follows

$$x_{k+1} = x_k - \frac{f(x_k)}{f'(x_k)} + \frac{f(x_k)f''(\xi)}{f'(x_k)^2}O(\delta).$$

The approximated Newton iteration is thus really an approximation of the usual Newton's method. For $\delta > 0$ small enough, good convergence can be achieved in practical applications. However, the correct choice of δ can be proven to be very difficult. Too small values of δ lead to truncation errors in the approximation of the derivative. Too large δ leads to a poor approximation of the derivatives. In Sect. 9.3, we will analyze the interplay between approximation of the derivative and rounding error using a simple example.

Example 8.33 (Approximated Newton Method) We again consider the function from Example 8.19 and determine the root

$$z_1 \approx -9.1635898.$$

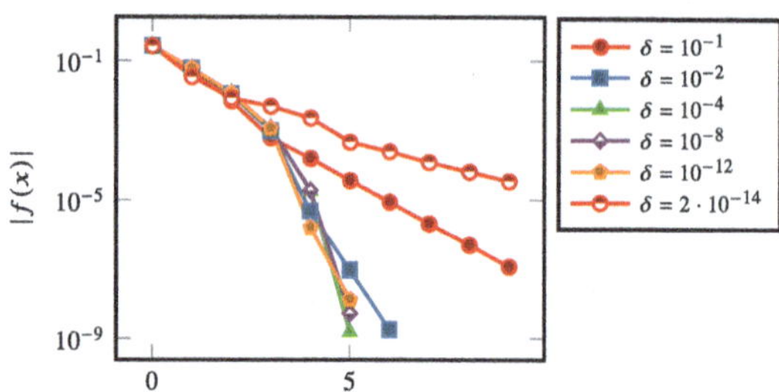

Fig. 8.5 Approximation of the Jacobian by finite differences in Newton's method. Optimal convergence is only achieved when δ is chosen neither too small nor too large

When applying the approximated Newton method, we particularly want to investigate the effect of the choice of stepsize δ. We test the method with values of δ between 10^{-1} and $2 \cdot 10^{-14}$. We show the results in Fig. 8.5. Here it becomes apparent that the correct choice of δ is crucial. If δ is too small, the method may even fail to converge.

A somewhat exotic variant of Newton's method is the *secant method*. It is closely related to the approximated Newton method. The derivative at x_k is approximated as a difference quotient between the two previous values

$$f'(x_k) \approx \frac{f(x_k) - f(x_{k-1})}{x_k - x_{k-1}}.$$

This leads to the iteration

$$x_{k+1} = x_k - \frac{x_k - x_{k-1}}{f(x_k) - f(x_{k-1})} f(x_k).$$

If during the approximation $x_k \to z$, then $x_k - x_{k-1} \to 0$, so the accuracy of the approximation constantly improves. This suggests good convergence of the method. Geometrically considered, Newton's method lays a tangent to the function, while the secant method uses a secant as an approximation to the derivative $f'(x)$.

Algorithm 8.34: Secant Method

Input: Function $f \in C[a, b]$, tolerance $\epsilon > 0$ and starting values $x_0, x_1 \in [a, b]$, $x_0 \neq x_1$.

1   $k = 1$
2   **while** $|f(x_k)| \geq \epsilon$ **do**
3      $x_{k+1} = x_k - f(x_k)(x_k - x_{k-1})/(f(x_k) - f(x_{k-1}))$
4      $k = k + 1$

Result: Approximate root $x_k \approx x$.

The secant method may be less common, but it is interesting as a curious example of a method with a non-integer order of convergence.

Theorem 8.35 (Secant Method) *Let $f \in C^2[a, b]$ with root $z \in (a, b)$ and*

$$m := \min_{x \in [a,b]} |f'(x)| > 0, \quad M := \max_{x \in [a,b]} |f''(x)|.$$

Further, let $\rho > 0$ be chosen such that

$$q := \frac{M}{2\,m}\rho < 1.$$

For arbitrary starting values

$$x_0, x_1 \in B_\rho(z) := \{x \in [a, b] : |x - z| < \rho\}$$

with $x_0 \neq x_1$, the secant method converges to the unique root. The convergence estimate holds

$$|x_k - z| \leq \frac{2\,m}{M} q^{\mu_k \gamma_k^k}$$

with a convergence order

$$\gamma_k \to \frac{1 + \sqrt{5}}{2} \approx 1.618 \quad and \quad \mu_k \to \frac{1 + \sqrt{5}}{2\sqrt{5}} \approx 0.723.$$

Proof *(i)* For the error $e_k := x_k - z$, it holds due to $f(z) = 0$ that

$$
\begin{aligned}
e_{k+1} = x_{k+1} - z &= e_k - \frac{f(x_k)}{f(x_k) - f(x_{k-1})}(x_k - x_{k-1}) \\
&= e_k \frac{x_k - x_{k-1}}{f(x_k) - f(x_{k-1})} \left(\frac{f(x_k) - f(x_{k-1})}{x_k - x_{k-1}} - \frac{f(x_k) - f(z)}{x_k - z} \right).
\end{aligned}
\tag{8.15}
$$

For any three arbitrary values x, y, z, it holds by applying the fundamental theorem of calculus twice

$$
\begin{aligned}
\frac{f(x) - f(y)}{x - y} - \frac{f(x) - f(z)}{x - z} &= \int_0^1 f'(x + s(y - x)) - f'(x + s(z - x))\,ds \\
&= \int_0^1 \int_0^1 f''(x + s(y - x) + rs(z - y))s(z - y)\,dr\,ds.
\end{aligned}
$$

It follows that

$$\left| \frac{f(x) - f(y)}{x - y} - \frac{f(x) - f(z)}{x - z} \right| \leq M \int_0^1 \int_0^1 s|z - y|\,dr\,ds = \frac{M}{2}|z - y|.$$

From (8.15), we infer

$$|e_{k+1}| \leq \frac{M}{2\,m}|e_k||e_{k-1}|$$

or, respectively,

$$\frac{M}{2m}|e_{k+1}| \leq \left(\frac{M}{2m}|e_k|\right)\left(\frac{M}{2m}|e_{k-1}|\right). \tag{8.16}$$

(ii) We show that the method is feasible. Let the initial values x_0, x_1 be such that

$$|e_0| < \rho, \quad |e_1| < \rho \quad \Rightarrow \quad \frac{M}{2m}|e_0| < 1, \quad \frac{M}{2m}|e_1| < 1.$$

Then from (8.16) it follows for all $k \in \mathbb{N}$ that the iterates remain within the domain of convergence.

Next, we check that we do not divide by zero. Suppose $f(x_k) = f(x_{k-1})$. Since $|f'| \geq m > 0$, it follows $x_k = x_{k-1}$. In this case, we have

$$x_{k-1} = x_k = x_{k-1} - \frac{f(x_{k-1})(x_{k-1} - x_{k-2})}{f(x_{k-1}) - f(x_{k-2})},$$

so necessarily $f(x_{k-1}) = f(x_k) = 0$, and the method terminates with the root. Thus, the secant method is always feasible under the given assumptions.

(iii) We derive the convergence estimate with (8.16)

$$q_k := \frac{M}{2m}|e_k| \quad \Rightarrow \quad q_{k+1} \leq q_k q_{k-1} \quad \Rightarrow \quad \log(q_{k+1}) \leq \log(q_k) + \log(q_{k-1}).$$

Logarithmizing is possible since $0 < q < 1$. We define the majorizing sequence

$$Q_0 = Q_1 = \log(q), \quad Q_{k+1} = Q_k + Q_{k-1}, \quad k = 1, 2, \ldots$$

This is just a scaled (and shifted by one) version of the Fibonacci sequence

$$f_1 = f_2 = 1, \quad f_{k+1} = f_k + f_{k-1}, \quad Q_k = \log(q) f_{k+1}.$$

For this, a closed representation exists [62]:

$$f_k = \frac{1}{\sqrt{5}}\left(\left(\frac{1+\sqrt{5}}{2}\right)^k - \left(\frac{1-\sqrt{5}}{2}\right)^k\right)$$

Since

$$\frac{1+\sqrt{5}}{2} \approx 1.618, \quad \frac{1-\sqrt{5}}{2} \approx -0.618,$$

asymptotically, we have

$$f_k \sim \frac{1}{\sqrt{5}}\left(\frac{1+\sqrt{5}}{2}\right)^k.$$

With this, we obtain for Q_k the estimate

$$Q_k = \log(q) f_{k+1} \approx \log(q) 0.723 \cdot 1.618^{k+1}.$$

The error of the secant method is finally estimated with

$$q_k \leq \exp\left(\log(q)\, Q_k\right) \leq q^{Q_k} \leq q^{0.723 \cdot 1.618^{k+1}},$$

as

$$|x_k - z| \leq \frac{2\,m}{M}\left(q^{1.17}\right)^{1.618^k}.$$

$\square$

The secant method is very efficient in terms of operations. In each step, only one new evaluation of the function $f(\cdot)$ must be performed. Newton's method further requires the evaluation of the derivative $f'(\cdot)$. Two steps of the secant method have a similar effort as one step of Newton's method, but combined, they deliver faster convergence with the order $1.618^2 \approx 2.618$. A problem of the secant method is the lack of stability. Cancellations may occur in the calculation of the difference quotient, especially if both iterates x_k and x_{k-1} lie on the same side of the root.

In combination with the bisection method in Sect. 8.2, we obtain the following stabilized method:

Algorithm 8.36: Regula Falsi

> **Input:** Function $f \in C[a, b]$, tolerance ϵ and initial values $x_0, x_1 \in [a, b]$ with
> $\qquad f(x_0) \cdot f(x_1) < 0$.
> 1 $k = 1$
> 2 **while** $|f(x_k)| \geq \epsilon$ **do**
> 3 $\qquad \tilde{x}_{k+1} := x_k - f(x_k) \cdot (x_k - x_{k-1})/(f(x_k) - f(x_{k-1}))$
> 4 $\qquad$ **if** $f(x_k)f(\tilde{x}_{k+1}) < 0$ **then**
> 5 $\qquad\qquad x_{k+1} = \tilde{x}_{k+1}$
> 6 $\qquad$ **else**
> $\qquad\qquad$ /* It holds $f(x_{k-1})f(\tilde{x}_{k+1}) < 0$ $\qquad\qquad\qquad\qquad$ */
> 7 $\qquad\qquad x_{k+1} = \tilde{x}_{k+1}$
> 8 $\qquad\qquad x_k = x_{k-1}$
> 9 $\qquad k = k + 1$
> **Result:** Approximate root $x_k \approx x$.

The *Regula falsi* method ensures that the last two approximations always lie on the opposite side of the root to avoid cancellations. However, the rapid convergence of the secant method is generally also lost. Examples can be constructed where the Regula falsi method converges more slowly than the simple bisection method. For details, we refer to the literature [36].

8.3.2 Convergence Concepts

So far, we have discovered different convergence speeds. The bisection method converges linearly with a fixed reduction factor $1/2$ in each step, whereas Newton's method has quadratic convergence.

Definition 8.37 (*Order of Convergence*) We call an iterative method generating the sequence $(x_k)_{k \in \mathbb{N}}$ with $x_k \to z$ convergent with the *order* $p \geq 1$, if

$$\|x_k - z\| \leq c\|x_{k-1} - z\|^p,$$

with a fixed constant $c > 0$. In the case $p = 1$, i.e., *linear* convergence, the *smallest* constant $c \in (0, 1)$ is called the *linear* convergence factor. If, in the case of linear convergence, $c_k \to 0$ for $k \to \infty$, i.e.,

$$\|x_k - z\| \leq c_k \|x_{k-1} - z\|, \quad c_k \to 0$$

we speak of *superlinear* convergence.

For the bisection method, it holds

$$|x_k - z| \leq \frac{b - a}{2^{k+1}}$$

i.e., the convergence factor is $\frac{1}{2}$ and the order of convergence is $p = 1$ (linear convergence).
  In the vicinity of a root, Newton's method has the convergence behavior

$$|x_k - z| \leq c\,|x_{k-1} - z|^2$$

and we say that the method converges *quadratically*, or, that it has convergence order two. Finally, the secant method converges with the order 1.618, i.e.,

$$|x_k - z| \leq c\,|x_{k-1} - z|^{1.618}.$$

In general, it holds (at least asymptotically) that superlinear convergent sequences converge faster than linear convergent sequences. More generally, methods of order $p + 1$ converge faster than methods of order p. The smaller the convergence factor c is, the faster the convergence. However, the order of convergence has a much greater influence on the convergence speed than the convergence factor.
  To abstract the concept of convergence, we now consider fixed-point problems of the form

$$x_{k+1} = g(x_k), \quad k = 0, 1, 2, \dots.$$

Assuming the mapping $g(\cdot)$ is continuously differentiable, the mean value theorem, Theorem 7.2, gives

$$\left| \frac{x_{k+1} - z}{x_k - z} \right| = \left| \frac{g(x_k) - g(z)}{x_k - z} \right| \rightarrow |g'(z)| \text{ for } k \rightarrow \infty. \tag{8.17}$$

From this, we infer that the asymptotic linear convergence factor for $k \rightarrow \infty$ is just equal to $|g'(z)|$. Accordingly, in the case $g'(z) = 0$ (at least) superlinear convergence is present. In the case $|g'(z)| > 1$, the fixed-point iteration generally does not converge.

From (8.17), we derive the following definition:

Definition 8.38 A fixed point z of a continuously differentiable mapping $g(\cdot)$ is called *attractive*, if $|g'(z)| < 1$ is. In the case $|g'(z)| > 1$, a fixed point is called *repelling*.

The following theorem provides a simple criterion for determining the order of convergence of a differentiable fixed-point iteration. It can also be used to construct methods of high order:

Theorem 8.39 (Order of Convergence of Fixed-Point Iterations) *Let $g(\cdot)$ be p times continuously differentiable in a neighborhood of the fixed point $z = g(z)$. The fixed-point iteration $x_{k+1} = g(x_k)$ has exactly the local order of convergence p, if*

$$g'(z) = \ldots = g^{(p-1)}(z) = 0 \quad \text{and} \quad g^{(p)}(z) \neq 0.$$

Proof "$\Leftarrow$": Let $g'(z) = \ldots = g^{(p-1)}(z) = 0$. Using a Taylor expansion, we have

$$x_{k+1} - z = g(x_k) - g(z) = \sum_{j=1}^{p-1} \frac{(x_k - z)^j}{j!} g^{(j)}(z) + \frac{(x_k - z)^p}{p!} g^{(p)}(\xi_k)$$

for $\xi_k \in (x_{k+1}, z)$. Thus, we obtain the estimate

$$|x_{k+1} - z| \leq \frac{1}{p!} \max |g^{(p)}| \, |x_k - z|^p.$$

"$\Rightarrow$": We assume the iteration is of order p, i.e.,

$$|x_{k+1} - z| \leq c|x_k - z|^p. \tag{8.18}$$

If there is a minimal $m \leq p - 1$ with $g^{(m)}(z) \neq 0$, but $g^{(j)}(z) = 0$, $j = 1, \ldots, m - 1$, then every iteration sequence $(x_k)_{k \in \mathbb{N}}$ with sufficiently small $|x_0 - z| \neq 0$ converges to with z with

$$|x_k - z| = \left| \frac{1}{m!} g^{(m)}(\xi_k) \right| \, |x_{k-1} - z|^m.$$

In other words, with order $m < p$, contradicting (8.18), i.e., the assumption that the iteration is of order p. Therefore, $g^{(j)}(z) = 0$ must hold for $j = 1, \ldots, p - 1$ and $g^{(p)}(z) \neq 0$. $\quad \Box$

We verify the theorem using the example of Newton's method, which we can write using

$$g(x) = x - \frac{f(x)}{f'(x)}$$

in the form of a fixed-point mapping. It holds

$$g'(x) = 1 - \frac{f'(x)^2 - f(x)f''(x)}{f'(x)^2} = f(x)\frac{f''(x)}{f'(x)^2} \quad \Rightarrow \quad g'(z) = 0,$$

and

$$g''(x) = \frac{f''(x)}{f'(x)} + \frac{f(x)f'''(x)}{f'(x)^2} - \frac{2f(x)f''(x)^2}{f'(x)^3} \quad \Rightarrow \quad g''(z) = \frac{f''(z)}{f'(z)},$$

which is not zero in the general case. Therefore, Newton's method converges—as we already know—quadratically. This analysis also shows that in the case of functions with $f''(z) = 0$ (but $f'(z) \neq 0$), at least third-order convergence is present. The function $f(x) = \sin(x)$ is such an example.

We now proceed in reverse. We search for

$$g(x) = x + h(x)f(x) \tag{8.19}$$

with a function $h(x)$ to be determined for a fixed point iteration with the highest possible order of convergence. First, it holds

$$g(z) = z + h(z)\underbrace{f(z)}_{=0} = z,$$

i.e., it is indeed a fixed-point iteration. For the first derivative, it holds

$$g'(z) := 1 + h'(z)\underbrace{f(z)}_{=0} + h(z)f'(z) \overset{!}{=} 0.$$

From this follows the condition on $h(x)$ in z

$$h(z) = -\frac{1}{f'(z)}.$$

We now choose $h(x)$ for all x according to this rule, so that

$$h(x) := -\frac{1}{f'(x)}.$$

Substituting into (8.19) yields the iteration function

$$g(x) := x - \frac{f(x)}{f'(x)}.$$

Thus, the iteration rule of Newton's method

$$x_{k+1} := g(x_k) = x_k - \frac{f(x_k)}{f'(x_k)}$$

is recovered.

8.4 Solving Nonlinear Systems

In this section, we consider nonlinear systems given by functions $f : \mathbb{R}^n \to \mathbb{R}^n$ with $n > 1$ and search for roots $z \in \mathbb{R}^n$. Every linear system $Ax = b$ can be seen as a special case of this problem by introducing the function

$$f(x) = b - Ax.$$

Since our iterates are vectors $x \in \mathbb{R}^n$, we use a superscript to denote the iteration index, i.e., $x^{(n)} \in \mathbb{R}$ and a subscript to denote the component $x_k^{(n)} \in \mathbb{R}$.

The bisection method does not have a counterpart in $\mathbb{R}^n$. However, we want to start with a correspondingly simple method and orient ourselves on the general convergence statement for fixed point iterations from Theorem 8.39. Let $f : D \subset \mathbb{R}^n \to \mathbb{R}^n$. Then we choose a matrix $C \in \mathbb{R}^{n \times n}$ and define the fixed point iteration

$$x^{(0)} \in \mathbb{R}^n, \quad x^{(k+1)} = g(x^{(k)}) := x^{(k)} + C^{-1} f(x^{(k)}), \quad k = 0, 1, 2, \ldots$$

This method converges if f is continuously differentiable on a ball $B_\rho(z) \subset \mathbb{R}^n$ and if

$$\sup_{\zeta \in B_\rho(z)} \|g'(\zeta)\| = \sup_{\zeta \in B_\rho(z)} \|I + C^{-1} Df(\zeta)\| =: q < 1$$

with the Jacobian $Df : \mathbb{R}^n \to \mathbb{R}^{n \times n}$.

8.4.1 Newton's Method in $\mathbb{R}^n$

Let $f : D \subset \mathbb{R}^n \to \mathbb{R}^n$. We are looking for a solution to

$$f(x) = (f_1(x), \ldots, f_n(x)) = 0.$$

Let $x^{(0)} \in D$ be an initial value. In every step, we linearize f with the Taylor approximation

$$f(x^{(k)} + w^{(k)}) = f(x^{(k)}) + \sum_{j=1}^{n} \frac{\partial f}{\partial x_j}(x^{(k)}) w_j^{(k)} + O(|w^{(k)}|^2).$$

The new iteration $x^{(k+1)} = x^{(k)} + w^{(k)}$ is defined as the root of this linearization. That is, we are looking for the solution $w^{(k)} \in \mathbb{R}^n$ of the linear system

$$\sum_{j=1}^{n} \frac{\partial f_i}{\partial x_j}(x^{(k)}) w_j^{(k)} = -f_i(x^{(k)}), \quad i = 1, \ldots, n.$$

With the Jacobian

$$Df = \begin{pmatrix} \frac{\partial f_1}{\partial x_1} & \frac{\partial f_1}{\partial x_2} & \cdots & \frac{\partial f_1}{\partial x_n} \\ \frac{\partial f_2}{\partial x_1} & \frac{\partial f_2}{\partial x_2} & \cdots & \frac{\partial f_2}{\partial x_n} \\ \vdots & & \ddots & \vdots \\ \frac{\partial f_n}{\partial x_1} & \frac{\partial f_n}{\partial x_2} & \cdots & \frac{\partial f_n}{\partial x_n} \end{pmatrix}$$

the linear system to be solved is written as

$$Df(x^{(k)}) w^{(k)} = -f(x^{(k)}).$$

With this and an initial value $x^{(0)} \in D$, the Newton iteration for $k = 0, 1, 2, \ldots$ is given by

$$Df(x^{(k)}) w^{(k)} = -f(x^{(k)}), \quad x^{(k+1)} = x^{(k)} + w^{(k)}. \tag{8.20}$$

The fact that in the multidimensional case, the derivative of $f(x)$ is a matrix is the crucial difference from the one-dimensional case. Instead of one derivative, n^2 derivatives must be calculated, and instead of a division by its derivative, a linear system with the coefficient matrix $Df(x_k) \in \mathbb{R}^{n \times n}$ has to be solved in each step of Newton's method.

The iteration (8.20) is formulated as a *residual correction scheme* (also known as defect correction method). We define analogously to Definition 3.51:

Definition 8.40 (*Residual*) Let $\tilde{x} \in \mathbb{R}^n$ be an approximation of the solution of $f(x) = y$. The *residual* is

$$d(\tilde{x}) = y - f(\tilde{x}).$$

A proof of convergence for the multidimensional Newton's method can be conducted as in the one-dimensional case with Taylor expansion. However, we give here the theorem of Newton-Kantorovich, which extends the result on the one hand and slightly weakens the requirements on the function f.

Theorem 8.41 (Newton-Kantorovich)   *Let $D \subset \mathbb{R}^n$ be an open and convex set and $f : D \subset \mathbb{R}^n \to \mathbb{R}^n$ be continuously differentiable. Furthermore, let the following hold:*

1. *The Jacobian Df is uniformly Lipschitz continuous for all $x, y \in D$:*

$$\|Df(x) - Df(y)\| \le L\|x - y\|, \quad x, y \in D, \tag{8.21}$$

   with some $L < \infty$.

2. *The Jacobian has a uniformly bounded inverse on D*

$$\|Df(x)^{-1}\| \le \beta, \quad x \in D, \tag{8.22}$$

for some $\beta < \infty$.

3. *For the starting point $x^{(0)} \in D$ let*

$$q := \alpha\beta L < \frac{1}{2}, \quad \alpha := \|Df(x^{(0)})^{-1} f(x^{(0)})\|. \tag{8.23}$$

4. *For $r := 2\alpha$ let the closed ball*

$$B_r(x^{(0)}) := \{x \in \mathbb{R}^n : \|x - x^{(0)}\| \le r\}$$

be contained in the set D.

Then the function f has exactly one root $z \in B_r(x^{(0)})$ and the Newton iteration

$$Df(x^{(k)})w^{(k)} = -f(x^{(k)}), \quad x^{(k+1)} = x^{(k)} + w^{(k)}, \quad k = 0, 1, 2, \ldots,$$

converges quadratically to this root z. Moreover, the a priori error estimate holds

$$\|x^{(k)} - z\| \le 2\alpha q^{2^k - 1}, \quad k = 0, 1, \ldots.$$

Proof We follow the proof from [75], closely related to the work of Kantorovich [3]. Alternative representations can be found, e.g., in [25].

As the proof of the theorem is by far more involved than the simple proof that we used to prove Theorem 8.16 in the one-dimensional case, we first provide an outline:

(i) Derivation of two auxiliary estimates.
(ii) We show: All iterates $x^{(k)}$ are in the ball $B_r(x^{(0)})$. Furthermore, the a priori error estimate holds.
(iii) We show that the iterates $(x^{(k)})_{k\in\mathbb{N}}$ form a Cauchy sequence and thus have a limit $z \in \mathbb{R}^n$.
(iv) We show that this limit must be a root of f.
(v) We show that there can only be one root in $B_r(x^{(0)})$.
(vi) We demonstrate the quadratic convergence.

(i) Let $x, y, z \in D$. Since D is convex, it holds

$$f(y) - f(x) = \int_0^1 Df\big(y + s(y - x)\big)(y - x)\mathrm{d}s.$$

Subtractinng $Df(z)(y - x)$ on both sides gives

$$f(y) - f(x) - Df(z)(y - x) = \int_0^1 \big((Df(x + s(y - x)) - Df(z)\big)(y - x)\mathrm{d}s.$$

Using the Lipschitz continuity of the Jacobian Df, we get

$$\|f(y) - f(x) - Df(z)(y - x)\| \le L\|y - x\| \int_0^1 \|s(x - z) + (1 - s)(y - z)\|ds$$

$$\le \frac{L}{2}\|y - x\|\big(\|x - z\| + \|y - z\|\big).$$

For the choice $z = x$, it follows

$$\|f(y) - f(x) - Df(x)(y - x)\| \le \frac{L}{2}\|y - x\|^2, \quad \forall x, y \in D. \tag{8.24}$$

Similarly for the choice $z = x^{(0)}$, we obtain

$$\|f(y) - f(x) - D(x^{(0)})(y - x)\| \le rL\|y - x\|, \quad \forall x, y \in B_r(x^{(0)}). \tag{8.25}$$

(ii) We now show that all iterates lie in $B_r(x^{(0)})$. In doing so, we can also prepare for the a priori error estimate. We conduct the proof by induction and will show that

$$\|x^{(k+1)} - x^{(0)}\| \le r, \quad \|x^{(k+1)} - x^{(k)}\| \le \alpha q^{2^k - 1}, \quad k = 1, 2, \dots . \tag{8.26}$$

We start with $k = 0$. For the Newton iteration $x^{(1)} - x^{(0)} = -Df(x^{(0)})^{-1} f(x^{(0)})$, we get from (8.23)

$$\|x^{(1)} - x^{(0)}\| = \|Df(x^{(0)})^{-1} f(x^{(0)})\| = \alpha = \frac{r}{2} < r,$$

i.e., $x^{(1)} \in B_r(x^{(0)})$, and the estimate

$$\|x^{(1)} - x^{(0)}\| \le \alpha = \alpha q^{2^0 - 1}$$

holds.

For $k \mapsto k + 1$, the two inequalities (8.26) are true for all $k \ge 0$. Thus, $x^{(k)} \in B_r(x^{(0)})$, so that the Newton iterate $x^{(k+1)}$ is well-defined. Then, using condition (8.22), the estimate (8.24) and the induction hypothesis (8.26) with the definition of q, the following chain of estimates holds:

$$
\begin{aligned}
\|x^{(k+1)} - x^{(k)}\| &= \|Df(x^{(k)})^{-1} f(x^{(k)})\| \\
&\le \beta \, \|f(x^{(k)})\| \\
&= \beta \| \underbrace{f(x^{(k)}) - f(x^{(k-1)}) - Df(x^{(k-1)})(x^{(k)} - x^{(k-1)})}_{=0} \| \\
&\le \frac{\beta L}{2} \underbrace{\|x^{(k)} - x^{(k-1)}\|^2}_{\text{Induction}} \le \frac{\beta L}{2}\left(\alpha q^{2^{(k-1)} - 1}\right)^2 \\
&= \frac{\alpha}{2} q^{2^{(k)} - 1} < \alpha q^{2^{(k)} - 1}
\end{aligned}
$$

Hereby, we obtain

$$\|x^{(k+1)} - x^{(0)}\| \le \|x^{(k+1)} - x^{(k)}\| + \ldots + \|x^{(1)} - x^{(0)}\|$$
$$\le \alpha\big(1 + q + q^3 + q^7 + \ldots + q^{2^{(k)}-1}\big)$$
$$\le \frac{\alpha}{1-q} \le 2\alpha = r,$$

i.e., it holds $x^{(k+1)} \in B_r(x^{(0)})$. This finalizes the induction step from $k \mapsto k+1$, i.e., the two inequalities (8.26) are valid for all k.

(iii) We now show that the $x^{(k)} \in B_r(x^{(0)})$ form a Cauchy sequence. Let $m > 0$. Since $q < \frac{1}{2}$, it holds

$$\|x^{(k)} - x^{(k+m)}\| \le \|x^{(k)} - x^{(k+1)}\| + \ldots + \|x^{(k+m-1)} - x^{(k+m)}\|$$
$$\le \alpha\big(q^{2^{(k)}-1} + q^{2^{(k+1)}-1} + \ldots + q^{2^{m+k-1}-1}\big)$$
$$= \alpha q^{2^{(k)}-1}\big(1 + q^{2^{(k)}} + \ldots + (q^{2^{(k)}})^{2^{m-1}-1}\big)$$
$$\le 2\alpha q^{2^{(k)}-1}. \tag{8.27}$$

By $q < \frac{1}{2}$ we conclude that $(x^{(k)})_{k\in\mathbb{N}} \subset D$ is a Cauchy sequence and the limit exists in the Banach space $\mathbb{R}^n$

$$z = \lim_{k\to\infty} x^{(k)} \in \mathbb{R}^n.$$

Going to the limit $k \to \infty$, we obtain with (8.26) that

$$\|z - x^{(0)}\| \le r \quad \Rightarrow \quad z \in B_r(x^{(0)}).$$

Taking the limit $m \to \infty$ in (8.27), we verify the error estimate

$$\|x^{(k)} - z\| \le 2\alpha q^{2^{(k)}-1}. \quad k = 0, 1, \ldots.$$

(iv) It remains to show that $z \in B_r(x^{(0)})$ is a root of f. The Newton iteration and condition (8.21) yield

$$\|f(x^{(k)})\| = \|Df(x^{(k)})(x^{(k+1)} - x^{(k)})\|$$
$$\le \|Df(x^{(k)}) - Df(x^{(0)}) + Df(x^{(0)})\| \, \|x^{(k+1)} - x^{(k)}\|$$
$$\le \Big(L\|x^{(k)} - x^{(0)}\| + \|Df(x^{(0)})\|\Big) \|x^{(k+1)} - x^{(k)}\| \to 0 \quad (k \to \infty).$$

Therefore, it holds
$$f(x^{(k)}) \to 0, \quad k \to \infty.$$

The continuity of f implies $f(z) = 0$.

(v) Finally, we must prove that $z \in B_r(x^{(0)})$ is unique. This is shown using the contraction property and the formulation of the Newton iteration as a fixed-point iteration. Every root of $f(\cdot)$ is a fixed point of the simplified Newton iteration

$$g(x) := x - Df(x^{(0)})^{-1} f(x).$$

The fixed point function $g(\cdot)$ is Lipschitz continuous, as (8.25) gives

$$\begin{aligned}
\|g(x) - g(y)\| &= \|x - y - Df(x^{(0)})^{-1} f(x) + Df(x^{(0)})^{-1} f(y)\| \\
&= \|Df(x^{(0)})^{-1} (f(y) - f(x) - Df(x^{(0)})(y - x))\| \\
&\leq \underbrace{\|Df(x^{(0)})^{-1}\|}_{\leq \beta} rL \|y - x\| \\
&\leq \beta Lr \|y - x\|,
\end{aligned}$$

for all $x, y \in B_r(x^{(0)})$. With $r = 2\alpha$ it follows $\beta Lr = 2\alpha\beta L \leq 2q < 1$. That is, g is a contraction. The Banach fixed point theorem says that there is only one fixed point, giving us uniqueness of the root.

(vi) By the definition of the Newton iteration (8.22) and the auxiliary estimate (8.24), we have

$$\begin{aligned}
\|x^{(n+1)} - z\| &= \|x^{(n)} - Df(x^{(n)})^{-1} f(x^{(n)}) - z\| \\
&\leq \|Df(x^{(n)})^{-1}\| \, \| \underbrace{f(z)}_{=0} - f(x^{(n)}) - Df(x^{(n)})(z - x^{(n)})\| \\
&\leq \beta \frac{L}{2} \|x^{(n)} - z\|^2.
\end{aligned}$$

$\square$

Remark 8.42 The Newton-Kantorovich theorem differs in some points from Theorem 8.16 on the one-dimensional Newton method. There, the existence of a root was one of the assumptions, while Newton-Kantorovich gives the root as a result. Another generalization lies in the regularity requirements; instead of twice differentiable, Newton-Kantorovich asks for Lipschitz continuity of the Jacobian. Finally, the domain of quadratic convergence no longer depends on the unknown root but on quantities that, in principle, could be evaluated. Of course, the theorem of Newton-Kantorovich also applies in the one-dimensional case.

From Theorem 8.41, we can derive a local convergence result

Corollary 8.43 *Let $D \subset \mathbb{R}^n$ be open and $f : D \subset \mathbb{R}^n \to \mathbb{R}^n$ be twice continuously differentiable. Assume that $z \in D$ is a root with a regular Jacobian $Df(z)$. Then Newton's method is locally convergent, i.e., there exists a neighborhood B around z, such that Newton's method converges for all initial values $x^{(0)} \in B$.*

Example 8.44 (Newton's Method in $\mathbb{R}^n$) We are looking for the roots of the function

$$f(x_1, x_2) = \begin{pmatrix} 1 - x_1^2 - x_2^2 \\ (x_1 - 2x_2)/(1/2 + x_2) \end{pmatrix}$$

with the Jacobian

$$Df(x) = \begin{pmatrix} -2x_1 & -2x_2 \\ \frac{2}{1+2x_2} & -\frac{4+4x_1}{(1+2x_2)^2} \end{pmatrix}.$$

The roots of f are given by

$$x \approx \pm(0.894427191, 0.447213595).$$

For the implementation in Python, we use our LU factorization with pivoting to solve the resulting linear systems. The code for this is

♣ **Implementation 8.45: Newton's Method in $\mathbb{R}^n$**

```python
def newton_vec(f, Df, x, n=10, tol=1e-10):
    b, y, w = np.zeros_like(x), np.zeros_like(x), np.zeros_like(x)
    for i in range(n):
        b[:] = - f(x)
        if np.linalg.norm(b) < tol:
            break
        jac = Df(x)
        pivot = lu_pivot(jac)
        for p in pivot:
            b[p] = b[[p[1], p[0]]]
        y[:] = forward(jac, b)
        w[:] = backward(jac, y)
        x[:] += w
    else:
        print(f"Newton's method did not converge after {n} iterations.", end=' ')
        print(f'||res||={np.linalg.norm(f(x)):.4e}')
    return i, x
```

To solve the above example with this, we still need to implement the function and Jacobian:

```python
def f(x):
    x, y = x[:]
    return np.array([1 - x**2 - y**2, (x - 2 * y) / (1 / 2 + y)], dtype=np.double)
```

```python
def Df(x):
    x, y = x[:]
    return np.array([[-2 * x, -2 * y],
                     [2 / (1 + 2 * y), - (4 + 4 * x)/ (1 + 2 * y)**2]],
                    ↪  dtype=np.double)
```

We start the iteration with $x^{(0)} = (1, 1)^T$,

```python
n, x = newton_vec(f, Df, x=np.array([1.0, 1.0]), n=15, tol=1e-10)
print(f'\nn = {n}, x = {x}')
```

and after five steps, we get the solution

```
n = 5, x = [0.89442719 0.4472136 ]
```

If we compare the solution in the individual steps,

```
x_1 = (1.1428571429, 0.3571428571)
x_2 = (0.9265873016, 0.4420634921)
x_3 = (0.8949352419, 0.4473485186)
x_4 = (0.8944273077, 0.4472136708)
x_5 = (0.8944271910, 0.4472135955)
```

we see that the solution was determined exactly to at least six digits. If we compare the solution with the exact value, we see that the solution was even determined exactly to eight significant digits.

In contrast to the one-dimensional case, the calculation of the derivative in $\mathbb{R}^n$ may be very complex, as not one, but n^2 partial derivatives

$$(Df)_{ij} = \frac{\partial f_i}{\partial x_j}, \quad i, j = 1, \ldots, n$$

must be calculated. The effort for this can be substantial. In the second step of Newton's method, a linear system has to be solved. This can be done, for example, with a factorization method in $O(n^3)$ operations followed by forward and backward solves requiring another $O(n^2)$ operations.

We have already learned about the quasi-Newton method in the one-dimensional case, see Algorithm 8.27. If the derivative $Df(x^{(k)})$ is not assembled in every iteration but only at a fixed value $c \in \mathbb{R}^d$, i.e., $Df(c)$ is used for all following iterations, convergence is still present (although only linear). In the multidimensional Newton's method, this simplification can fully play out its strength, e.g., when the linear system is solved with the LU factorization:

Algorithm 8.46: Simplified Newton Method in $\mathbb{R}^n$

> **Input:** Function $f \in C^1(D)$, initial value $x^{(0)} \in D$, $c \in D$ and tolerance $\epsilon > 0$.
> 1 Calculate the Jacobian $Df(c)$
> 2 Create the LU factorization $Df(c) = L(c)U(c)$
> 3 $k = 0$
> 4 **while** $\|f(x_k)\| \geq \epsilon$ **do**
> 5 $L(c)U(c)w^{(k)} = -f(x^{(k-1)})$
> 6 $x^{(k)} = x^{(k-1)} + w^{(k)}$
> **Result:** Approximate root $x_k \approx x \in D$.

Example 8.47 (Simplified Newton Method in $\mathbb{R}^n$) We apply the simplified Newton method to the function from Example 8.44. For this, we first have to adjust our vector-valued implementation by calculating the LU factorization only once outside the loop.

♣ **Implementation 8.48: Simplified Newton Method in $\mathbb{R}^n$**

```python
def newton_simpl_vec(f, Df, x, n=10, tol=1e-10):
    b, y, w = np.zeros_like(x), np.zeros_like(x), np.zeros_like(x)
    jac = Df(x)
    pivot = lu_pivot(jac)
    for i in range(n):
        b[:] = - f(x)
        if np.linalg.norm(b) < tol:
            break
        for p in pivot:
            b[p] = b[[p[1], p[0]]]
        y[:] = forward(jac, b)
        w[:] = backward(jac, y)
        x[:] += w
    else:
        print(f'The simplified Newton method did not converge after {n}
        ↪ iterations.', end=' ')
        print(f'||res||={np.linalg.norm(f(x)):.4e}')
    return i, x
```

If we now apply this code with the better initial value $x^{(0)} = (1, 0.5)^T$, we see that we reached the root after 10 steps

```python
n, x = newton_simpl_vec(f, Df, x=np.array([1.0, 0.5]), n=15)
print(f'n = {n}, x = {x}')
```

```
n = 10, x = [0.89442719 0.4472136 ]
```

With the starting vector $x^{(0)} = (1, 1)^T$, the simplified Newton method takes 670 steps to determine the root to an residual tolerance $< 10^{-10}$. Therefore, we always have to weight up between the cost of matrix factorization in Newton's method and the slower convergence in the simplified Newton method.

The two expensive steps of Newton's method, setting up and factorizing the Jacobian, are each performed only once. The actual iteration can be efficiently performed with forward and backward substitution. This method can be refined by not statically choosing the linearization point $c \in \mathbb{R}^n$, but regularly updating it. An update can either be performed after a fixed number of steps $c = x^{(3)}$, $c = x^{(6)}, \ldots$ or whenever the method converges poorly. The residual of two consecutive iterations

$$\rho_k = \frac{\|f(x^{(k)})\|}{\|f(x^{(k-1)})\|}$$

is suitable for controlling the convergence. If $\rho_k > \rho_0$ with a limit $\rho_0 \approx 0.1$ applies, i.e., if the convergence is poor, an update of the Jacobian is performed with $c = x^{(k)}$.

Another variant of Newton's method is the *inexact method*. Here we assume that the linear system $Df(x^{(k)})w^{(k)} = -f(x^{(k)})$ is not solved exactly but only approximated with an iterative linear solver up to some accuracy $\epsilon > 0$. It can be shown that this inexact version of Newton's method is still converging. The convergence rate, however, will be limited by the accuracy of the linear residuals, and quadratic convergence is only gained if this linear accuracy is increased in every step. For large systems, where a direct solution of the linear problems is not feasible, the inexact Newton method is the standard.

8.4.2   Globalization of Newton's Method

The great challenge in the practical implementation of Newton's method is the choice of a good starting value. A globalization, i.e., an enlargement of the convergence range, can be achieved by damping. To this end, line 6 in Algorithm 8.46 is replaced by

$$x^{(k)} = x^{(k-1)} + \omega_k w^{(k)}$$

with a damping parameter $\omega_k \in (0, 1]$, or equivalently

$$x^{(k+1)} = x^{(k)} - \omega_k Df(x^{(k)})^{-1} f(x^{(k)}).$$

Theorem 8.49 (The Damped Newton Method) *Let the prerequisites of Theorem 8.41 hold, except for the condition for the starting value, (8.23). The damped Newton iteration*

$$Df(x^{(k)})w = -f(x^{(k)}), \quad x^{(k+1)} = x^{(k)} + \omega_k w$$

with

$$\omega_k := \min\{1, \frac{1}{\alpha_k \beta L}\}, \quad \alpha_k := \|Df(x^{(k)})^{-1} f(x^{(k)})\|$$

generates a sequence $(x^{(k)})_{k \in \mathbb{N}}$, *for which after* k_* *steps*

$$q_* := \alpha_{k_*} \beta L < \frac{1}{2}$$

is fulfilled. For $k > k_*$, $x^{(k)}$ *converges quadratically and the a priori error estimate*

$$\|x^{(k)} - z\| \leq \frac{\alpha}{1 - q_*} q_*^{2^k - 1}, \quad k \geq k_*$$

holds.

Proof We want to show monotone convergence in the sense of $\|f(x^{(k+1)})\| \leq \|f(x^{(k)})\|$ of the damped iteration. For $x^{(0)}$, the set

$$D_0 := \{x \in D : \|f(x)\| \leq \|f(x^{(0)})\|\}$$

is closed and non-empty. Further, we define the damped iteration

$$x_\omega = g_\omega(x) := x - \omega Df(x)^{-1}f(x).$$

For an $x \in D_0$, let $\omega_{max} \leq 1$ be the maximum damping parameter, such that:

$$f(x_\omega) \leq f(x^{(0)}), \quad 0 \leq \omega \leq \omega_{max} \leq 1.$$

We define

$$h(\omega) := f(x_\omega)$$

with $h(0) = f(x)$. For $h(\cdot)$, it holds

$$h'(\omega) = \frac{d}{d\omega} f(x - \omega Df(x)^{-1}f(x)) = -Df(x_\omega)Df(x)^{-1}f(x).$$

With $h'(0) = -f(x) = -h(0)$, it follows

$$f(x_\omega) - f(x) = h(\omega) - h(0) = \int_0^\omega h'(s)ds = -\int_0^\omega Df(x_s)Df(x)^{-1}f(x)ds =$$

$$-\int_0^\omega \left((Df(x_s) - Df(x))Df(x)^{-1}f(x) + f(x)\right)ds.$$

Therefore,

$$f(x_\omega) = \left(-\int_0^\omega (Df(x_\omega) - Df(x))Df(x)^{-1}ds\right)f(x) + (1 - \omega)f(x)$$

and consequently

$$\|f(x_\omega)\| \leq (1 - \omega)\|f(x)\| + \beta L \int_0^\omega \|x_s - x\|ds\|f(x)\|$$

$$= (1 - \omega)\|f(x)\| + \beta L \int_0^\omega s\|Df(x)^{-1}f(x)\|ds\|f(x)\|$$

$$\leq \left(1 - \omega + \alpha_x \beta L \frac{\omega^2}{2}\right)\|f(x)\|,$$

with $\alpha_x = \|Df(x)^{-1}f(x)\|$. We now determine $0 \leq \omega \leq \omega_{max} \leq 1$ such that the expression

$$\rho(\omega) := \left|1 - \omega + \alpha_x \beta L \frac{\omega^2}{2}\right|$$

becomes minimal. The minimum is realized at

$$\omega_{min} = \min\left\{1, \frac{1}{\alpha_x \beta L}\right\}.$$

Then, it holds

$$1 - \omega_{min} + \frac{\omega_{min}^2}{2}\alpha_x \beta L \leq 1 - \frac{1}{2\alpha_x \beta L} < 1.$$

With this choice of the damping parameter, monotone convergence

$$\|f(x^{(k+1)})\| \leq \left(1 - \frac{1}{2\alpha_k \beta L}\right)\|f(x^{(k)})\|$$

is achieved. After a finite number of steps, it holds

$$\frac{\alpha_k \beta L}{2} \leq \frac{\beta^2 L}{2}\|f(x^{(k)})\| < 1$$

and the domain of quadratic convergence is reached. $\square$

Example 8.50 (Damped Newton Method) If we choose the starting vector poorly, then Newton's method may converge poorly or not at all. For example, if we take the starting vector $x = (0, -0.49999)$ (where the Jacobian is nearly singular), we do not get convergence after 35 steps:

```
1 newton_vec(f, Df, x=np.array([0, -0.49999]), n=35, tol=1e-10)
    it  ||f(x)||
    ---------------
    00  9.9998e+04
    01  5.6255e+09
    02  1.4064e+09
    ...
    19  3.0457e+00
    20  6.8497e+03
    21  1.5718e+04
    ...
    32  3.1562e+00
    33  1.8035e+00
    34  2.2157e+00
```

To achieve convergence, we incorporate damping. To implement damping in our Newton's method, we need to give the function a list of damping parameters omega and adjust the correction line as follows

```
12        x[:] += omega[i] * w
```

If we now apply this, we get

```
1 newton_damped_vec(f, Df, x=np.array([0, -0.49999]), omega=[0.88] * 50, n=50,
2                    tol=1e-10)
   it  ||f(x)||
   --------------
   00  9.9998e+04
   01  4.3564e+09
   02  1.3662e+09
   ...
   19  3.7163e+00
   20  1.0136e+00
   21  6.6323e-01
   22  5.5081e-01
   23  1.1150e-02
   ...
   29  3.6487e-08
   30  4.3784e-09
   31  5.2541e-10
   32  6.3050e-11
```

Even after we are in the domain, the method only converges linearly. This is clear, as a damping parameter < 1 is always used. The goal must be to automatically end the damping when the domain of quadratic convergence is found.

Example 8.51 (Damping in the Simplified Newton Method) We can also incorporate damping in the simplified Newton method. With the starting vector $x^{(0)} = (1, 1)^T$ and $\omega = 0.74$, the damped simplified Newton method only needs 27 steps. In comparison, the undamped simplified Newton method needed 670 steps, see Example 8.47.

Remark 8.52 (Globalization and Transition to Quadratic Convergence) The damped Newton method includes a globalization strategy. By choosing a damping parameter, the convergence domain of Newton's method can be enlarged. The initial value condition

$$x^{(0)} \in D : \quad \underbrace{\|Df(x^{(0)})^{-1} f(x^{(0)})\|}_{=\alpha_0} \beta L < \frac{1}{2}$$

does not have to be fulfilled. If this product is too large, the damped iteration is iterated until the product $\alpha_k \beta L$ with $\alpha_k = \|Df(x^{(k)})^{-1} f(x^{(k)})\|$ fulfills the condition. Globalization strategies increase the convergence range. However, only slow convergence is present initially. We have proven the convergence factor

$$\rho \le 1 - \frac{1}{2\alpha_k \beta L}$$

above. For large α_k, ρ can be arbitrarily close to 1.

An essential feature of *efficient* globalization strategies is to automatically switch to fast Newton convergence when the iteration is in the domain of quadratic convergence. This is given here, since

$$\alpha_k \beta L < \frac{1}{2} \quad \Rightarrow \quad \omega_{min} = \min\left\{1, \frac{2}{\alpha_x \beta L}\right\} = 1.$$

A problem with the application of the damped Newton's method is the practical determination of the damping parameter ω_k. The quantities

$$\alpha_k = \|Df(x^{(k)})^{-1}f(x^{(k)})\|, \quad \beta = \max_{x \in D} \|Df(x)^{-1}\|,$$

as well as the Lipschitz constant of $Df(x)$ are generally not computable. A common strategy for finding a good damping parameter is the *Line-Search Algorithm* (the best solution is sought along a line):

Algorithm 8.53: Line-Search

> **Input:** Function $f \in C^1(D)$, initial value $x^{(0)} \in D$, tolerance $\epsilon > 0$, $\sigma \in (0,1)$ and
> maximum number of Line-Search steps $L_{max} \in \mathbb{N}$.
> 1 $k = 0$
> 2 **while** $\|f(x^{(k)})\| \geq \epsilon$ **do**
> 3 $Df(x_k)w_{(k)} = -f(x^{(k)})$
> 4   Set $\omega_0^k = 1$
> 5   **for** $l = 0$ **to** L_{max} **do**
> 6 $x^{(k+1)} = x^{(k)} + \omega_l w^{(k)}$
> 7   **if** $\|f(x^{(k+1)})\| < \|f(x^{(k)})\|$ **then**
> 8   Break
> 9 $\omega_{l+1}^{(k)} = \sigma \omega_l^{(k)}$
> 10 $k = k + 1$

In the Line-Search method, monotone convergence is enforced in each step. Initially, a full Newton step with $\omega = 1$ is attempted. As long as $\|f(x^{(k+1)})\| > \|f(x^{(k)})\|$, ω is reduced. Common values for σ are $\sigma = \frac{1}{2}$ or $\sigma = \frac{1}{4}$. There are cases where monotone convergence is not achievable. Then it makes sense to perform only a few iterations of Line-Search. After, e.g., $L = 4$ steps, the new approximation is accepted, even if the residual in this Newton step has increased, and a new search direction is computed.

Example 8.54 (Globalized Newton Method in $\mathbb{R}^n$) In Example 8.47, we saw that the quasi-Newton method required 670 steps, with an initial value at which Newton's method converged after five steps. In Example 8.51, we then saw that the damped, simplified Newton method required only 27 steps with the correct choice of the damping parameter. However, it is not always clear what a good choice for the damping parameter is, so we want to automate this choice. In addition, we saw in Example 8.50 that the damped Newton's method only converges linearly, so we want to use the full Newton's method when we are in the range of quadratic convergence.

To improve the convergence and the convergence radius, while keeping the number of matrix factorizations minimal, we update the Jacobian whenever the residual has not decreased by at least a factor of 0.3, and implement the Line-Search algorithm to automate the choice of damping.

♣ Implementation 8.55: Globalized Newton Method

```python
def newton_global_vec(f, D, x, sigma=0.5, Lmax=10, n=10, tol=1e-10):
    x0, b, y, w = [np.zeros_like(x) for _ in range(4)]
    b[:] = -f(x)
    res0, res1 = np.linalg.norm(b), float('nan')
    if res0 < tol:
        return 0, x

    for i in range(n):
        if i == 0 or res0 / res1 > 0.3:
            print(f'i = {i}: Jacobian updated')
            jac = D(x)
            pivot = lu_pivot(jac)
        for p in pivot:
            b[p] = b[[p[1], p[0]]]
        y[:] = forward(jac, b)
        w[:] = backward(jac, y)
        for l in range(Lmax):
            x0[:] = x + sigma**l * w
            b[:] = -f(x0)
            res2 = np.linalg.norm(b)
            if res2 < res0:
                if l > 0:
                    print(f'i = {i}: {l} line-search steps necessary')
                x[:] = x0
                break
            else:
                print(f'Step {i}: {l} line search failed, ||res|| = {res2:.4e}')
            x[:] = x0
        res1 = res0
        res0 = res2
        if res0 < tol:
            break
    else:

        print('The globalized Newton method did not converge after', end=' ')
        print(f' {n} iterations. ||res|| = {res0:.4e}')
    return i, x
```

Applied to the function from Example 8.47, we get

```python
n, x = newton_global_vec(f, Df, x=np.array([1.0, 1.0]), n=20, tol=1e-10)
print(f'\nn = {n}, x = {x}')
    i = 0: Jacobian updated
    i = 1: Jacobian updated

    n = 15, x = [0.89442719 0.4472136 ]
```

So, we only needed one update of the Jacobian for the method to calculate the root after 15 instead of 670 steps. If we take the initial value $x^{(0)} = (-1, 1)^T$, at which the quasi-Newton method diverges, we now have convergence after 14 steps, where the Jacobian was reassembled six times. A line search was necessary four times to reduce the residual in each step.

```
1 n, x = newton_global_vec(f, Df, x=np.array([-1.0, 1.0]), n=50)
2 print(f'\nn = {n}, x = {x}')
        i = 0: Jacobian updated
        i = 0: 3 line-search steps necessary
        i = 1: Jacobian updated
        i = 1: 3 line-search steps necessary
        i = 2: Jacobian updated
        i = 2: 2 line-search steps necessary
        i = 3: Jacobian updated
        i = 3: 1 line-search steps necessary
        i = 4: Jacobian updated
        i = 6: Jacobian updated

        n = 14, x = [0.89442719 0.4472136 ]
```

The line search method changes the length of the step, which is given by $w^{(k)} = -Df(x^{(k)})^{-1}f(x^{(k)})$, but not the *direction* $w^{(k)}$ itself. We now consider the scalar function

$$f(x) = x + \sin(2x)$$

with the only root $f(0) = 0$. Newton's method is only applicable locally, e.g., for the initial value $x_0 = 2$, we have $f'(x_0) \approx -0.31$ and the first Newton step is

$$x_0 + \omega w^{(1)} = x_0 - \frac{f(x_0)}{f'(x_0)} \approx 2 + 4.05\omega.$$

For $\omega \in (0, 1]$, we can certainly not expect convergence. The search direction $w^{(1)}$ is thus pointing in the wrong direction. Another class of globalization methods, which also change the search direction, are *trust-region methods*. The idea is that Newton's method approximates the function $f(x)$ in each step by a linear function. For tiny steps, it can be assumed that this approximation is good and ultimately leads to quadratic convergence. Suppose the iteration (or the initial value) is still far from the desired solution. In that case, we generally cannot assume that the function can be well represented globally as a linear function. The line-search approach was to reduce the step size. Trust-region methods, on the other hand, modify the Jacobian of the function so that it is more similar to a linear function, e.g., by simply adding the weighted unit matrix

$$D_\gamma f(x^k) := \tilde{D}f(x^k) := Df(x^{(k)}) + \gamma I.$$

The trust-region methods come from the field of optimization, where Newton's method is used to find stationary points, i.e., roots of the derivative, for details we refer to [70].

Newton's method is one of the most important methods in numerical mathematics. If all conditions are met and if an initial value in the domain of quadratic convergence is known, then the root is quickly found. Of course, determining a good initial value is often not easy. Therefore, Newton's method is an active and large field of research in various application fields, e.g., in the consideration of minimization problems (search for stationary points, i.e., roots of the first derivative) or in the approximation of non-linear differential equations. For a detailed presentation, we refer to [25].

8.5   Excursus: Root Finding in the Complex Plane

The *Fundamental Theorem of Algebra* states that a polynomial of degree n has exactly n roots (counting their multiplicity), see [5, 62] or [35]. These roots, however, can be complex. The polynomial

$$p(x) = x^3 - 1$$

has exactly one simple real root $x = 1$ in $\mathbb{R}$. If we consider the polynomial as a function of complex numbers

$$p : \mathbb{C} \to \mathbb{C}, \quad p(z) = z^3 - 1,$$

then the three roots are the *complex roots of unity*

$$z_1 = \mathrm{e}^{0\mathrm{i}} = 1, \quad z_2 = \mathrm{e}^{\frac{2}{3}\pi\mathrm{i}} \approx -0.5 + 0.866\mathrm{i}, \quad \mathrm{e}^{\frac{4}{3}\pi\mathrm{i}} \approx -0.5 - 0.866\mathrm{i}.$$

In Fig. 8.6, we illustrate these three solutions.

In this excursus, we want to explore the extent to which methods for finding roots can be transferred to complex functions in complex numbers. We begin by recalling some of the most critical facts of complex numbers.

Definition 8.56 (*Complex Numbers*) The set $\mathbb{C}$ of *complex numbers* is a commutative field $(\mathbf{C}, +, -)$, which contains the field of real numbers $\mathbb{R}$ as a subfield. There exists an element $\mathrm{i} \in \mathbb{C}$ with the property

$$\mathrm{i}^2 = -1.$$

This element is called *imaginary unit*. Every complex number $z \in \mathbb{C}$ can be uniquely written with two real numbers $a, b \in \mathbb{R}$ as

$$z = a + \mathrm{i}b,$$

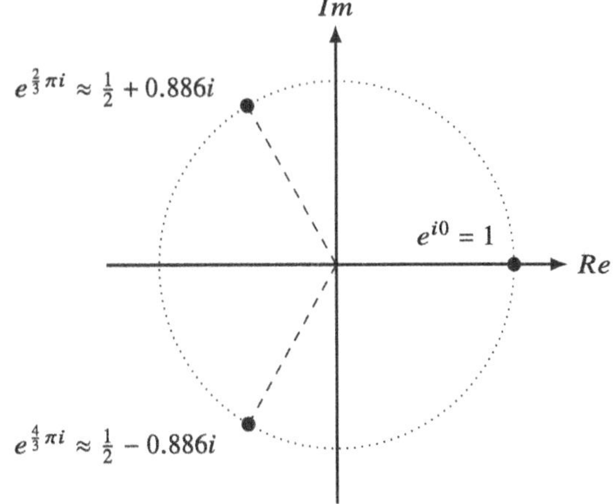

Fig. 8.6 Solutions of the equation $z^3 - 1 = 0$ in the complex plane. The three roots $z_1, z_2, z_3 \in \mathbb{C}$ are the *third roots of unity*. The n-th roots of unity can be given as $z_k^n = \mathrm{e}^{2k\pi/n\mathrm{i}}$

where $a = Re(z)$ is called the *real part* and $b = Im(z)$ the *imaginary part* of z. That means if $b = 0$, we are working on the real axis with the known real numbers.

For a complex number $z = a + ib$, $\bar{z} = a - ib$ is called the *complex conjugate*. For two complex numbers $z_{1/2} = a_{1/2} + ib_{1/2}$, addition and multiplication are given as

$$z_1 + z_2 = (a_1 + a_2) + i(b_1 + b_2),$$
$$z_1 \cdot z_2 = (a_1 \cdot a_2 - b_1 \cdot b_2) + i(a_1 b_2 + a_2 b_1).$$

For division, it holds

$$\frac{z_1}{z_2} = \frac{a_1 a_2 + b_1 b_2}{|z_2|^2} + i\frac{b_1 a_2 - a_1 b_2}{|z_2|^2}.$$

The *magnitude of a complex number* $z = a + ib$ is defined as

$$|z| = \sqrt{z\bar{z}} = \sqrt{a^2 + b^2}.$$

By considering $i^2 = -1$, the calculation rules from $\mathbb{R}$ are transferred to $\mathbb{C}$. The field of complex numbers $\mathbb{C}$ is isomorphic to the vector space $\mathbb{R}^2$. Complex numbers are therefore represented as points in the *complex plane*, see Fig. 8.6. Furthermore, the *Euler formula* holds for $b \in \mathbb{R}$

$$e^{ib} = \cos(b) + i\sin(b) \quad \Rightarrow \quad |e^z| = |e^{a+ib}| = e^a.$$

Definition 8.57 (*Complex Function*) A mapping $f : D \subset \mathbb{C} \to \mathbb{C}$ is called a *complex function*. Using $z = x + iy$ and with $f = u + iv$ for two functions $u, v : D \to \mathbb{R}$, every complex function can be interpreted as a function $f : \mathbb{R}^2 \to \mathbb{R}^2$:

$$f(z) = u(z) + iv(z), \quad f(x, y) = u(x, y) + iv(x, y).$$

A *root* $z \in \mathbb{C}$ is a complex number, in which the function f takes the value zero, i.e.,

$$f(z) = u(z) + iv(z) = 0.$$

Thus, it holds $Re(f(z)) = Im(f(z)) = 0$.

The analogy to functions in $\mathbb{R}^2$ allows a first classification of the complex root problem. It is a two-dimensional task. Concepts like the bisection method cannot work. However, Newton's method can be directly transferred:

Theorem 8.58 (Real Newton's Method for Complex Functions) *Let $f : \mathbb{C} \to \mathbb{C}$ be a complex function, which—considered as a real-valued function $f : \mathbb{R}^2 \to \mathbb{R}^2$ - fulfills the prerequisites of Theorem 8.41 (Newton-Kantorovich). Then Newton's method can be applied accordingly.*

Example 8.59 We consider the function

$$p(z) = z^3 - 1.$$

With $z = a + ib$, we have

$$p(x, y) = p(a + ib) = (a + ib)^3 - 1 = (a^3 - 3ab^2 - 1) + i(3a^2 b - b^3),$$

so we will examine in $\mathbb{R}^2$ the function

$$f(x, y) = \begin{pmatrix} x^3 - 3xy^2 - 1 \\ 3x^2 y - y^3 \end{pmatrix}.$$

We are looking for x, y such that $f(x, y) = 0$ holds. This function is in $C^\infty(\mathbb{R}^2)$, so Newton's method is applicable. For the Jacobian, we have

$$Df(x, y) = \begin{pmatrix} 3x^2 - 3y^2 & -6xy \\ 6xy & 3x^2 - 3y^2 \end{pmatrix}.$$

Because

$$\det(Df(x, y)) = 9(x^2 + y^2)^2$$

the Jacobian is regular away from the origin $(0, 0)$. We apply the two-dimensional Newton's method for various initial values (x_0, y_0) until the first three digits are exactly determined.

k	(x_k, y_k)			
1	(1.000, 0.000)	(0.000, 1.000)	(−1.000, 0.000)	(0.000, −1.000)
2		(−0.333, 0.667)	(−0.333, 0.000)	(−0.333, −0.667)
3		(−0.582, 0.924)	(2.778, 0.000)	(−0.582, −0.924)
4		(−0.508, 0.868)	(1.895, 0.000)	(−0.508, −0.868)
5		(−0.500, 0.866)	(1.356, 0.000)	(−0.500, −0.866)
6			(1.085, 0.000)	
7			(1.007, 0.000)	
8			(1.000, 0.000)	

The method converges as expected, and we find the three complex unit roots from Fig. 8.6. Here, we have considered the problem completely as a problem in real numbers. This way, we can rely on known facts, but we cannot use, perhaps, more efficient, complex arithmetic.

In the following, we will deal exclusively with the zeros (real and complex) of real polynomials. We summarize some statements:

Theorem 8.60 (Roots of Polynomials With Real Coefficients) *The polynomial*

$$p(z) = z^n + \alpha_{n-1}z^{n-1} + \cdots + \alpha_0, \quad \alpha_i \in \mathbb{R}, \tag{8.28}$$

has exactly n (possibly complex) zeros $z_1, \ldots, z_n \in \mathbb{C}$ (multiple zeros are possible) and allows the factorization into linear factors

$$p(z) = \prod_{k=1}^{n}(z - z_i).$$

If z_i is a zero with imaginary part $Im(z_i) \neq 0$, then $\bar{z}_i$ is another zero. All zeros lie in the circle

$$z_i \in B_R(0) := \left\{ z \in \mathbb{C}, \ |z| < R := \max\left\{ 1, \sum_{k=0}^{n-1} |\alpha_k| \right\} \right\}.$$

Proof The proofs for the individual statements can be found in [35] or similar books on complex analysis. $\qquad\square$

The Bairstow Method

Theorem 8.60 contains some essential statements for the numerical search for roots: First, all roots lie in a circle around the origin. This gives us a first idea for the area in which we have to search for roots. Furthermore, complex roots occur in pairs:

$$p(a + ib) = 0 \quad \Rightarrow \quad p(a - ib) = 0$$

Suppose that the complex roots are given by $z_1, \ldots, z_{2k} \in \mathbb{C} \setminus \mathbb{R}$ with $2k \leq n$ and $z_{2i} = \bar{z}_{2i-1}$ with $z_{2i} = a_{2i} + ib_{2i}$ and $z_{2i-1} = a_{2i} - ib_{2i}$. Let $z_{2k+1}, \ldots, z_n \in \mathbb{R}$ be the remaining $n - 2k$ real roots of $p(\cdot)$. Thus, the factorization follows

$$p(z) = \left(\prod_{i=1}^{k}(z - z_{2i})(z - \bar{z}_{2i}) \right) \prod_{i=2k+1}^{n} (z - z_i)$$

$$= \left(\prod_{i=1}^{k}((z - a_{2i})^2 + b_{2i}^2) \right) \prod_{i=2k+1}^{n} (z - z_i).$$

The polynomial can thus be divided into linear factors with real roots and quadratic factors with purely complex roots.

The *Bairstow method* is based on this factorization; see [36]. Instead of a direct search for all roots, we first look for quadratic factors of the type

$$a(x) := (x - z_i)(x - \bar{z}_i) = (x - a_i)^2 + b_i^2 = x^2 - 2xa_i + (a_i^2 + b_i^2).$$

Once such a factor is found, the two roots can be easily determined as

$$z_{i,1/2} = a_i \pm ib_i.$$

Subsequently, the polynomial $p \in P_n$ is divided by the found factor $a \in P_2$ and the process continues with a polynomial from P_{n-2}.

In the following, we describe the identification of these quadratic factors. We aim to identify the quadratic polynomial

$$a(x) = x^2 - \alpha x + \beta,$$

with $\alpha, \beta \in \mathbb{R}$ and $\beta > 0$, such that the polynomial division

$$p(x) = a(x)b(x) + r(x) \tag{8.29}$$

is without remainder, i.e., with $r \equiv 0$. The coefficients α, β of $a(x)$ are sought iteratively. With $p \in P_n$ and for any $a \in P_2$ it follows that $b \in P_{n-2}$ and $r \in P_1$, i.e., the remainder is zero, a constant or a linear polynomial. We make the general approach

$$b(x) = \sum_{k=-2}^{n} b_k x^k, \tag{8.30}$$

where we have introduced $b_n = b_{n-1} = b_{-2} = b_{-1} = 0$ to ease the notation. We calculate the general product

$$a(x)b(x) = \sum_{k=0}^{n} (b_{k-2} - \alpha b_{k-1} + b_k \beta)x^k.$$

A comparison of coefficients from (8.29) yields

$$a_k = b_{k-2} - \alpha b_{k-1} + b_k \beta, \quad k = n, n-1, \ldots, 0. \tag{8.31}$$

These are $n + 1$ equations for only $n - 1$ coefficients $b_0, \ldots, b_{n-2}$, as $b_n = 0$ and $b_{n-1} = 0$ were introduced in (8.30) only to simplify the notation. Suppose we determine the coefficients $b_0, \ldots, b_{n-2}$ from the (8.31), i.e.,

$$b_{k-2} = a_k + \alpha b_{k-1} - \beta b_k, \quad k = 2, \ldots, n.$$

Then the remainder is given by

$$r(x) = p(x) - a(x)b(x) = (a_1 + \alpha b_0 - \beta b_1)x + (a_0 - \beta b_0).$$

The goal now is to determine $\alpha, \beta \in \mathbb{R}$, such that the remainder is as small as possible. The coefficients b_0 and b_1 depend on α and β and can be easily calculated:

Algorithm 8.61: Calculation of the Bairstow Coefficients

Input: Polynomial $p(x) = \sum_{k=0}^{n} a_k x^k$.
1 Choose initial guesses $\alpha, \beta \in \mathbb{R}$
2 Set $b_0 = 1, b_1 = 0$
3 **for** $k = n - 1$ **to** 2 **do**
4      $b_2 = b_1$
5      $b_1 = b_0$
6      $b_0 = a_k + \alpha b_1 - \beta b_2$

The coefficients b_0 and b_1 implicitly depend on the values $\alpha, \beta \in \mathbb{R}$. To determine these optimally, we apply Newton's method to the function

$$F(\alpha, \beta) = \begin{pmatrix} a_1 + \alpha b_0(\alpha, \beta) - \beta b_1(\alpha, \beta) \\ a_0 - \beta b_0(\alpha, \beta) \end{pmatrix} = 0.$$

Given an approximation $\alpha^{(l)}, \beta^{(l)}$, the residual, i.e., $F(\alpha^{(l)}, \beta^{(l)})$, can be calculated with Algorithm 8.61. The Jacobian given as

$$DF(\alpha, \beta) = \begin{pmatrix} b_0 + \alpha D_\alpha b_0 - \beta D_\alpha b_1 & \alpha D_\beta b_0 - b_1 - \beta D_\beta b_1 \\ -\beta D_\alpha b_0 & -b_0 - \beta D_\beta b_0 \end{pmatrix},$$

where $b_0(\cdot)$ and $b_1(\cdot)$, as well as their derivatives in the direction of α and $q = \beta$ naturally depend on α and β. To calculate the derivative, Algorithm 8.61 could be formally derived with respect to α and β. However, by a clever transformation, the calculation of the derivatives can be made much more efficient. We refer to the literature [36] and consider a simple example.

Example 8.62 (Bairstow Method to Determine the Third Roots of Unity) We consider the polynomial

$$p(x) = x^3 - 1, \quad a_3 = 1, \ a_2 = 0, \ a_1 = 0, \ a_0 = 1.$$

Using the Bairstow method, we are looking for a quadratic factor of this polynomial. Division of $p(x) = x^3 - 1$ by $a(x) = x^2 - px + q$ yields the remainder

$$p(x) = (x^2 - \alpha x + \beta)(x + \alpha) - \left((\beta - \alpha^2)x + \alpha\beta + 1\right), \tag{8.32}$$

i.e.,

$$r(x) = (\beta - \alpha^2)x + \alpha\beta + 1.$$

We are looking for $\alpha, \beta \in \mathbb{R}$, such that

$$F(\alpha, \beta) = \begin{pmatrix} \beta - \alpha^2 \\ \alpha\beta + 1 \end{pmatrix} \overset{!}{=} 0.$$

The Jacobian and its inverse are

$$DF(\alpha, \beta) = \begin{pmatrix} -2\alpha & 1 \\ \beta & \alpha \end{pmatrix}, \quad DF(\alpha, \beta)^{-1} = \frac{1}{2\alpha^2 + \beta} \begin{pmatrix} -\alpha & 1 \\ \beta & 2\alpha \end{pmatrix}.$$

The solution is then quickly identified with a few iterations of Newton's method as $\alpha = -1$ and $\beta = 1$, such that the quadratic term gets

$$p(x) = x^2 + x + 1$$

with remainder $r(x) = 0$. The two complex roots are given as

$$z_{1/2} = \frac{\alpha}{2} \pm \sqrt{\frac{\alpha^2}{4} - \beta} = -\frac{1}{2} \pm \frac{\sqrt{3}}{2}i.$$

According to (8.32), the remaining linear factor is

$$b(x) := x + p = x - 1,$$

i.e., the third (real) root is

$$z_3 = 1.$$

The great advantage of the Bairstow method is that it can be carried out entirely in real arithmetic. Since it is based on Newton's method for determining the coefficients $\alpha, \beta \in \mathbb{R}$, it is extremely fast and efficient.

The Complex Newton Method

Finally, we discuss the implementation of Newton's method in complex arithmetic. First, we introduce the concept of a *complex derivative*:

Definition 8.63 (*Complex Differentiability*) A function $f : D \subset \mathbb{C} \to \mathbb{C}$ is called *complex differentiable* at $z_0 \in \mathbb{C}$, if the limit

$$f'(z_0) := \lim_{h \to 0} \frac{f(z_0 + h) - f(z_0)}{h}$$

exists for any $h \in \mathbb{C}$. The function is called *holomorphic at z_0*, if there exists a neighborhood of z_0 in which f is complex differentiable. If f is complex differentiable on all of D, then f is called a *holomorphic function*. If f is complex differentiable on $\mathbb{C}$, then f is called an *entire function*.

A complex function $f = u + iv$ is complex differentiable if and only if its real counterpart

$$f = u + iv, \quad f(x, y) = \begin{pmatrix} u(x, y) \\ v(x, y) \end{pmatrix}$$

is totally differentiable in the real sense and if the *Cauchy-Riemann differential equations* hold

$$\frac{d}{dx}u(x, y) = \frac{d}{dy}v(x, y), \quad \frac{d}{dy}u(x, y) = -\frac{d}{dx}v(x, y). \tag{8.33}$$

The rules for determining derivatives then correspond to the known rules of real derivatives. For example, the polynomial $p(z)$ is complex differentiable everywhere, and it holds

$$p'(z) = nz^{n-1} + a_{n-1}(n - 1)z^{n-2} + \cdots + 2a_2 z + a_1.$$

Complex conjugation $f(z) = \bar{z}$ is not complex differentiable, because for $h \in \mathbb{R}$ with $h \to 0$ it holds

$$\lim_{h \to 0} \frac{\overline{z + h} - \bar{z}}{h} = \lim_{h \to 0} \frac{h}{h} = 1, \quad \lim_{h \to 0} \frac{\overline{z + ih} - \bar{z}}{ih} = \lim_{h \to 0} \frac{-ih}{ih} = -1.$$

On the other hand, the complex exponential function $\exp(z)$ and the complex trigonometric functions $\sin(z)$ and $\cos(z)$ are complex differentiable everywhere, thus *entire functions*. In complex analysis, the astonishing connection applies:

Theorem 8.64 (Holomorphic Functions) *Let $f : D \subset \mathbb{C} \to \mathbb{C}$ be a holomorphic function. Then f is arbitrarily often complex, continuously differentiable on D, and can be developed into a power series at any point.*

For a proof, we refer to [35]. Holomorphic functions can be represented locally as a Taylor series.

With these preparations, we can define Newton's method on the complex numbers:

Theorem 8.65 (Complex Newton's Method) *Let $f : D \subset \mathbb{C} \to \mathbb{C}$ be a holomorphic function with a simple zero $\hat{z} \in D$. Then there exists some $\rho > 0$ and a neighborhood $B_\rho(\hat{z})$, such that the complex Newton method converges for all initial values $z_0 \in B_\rho(\hat{z})$ quadratically to the root*

$$z_{k+1} = z_k - \frac{f(z_k)}{f'(z_k)}, \quad k = 0, 1, 2, \dots .$$

Proof The function f is arbitrarily often complex, continuously differentiable, and it holds $f'(\hat{z}) \neq 0$. Therefore, there exists a neighborhood $B_{\rho'}(\hat{z})$, on which $|f'(\hat{z})| \geq m > 0$ is bounded from below. We show in the following that Newton's method with complex arithmetic is equivalent to the application of the vector-valued Newton's method to the function $f(z) = f(x, y)$.

For the holomorphic function $f = u + iv$, the Cauchy-Riemann differential equations hold, and we have

$$f'(z) = u_x(z) + iv_x(z) = u_x(z) - iu_y(z).$$

With this, Newton's method can be written as

$$\begin{aligned}
z_{k+1} &= z_k - \frac{u(z_k) + iv(z_k)}{u_x(z_k) - iu_y(z_k)} \\
&= z_k - \frac{(u(z_k) + iv(z_k))(u_x(z_k) + iu_y(z_k))}{u_x(z_k)^2 + u_y(z_k)^2}, \\
&= z_k - \frac{u_x(z_k)u(z_k) - u_y(z_k)v(z_k) + i(u_x(z_k)v(z_k) + u_y(z_k)u(z_k))}{u_x(z_k)^2 + u_y(z_k)^2}.
\end{aligned}$$

For $z_k = a_k + ib_k$, it then holds

$$\begin{aligned}
a_{k+1} &= a_k - \frac{u_x(z_k)u(z_k) - u_y(z_k)v(z_k)}{u_x(z_k)^2 + u_y(z_k)^2}, \\
b_{k+1} &= b_k - \frac{u_x(z_k)v(z_k) + u_y(z_k)u(z_k)}{u_x(z_k)^2 + u_y(z_k)^2}.
\end{aligned} \tag{8.34}$$

Now, we consider the real-valued Newton method applied to the function

$$f(x, y) = \begin{pmatrix} u(x, y) \\ v(x, y) \end{pmatrix}.$$

It results in

$$\begin{pmatrix} x_{k+1} \\ y_{k+1} \end{pmatrix} = \begin{pmatrix} x_k \\ y_k \end{pmatrix} - \begin{pmatrix} u_x(x_k, y_k) & u_y(x_k, y_k) \\ v_x(x_k, y_k) & v_y(x_k, y_k) \end{pmatrix}^{-1} \begin{pmatrix} u(x_k, y_k) \\ v(x_k, y_k) \end{pmatrix}.$$

Since the function f is holomorphic, the Cauchy-Riemann differential equations apply, so

$$\begin{aligned}
\begin{pmatrix} x_{k+1} \\ y_{k+1} \end{pmatrix} &= \begin{pmatrix} x_k \\ y_k \end{pmatrix} - \begin{pmatrix} u_x(x_k, y_k) & u_y(x_k, y_k) \\ -u_y(x_k, y_k) & u_x(x_k, y_k) \end{pmatrix}^{-1} f(x_k, y_k) \\
&= \begin{pmatrix} x_k \\ y_k \end{pmatrix} - \frac{1}{u_x(x_k, y_k)^2 + u_y(x_k, y_k)^2} \begin{pmatrix} u_x(x_k, y_k) & -u_y(x_k, y_k) \\ u_y(x_k, y_k) & u_x(x_k, y_k) \end{pmatrix} f(x_k, y_k).
\end{aligned}$$

This is the same iteration as (8.34).

Thus, for a sufficiently small domain $B_\rho(\hat{z})$, all prerequisites of Theorem 8.41 are fulfilled. $\qquad\square$

Example 8.66 (Complex Newton's Method to Determine the Unit Roots) We again consider

$$p(z) = z^3 - 1.$$

The complex Newton method results in the iteration

$$z_{k+1} = z_k - \frac{p(z_k)}{p'(z_k)} = \frac{2}{3}z_k + \frac{1}{3z_k^2}.$$

With the notation $z_k = a_k + ib_k$ this leads to

$$a_{k+1} + ib_{k+1} =$$
$$\frac{2x_k^5 + 4x_k^3 y_k^2 + x_k^2 + 2y_k^4 x_k - y_k^2}{3(x_k^2 + y_k^2)^2} + i\,\frac{2x_k^4 y_k + 4y_k^3 x_k^2 - 2x_k y_k + y_k^5}{3(x_k^2 + y_k^2)^2}.$$

It is easy to verify that this is precisely the iteration formula from Example 8.59. We, therefore, refrain from performing the procedure again. Instead, we briefly deal with the question of the domain of convergence. Instead of an analysis, we determine the domain of convergence of the three zeros numerically and perform the procedure for different values, each from $x_0, y_0 \in [-8, 8]$. In Fig. 8.7, we show the convergence behavior for all initial values

$$z_0 = hk + ihl, \quad k, l = -250, \ldots, 250, \quad h = \frac{8}{250}.$$

Here, a red point means that for the corresponding initial value the root $z_1 = 1$ is reached, a green initial value leads to convergence to $z_2 = e^{\frac{2}{3}\pi i}$ and a blue initial value to convergence to $z_3 = e^{\frac{4}{3}\pi i}$. Gray values mean that the procedure diverges. More precisely, the iteration is aborted as soon as $|z_k| > 10$. Then the shade of gray value indicates the index k at which the iteration is aborted. Dark values indicate immediate termination. In white areas, it could not be decided within the allowed 255 iterations whether Newton's method converges or not. The set of complex numbers at which the convergence behavior suddenly changes is called *Julia set* or *Newton fractal*.

8.6 Excursus: Fast Inverse Square Root

In the 1999 released computer game QUAKE III ARENA[1] Algorithm 8.67 was used to approximate the square root, more precisely, to approximate the inverse of the square root

$$y = \frac{1}{\sqrt{x}}.$$

The authorship of the method is precisely documented,[2] literature can be found in abundance [68].

---

[1] A computer game developed by id Software in 1999.
[2] See the discussion at https://www.beyond3d.com/content/articles/8/.

Fig. 8.7 The domain convergence of the complex Newton method forms a *Julia set*. From top left to bottom right, the initial values are chosen from increasingly smaller intervals. Top left applies $z_0 = a_0 + ib_0$ with $a_0, b_0 \in [-8, 8]$. The subintervals are each excerpts from the previous representation

Algorithm 8.67: Fast Inverse Square Root

```
1  float Q_rsqrt( float number )
2  {
3    int i;
4    float x2, y;
5    const float threehalfs = 1.5F;
6
7    x2 = number * 0.5F;
8    y  = number;
9    i  = * ( int * ) &y;
10   i  = 0x5f3759df - ( i >> 1 );
11   y  = * ( float * ) &i;
12   y  = y * ( threehalfs - ( x2 * y * y ) );
13   // y  = y * ( threehalfs - ( x2 * y * y ) );
14
15   return y;
16 }
```

A first numerical test, which we present in Fig. 8.8, shows the high relative accuracy of the method with a relative error of a maximum of 0.18% for any input values $x \in [10^{-3}, 10^2]$. The method must be discussed in its historical context. In 1999, the PENTIUM II was a common processor. This processor did have a floating point unit, but also had some limitations (see [34]):

- Calculations with integers were faster than the floating-point arithmetic.
- Additions and multiplications were by far faster than divisions on this processor. Instead of a division, about 20 multiplications or 40 additions could be performed.
- The calculation of a root was performed in the floating point unit, but was not very efficient and required about 35 multiplications or 70 additions. Calculating the inverse of a root thus had a similar effort as 100 additions or 50 multiplications.

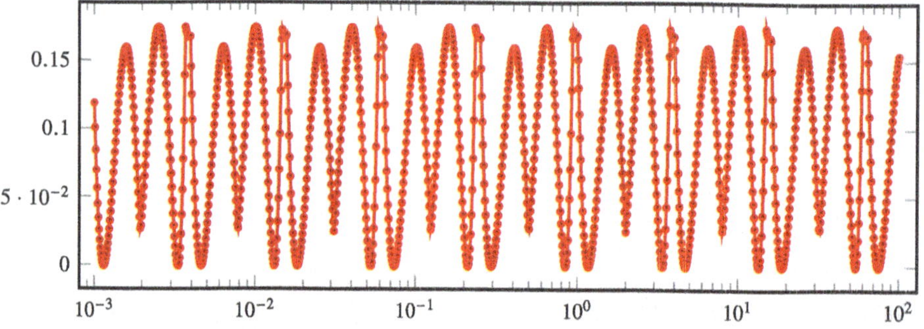

Fig. 8.8 Relative error of the function *Fast Inverse Square Root* from Algorithm 8.67 for $x \in [10^{-3}, 10^2]$. It shows that the inverse of the root for all numbers is calculated with a relative error smaller than 0.2%

Only the PENTIUM III processor, introduced in 1999, had an efficient function for approximating the root and its inverse with the SSE extension. It could perform these (as well as the division) in the same time it takes to perform a multiplication.

Shading—Fast Calculation of Normal Vectors

QUAKE III was one of the first games to popularize fast, realistic-looking 3D graphics. To create a spatial impression in a two-dimensional projection on the computer screen, lighting effects are significant. Depending on the position of the viewer and the position of the light source, objects appear differently bright. The situation is shown in Fig. 8.9. We want to determine the brightness at point $x \in \mathbb{R}^3$. The viewer (the camera) is at point $C \in \mathbb{R}^3$ and the light source is at $L \in \mathbb{R}^3$. The brightness is determined by the simple rule "angle of incidence equals angle of reflection", i.e., it depends on the angle between the line from the viewer to point x, i.e., $\mathbf{c} := \overline{Cx}$ and the outgoing light ray $\mathbf{l}_\perp$. For the latter, the projection

$$\mathbf{l}_\perp = \mathbf{l} - 2\langle \mathbf{n}, \mathbf{l}\rangle \mathbf{n}$$

holds, where $\mathbf{l} := \overline{Lx}$ is the line connecting the light source and point x, and $\mathbf{n}$ is the normal vector at x on the plane. The angle $\alpha = \sphericalangle(\mathbf{c}, \mathbf{l}_\perp)$ can now be determined using

$$\cos(\alpha) = \frac{\langle \mathbf{l}_\perp, \mathbf{c}\rangle}{|\mathbf{l}_\perp|\,|\mathbf{c}|}.$$

To calculate the scalar products and projections, some multiplications and additions are necessary. These can all be performed quickly and efficiently. In addition, however, the result must be scaled with the length of the vectors, i.e.,

$$\frac{1}{|\mathbf{c}|} = \frac{1}{\sqrt{c_1^2 + c_2^2 + c_3^3}}.$$

In this step, the calculation of the root or its inverse is necessary. In the application of computer games, an approximation is sufficient. In complex games like QUAKE III, these calculations have to be performed so frequently that it is worth considering a fast algorithm.

Newton's Method for Calculating the Inverse Root

We first consider lines 12 and 13 in Algorithm 8.67,

```
12    y  = y * ( threehalfs - ( x2 * y * y ) );
13 // y  = y * ( threehalfs - ( x2 * y * y ) );
```

This appears to be two iterations of a method, where the second iteration is not used[3]. We can formulate this iteration as

$$y_{n+1} = y_n \left(\frac{3}{2} - \frac{y_n^2 x}{2} \right) =: g(y_n),\qquad (8.35)$$

where x is the number whose inverse root is to be determined, and which is passed in line 1 as `number`, for which `threehalfs` was defined as $3/2$ (line 5) and `x2` as $x/2$. This iteration has a fixed point

$$y \overset{!}{=} g(y) \quad \Leftrightarrow \quad y^2 = \frac{1}{x},$$

which is the desired result. In the sense of Theorem 8.39, we determine the order of convergence of this fixed point iteration by analyzing the derivatives. It holds

$$g(y) = \frac{3y}{2} - \frac{y^3 x}{2}, \quad g'(y) = \frac{3}{2} - \frac{3}{2}y^2 x, \quad g''(y) = -3yx,$$

so for $y = x^{-\frac{1}{2}}$, we have

$$g(x^{-\frac{1}{2}}) = x^{-\frac{1}{2}}, \quad g'(x^{-\frac{1}{2}}) = 0, \quad g''(x^{-\frac{1}{2}}) = -3x^{\frac{1}{2}}.$$

Therefore, we have a fixed point iteration with quadratic convergence. The iteration can be written as an application of Newton's method for the function

$$f(y) = x - \frac{1}{y^2}$$

which is given by

$$y_{n+1} = y_n - \frac{f(y_n)}{f'(y_n)} = y_n - \frac{x - y_n^{-2}}{2y_n^{-3}} = g(y_n).$$

We, therefore, understand the mathematics behind the first two lines of code. Within the required precision of QUAKE III, a single Newton step seems to suffice.

The investigation of lines $8 - 11$ of the algorithm remains. These four lines must represent a good approximation of the initial value, so that in the subsequent Newton's method, there is indeed quadratic convergence. The expression `0x5f3759df` will be of particular importance, which enters the procedure as a constant. Much has already been written about this number. Constants, whose immediate purpose is not immediately apparent, are sometimes called *magic numbers* in computer science. However, we will see through mathematical analysis that the constant chosen here is the optimal choice.

First, we analyze Newton's method for the inverse root in detail. Let $y \in \mathbb{R}_+$ be a first approximation, then with (8.35) for $\bar{y} = g(y)$, we get with polynomial division

---

[3] The leading slashes "//" start for a *comment*, the program text is not executed.

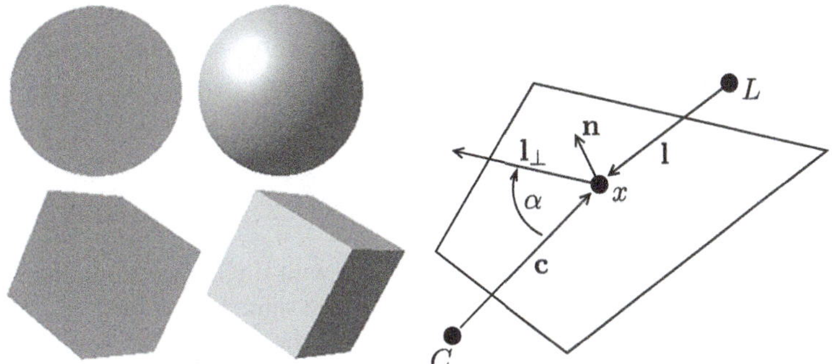

Fig. 8.9 Left: 3D through light effects. Right: Calculation of the brightness of a surface with normal vector **v**. The observer is at point C, the light source is at point L

$$\bar{y} - \frac{1}{\sqrt{x}} = -\frac{x}{2}\left(y + \frac{2}{\sqrt{x}}\right)\left(y - \frac{1}{\sqrt{x}}\right)^2. \tag{8.36}$$

Here, we can read off both quadratic convergence and

$$\bar{y} = g(y) \leq \frac{1}{\sqrt{x}} \quad \forall y > 0.$$

The method then converges monotonically increasing from below, since

$$y < \frac{1}{\sqrt{x}} \quad \Rightarrow \quad g(y) = \frac{y}{2}(3 - y^2 x) > \frac{y}{2}(3 - 1) = y$$

holds.

Determining the Initial Value

The task is now to find a good initial value for Newton's method. In Algorithm 8.67, this is done in lines $9 - 11$ which are initially difficult to understand:

```
 9  i  = * ( int * ) &y;
10  i  = 0x5f3759df - ( i >> 1 );
11  y  = * ( float * ) &i;
```

We will have to delve a little deeper into the binary number representation of computers and also refer to Sect. 1.4. In current C/C++ versions[4] the data type int describes an integer, for which 32 bits are available; one for the sign and 31 for the digits

---

[4] This applies, for example, to gcc version 14. At the time of Quake III, a variable of type int only had 16 bits available. Therefore, in the original algorithm [96], the data type long was used, an integer with 32 bits, corresponding to the (already used at that time) 32 bits in the floating point type float.

$$x_{\text{int}} = (-1)^{b_{32}} \sum_{i=1}^{31} b_i \cdot 2^{i-1}, \quad b_1, \ldots, b_{32} \in \{0, 1\}.$$

We will write integers shorter as

$$x_{\text{int}} = [\pm^{b_{32}} b_{31} \ldots b_1].$$

The data type **float** describes a floating point number with single precision. More precisely, a number stored using 32 bits, of which one, b_{32}, for the sign, the eight bits $b_{31}, \ldots, b_{24}$ for the exponent and the remaining 23 bits $b_{23}, \ldots, b_1$ for the mantissa are available:

$$x_{\text{float}} = (-1)^{b_{32}} \cdot 2^{\left(\sum_{i=0}^{7} b_{24+i} 2^i - 127\right)} \cdot \left(1 + 2^{-24} \sum_{i=1}^{23} b_i 2^i\right), \tag{8.37}$$

where $b = 127$ gives the *bias* of the exponent representation. We write such a number in short notation as

$$x_{\text{float}} = [(-1)^{b_{32}} \cdot 2^{b_{31} \ldots b_{24} - b} 1.b_{23} \ldots b_1]_2 = (-1)^S 1.m \cdot 2^{E-b}.$$

The leading 1 in the mantissa serves for normalization, so that $m \in [0, 1)$ and one digit is saved. In the following, we only consider positive numbers, so we always assume $S = 0$ and do not list the sign separately. Lines 9 and 11 in the algorithm have a special meaning. The number representations are "reinterpreted". For example, in

```
9  i = * (int *) & y;
```

there is no rounding of the floating point number y_{float} to an integer i_{int}, but a reinterpretation of the individual bits occurs. The floating-point number

$$x_{\text{float}} = 1.m \cdot 2^{E-b}$$

becomes the integer

$$i_{\text{int}} = 2^{23}(E + m) =: LE + M,$$

with

$$L = 2^{23}, \quad M = Lm.$$

Let us consider an example. The number $x = 3.1415$ has the floating point representation

$$3.1415 = 1.57075 \cdot 2^{128-127}$$
$$= [2^{10000000 - 127}{}_{10} 1.10010010000111010101010110]_2.$$

In the corresponding binary representation, the integer with the same bit sequence represents a completely different decimal number:

$$i_{\text{int}} = [1000000010010010000111010101010110]_2 = 1\,078\,529\,622 \tag{8.38}$$

This has no relation to $x_{\text{float}} = 3.1415$. We must therefore explain the meaning of lines 9 and 11 differently.

Another special feature is the second part of line 10

```
10  i  = 0x5f3759df - ( i >> 1 );
```

is the operation $i \gg 1$. The "$\gg$" operation shifts all bits in the binary representation of a number one place to the right. We stick with the above example. From (8.38) this results in

$$i_{\text{int}} = [10000000010010010000111010101010110]_2 = [1\,078\,529\,622]_{[10]},$$

$$i_{\text{int}} \gg 1 = [01000000001001001000011101010101011]_2 = [539\,264\,811]_{[10]}.$$

This means that the original number was just divided by two. The shifting of bits can be performed very quickly by CPUs. Applied to floating point numbers, the "$\gg$" operation makes little sense, as no regard is taken for the meaning of the bits. The sign bit is shifted into the exponent, and the last exponent bit into the mantissa.

The abbreviation $0x$ in the first part of line 10 in Algorithm 8.67 stands for the specification of a number in hexadecimal, i.e., base 16 system, where a is the 10, b is the 11, and finally f is the 15th place. The number $0x5f3759df$ is therefore the number

$$5f3759df_{16} = 1\,597\,463\,007,$$

in binary representation given as

$$5f3759df_{16} = [1011111100110111010110011101011111]_2. \tag{8.39}$$

In summary, lines $9-11$ in the algorithm represent an operation

$$Y = Z - \frac{X}{2},$$

where Z is the *magic number* and Y and X represent the reinterpretations of the input $x \in \mathbb{R}_+$ and the initial value $y_0 \in \mathbb{R}_+$ as integers.

To understand this trick, one uses a reformulation of the actual task. It holds

$$y = x^{-\frac{1}{2}} \quad \Leftrightarrow \quad \log_2(y) = -\frac{1}{2}\log_2(x).$$

Usually, the logarithm is a costly operation. Approximating the logarithm to base 2, however, is straightforward in floating point notation (8.37). For $x_{\text{float}} = \mathbf{1}.m \cdot 2^{E-b}$ it holds:

$$\log_2(y) = -\frac{1}{2}\log_2(x_{\text{float}}) = -\frac{1}{2}\log_2(\mathbf{1}.m) - \frac{1}{2}\log_2(2^{E-b})$$

$$= -\frac{1}{2}\log_2(1+m) - \frac{E-b}{2}$$

The first part is the logarithm of a number between 1 and 2, the second part can be calculated as the division of the exponent by 2, i.e., by bit shifting. We are now looking for an approximation to y, written as

$$y_{\text{float}} = \mathbf{1}.m_y \cdot 2^{E_y - b},$$

so that

$$\log_2(y_{\text{float}}) = \log_2(1 + m_y) + E_y - b \stackrel{!}{=} -\frac{1}{2}\log_2(1 + m) - \frac{E - b}{2}. \tag{8.40}$$

From the series expansion of $\log(1 + x)$ it follows that $\log_2(1 + m) \approx z + m$ is a good approximation for $m \in [0, 1]$, where we will optimally determine z, see Fig. 8.10. We insert this into (8.40) and write

$$\log_2(y_{\text{float}}) \approx z + m_y + E_y - b \stackrel{!}{\approx} -\frac{z + m}{2} - \frac{E - b}{2} \approx -\frac{1}{2}\log_2(x_{\text{float}}). \tag{8.41}$$

We are looking for a simple and good approximation for all arguments $m \in [0, 1]$ and for all exponents. A natural assignment seems to be

$$E_y - b \approx -\frac{E - b}{2}, \quad z + m_y \approx -\frac{z + m}{2}$$

However, this leads to

$$m_y \approx -\frac{3z + m}{2} < 0,$$

contrary to the convention $m_y \in [0, 1)$. Therefore, we introduce a number $1 + \sigma \in \{1, 2, \ldots\}$ always to obtain a valid range of m_y. We will determine the exact choice of σ later. It also depends on the yet to be determined number $z \in \mathbb{R}$. So we write (8.41) as

$$\log_2(y_{\text{float}}) = z + m_y + E_y - b \stackrel{!}{\approx} \left(-\frac{z + m}{2} + \frac{\sigma + 1}{2}\right) - \frac{E - b + (\sigma + 1)}{2}.$$

Now, we determine

$$\begin{aligned}
z + m_y &= -\frac{z + m}{2} + \frac{\sigma + 1}{2}, & m_y &= \frac{\sigma + 1 - 3z - m}{2}, \\
E_y - b &= -\frac{E - b + \sigma + 1}{2}, & E_y &= \frac{3b - 1 - \sigma - E}{2} = 190 - \frac{E + \sigma}{2},
\end{aligned} \tag{8.42}$$

where we used $b = 127$. First, we can see that the still unknown number $z \in \mathbb{R}$ does not enter into the determination of the exponent E_y. Next, we can determine the value $\sigma \in \mathbb{N}$. In order for the new exponent E_y to be calculated exactly by bit shifting, we choose σ to be even if E is even, otherwise we choose σ to be odd.

The exponent E_y and mantissa m_y determine the approximation

$$y_0 = (1 + m_y) \cdot 2^{E_y - b},$$

which we can also interpret as an integer. With $L = 2^{23}$, we have

$$Y_0 = Lm_y + LE_y = M_y + LE_y.$$

Using (8.42) for m_y and E_y, we get

$$Y_0 = L\frac{\sigma + 1 - 3z}{2} + 190L - \frac{\sigma L}{2} - \overbrace{\frac{Lm + LE}{2}}^{=X}. \tag{8.43}$$

We define

$$Z = \lfloor L\frac{\sigma + 1 - 3z}{2} + 190L - \frac{\sigma L}{2} - \overbrace{\frac{Lm + LE}{2}}^{=X} \rfloor \quad \Rightarrow \quad Y_0 = Z - \frac{X}{2}$$

and the integer Z is just the *magic number* used in the method. In the following, we will optimally determine the number Z.

Approximation of the Exponent

The above already explains the second part of line 10. Using integer arithmetic, it is

$$E_y = 190 - (E + \sigma) \gg 1$$

and in binary representation

$$E_y = [10111110]_2 - (E + \sigma) \gg 1. \tag{8.44}$$

The digits of the binary representation of the number 190 are already found in the constant `0x5f3759df`, see (8.39):

$$5f3759df_{16} = [\underline{10111110}01101110101100111011111]_2$$

If we interpret this integer again as a floating point number, as is done in line 12 of Algorithm 8.67, then $190 = 10111110_2$ is just the binary representation of the exponent, i.e., the bits $b_{31}, \ldots, b_{24}$.

Approximation of the Mantissa

We still must approximate the mantissa $\mathbf{1}.m$ in (8.41), well and efficiently. Depending on $m \in [0, 1)$ and $\sigma \in \mathbb{N}$, we are looking for a $z \in \mathbb{R}$ such that

$$m_y = \frac{\sigma + 1 - 3z - m}{2} = \frac{\sigma + 1 - 3z}{2} - \frac{m}{2}$$

is approximated optimally. We have to distinguish two cases, σ even and σ odd. The first case stands for even exponents E in the representation of the number x_{float}. From the analysis of Fig. 8.10, we expect

$$0 \le z \le \frac{1}{10}$$

and we only consider this range. First, let E be odd, e.g., $E = 127$, and thus $\sigma = 1$. Then,

$$0.85 = \frac{1 + 1 - 3 \cdot 0.1}{2} - \frac{0}{2} \le \underbrace{\frac{1 + 1 - 3z}{2} - \frac{m}{2}}_{=m_y} \le \frac{1 + 1 - 3 \cdot 0}{2} - \frac{0}{2} \le 1,$$

i.e., m_y is always in the valid range. With (8.44) the initial value is

$$y_0 = \left(1 + \frac{2 - 3z}{2} - \frac{m}{2}\right) \cdot 2^{190 - \frac{E+1}{2}}.$$

We recall that for the exact result in the case $1 + \sigma = 2$, it holds:

$$y = 2^{-\frac{1}{2}\log_2(1+m)+1} \cdot 2^{190 - \frac{E+1}{2}}$$

Thus, the relative error of the initial value (still depending on z) is

$$y_0 = x^{-\frac{1}{2}} \frac{2 - \frac{3z+m}{2}}{2^{-\frac{1}{2}\log_2(1+m)+1}} =: x^{-\frac{1}{2}} \gamma(z, m).$$

We can now determine z such that $\gamma(z, m)$ is as close as possible to 1 for all $m \in [0, 1]$. However, we want to directly predict the result of the first Newton iteration (8.35) and calculate

$$y_1 = g(y_0) = x^{-\frac{1}{2}} \frac{3\gamma(z, m) - \gamma(z, m)^3}{2} =: x^{-\frac{1}{2}} \delta(z, m).$$

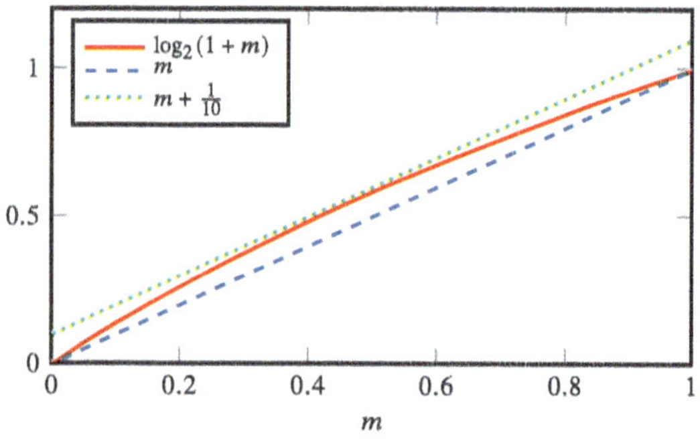

Fig. 8.10 Approximation of the logarithm by a linear function $\log_2(1 + m) \approx \sigma + m$ with slope 1—here, $\sigma = 0$ and $\sigma = \frac{1}{10}$ as enclosing functions

Now, we need to determine $z \in \mathbb{R}$ such that the relative error after one Newton step is minimized:

$$\max_{x \in [0,1]} |1 - \delta(z,m)| \to \min$$

Since the Newton iterations satisfy $y_n \leq x^{-\frac{1}{2}}$, this is equivalent to

$$\min_{x \in [0,1]} \delta(z,m) \to \max . \tag{8.45}$$

Possible extremal points can be found at the boundary, i.e., for $m = 0$ and $m = 1$

$$\begin{aligned}
\delta_1(z) &:= \delta(z,0) = \frac{27}{128}z^3 - \frac{27}{32}z^2 + 1, \\
\delta_2(z) &:= \delta(z,1) = \frac{9\sqrt{2}}{64}\left(3z^3 - 9z^2 + z + 5\right),
\end{aligned} \tag{8.46}$$

as well as inside at the zeros of the first derivative, i.e.,

$$\frac{\partial}{\partial m}\delta(z,m) \stackrel{!}{=} 0.$$

A real zero results at

$$m' = \frac{2}{3} - z,$$

and here it holds

$$\delta_3(z) := \delta(z,m') = -\frac{\sqrt{3(5-3z)}}{3888}\left(81z^4 - 540z^3 + 1350z^2 - 528z - 995\right). \tag{8.47}$$

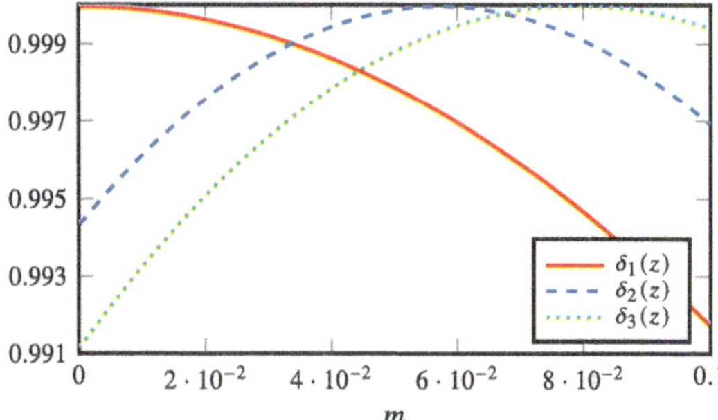

Fig. 8.11 Maximizing the minimum of $\delta_1(z)$, $\delta_2(z)$ and $\delta_3(z)$. The optimal result is obtained at the intersection of $\delta_1(z)$ and $\delta_3(z)$

The task now is to determine the value z such that

$$\delta_{min}(z) := \min\{\delta_1(z), \delta_2(z), \delta_3(z)\} \to \max \quad \forall z \in [0, 0.1].$$

A graphical analysis in Fig. 8.11 shows that this happens exactly at the intersection of $\delta_1(z)$ and $\delta_3(z)$. In the interval $z \in [0, 0.1]$ there is only one intersection point

$$\delta_1(z) = \delta_3(z) \quad \Rightarrow \quad z \approx 0.044513572756643.$$

We thus approximate the initial value as

$$y_0 = \left(2 - \frac{3z}{2} - \frac{m}{2}\right) 2^{190 - \frac{E+1}{2}}.$$

From this value for z, we can use (8.43) with $\sigma = 1$ to determine the optimal constant as

$$Z = \lfloor L\frac{1 + 1 - 3z}{2} + 190L - \frac{1}{2}\frac{L}{}\rfloor = 1\,597\,469\,713 = 5f377411_{16}.$$

This number deviates slightly from the constant $\texttt{0x5f3759df}$ used in the algorithm, but so far we have only considered the case of odd exponents E. Nevertheless, we determine the value z', which belongs to the number $\texttt{0x5f3759df}$. It holds

$$Z' = 5f3759df_{16} = 1\,597\,463\,007 = \lfloor L\frac{1 + 1 - 3z'}{2} + 190L - \frac{L}{2}\rfloor$$

$$\Leftrightarrow \quad z' = 0.04504656792.$$

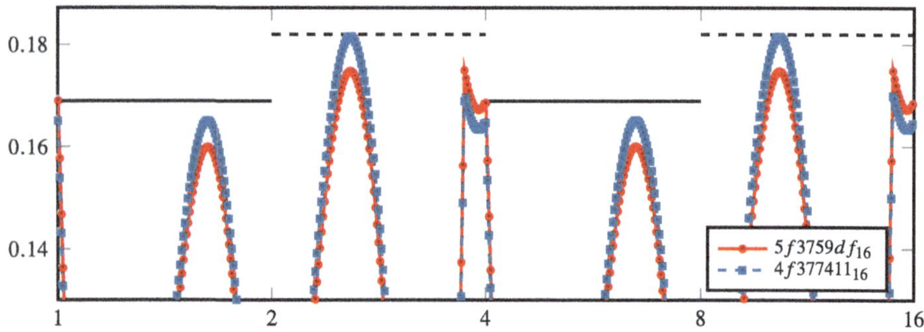

Fig. 8.12 Comparison of the *Fast Inverse Square Root* for different choices of the *magic number*. The solid (red) line represents the constant from the original algorithm. The dashed line comes from the optimization in the case E is odd. In these intervals $[1, 2]$ ($1.M \cdot 2^0 = 2^{127-127}$) and $[4, 8]$ ($1.M \cdot 2^2 = 2^{129-127}$), the error when using $\texttt{0x5f377411}$ is smaller. Considering all possible inputs $x \in \mathbb{R}_+$, the constant from the original method is slightly better

In Fig. 8.12, we present the result of the inverse root algorithm for both *magic numbers* `0x5f3759df` and `0x5f377411`. It turns out that the constant found here produces smaller errors if the input comes from an interval with an odd exponent E,

$$x_{\text{float}} = (1 + m) \cdot 2^{E-127},$$

that is

$$x_{\text{float}} \in [1, 2], \quad x_{\text{float}} \in [4, 8],$$

with $E = 127$ or with $E = 129$. On intervals with even E, the constant of the original method is superior. However, the deviation is small. A corresponding analysis for this case can be carried out as described above.

In this chapter, we will deal with the question of how a series of data points (e.g., from physical measurements, experimental observations, stock market data, etc.) can be approximated by a *simple* function $p(x)$ (e.g., polynomials). On the other hand, we want to clarify how *complex* functions can be approximated by simpler functions.

Interpolation means that the approximating function p is constructed in such a way that it is exact at the given discrete nodes x_k and values y_k:

$$p(x_k) = y_k, \quad k = 0, 1, 2, 3, \ldots, n$$

The support points (x_k, y_k) are either discrete data values (e.g., from experiments) or determined by functions $y_k = f(x_k)$. With the help of the interpolation $p(x)$, hitherto unknown values at intermediate points $\xi \in (x_k, x_{k+1})$ or the integral or the derivative of $p(x)$ can be determined. Moreover, interpolation has a significant importance for the development of further numerical methods, such as numerical quadrature, differentiation, or for the development of finite elements.

Approximation[1] is more generally defined. Again, a simple function $p(x)$ (i.e., a polynomial) is sought, which approximates discrete or function-determined data values (x_k, y_k) as well as possible. In contrast to interpolation, however, it is not necessarily required that $p(x_k) = y_k$ is exact at distinguished points x_k. What characterizes the approximation is decided on a case-by-case basis. For example, in the approximation of a function $f \in V$, the best approximation with respect to a norm within a subspace $P \subset V$ is defined as

---

[1] Approximation [27] (approximation): "Representation of a number, a function, a curve, or another mathematical object by a simpler number, function, curve, object, while keeping the error as small as possible."

T. Richter et al., *Introduction to Numerical Mathematics*, Mathematics Study Resources 25, https://doi.org/10.1007/978-3-662-72546-7_9

$$p \in P: \quad \|p - f\| = \min_{q \in P} \|q - f\|.$$

Here, P is the class of considered simple functions (i.e., all quadratic polynomials). An application of approximation is the "fitting" of discrete data values given by experiments. Often, many thousands of measured values are taken into account, which, due to statistical measurement errors, should not be exactly fulfilled in the sense of interpolation, but only approximately. A second application is the simplest possible representation of given functions.

The representation of discrete data as a simple function or the approximation of a general function by, for example, a polynomial has the great advantage that elementary operations such as differentiation and integration can be performed much more easily. As function spaces for approximation, the simplest possible functions are used. Examples are the polynomial spaces with functions

$$p(x) = \sum_{k=0}^{n} \alpha_k x^k$$

or the spaces of trigonometric functions

$$q(x) = \sum_{k=0}^{n} \alpha_k \cos(k\pi x) + \beta_k \sin(k\pi_x).$$

The *Weierstrass approximation theorem*, Theorem 7.6, states that for every continuous function $f : [a, b] \to \mathbb{R}$ and any given accuracy ϵ, there is a polynomial p such that $|f(x) - p(x)| < \epsilon$ for all $x \in [a, b]$. A proof of the Weierstrass approximation theorem is constructive and generates a sequence of so-called *Bernstein polynomials* that converge to f. However, this convergence is slow and is not suitable as a basis for an efficient numerical method [49].

Another approach to approximating functions with polynomials is the Taylor expansion, which we described in Theorem 7.5. This requires high regularity of the function f and then provides (locally) very good convergence. However, the Taylor expansion is also not suitable as the basis for an efficient and stable method. Taylor approximations will, however, be an essential tool in the upcoming analysis.

Lately, approximation theory also plays a significant role in connection with *machine learning*, especially in the field of *artificial neural networks*. These are special functions that are built from elementary building blocks but have many free parameters and are capable of representing very complex relationships well. We will discuss this in Sect. 11.5.

9.1 Polynomial Interpolation

We denote by P_n the vector space of polynomials up to degree $n \in \mathbb{N}$:

$$P_n := \left\{ p(x) = \sum_{k=0}^{n} a_k x^k, \ a_k \in \mathbb{R}, \ k = 0, \dots, n \right\}$$

Definition 9.1 (*Lagrange Interpolation Problem*) The Lagrange interpolation problem consists of determining a polynomial $p \in P_n$ for $n + 1$ pairwise different support points (nodes) $x_0, \dots, x_n \in \mathbb{R}$ and corresponding given support values $y_0, \dots, y_n \in \mathbb{R}$, such that the interpolation condition

$$p(x_k) = y_k, \quad k = 0, \dots, n,$$

is satisfied.

Theorem 9.2 *The Lagrange interpolation problem is uniquely solvable.*

Proof *(i)* We first prove uniqueness. Suppose $p_1, p_2 \in P_n$ are two solutions, given as

$$p_i(x) = \sum_{k=0}^{n} a_k^{(i)} x^k.$$

For the difference $q := p_1 - p_2 \in P_n$, we have $q(x_i) = 0$ at the $n + 1$ pairwise different points. According to Rolle's theorem, Theorem 7.3 in Chap. 7, the derivative q' then has n pairwise different zeros $x_i^{(1)}$, each in the intervals

$$q'(x_i^{(1)}) = 0, \quad x_i^{(1)} \in (x_i, x_{i+1}), \quad i = 0, \dots, n - 1.$$

We can repeatedly apply Rolle's theorem until we finally obtain one zero $x_0^{(n)}$ of the n-th derivative of $q(x)$, i.e., $q^{(n)}(x)$. But the polynomial $q^{(n)}(x)$ has the representation

$$q^{(n)}(x) = n!(a_n^{(1)} - a_n^{(2)}),$$

so it follows that $a_n^{(1)} = a_n^{(2)}$ and $q^{(n)} \equiv 0$. For the $n - 1$-th derivative, we have

$$q^{(n-1)}(x) = (n - 1)!(a_{n-1}^{(1)} - a_{n-1}^{(2)}),$$

and since this has two zeros, it follows again $a_{n-1}^{(1)} = a_{n-1}^{(2)}$, so $q^{(n-1)} \equiv 0$. Finally, $q \equiv 0$, so $p_1 = p_2$.

(ii) To prove existence, we consider the determining equations

$$p(x_k) = y_k, \quad k = 0, \dots, n.$$

This can be viewed as a linear system with $n + 1$ equations for the $n + 1$ unknown coefficients $a_0, \ldots, a_n$ of the polynomial $p \in P_n$. From the uniqueness of the solution (i.e., injectivity) also follows the solvability (i.e., the surjectivity). $\qquad\square$

The given support values y_k can be values of a given function f, i.e.,

$$f(x_k) = y_k, \quad k = 0, 1, \ldots, n,$$

or any discrete data values

$$(x_k, y_k) \in \mathbb{R} \times \mathbb{R}, \quad k = 0, 1, \ldots, n.$$

The interpolation problem can be formulated as a linear system

$$a_0 + a_1 x_0 + a_2 x_0^2 + \cdots + a_n x_0^n = y_0$$
$$a_0 + a_1 x_1 + a_2 x_1^2 + \cdots + a_n x_1^n = y_1$$
$$\vdots$$
$$a_0 + a_1 x_n + a_2 x_n^2 + \cdots + a_n x_n^n = y_n,$$

or in matrix notation

$$\begin{pmatrix} 1 & x_0 & x_0^2 & \cdots & x_0^n \\ 1 & x_1 & x_1^2 & \cdots & x_1^n \\ \vdots & \vdots & \ddots & \ddots & \vdots \\ 1 & x_n & x_n^2 & \cdots & x_n^n \end{pmatrix} \begin{pmatrix} a_0 \\ a_1 \\ \vdots \\ a_n \end{pmatrix} = \begin{pmatrix} y_0 \\ y_1 \\ \vdots \\ y_n \end{pmatrix}.$$

Creating the interpolation polynomial, therefore, requires solving a quadratic linear system with $n + 1$ equations. The matrix is the so-called *Vandermonde matrix*, see [33]. In general, this matrix is very poorly conditioned with an exponentially increasing condition number. This direct approach is therefore not suitable for practically solving the interpolation problem.

9.1.1 Lagrange Basis Polynomials

To handle the interpolation problem numerically, we will define a basis of the space P_n which allows for an immediate construction of the interpolation polynomial. We define:

Theorem 9.3 (Lagrange Basis Polynomials) *Let $x_k \in \mathbb{R}$ for $k = 0, \ldots, n$ be pairwise different support points. The Lagrange polynomials*

$$L_k^{(n)}(x) := \prod_{j=0, j \neq k}^{n} \frac{x - x_j}{x_k - x_j} \in P_n, \quad k = 0, 1, \ldots, n,$$

define a basis of P_n. Furthermore, we have

$$L_k^{(n)}(x_l) = \delta_{kl}. \tag{9.1}$$

Proof *(i)* We first show the property (9.1). We have

$$L_k^{(n)}(x_k) = \prod_{j=0,\,j\neq k}^{n} \frac{x_k - x_j}{x_k - x_j} = 1.$$

Furthermore, for $l \neq k$

$$L_k^{(n)}(x_l) = \prod_{j=0,\,j\neq k}^{n} \frac{x_l - x_j}{x_k - x_j} = \frac{x_l - x_l}{x_k - x_l} \prod_{j=0,\,j\neq k,\,j\neq l}^{n} \frac{x_l - x_j}{x_k - x_j} = 0.$$

(ii) By construction, $L_k^{(n)} \in P_n$ holds for $k = 0, \ldots, n$. Furthermore, all polynomials $L_k^{(n)}$ are linearly independent. Otherwise, there would be coefficients $\alpha_1, \ldots, \alpha_n$ with

$$L_0^{(n)}(x) = \sum_{i=1}^{n} \alpha_i L_i^{(n)}(x).$$

For $x = x_0$, a contradiction arises. $\qquad\square$

Theorem 9.4 (Lagrange Representation) *Let $x_0, \ldots, x_n \in \mathbb{R}$ be pairwise distinct support points and $y_0, \ldots, y_n \in \mathbb{R}$. The (unique) solution of the Lagrange interpolation problem $p(x_k) = y_k$ for $k = 0, \ldots, n$ is given by*

$$p(x) := \sum_{k=0}^{n} y_k L_k^{(n)}(x).$$

Proof This statement follows directly from the just-proven properties of the Lagrange basis polynomials. $\qquad\square$

The Lagrange representation of the interpolation polynomial impresses with its simplicity. However, it has the significant disadvantage that each basis polynomial $L_k^{(n)}(x)$ depends on all support points $x_0, x_1, \ldots, x_n$. Suppose we want to increase the accuracy by adding a support point x_{n+1}. Then, all basis polynomials must be replaced.

Example 9.5 Given are the points $(x_0, y_0) = (1, 3)$ and $(x_1, y_1) = (4, 8)$. We construct the Lagrange basis polynomials and the corresponding linear interpolation polynomial. We obtain

$$L_0^{(1)}(x) = \frac{x - x_1}{x_0 - x_1} = \frac{x - 4}{-3} \in P_1,$$

$$L_1^{(1)}(x) = \frac{x - x_0}{x_1 - x_0} = \frac{x - 1}{3} \in P_1.$$

In particular, the interpolation conditions, i.e., the Kronecker delta property, hold in the corresponding interpolation points:

$$L_0^{(1)}(x_0) = \frac{1 - 4}{-3} = 1, \quad L_0^{(1)}(x_1) = \frac{4 - 4}{-3} = 0$$

The interpolation polynomial is

$$p(x) = y_0 L_0^{(1)}(x) + y_1 L_1^{(1)}(x) = 3\frac{x - 4}{-3} + 8\frac{x - 1}{3}.$$

◀

9.1.2 Newton Basis Polynomials

The Newton representation is based on a basis that allows a successive extension of the interpolation polynomial by new support points.

Theorem 9.6 (Newton Basis Polynomials) *Let $x_0, \ldots, x_n \in \mathbb{R}$ be pairwise distinct support points. The polynomials $N_0, \ldots, N_n \in P_n$ defined as*

$$N_0(x) := 1,$$

$$N_k(x) := \prod_{j=0}^{k-1}(x - x_j), \quad k = 1, \ldots, n,$$

are a basis of P_n. They are called the Newton basis.

Proof The basis property can be easily proven again. First, $N_k \in P_n$ holds for $k = 0, \ldots, n$. Furthermore, the basis polynomials are linearly independent. This follows from the property

$$N_k(x_l) = \begin{cases} 0 & k > l, \\ \neq 0 & k \le l. \end{cases}$$

A polynomial $N_k \in P_k \setminus P_{k-1}$ cannot be represented by a linear combination of polynomials P_l with $l < k$. $\qquad\square$

The basis polynomial $N_k(x)$ depends only on the support points $x_0, \ldots, x_k$. When adding a support point x_{k+1}, the first basis polynomials do not need to be changed. The interpolation polynomial is determined by the approach

$$p(x) = \sum_{k=0}^{n} a_k N_k(x).$$

For the support point x_k, $N_l(x_k) = 0$ holds for all $l > k$. We proceed recursively:

$$y_0 \overset{!}{=} p(x_0) = a_0$$

$$y_1 \overset{!}{=} p(x_1) = a_0 + a_1(x_1 - x_0)$$

$$\vdots$$

$$y_n \overset{!}{=} p(x_n) = a_0 + a_1(x_n - x_0) + \ldots + a_n(x_n - x_0) \cdots (x_n - x_{n-1})$$

In contrast to the previously learned Lagrange representation, an additional data pair (x_{n+1}, y_{n+1}) can be easily added. If the interpolation polynomial $p_n \in P_n$ is given, a support point x_{n+1} can be easily added:

$$y_{n+1} \overset{!}{=} p_{n+1}(x_{n+1}) = p_n(x_{n+1}) + a_{n+1} N_{n+1}(x_{n+1})$$

$$\Rightarrow \quad a_{n+1} = \frac{y_{n+1} - p_n(x_{n+1})}{N_{n+1}(x_{n+1})} \tag{9.2}$$

In practice, the coefficients a_k are calculated numerically stable by the following result:

Theorem 9.7 *Let $x_k \in \mathbb{R}$ for $k = 0, \ldots, n$ be pairwise different support points and $y_k \in \mathbb{R}$. The Lagrange interpolation polynomial with respect to the Newton polynomial basis is given as*

$$p(x) = \sum_{k=0}^{n} y[x_0, \ldots, x_k] N_k(x).$$

The notation $y[x_0, \ldots, x_k]$ denotes the divided differences, *which are defined by the following recursive rule:*

> **Input:** *Support points (x_k, y_k) for $k = 1, \ldots n$.*
> 1 **for** $k = 0$ **to** n **do**
> 2 $y[x_k] := y_k$
> 3 **for** $l = 1$ **to** n **do**
> 4   **for** $k = 0$ **to** $n - l$ **do**
> 5 $y[x_k, \ldots, x_{k+l}] := \frac{y[x_{k+1}, \ldots, x_{k+l}] - y[x_k, \ldots, x_{k+l-1}]}{x_{k+l} - x_k}$
> **Result:** *The coefficients of the Lagrange interpolation polynomial with respect to the Newton polynomial basis.*

Proof We denote by $p_{k,k+l} \in P_l$ the polynomial, which represents the interpolation to the $l+1$ points $(x_k, y_k), \ldots, (x_{k+l}, y_{k+l})$. In particular, the desired polynomial $p \in P_n$ is given by $p_{0,n}$. We show that the following statement holds for all $l = 0, \ldots n$ and $k = 0, \ldots, n - l$:

$$p_{k,k+l}(x) = y[x_k] + y[x_k, x_{k+1}](x - x_k)$$
$$+ \cdots + y[x_k, \ldots, x_{k+l}](x - x_k) \cdots (x - x_{k+l-1}) \quad (9.3)$$

We prove by induction on the polynomial degree l. For $l = 0$ we have $p_{k,k}(x) = y[y_k] = y_k$. Assume the claim is correct for $l - 1 \geq 0$. This means in particular, the polynomial $p_{k,k+l-1}$ is given in representation (9.3) and interpolates the points $(x_k, y_k), \ldots,$ (x_{k+l-1}, y_{k+l-1}), and the polynomial $p_{k+1,k+l}$ interpolates the points $(x_{k+1}, y_{k+1}), \ldots,$ (x_{k+l}, y_{k+l}). Then by

$$q(x) = \frac{(x - x_k) p_{k+1,k+l}(x) - (x - x_{k+l}) p_{k,k+l-1}(x)}{x_{k+l} - x_k} \quad (9.4)$$

an interpolation polynomial through the points $(x_k, y_k), \ldots, (x_{k+l}, y_{k+l})$ is given. This becomes clear by substituting x_i into $q(x)$. For inner points $i = k + 1, \ldots, k + l - 1$ we have $p_{k,k+l-1}(x_i) = p_{k+1,k+l}(x_i) = y_i$, for x_k and x_{k+l} one of the two factors is equal to zero. Thus, the desired interpolation polynomial is $p_{k,k+l} = q$. However, according to construction (9.2) the following representation also holds:

$$p_{k,k+l}(x) = p_{k,k+l-1}(x) + a(x - x_k) \cdots (x - x_{k+l-1})$$

Comparing the leading monomial x^n of this representation with (9.4), using (9.3) for $p_{k,k+l-1}$ and $p_{k+1,k+l}$ yields

$$a = \frac{y[x_{k+1}, \ldots, x_{k+l}] - y[x_k, \ldots, x_{k+l-1}]}{x_{k+l} - x_k} = y[x_k, \ldots, x_{k+l}].$$

$\square$

Example 9.8 We continue with Example 9.5. Given are the points $(x_0, y_0) = (1, 3)$ and $(x_1, y_1) = (4, 8)$. In the Newton representation, we have the basis polynomials

$$N_0(x) = 1, \quad N_1(x) = (x - x_0) = (x - 1).$$

The coefficients are determined by

$$3 = y[x_0], \quad 8 = y[x_1]$$

and the first divided difference is

$$y[x_0, x_1] = \frac{y[x_1] - y[x_0]}{x_1 - x_0} = \frac{8 - 3}{3} = \frac{5}{3}.$$

Thus, we obtain the interpolation polynomial

$$p(x) = y[x_0]N_0(x) + y[x_0, x_1]N_1(x) = 3 \cdot 1 + \frac{5}{3}(x - 1).$$

We do a small test to see if we can recover the interpolation conditions:

$$p(x_0) = 3 + \frac{5}{3}(1 - 1) = 3$$

$$p(x_1) = 3 + \frac{5}{3}(4 - 1) = 3 + 5 = 8$$

The test is successful.   ◄

This recursive construction principle immediately suggests an algorithm for evaluating the interpolation at a point $\xi \in \mathbb{R}$ without having to explicitly determine the desired interpolation polynomial $p(x)$ beforehand. We formulate the following algorithm:

Algorithm 9.9: Neville Scheme

> **Input:** Support points (x_k, y_k) for $k = 0, \dots n$ and an evaluation point ξ.
> 1 **for** $k = 0$ **to** n **do**
> 2 $p_{k,k} := y_k$
> 3 **for** $j = 1$ **to** n **do**
> 4   **for** $k = 0$ **to** $n - j$ **do**
> 5 $p_{k,k+j} := p_{k,k+j-1} + (\xi - x_k)\frac{p_{k+1,k+j} - p_{k,k+j-1}}{x_{k+j} - x_k}$
> **Result:** $p_{0,n}$ is the value of the interpolation polynomial evaluated at the point ξ.

♣ **Implementation 9.10: Neville Scheme**

```python
def neville(data, xi):
    n = data.shape[0]
    x = data[:, 0]
    p = np.diag(data[:, 1])

    for j in range(1, n):
        for k in range(n - j):
            p[k, k + j] = p[k, k + j - 1] + (xi - x[k]) * (p[k + 1, k + j] - p[k, k
            ↪ + j - 1]) / (x[k + j] - x[k])
    return p
```

When executing the Neville scheme, we also obtain all intermediate interpolations $p_{k,l}$ through the points (x_k, y_k) to (x_{k+l}, y_{k+l}).

Remark 9.11 (Divided Differences) The divided differences $y[x_0, \dots, x_n]$ initially appear unclear in their purpose, but they are ideally suited for algorithmic use, as they can be calculated very easily. We assume that the support values are taken from an analytical function $f \in C^\infty$ such that

$$y_k = f(x_k)$$

holds. Furthermore, let the nodes be evenly distributed with $x_k = x_0 + hk$ and a $h \in \mathbb{R}$. Then, with Taylor expansion

$$y[x_k, x_{k+1}] = \frac{f(x_k + h) - f(x_k)}{h} = f'(x_{k+\frac{1}{2}}) + \sum_{j=1}^{\infty} \frac{f^{(2j+1)}(x_{k+\frac{1}{2}})}{(2j+1)!} h^{2j}.$$

The first divided difference is thus an approximation to the first derivative of the function f. Then it further holds

$$\begin{aligned} y[x_k, x_{k+1}, x_{k+2}] &= \frac{y[x_{k+1}, x_{k+2}] - y[x_{k+1}, x_{k+2}]}{2h} \\ &= \frac{f'(x_{k+\frac{3}{2}}) - f'(x_{k+\frac{1}{2}}) + \mathcal{R}_{k,k+2}}{2h} \\ &= \frac{1}{2} f''(x_{k+1}) + \sum_{j=1}^{\infty} \frac{f^{(2j+1)}}{2(2j+1)!} h^{2j} \end{aligned}$$

with a remainder

$$\frac{\mathcal{R}_{k,k+2}}{2h} = \sum_{j=1}^{\infty} \frac{f^{(2j+1)}(x_{k+\frac{3}{2}}) - f^{(2j+1)}(x_{k+\frac{1}{2}})}{2(2j+1)!} h^{2j-1}.$$

We only consider the first critical term for $j = 1$, as the remainder is only of first order. It holds

$$\frac{f'''(x_{k+\frac{3}{2}}) - f'''(x_{k+\frac{1}{2}})}{12} h = \sum_{j=1}^{\infty} \frac{f^{(2j+2)}(x_{k+1})}{12(2j)!} h^{2j}.$$

Thus, it again holds

$$y[x_k, x_{k+1}, x_{k+2}] = \frac{1}{2} f''(x_{k+1}) + O(h^2).$$

In general, for analytical functions, it can be shown that

$$y[x_k, x_{k+1}, \ldots, x_{k+l}] = \frac{f^{(l)}\left(\frac{x_k + x_{k+l}}{2}\right)}{l!} + O(h^2).$$

The divided differences are approximations of the Taylor coefficients. ♦

Example 9.12 (Neville Scheme) We consider the support points

$$\{(x_k, y_k)_k, \; k = 0, 1, 2, 3\} = \{(0, 0), (1, 1), (2, 8), (3, 27)\},$$

which are taken from the function $f(x) = x^3$. We execute the Neville scheme to calculate the interpolation at the point $\xi = 0.5$ recursively and obtain the following results with our code:

```
[[ 0.    0.5   -0.25   0.125]
 [ 0.    1.    -2.5    2.   ]
 [ 0.    0.     8.    -20.5  ]
 [ 0.    0.     0.     27.   ]]
```

The final approximation $p_{03} = 0.125 = 0.5^3$ is exact. This is to be expected, as $f \in P_3$. As a test, we consider the very poor approximation p_{23}, which is obtained by the linear interpolation polynomial $p_{2,3}(x)$ through the support points $(2, 8)$ and $(3, 27)$, thus

$$p_{2,3}(x) = 8 + \frac{27 - 8}{3 - 2}(x - 2) = 19x - 30.$$

It holds $p_{2,3}(0.5) = -20.5$.   ◄

9.1.3 Interpolation of Functions and Error Estimates

In this section, we discuss the interpolation of functions. The points are no longer given by a data set, but by evaluating a given function f on $[a, b]$:

$$y_k = f(x_k), \quad x_k \in [a, b], \ k = 0, \ldots, n.$$

The feasibility, i.e., existence and uniqueness of an interpolation polynomial, has already been answered in the preceding sections. Regarding the interpolation of functions, the question arises of *how well* the interpolation polynomial $p \in P_n$ approximates the function f on $[a, b]$.

Theorem 9.13 (Interpolation Error With Differential Remainder) *Let $f \in C^{n+1}[a, b]$ and $p \in P_n$ be the interpolation polynomial to f at the $n + 1$ pairwise different nodes $x_0, \ldots, x_n$. Then for each $x \in [a, b]$ there exists a $\xi \in (a, b)$ such that*

$$f(x) - p(x) = \frac{f^{(n+1)}(\xi)}{(n + 1)!} \prod_{j=0}^{n}(x - x_j). \tag{9.5}$$

In particular, it holds

$$|f(x) - p(x)| \leq \frac{\max\limits_{\xi \in (a,b)} |f^{(n+1)}(\xi)|}{(n + 1)!} \prod_{j=0}^{n} |x - x_j|. \tag{9.6}$$

Proof If x coincides with a node, i.e., $x = x_k$ for a $k \in \{0, \ldots, n\}$, then the error and remainder vanish, and we are done. Let therefore $x \neq x_k$ for all $k = 0, 1, \ldots, n$ and $F \in C^{n+1}[a, b]$ be defined as

$$F(t) := f(t) - p(t) - K(x) \prod_{j=0}^{n}(t - x_j).$$

Here, $K(x)$ is determined such that $F(x) = 0$. This is possible, since

$$\prod_{j=0}^{n}(x - x_j) \neq 0 \quad \Rightarrow \quad K(x) = \frac{f(x) - p(t)}{\prod_{j=0}^{n}(x - x_j)}.$$

Then $F(t)$ has at least $n + 2$ different zeros $x_0, x_1, \ldots, x_n, x$ in $[a, b]$. By repeated application of Rolle's theorem, the derivative $F^{(n+1)}$ has at least one zero $\xi \in (a, b)$. With

$$0 = F^{(n+1)}(\xi) = f^{(n+1)}(\xi) - p^{(n+1)}(\xi) - K(x)(n + 1)!$$
$$= f^{(n+1)}(\xi) - 0 - K(x)(n + 1)!,$$

the claim follows from

$$K(x) = \frac{f^{(n+1)}(\xi)}{(n + 1)!} \quad \Rightarrow \quad f(x) - p(x) = \frac{f^{(n+1)}(\xi)}{(n + 1)!} \prod_{j=0}^{n}(x - x_j).$$

$\square$

The proof of an integral remainder of the interpolation is somewhat more involved. For this, we first formulate the *Hermite relation*.

Theorem 9.14 (Hermite Relation) *For pairwise different nodes $x_0, \ldots, x_{m+1} \in [a, b]$ it holds:*

$$f[x_0, \ldots, x_{l-1}, x] = f[x_0, \ldots, x_{l-1}, x_l] + f[x_0, \ldots, x_{l-1}, x_l, x](x - x_l)$$

with $1 \leq l \leq m + 1$.

Proof On the one hand, for $x = x_l$ we have

$$f[x_0, \ldots, x_{l-1}, x_l] + f[x_0, \ldots, x_{l-1}, x_l, x](x - x_l)$$
$$= f[x_0, \ldots, x_{l-1}, x_l] + f[x_0, \ldots, x_{l-1}, x_l, x_l](x_l - x_l)$$
$$= f[x_0, \ldots, x_{l-1}, x].$$

On the other hand, let $x \neq x_l$. We define

$$F(t_{l+1}) := f^{(l)}(x_0 + \ldots + t_l(x_l - x_{l-1}) + t_{l+1}(x - x_l)).$$

Its derivative is

$$\frac{\mathrm{d}}{\mathrm{d}t_{l+1}} F(t_{l+1}) := f^{(l+1)}(x_0 + \ldots + t_l(x_l - x_{l-1})t_{l+1}(x - x_l)) \cdot (x - x_l).$$

With this, we conclude

$$(x - x_l) f[x_0, \ldots, x_l, x]$$

$$= (x - x_l) \int_0^1 dt_1 \ldots \int_0^{t_{l-1}} dt_l \ldots$$

$$\ldots \int_0^{t_l} dt_{l+1} f^{(l+1)}(x_0 + t_1(x_1 - x_0) + \ldots + t_l(x_l - x_{l-1}) + t_{l+1}(x - x_l))$$

$$= (x - x_l) \frac{1}{x - x_l} \int_0^1 dt_1 \ldots \int_0^{t_{l-1}} dt_l \ldots \int_0^{t_l} dt_{l+1} \frac{d}{dt_{l+1}} F(t_{l+1})$$

$$= \int_0^1 dt_1 \ldots \int_0^{t_{l-1}} dt_l \, F(t_{l+1})|_0^{t_l}$$

$$= \int_0^1 dt_1 \ldots \int_0^{t_{l-1}} dt_l \, f^{(l)}(x_0 + \ldots + t_l(x - x_{l-1}))$$

$$- \int_0^1 dt_1 \ldots \int_0^{t_{l-1}} dt_l \, f^{(l)}(x_0 + \ldots + t_l(x_l - x_{l-1}))$$

$$= f[x_0, \ldots, x_{l-1}, x] - f[x_0, \ldots, x_{l-1}, x_l].$$

The statement is thus proven. $\qquad\square$

Theorem 9.15 (Interpolation Error With Integral Remainder) *Let $f \in C^{n+1}[a, b]$. Then for $x \in [a, b] \setminus \{x_0, \ldots, x_n\}$ the representation*

$$f(x) - p(x) = f[x_0, \ldots, x_n, x] \prod_{j=0}^{n} (x - x_j)$$

holds with the interpolation conditions $f[x_i, \ldots, x_{i+k}] := y[x_i, \ldots, x_{i+k}]$ and

$$f[x_0, \ldots, x_n, x] =$$

$$\int_0^1 \int_0^{t_1} \ldots \int_0^{t_n} f^{(n+1)}(x_0 + t_1(x_1 - x_0) + \ldots + t(x - x_n)) dt \, dt_n, \ldots, dt_1.$$

Proof The goal is to show that the integral representation matches the divided differences from Theorem 9.7. So if $x_0, \ldots, x_n$ are pairwise different, then we want to show that:

$$y[x_0, \ldots, x_n] = f[x_0, \ldots, x_n]$$

We justify this in the following. The Newton interpolation polynomial p for f and pairwise different $x_0, \ldots, x_n$ can be represented as

$$p(x) = y[x_0] + y[x_0, x_1](x - x_0) + \ldots + y[x_0, \ldots, x_n](x - x_0) \cdots (x - x_{n-1}).$$

Thus, $f(x_k) = p(x_k)$ for $k = 0, \ldots, n$. On the other hand, from the Hermite relation, Theorem 9.14, we get

$$\begin{aligned}
f(x) &= f[x_0] + (x - x_0) f[x_0, x] \\
&= f[x_0] + (x - x_0)(f[x_0, x_1] + (x - x_1) f[x_0, x_1, x]) \\
&= f[x_0] + (x - x_0) f[x_0, x_1] + (x - x_0)(x - x_1) f[x_0, x_1, x] \\
&= f[x_0] + (x - x_0) f[x_0, x_1] + (x - x_0)(x - x_1)(f[x_0, x_1, x_2] \\
&\quad + (x - x_2) f[x_0, x_1, x_2, x]) \\
&= \ldots \\
&= f[x_0] + (x - x_0) f[x_0, x_1] + \ldots \\
&\quad + (x - x_0) \cdots (x - x_{n-1}) f[x_0, \ldots, x_{n-1}, x].
\end{aligned}$$

Above we already had $f(x_k) = p(x_k)$ for $k = 0, \ldots, n$ and therefore

$$y[x_0] = p(x_0) = f(x_0) = f[x_0].$$

Since the nodes are pairwise different, in particular $x_1 - x_0 \neq 0$, we get for x_1

$$p(x_1) = y[x_0] + y[x_0, x_1](x_1 - x_0)$$
$$\text{and } f(x_1) = f[x_0] + f[x_0, x_1](x_1 - x_0).$$

By comparison, it is
$$f[x_0, x_1] = y[x_0, x_1].$$

With induction, this relation can be shown for $k = 1, \ldots, n$:

$$\begin{aligned}
p(x_k) &= y[x_0] + y[x_0, x_1](x_1 - x_0) + \ldots \\
&\quad + y[x_0, \ldots, x_k](x_k - x_0) \cdots (x_k - x_{k-1}) \\
\text{and } f(x_k) &= f[x_0] + f[x_0, x_1](x_1 - x_0) + \cdots \\
&\quad + f[x_0, \ldots, x_k](x_k - x_0) \cdots (x_k - x_{k-1}).
\end{aligned}$$

By comparison, we get:
$$f[x_0, \ldots, x_k] = y[x_0, \ldots, x_k]$$

for $k = 1, \ldots, n$. $\qquad\square$

Analyzing the different contributions to the Lagrangian interpolation remainder (9.5), we first note that the term $\frac{1}{(n+1)!}$ gets very small for large n. The role of the product $\prod_{j=0}^{n}(x - x_j)$ is more difficult to understand. It gets small if the nodes are close together. However, if the majority of the factors $|x - x_j|$ are large, the product will also most likely be larger. If all derivatives of f are uniformly bound in $[a, b]$, it holds that

$$\max_{a \leq x \leq b} |f(x) - p(x)| \to 0, \quad n \to \infty,$$

if the nodes are uniformly distributed. Whenever the derivatives of the function grow quickly, e.g.,

$$f(x) = (1 + x^2)^{-1}, \quad |f^{(n)}(x)| \approx 2^n n! O(|x|^{-2-n}),$$

the interpolant does not converge uniformly on $[a, b]$.

Example 9.16 The function $f(x) = |x|, x \in [-1, 1]$ is interpolated using the Lagrange interpolation at the nodes

$$x_k = -1 + kh, \quad k = 0, \ldots, 2m, \quad h = 1/m, \quad x \neq x_k.$$

This results in the global behavior

$$p_m(x) \not\to f(x), \quad m \to \infty.$$

Although f is not differentiable in this example, the behavior is typical for the Lagrange interpolation and often observed, for instance, for the smooth function

$$f(x) = (1 + x^2)^{-1}, \quad x \in [-5, 5].$$

◄

We summarize the results so far:

Remark 9.17 The approximation theorem of Weierstrass states that any function $f \in C([a, b])$ can be approximated arbitrarily well by a polynomial. However, analysis does not provide any guidance on how to carry out the approximation in practice. The Lagrange interpolation is one way to approximate. The quality of this approximation is primarily determined by the regularity of the data, i.e., by f. A uniform approximation of functions with Lagrange interpolation polynomials is, in general, not possible.   ◆

The Lagrange interpolation "suffers" from the same limitations as the Taylor expansion, Theorem 7.5. We are still far from the possibility of approximating a merely continuous function $f \in C([a, b])$ arbitrarily well.

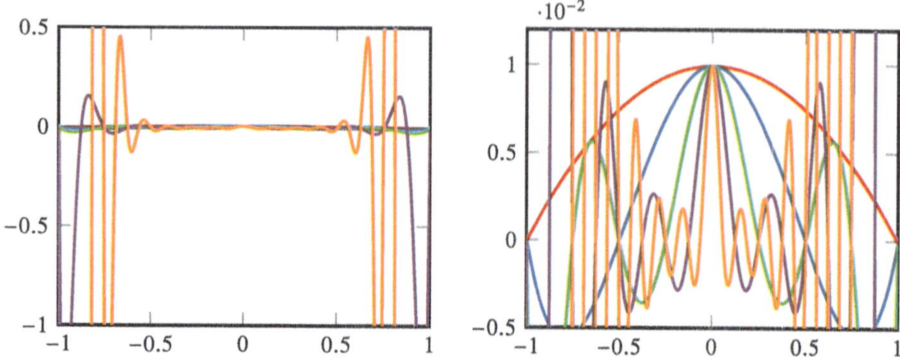

Fig. 9.1 Interpolation polynomials $\tilde{p}_m(x) \in P_{2m}$ for $m = 1, 2, 4, 8$. Left: the complete interpolation interval. Right: Detail near $y = 0$

A second disadvantage of Lagrange interpolation is the lack of locality. A disturbance in a support point $(\tilde{x}_k, \tilde{y}_k)$ affects all Lagrange polynomials and, in particular, the entire interpolation interval. We consider an example for this:

Example 9.18 (Global Error Influence) We are looking for the interpolation polynomial for the function $f(x) = 0$ at the $2m + 1$ nodes

$$x_k = kh, \quad k = -m, \ldots, m, \quad h = \frac{1}{m}.$$

The exact interpolation polynomial $p \in P_{2m}$ is naturally given by $p = 0$. We assume that the function evaluation is disturbed at $x = 0$ with a small error $\epsilon > 0$:

$$y_k = \begin{cases} 0 & k \neq 0 \\ \epsilon & k = 0. \end{cases}$$

The disturbed interpolation polynomial is given by

$$\tilde{p}(x) = \epsilon \prod_{i=-m, i \neq 0}^{m} \frac{x - x_i}{x_i}.$$

In Fig. 9.1, we show $\tilde{p}_m \in P_{2m}$ for the cases $m = 1, 2, 4, 8$ with a disturbance $\epsilon = 0.01$. Despite this small disturbance at the point $x = 0$, the interpolation polynomials deviate significantly from $y_k = 0$ at the interval boundaries. In the right figure, one can see that for small polynomial degrees, i.e., $m = 1$ and $m = 2$, the maximum error on the interval is not larger than the initial disturbance $\epsilon = 0.01$. The Lagrange interpolation is unstable for large polynomial degrees.   ◀

The unfavorable behavior of the interpolation at the interval boundaries is called *Runge's phenomenon*. It is explained by the fact that every polynomial $p(x)$ for $x \to \pm\infty$ tends to $p \to \pm\infty$. A function, which does not have this property, such as

$$r(x) = \frac{1}{1 + x^2},$$

for which $r \to 0$ for $x \to \pm\infty$, can generally only be represented poorly with polynomials globally.

9.1.4 Hermite Interpolation

One generalization of the Lagrangian interpolation is to also consider derivative values as conditions, such as $p^{(i)}(x_k) = f^{(i)}(x_k)$ for $i = 1, 2, \ldots$. This approach is called *Hermite interpolation*.

Theorem 9.19 (Hermite Interpolation) *Let $f \in C^{(n+1)}([a, b])$. Let $x_0, \ldots, x_m$ be pairwise different nodes and $\mu_k \in \mathbb{N}$ for $k = 0, \ldots, m$ a derivative index. Furthermore, let $n = m + \sum_{k=0}^{m} \mu_k$. The Hermite interpolation $p \in P_n$ satisfying*

$$k = 0, \ldots, m : \quad p^{(i)}(x_k) = f^{(i)}(x_k), \quad i = 0, \ldots, \mu_k,$$

is uniquely determined and for every $x \in [a, b]$ there exists an intermediate point $\xi \in [a, b]$ such that:

$$f(x) - p(x) = f[\underbrace{x_0, \ldots, x_0}_{\mu_0+1\text{-}times}, \ldots, \underbrace{x_m, \ldots, x_m}_{\mu_m+1\text{-}times}, x] \prod_{k=0}^{m} (x - x_k)^{\mu_k+1}$$

$$= \frac{1}{(n+1)!} f^{(n+1)}(\xi) \prod_{k=0}^{m} (x - x_k)^{\mu_k+1}$$

For the proof and further details on practical calculation, we refer to the literature [36].

In the case $\mu_k = 0$ for all k, Hermite and Lagrange interpolation coincide. In the case of a single support point $m = 0$ with $\mu_0 = n$, the Hermite interpolation exactly corresponds to the Taylor sum, i.e.,

$$p(x) = \sum_{i=0}^{k} \frac{(x - x_0)^k}{k!} f^{(k)}(x_0).$$

9.2   Piecewise Interpolation

A significant defect of the *global* interpolation is that the interpolating polynomials tend to strong oscillations between the nodes. The reason is the generic stiffness given by the *implicit* requirement of C^∞-transitions at the nodes. The stiffness can be reduced by composing the global function as *piecewise* polynomials with respect to the partition $a = x_0 < x_1 < \ldots < x_n = b$. At the nodes x_i, suitable differentiability properties such as continuity or once differentiability may be required.

The term *spline* will play a central role and is usually used in the literature for the *cubic* spline, see Definition 9.24. More generally, a spline refers to piecewise interpolants that have a given smoothness property at the transition points. Simple piecewise interpolants, such as piecewise linear functions, have a prominent role in the finite element method for approximating the solution of differential equations.

The global interval $I = [a, b]$ is divided into subintervals $I_i = [x_{i-1}, x_i]$ with the length $h_i = x_i - x_{i-1}$. The fineness of the entire interval subdivision is characterized by $h := \max_{i=1,\ldots,n} h_i$, see Fig. 9.2. To define the spline, the vector spaces of piecewise polynomial functions are given as follows:

$$S_h^{k,r}[a, b] := \{s \in C^r[a, b], \ s|_{I_i} \in P_k(I_i)\}.$$

Here, $k \in \mathbb{N}$ is the local polynomial degree and $r < k$ is the global regularity of the functions. For a set of support values (either data points or taken from a function) in the total interval I, an interpolant $p \in S_h^{k,r}[a, b]$ is determined using suitable interpolation conditions.

We discuss some examples, focusing on the idea of the whole process rather than on proving questions of existence, uniqueness, and error estimates.

Example 9.20 (Piecewise Linear Interpolation) In Fig. 9.2, we show the piecewise linear Lagrange interpolant (i.e., $k = 1, r = 0$) for approximating a given function f on $[a, b]$ at the nodes x_i, $i = 0, \ldots, n$:

$$s \in S_h^{1,0}[a, b] = \{s \in C[a, b], \ s|_{I_i} \in P_1(I_i)\}$$

with the interpolation conditions

Fig. 9.2 Piecewise linear interpolation s (dashed) of a function f (solid)

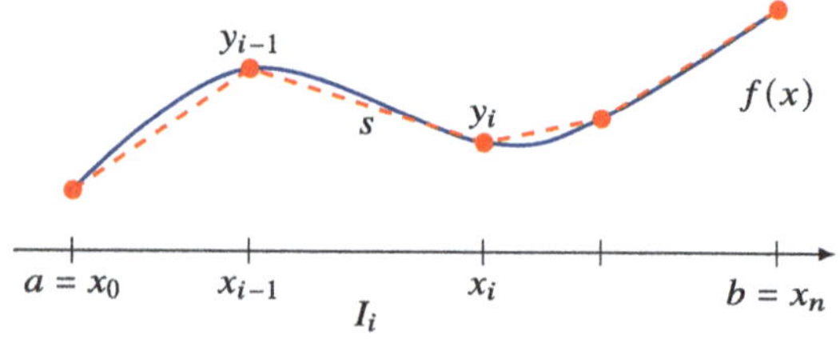

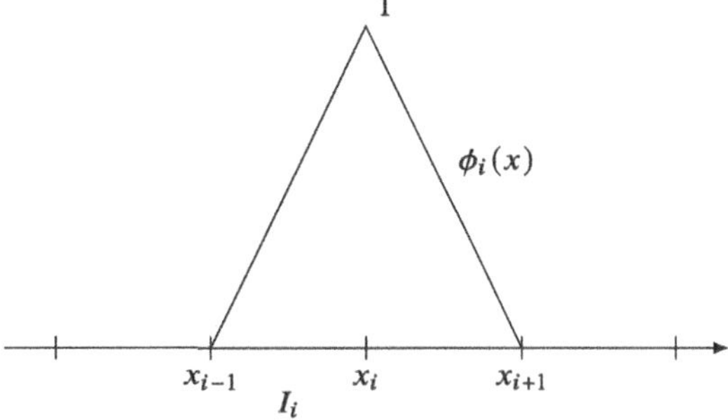

Fig. 9.3 Piecewise linear basis function, also known as hat function

$$s(x_i) = f(x_i), \quad i = 0, \ldots, n.$$

The application of the a priori error estimate for the Lagrange interpolation separately on each I_i yields the global a priori error estimate

$$\max_{x \in [a,b]} |f(x) - p(x)| \le \frac{1}{2} h^2 \max_{x \in [a,b]} |f''(x)|. \tag{9.7}$$

For small step sizes, it holds that

$$\max_{x \in [a,b]} |f(x) - s(x)| \to 0 \quad (h \to 0)$$

uniformly over the entire interval. ◀

The interpolant of the previous example is constructed using the so-called nodal basis of $S_h^{1,0}([a, b])$, consisting of the *hat functions* $\phi_i \in S_h^{1,0}[a, b]$, $i = 0, \ldots, n$, which are uniquely determined by the condition

$$\phi_i(x_j) = \delta_{ij}, \quad i, j = 0, \ldots, n.$$

The interpolant s of f then allows the representation in the form of

$$s(x) = \sum_{i=0}^{n} f(x_i)\phi_i(x).$$

This construction establishes the analogy to the Lagrangian representation of the Lagrange interpolation polynomial. In contrast to this *global* perspective, we work *locally* with the splines, since the hat functions ϕ_i are only non-zero in the subintervals directly adjacent to the respective support point x_i (see Fig. 9.3).

However, if the local polynomial degree is increased on each subinterval, higher smoothness can also be achieved globally using the interpolation conditions. This leads to the cubic splines, i.e., $k = 3$.

Example 9.21 (Piecewise Cubic Interpolation) Let a cubic polynomial be prescribed on each subinterval I_i, i.e., $k = 3$ with $r = 0$, i.e., so we $S_h^{3,0}[a, b]$. The interpolation conditions for the case $r = 0$ (i.e., global continuity) are

$$s(x_i) = f(x_i).$$

To uniquely determine a cubic polynomial, four conditions are necessary in each I_i, so that two additional interpolation points are given:

$$s(x_{ij}) = f(x_{ij}),$$

where $x_{ij} \in I_i, i = 1, \ldots, n$ and $j = 1, 2$. This piecewise cubic Lagrange interpolation uniquely defines a globally continuous function $s \in S_h^{3,0}[a, b]$.   ◄

Instead of enriching the interpolation condition by intermediate values $x_{ij} \in (x_{i-1}, x_i)$, it is also possible to specify derivative values:

Example 9.22 (Piecewise Cubic Hermite Interpolation) Again, let a cubic polynomial be prescribed on each subinterval I_i, i.e., $k = 3$, and this time with $r = 1$, i.e., we consider $S_h^{3,1}[a, b]$. The interpolation conditions for the case $r = 1$ (i.e., global continuity and on times globally differentiable) are

$$s(x_i) = f(x_i), \quad s'(x_i) = f'(x_i).$$

These four conditions uniquely determine $s \in S_h^{3,1}[a, b]$.   ◄

Theorem 9.23 (Piecewise Cubic Interpolation) *The piecewise cubic interpolant s (i.e., $k = 3$) with $r = 0$ or $r = 1$ to approximating the function f satisifies the following global a priori error estimate:*

$$\max_{x \in [a,b]} |f(x) - s(x)| \le \frac{1}{4!} h^4 \max_{x \in [a,b]} |f^{(4)}(x)|$$

Proof On each interval $I_i = [x_{i-1}, x_i]$ with $h_i = x_i - x_{i-1}$, the estimate for the Lagrange or Hermite interpolation can be applied:

$$\left. f(x) - s(x) \right|_{I_i} \le \frac{\max\limits_{I_i} |f^{(4)}(x)|}{4!} h_i^4$$

□

To uniquely determine a piecewise cubic function, four conditions are required on each interval I_i. However, the sets of conditions in the two considered examples differ. In the piecewise cubic Lagrange interpolation in Example 9.21, four conditions $s(x_{ij}) = f(x_{ij})$ are set in each of the n intervals, where the same interpolation condition appears twice at the interval ends x_i. In total, the piecewise cubic Lagrange interpolation is given by $3n + 1$ conditions. The piecewise Hermite interpolation in Example 9.22 also has four conditions per interval. At the interval ends, however, two conditions appear twice, resulting in a total of $2n + 2$ global conditions.

If the global regularity of the interpolation is to be further increased, the trivial approach of additionally requiring $s''(x_i) = f''(x_i)$ fails, as this would lead to six conditions per subinterval (with only four unknowns of a cubic polynomial). Instead of prescribing function values for the derivative and second derivative at each support point x_i, we only require the continuity of the derivatives:

$$\lim_{h \downarrow 0} s'(x_i + h) = \lim_{h \downarrow 0} s'(x_i - h), \quad \lim_{h \downarrow 0} s''(x_i + h) = \lim_{h \downarrow 0} s''(x_i - h).$$

We count the conditions and unknowns. The space $S_h^{(3,2)}$ has on each of the n intervals $I_i = [x_{i-1}, x_i]$ four unknown parameters

$$s(x)\big|_{I_i} = \alpha_0^{(i)} + \alpha_1^{(i)} x + \alpha_2^{(i)} x^2 + \alpha_3^{(i)} x^3.$$

The interpolation condition $s(x_i) = y_i$ provides a total of $n + 1$ conditions. Due to the required global continuity of function value, derivative, and second derivative, on each of the $n - 1$ interval transitions, three further conditions are added. Thus, there are $4n - 2$ conditions for the $4n$ unknown parameters. The problem is closed by specifying additional boundary conditions at points a and b. By specifying

$$s''(a) = s''(b) = 0$$

the *natural cubic spline* is defined.

The name *spline* comes from a flexible strip of wood or metal that shipbuilders historically used to draw smooth curves through a set of points. The tool was called spine. Imagine a flexible rod that is held at specific points but is allowed to take a free form in between. According to physical principles, this form will minimize the energy of the rod. The curvature of the rod gives the energy. The natural spline can be described as the energy-minimizing function, which is twice globally differentiable and fulfils the interpolation property at the nodes:

$$\min_{s \in S[a,b]} E(s)$$

with

$$E(s) := \int_a^b |s''(x)|^2 \mathrm{d}x, \quad S := \{\phi \in S_h^{3,2}, \ \phi(x_i) = y_i \text{ for } i = 0, \ldots, n\}.$$

This form of spline is the most commonly used spline interpolation task and has applications in computer graphics.

Definition 9.24 (*Cubic Spline*)　A function $s : [a, b] \to \mathbb{R}$ is called a *cubic* spline with respect to partition $a = x_0 < x_1 < \ldots < x_n = b$ if:

(i) $s \in C^2[a, b]$
(ii) $s|_{[x_{i-1}, x_i]} \in P_3, \quad i = 1, \ldots n.$

If additionally

(iii) $s''(a) = s''(b) = 0$

holds, the spline is called *natural*.

Theorem 9.25　(Natural Cubic Splines) *Let* $x_0, \ldots, x_n \in \mathbb{R}$ *be pairwise different nodes and* $y_k \in \mathbb{R}$ *be given. The natural cubic spline* $s \in S_h^{3,2}$ *described by the interpolation rule*

$$s(x_i) = y_i \in \mathbb{R}, \quad i = 0, \ldots, n \tag{9.8}$$

exists and to unique.

For every function $g \in S_h^{3,2}$ *with* $g(x_k) = y_k$, *for* $k = 0, \ldots, n$, *it holds*

$$\int_a^b |s''(x)|^2 dx \le \int_a^b |g''(x)|^2 dx.$$

In the case $y_k = f(x_k)$ *for a function* $f \in C^4[a, b]$ *the a priori error estimate holds*

$$\max_{a \le x \le b} |f(x) - s(x)| \le \frac{1}{2} h^4 \max_{a \le x \le b} |f^{(4)}(x)|.$$

Proof (*i*) *Determination of the spline*: We show that the spline is given as solution of a linear system. On each interval $I_i := [x_{i-1}, x_i]$ we write

$$s(x)\Big|_{I_i} = a_0^{(i)} + a_1^{(i)}(x - x_i) + a_2^{(i)}(x - x_i)^2 + a_3^{(i)}(x - x_i)^3, \quad i = 1, \ldots, n.$$

It is necessary to determine the $4n$ coefficients $a_j^{(i)}$. The first coefficient can be read directly as $a_0^{(i)} = y_i$:

$$y_i = s(x_i) = a_0^{(i)}, \quad i = 1, \ldots, n$$

For the $3n$ unknown coefficients, the conditions remain

$$s(x_{i-1}) = y_{i-1} \qquad y_{i-1} = a_0^{(i)} - a_1^{(i)}h_i + a_2^{(i)}h_i^2 - a_3^{(i)}h^3 \qquad i = 1,\ldots,n$$

$$\left. s'(x_i)\right|_{I_i} = \left. s'(x_i)\right|_{I_{i+1}} \qquad a_1^{(i)} = a_1^{(i+1)} - 2a_2^{(i+1)}h_i + 3a_3^{(i+1)}h_i^2 \qquad i = 1,\ldots,n-1$$

$$\left. s''(x_i)\right|_{I_i} = \left. s''(x_i)\right|_{I_{i+1}} \qquad 2a_2^{(i)} = 2a_2^{(i+1)} - 6a_3^{(i+1)}h_i \qquad i = 1,\ldots,n-1$$

$$s''(x_0) = 0 \qquad 0 = a_2^{(1)} - 2a_3^{(1)}h_1 \qquad i = 0$$

$$s''(x_n) = 0 \qquad 0 = a_2^{(n)} \qquad i = n,$$

i.e., $n + 2(n-1) + 2 = 3n$ equations. With appropriate numbering, the system can be written as a band matrix with bandwidth 3. For efficient determination of the spline, the system can be further reduced by reformulating the conditions, resulting in a tridiagonal system.

(ii) Existence and uniqueness: Since the spline is determined as a solution of a system of linear equations, it is sufficient to show uniqueness. We consider the space

$$N := \{\phi \in S_h^{3,2}, \ \phi(x_i) = 0, \ i = 0, \ldots, n\}.$$

On this space, a scalar product is given by

$$(u, v)_N := \int_a^b u''(x)v''(x)\mathrm{d}x.$$

Linearity, symmetry, and positivity are clear. Homogeneity follows, as from $(v, v)_N = 0$ it immediately follows pointwise that $v'' = 0$. This means v is a linear function. From $v(a) = v(b) = 0$ it then follows that $v = 0$.

Let s_1, s_2 be two splines to the support points $\{(x_i, y_i), \ i = 0, \ldots, n\}$. Then $s = s_1 - s_2 \in N$ and it holds for $w \in N$ with integration by parts (note that $s \in C^\infty(I_i)$, since polynomial)

$$\begin{aligned}
(s, w)_N &= \int_a^b s''w''\mathrm{d}x = \sum_{i=1}^{N}\left\{\left. s''w'\right|_{x_{i-1}}^{x_i} - \int_{I_i} s'''(x)w'(x)\mathrm{d}x\right\} \\
&= \sum_{i=1}^{N}\left\{\left. s''w'\right|_{x_{i-1}}^{x_i} - s'''\underbrace{\left. w\right|_{x_{i-1}}^{x_i}}_{=0} + \int_{I_i} \underbrace{s^{(4)}(x)}_{=0}\, w(x)\mathrm{d}x\right\} \\
&= \sum_{i=1}^{n}\left\{s''(x_i)w'(x_i) - s''(x_{i-1})w'(x_{i-1})\right\} \\
&= s''(b)w'(b) - s''(a)w'(a) = 0,
\end{aligned} \tag{9.9}$$

since $s''(a) = s_1''(a) - s_2''(a) = 0$. This argument applies analogously for b. From this follows

$$(s, s)_N = 0 \quad \Rightarrow \quad s'' = 0 \quad \Rightarrow \quad s = 0,$$

and thus the uniqueness $s_1 = s_2$ and hence also the existence of a solution.

(iii) Energy minimization: The relation

$$(s, w)_N = 0$$

does not only hold for all functions $s \in N$ but for all natural splines satisfying $s''(a) = s''(b) = 0$ and that satisfy (9.8). Thereby, the space N is orthogonal to the spline s. Let $f \in S_h^{3,2}$ with $y_i = f(x_i)$ for $i = 0, \ldots, n$. Then $w := s - f \in N$ and

$$0 = (s, s - f)_N = (s, s)_N - (s, f)_N,$$

thus

$$\int_a^b s''(x)^2 \mathrm{d}x = \int_a^b s''(x) f''(x) \mathrm{d}x \leq \left(\int_a^b s''(x)^2 \mathrm{d}x \right)^{\frac{1}{2}} \left(\int_a^b f''(x)^2 \mathrm{d}x \right)^{\frac{1}{2}},$$

from which the energy minimization follows.

(iv) Error bound: The proof of the a priori error estimate is involved, and we refer to the literature [84]. $\qquad\square$

The minimization of the second derivative can be understood as a physical principle. However, for numerical purposes, it is relevant that the minimization of the second derivatives also suppresses oscillations. The natural spline has excellent stability properties. To calculate the spline, a linear system must be solved. The equations are given in the proof of Theorem 9.25. The values $a_0^{(i)} = y_i$ can be determined immediately. Furthermore, in a preparatory step, the unknowns $a_1^{(i)}$ and $a_3^{(i)}$ can be expressed through the $a_2^{(i)}$. This leaves a linear system in the n unknowns $a_2^{(1)}, \ldots, a_2^{(n)}$. The matrix has a tridiagonal shape and can be solved very efficiently.

9.3　Numerical Differentiation

In this section, we deal with the numerical approximation of the derivatives of a function $f : [a, b] \to \mathbb{R}$. We assume that the analytical evaluation of the derivative is either not possible or very costly, which might be the case if evaluating $f(x_0)$ might already be based on a complex algorithm. The following procedure describes the basic idea:

Algorithm 9.26: Numerical Differentiation

> **Input:** Function $f \in C(a, b)$.
> 1 Interpolate $f(x)$ by polynomial $p \in P_m$ with $m \geq n$
> 2 Calculate $p^{(n)}(x_0)$
> **Result:** Approximation $f^{(n)}(x_0) \approx p^{(n)}(x_0)$

In the following, we develop some simple methods for approximating derivatives, which are based on interpolation.

9.3.1 Approximation of the First Derivative

We interpolate a function $f(x)$ linearly in the two nodes x_0 and x_1

$$p_1(x) = \frac{x - x_1}{x_0 - x_1} f(x_0) + \frac{x - x_0}{x_1 - x_0} f(x_1)$$

and approximate the derivative $f'(x)$ by the corresponding derivative of the polynomial

$$p_1'(x) = \frac{f(x_1) - f(x_0)}{x_1 - x_0}.$$

In the followineg we study, how well $p_1'(x)$ approximates the true derivative $f'(x)$. For the linear interpolation polynomial, the derivative is a constant function. Thus, we obtain the same approximation to the derivative for all values x.

Theorem 9.27 (Difference Quotient for the First Derivative) *Let $f \in C^2[a, b]$. For arbitrary points $x_0, x_1 \in [a, b]$ with $h := x_1 - x_0 > 0$ the one-sided difference quotients (forward and backward) satisfies*

$$\frac{f(x_1) - f(x_0)}{h} = f'(x_0) + \frac{1}{2}hf''(x_0) + O(h^2),$$

$$\frac{f(x_1) - f(x_0)}{h} = f'(x_1) - \frac{1}{2}hf''(x_0) + O(h^2).$$

In the case $f \in C^4[a, b]$, for $x_0, x_0 - h, x_0 + h \in [a, b]$ the central difference quotient yields the improved estimate

$$\frac{f(x_0 + h) - f(x_0 - h)}{2h} = f'(x_0) + \frac{1}{6}h^2 f'''(x_0) + O(h^4).$$

Proof *(i)* With Taylor expansion, we have

$$f(x_0 \pm h) = f(x_0) \pm hf'(x_0) + \frac{1}{2}h^2 f''(x_0) \pm \frac{1}{6}h^3 f'''(x_0) + \frac{1}{24}h^4 f^{(4)}(\xi)$$

with a $\xi \in [a, b]$. From this follows, for example, for the forward difference quotient

$$\frac{f(x_0 + h) - f(x_0)}{h} = \frac{hf'(x_0) + \frac{1}{2}h^2 f''(x_0) + O(h^3)}{h}$$

$$* = f'(x_0) + \frac{1}{2}hf''(x_0) + O(h^2).$$

The same is easily shown for the backward difference quotient.

(ii) Symmetry effects in the Taylor extension of $f(x_0 + h)$ and $f(x_0 - h)$ leads to cancellation of all even powers in h

$$f(x_0 + h) - f(x_0 - h) = 2hf'(x_0) + \frac{1}{3}h^3 f'''(x_0) + O(h^5).$$

$\square$

Numerical methods that converge better in individual points—or in special situations—than in the general case, are called *superconvergent*. The central difference quotient exhibits superconvergent behavior, which can be explained by choosing the midpoint with symmetric Taylor expansions to the left and right. A "central approximation" in x_0 using the two points $x_0 - h$ and $x_0 + 2h$ is not of second order:

$$\frac{f(x_0 + 2h) - f(x_0 - h)}{3h}$$

$$= f'(x_0) + \left(2hf''(x_0) + \frac{8}{6}h^2 f'''(x_0)\right) - \left(\frac{1}{2}hf''(x_0) - \frac{1}{6}h^2 f'''(x_0)\right) + O(h^3)$$

$$= f'(x_0) + \frac{3}{2}hf''(x_0) + O(h^2).$$

Terms of higher order in h do not cancel each other out, and the central, but asymmetric difference quotient (which is actually not really central) converges with order one only.

9.3.2 Approximation of the Second Derivative

We interpolate $f(x)$ using the three nodes $x_0 - h, x_0, x_0 + h$ through a quadratic polynomial:

$$p_2(x) = f(x_0 - h)\frac{(x_0 - x)(x_0 + h - x)}{2h^2}$$

$$+ f(x_0)\frac{(x - x_0 + h)(x_0 + h - x)}{h^2} + f(x_0 + h)\frac{(x - x_0 + h)(x - x_0)}{2h^2}$$

In $x = x_0$, the first and second derivatives are

$$p_2'(x_0) = \frac{f(x_0 + h) - f(x_0 - h)}{2h}, \quad p_2''(x_0) = \frac{f(x_0 + h) - 2f(x_0) + f(x_0 - h)}{h^2}.$$

Using a quadratic approximation does not improve the first-order quotient. Instead, the known symmetric approximation scheme is recovered. For the second derivative, the Taylor expansion gives

Table 9.1 Difference approximation of $f'(1/2)$ (two tables to the left) and $f''(1/2)$ (right) of the function $f(x) = \tanh(x)$. In each case, the one-sided or the central difference quotient was used

h	$\dfrac{f(\frac{1}{2}+h)-f(\frac{1}{2})}{h}$	$\dfrac{f(\frac{1}{2}+h)-f(\frac{1}{2}-h)}{2h}$	$\dfrac{f(\frac{1}{2}+2h)-2f(\frac{1}{2}+h)+f(\frac{1}{2})}{h^2}$	$\dfrac{f(\frac{1}{2}+h)-2f(\frac{1}{2})+f(\frac{1}{2}-h)}{h^2}$
$\frac{1}{2}$	0.598954	0.761594	-0.623692	-0.650561
$\frac{1}{4}$	0.692127	0.780461	-0.745385	-0.706667
$\frac{1}{8}$	0.739861	0.784969	-0.763733	-0.721740
$\frac{1}{16}$	0.763405	0.786079	-0.753425	-0.725577
$\frac{1}{32}$	0.775004	0.786356	-0.742301	-0.726540
$\frac{1}{64}$	0.780747	0.786425	-0.735134	-0.726782
Exact	0.786448	0.786448	-0.726862	-0.726862

$$p_2''(x_0) = \frac{-2f(x_0) + f(x_0 - h) + f(x_0 + h)}{h^2} = f''(x_0) + \frac{1}{12}h^2 f^{(4)}(x_0) + O(h^4),$$

which shows second order..

We can also derive a one-sided difference quotient for the second derivative based on p_2. This is done by approximating $f''(x_0 - h) \approx p_2''(x_0 - h)$. Taylor expansion shows that this approximation is only first order

$$p''(x_0 - h) = f''(x_0 - h) + hf'''(x_0 - h) + O(h^2).$$

Example 9.28 Let

$$f(x) = \tanh(x).$$

We are looking for an approximation of

$$f'\left(\frac{1}{2}\right) \approx 0.7864477329659274, \quad f''\left(\frac{1}{2}\right) \approx -0.7268619813835874.$$

For approximation, we use the four difference quotients discussed so far for different step sizes $h > 0$, see Table 9.1.

In Fig. 9.4, we plot the errors of the approximations against the step size. Here, the difference between linear and quadratic order in h is clearly visible. ◄

9.3.3 Stability of Numerical Differentiation

Finally, we examine the stability of the difference approximation. The coefficients of the various formulas change sign, so there is a risk of cancellation. As an example, we perform the stability analysis for the central difference approximation to determine the first derivative.

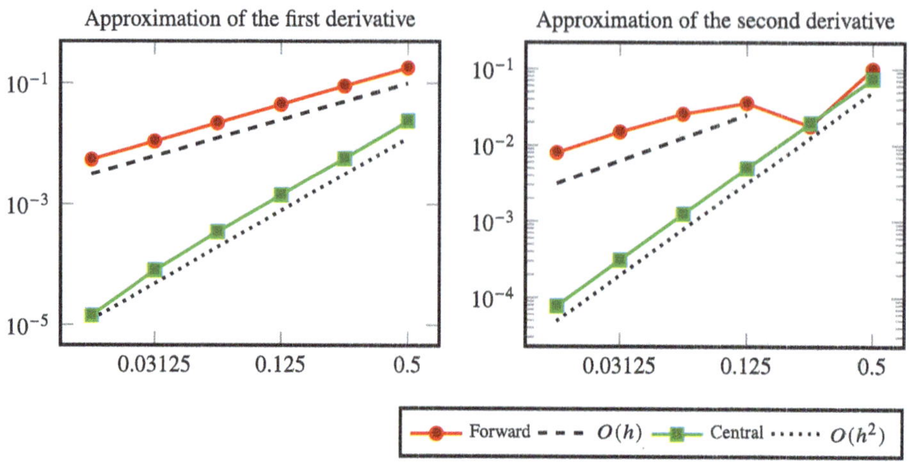

Fig. 9.4 Error in the difference approximation of the first (left) and second (right) derivative of $f(x) = \tanh(x)$ at the point $x_0 = \frac{1}{2}$

We assume that the function evaluation is distorted with a relative error $|\epsilon_i| \le \epsilon$

$$\tilde{p}'(x_0) = \frac{f(x_0 + h)(1 + \epsilon_1) - f(x_0 - h)(1 + \epsilon_2)}{2h}(1 + \epsilon_3)$$

$$= \frac{f(x_0 + h) - f(x_0 - h)}{2h}(1 + \epsilon_3) + \frac{\epsilon_1 f(x_0 + h) - \epsilon_2 f(x_0 - h)}{2h} + O(\epsilon^2).$$

The relative error is computed as

$$\left| \frac{\tilde{p}'(x_0) - p'(x_0)}{p'(x_0)} \right| \le \epsilon + \frac{|f(x_0 + h)| + |f(x_0 - h)|}{|f(x_0 + h) - f(x_0 - h)|}\epsilon + O(\epsilon^2).$$

and especially $f(x_0 + h) \approx f(x_0 - h)$, i.e., $f'(x_0) \approx 0$ can lead to an arbitrary amplification of the error. The smaller we choose h, the larger this effect becomes, as it holds

$$f(x_0 + h) - f(x_0 - h) = 2hf'(x_0) + O(h^3).$$

This rounding error counteracts the approximation error, as seen in Theorem 9.27. The total error is the sum of both components

$$\frac{|f'(x_0) - \tilde{p}'(x_0)|}{|f'(x_0)|} \le \frac{|f'(x_0) - p'(x_0)|}{|f'(x_0)|} + \frac{|p'(x_0) - \tilde{p}'(x_0)|}{|f'(x_0)|}$$

$$= \frac{|f'(x_0) - p'(x_0)|}{|f'(x_0)|} + \frac{|p'(x_0) - \tilde{p}'(x_0)|}{|p'(x_0)|} \cdot \frac{|p'(x_0)|}{|f'(x_0)|}$$

$$\le h^2 \frac{|f'''(x_0)|}{6|f'(x_0)|} + O(h^4) + \frac{\max_\xi |f(\xi)|}{|f'(x_0)|h}\epsilon + O(\epsilon^2).$$

Best performance is given for balanced error contributions, i.e., in the case

$$h^2 \approx \frac{\epsilon}{h} \quad \Rightarrow \quad h \approx \sqrt[3]{\epsilon}.$$

For $\epsilon \approx 10^{-16}$, step sizes $h < 10^{-5}$ are therefore not sensible. In general, numerical differentiation is less stable than numerical integration.

9.4 Richardson Extrapolation

An important application of interpolation is the *extrapolation to the limit*. The idea can be explained most easily using an example. Assume, we want to calculate the derivative $f'(x_0)$ of a function f at the point x_0 using the one-sided difference quotient:

$$a(h) := \frac{f(x_0 + h) - f(x_0)}{h}$$

The step size parameter h determines the quality of the approximation $a(h) \approx f'(x_0)$. For $h \to 0$ (neglecting rounding errors) $a(h) \to f'(x_0)$. However, $a(h)$ cannot be evaluated at the point $h = 0$. The idea of extrapolation to the limit is to construct an interpolation polynomial $p(x)$ through the support points $(h_i, a(h_i))$ for a sequence of step sizes $h_0, h_1, \ldots, h_n$ and to use the value $p(0)$ as an approximation for $a(0)$.

The fundamental question arises whether the interpolating function at the nodes $h_0, \ldots, h_n$ also has significance outside the interval $I := [\min_i\{h_i\}, \max_i\{h_i\}]$. Example 9.18 does not initially suggest this. Here, the approximation property was already strongly disturbed at the boundary of the interval I due to oscillations in points between the nodes. Nevertheless, we consider a simple example:

Example 9.29 (Extrapolation of the Forward Difference Quotient) Let $f(x) = \tanh(x)$ and we want to evaluate the derivative at the point $x_0 = 1/2$. The exact value is $f'(1/2) \approx 0.786448$. For this, we define the difference quotient

$$a(h) = \frac{\tanh(x + h) - \tanh(x)}{h}$$

and evaluate for the step sizes $h = 2^{-1}, 2^{-2}, 2^{-3}, 2^{-4}$. The results are shown in Table 9.2 on the left side.

Formally, we can construct an interpolation polynomial $p(h)$ through the points $h_i = 2^{-i}$ and $a_i = a(h_i)$ and evaluate this at the point $h = 0$. This is done efficiently with the Neville scheme for the point $h = 0$:

$$a_{k,k} = a_i, \qquad\qquad\qquad\qquad i = 1, \ldots, n,$$

$$a_{k,k+l} = a_{k,k+l-1} - h_k \frac{a_{k+1,k+l} - a_{k,k+l-1}}{h_{k+l} - h_k}, \quad l = 1, \ldots, n-1, \; k = 1, \ldots, n-l$$

Table 9.2 Extrapolation of the one-sided difference quotient to approximate $\tanh'(1/2)$. By $a(h)$ we denote the value of the difference quotient, and on the right we show all approximations given by the Neville extrapolation scheme

h_i	$a(h)$	$a_i = p_{ii}$	$p_{i,i+1}$	$p_{i,i+2}$	$p_{i,i+3}$
2^{-1}	0.5989	0.5989	0.7853	0.78836	0.7865
2^{-2}	0.6921	0.6921	0.7976	0.78673	
2^{-3}	0.7399	0.7399	0.7869		
2^{-4}	0.7634	0.7634			

The results of extrapolation are shown in the right part of Table 9.2 and the exact digits are underlined. Thus, we see that the accuracy can indeed be improved by extrapolation. ◄

We want to examine this example for the simplest case of a linear interpolation. For $f \in C^3([a, b])$, let

$$a(h) = \frac{f(x_0 + h) - f(x_0)}{h} = f'(x_0) + \frac{h}{2} f''(x_0) + \frac{h^2}{6} f'''(\xi_{x_0,h}) \tag{9.10}$$

be the one-sided approximation of the first derivative. We define the linear interpolation polynomial through the support points $(h, a(h))$ and $(h/2, a(h/2))$ as

$$p(t) = \left(\frac{t - \frac{h}{2}}{h - \frac{h}{2}} \right) a(h) + \left(\frac{t - h}{\frac{h}{2} - h} \right) a\left(\frac{h}{2} \right)$$

and evaluate this at the point $t = 0$ as an approximation to $a(0)$:

$$a(0) \approx p(0) = 2a\left(\frac{h}{2} \right) - a(h)$$

For $a(h/2)$ and $a(h)$, we insert the Taylor expansion (9.10) and obtain with

$$p(0) = 2\left(f'(x_0) + \frac{h}{4} f''(x_0) + \frac{h^2}{24} f'''(\xi_{x_0,h/2}) \right)$$
$$- \left(f'(x_0) + \frac{h}{2} f''(x_0) + \frac{h^2}{6} f'''(\xi_{x_0,h}) \right)$$
$$= f'(x_0) + O(h^2),$$

which is an approximation of the first derivative of second order in the step size h.

Higher-order methods can be generated by extrapolation. Such a procedure, in which the results of a numerical method are further processed, is called *postprocessing*.

Theorem 9.30 (Simple Extrapolation) *Let $a(h) : \mathbb{R}_+ \to \mathbb{R}$ be a $(n+1)$-times continuously differentiable function with the expansion*

$$a(h) = a_0 + \sum_{j=1}^{n} a_j h^j + a_{n+1}(h)h^{n+1}$$

with coefficients $a_j \in \mathbb{R}$ and $a_{n+1}(h_{k+i}) = a_{n+1} + o(1)$. Furthermore, let $(h_k)_{k=0,1,\ldots}$ with $h_k \in \mathbb{R}_+$ be a monotonically decreasing step size sequence with the property

$$0 < \frac{h_{k+1}}{h_k} \le \rho < 1. \tag{9.11}$$

Let $p_n^{(k)} \in P_n$ be the polynomial interpolating

$$(h_k, a(h_k)), \ldots, (h_{k+n}, a(h_{k+n})).$$

Then it holds

$$a(0) - p_n^{(k)}(0) = O(h_k^{n+1}) \quad (k \to \infty).$$

Proof In Lagrange representation, it holds

$$p_n^{(k)}(t) = \sum_{i=0}^{n} a(h_{k+i})L_{k+i}^{(n)}(t), \quad L_{k+i}^{(n)} = \prod_{l=0,l\neq i}^{n} \frac{t - h_{k+l}}{h_{k+i} - h_{k+l}}. \tag{9.12}$$

We insert the expansion of $a(h)$ into the polynomial representation (9.12) and obtain at $t = 0$

$$p_n^{(k)}(0) = \sum_{i=0}^{n} \left\{ a_0 + \sum_{j=1}^{n} a_j h_{k+i}^j + a_{n+1}(h_{k+i})h_{k+i}^{n+1} \right\} L_{k+i}^{(n)}(0).$$

The Lagrange basis polynomials satisfy

$$\sum_{i=0}^{n} h_{k+i}^r L_{k+i}^{(n)}(0) = \begin{cases} 1 & r = 0, \\ 0 & r = 1, \ldots, n, \\ (-1)^n \prod_{i=0}^{n} h_{k+i} & r = n+1. \end{cases} \tag{9.13}$$

This is shown by analyzing the interpolation of the polynomial $p(x) = x^r$ with

$$p(0) = \begin{cases} 1 & r = 0 \\ 0 & r > 0 \end{cases}$$

for various exponents r. With (9.13) and $a_{n+1}(h_{k+i}) = a_{n+1} + o(h_{k+i})$ follows

$$p_n^{(k)}(0) = a_0 + a_{n+1}(-1)^n \prod_{i=0}^{n} h_{i+k} + \sum_{i=0}^{n} o(1) h_{k+i}^{n+1} L_{k+i}^{(n)}(0).$$

With the step size condition (9.11) $h_{k+1} \leq \rho h_k$ and $\rho < 1$, the Lagrange basis polynomials at $t = 0$ satisfy

$$|L_{k+i}^{(n)}(0)| = \prod_{l=0, l \neq i}^{n} \left| \frac{1}{\frac{h_{k+i}}{h_{k+l}} - 1} \right| \leq c(\rho, n), \tag{9.14}$$

since $|h_{k+1}/h_{k+l}|$ is bounded away from 1. Overall, with $h_{k+i} \leq h_k$, it holds

$$\left| p_n^{(k)} - a(0) \right| = O(h_k^{n+1}).$$

$$\square$$

The step size condition is necessary for estimating $|L_{k+i}^{(n)}(0)| = O(1)$ and prevents strong oscillations of the basis functions at $t = 0$ as observed in Example 9.18. An admissible sequence of steps for extrapolation is $h_i = \frac{h_0}{2^i}$. The simple choice $h_i = h_0/i$ is not permissible, as it gives $h_{k+1}/h_k \to 1$ and an estimate in step (9.14) does not hold uniform in k.

To apply extrapolation to a given method, the approximations $p_n^{(k)}(0)$ are calculated using the Neville scheme from Algorithm 9.9. We extend Example 9.29.

Example 9.31 (Extrapolation of the Central Difference Quotient) We approximate the derivative of $f(x) = \tanh(x)$ at the point $x_0 = 0.5$ and calculate the approximations with the central difference quotient. The exact value is $\tanh'(0.5) \approx 0.78645$. We obtain

h	$a(h) = a_{kk}$	$a_{k,k+1}$	$a_{k,k+2}$	$a_{k,k+3}$
2^{-1}	**0.76**16	**0.7**993	**0.786**20	**0.78645**93
2^{-2}	**0.78**05	**0.78**95	**0.78643**	
2^{-3}	**0.78**50	**0.78**72		
2^{-4}	**0.7861**			

The approximation is improved. However, it turns out that in the first step, i.e., for all values $a_{k,k+1}$, which result from linear interpolation of two values, no improvement can be achieved. ◀

In the preceding example, the central difference quotient is extrapolated. Given $f \in C^\infty$, the series expansion is

$$a(h) = \frac{f(x+h) - f(x-h)}{2h}$$

$$= \frac{1}{2h}\left(\sum_{k=0}^{\infty}\frac{f^{(k)}(x)}{k!}h^k - \sum_{k=0}^{\infty}\frac{f^{(k)}(x)}{k!}(-h)^k\right) = \sum_{k=0}^{\infty}\frac{f^{(2k+1)}(x)}{(2k+1)!}h^{2k}.$$

That is, $a(h)$ has an expansion in even powers of h. This explains why interpolation with linear polynomials does not yield any gain. Instead, we will exploit the known power expansion of $a(h)$ and look for interpolation polynomials in h^2

$$p(h) = a_0 + a_1 h^2 + a_2 h^4 + \cdots .$$

Example 9.32 (Extrapolation of the Central Difference Quotient With Quadratic Polynomials) We approximate the derivative of $f(x) = \tanh(x)$ and interpolate the already calculated values $(a(h_3), h_3)$ and $(a(h_4), h_4)$ using a linear polynomial in h^2. It holds

$$p(h) = a(h_3)\frac{h^2 - h_4^2}{h_3^2 - h_4^2} + a(h_4)\frac{h^2 - h_3^2}{h_4^2 - h_3^2} \quad\Rightarrow\quad p(0) = \frac{h_3^2 a(h_4) - h_4^2 a(h_3)}{h_3^2 - h_4^2}.$$

With the values from Example 9.31, we get

$$p(0) = \mathbf{0.7864}493,$$

i.e., after only one step with the correct order, the first four digits are already exact.   ◀

The Richardson extrapolation fully plays out its strength when the underlying function $a(h)$ has a special series expansion that contains, for example, only even powers of h. Given sufficient regularity of f, we have

$$a(h) := \frac{f(x_0 + h) - f(x_0 - h)}{2h} = f'(x_0) + \sum_{i=1}^{n}\frac{f^{(2i+1)}(x_0)}{(2i+1)!}h^{2i} + O(h^{2n+1}).$$

The idea is to exploit this expansion in interpolation and only consider polynomials in $1, h^2, h^4, \ldots$. The central difference quotient is not an isolated case. In numerical quadrature, we will get to know the Romberg method, which is also based on the extrapolation of a method with error expansion in h^2. In the context of ordinary differential equations, Gragg's extrapolation method is also based on this principle.

Theorem 9.33 (Richardson Extrapolation to the Limit) *Let $a : \mathbb{R}_+ \to \mathbb{R}$ be a function that is $q(n+1)$ times continuously differentiable for $q > 0$ and that allows for the expansion*

$$a(h) = a_0 + \sum_{j=1}^{n} a_j h^{qj} + a_{n+1}(h)h^{q(n+1)}$$

Table 9.3 Extrapolation of the central difference for $\tanh'(0.5)$

h	$a(h) = p_{kk}$	$p_{k,k+1}$	$p_{k,k+2}$	$p_{k,k+3}$
2^{-1}	**0.7**61594156	**0.7867**493883	**0.7864**537207	**0.7864477**443
2^{-2}	**0.78**0460580	**0.78647**21999	**0.7864478**377	
2^{-3}	**0.784**969295	**0.78644**93603		
2^{-4}	**0.786**079344			

with coefficients $a_j \in \mathbb{R}$ and $a_{n+1}(h_{k+i}) = a_{n+1} + o(1)$. Furthermore, let $(h_k)_{k=0,1,\ldots}$ be a monotonically decreasing sequence of step sizes $h_k > 0$ with the property

$$0 < \frac{h_{k+1}}{h_k} \le \rho < 1. \tag{9.15}$$

Let $p_n^{(k)} \in P_n$ (in h^q) be the interpolation to

$$(h_k^q, a(h_k)), \ldots, (h_{k+n}^q, a(h_{k+n})).$$

Then it holds

$$a(0) - p_n^{(k)}(0) - O(h_k^{q(n+1)}) \quad (k \to \infty).$$

Proof The proof is a simple modification of Theorem 9.30 and can essentially be completed using the substitution $\tilde{h}_k = h_k^q$. For details, we refer to [75]. $\qquad\square$

In principle, Richardson extrapolation allows for an arbitrary increase in the order of the method given sufficient regularity. However, usually only a few extrapolation steps with $h_k, h_{k+1}, \ldots, h_{k+n}$ are used. The extrapolation is most easily performed with the Neville scheme at the point $\xi = 0$, see Algorithm 9.9.

We conclude the Richardson extrapolation with two another examples:

Example 9.34 (Extrapolation of the Central Difference Quotient) We approximate the derivative of $f(x) = \tanh(x)$ at the point $x = 0.5$ with the central difference quotient and extrapolation. The exact value is $\tanh'(0.5) \approx 0.7864477329$. The results are shown in Table 9.3. Already, the first extrapolation $p_{k,k+1}$ provides far better results than the extrapolation of the forward difference quotient. The best approximation $p_{k,k+3}$ has seven correct significant digits. ◀

Table 9.4 Extrapolation of the central difference approximation with suboptimal order $q = 1$

h	$a(h_k) = p_{kk}$	$a(h_k) - f'(1)$	$p_{k,k+1}$	$p_{k,k+1} - f'(1)$	$p_{k,k+2}$	$p_{k,k+2} - f'(1)$
2^{-2}	60.03	1.29×10^1	78.61	5.68×10^0	73.36353	4.36×10^{-1}
2^{-3}	69.32	3.60×10^0	74.68	1.75×10^0	**72.95**825	3.12×10^{-2}
2^{-4}	**72.**00	9.28×10^{-1}	73.39	4.61×10^{-1}	**72.92**908	2.02×10^{-3}
2^{-5}	**72.**69	2.34×10^{-1}	73.04	1.17×10^{-1}	**72.927**19	1.27×10^{-4}
2^{-6}	**72.**87	5.85×10^{-2}	**72.9**6	2.93×10^{-2}		
2^{-7}	**72.9**1	1.46×10^{-2}				

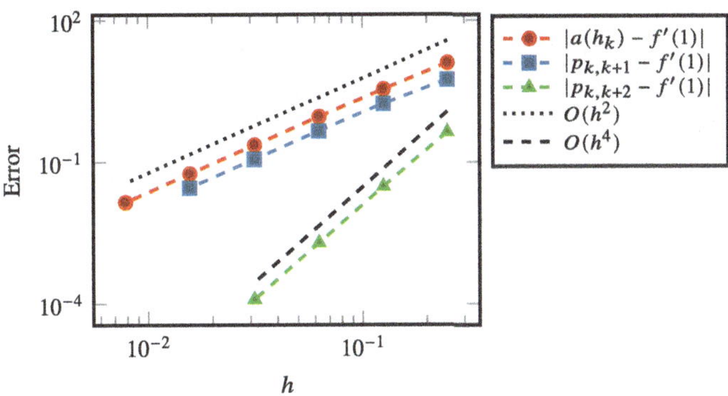

Fig. 9.5 Convergence of the extrapolation of the central difference approximation with suboptimal order $q = 1$

Example 9.35 (Extrapolation with Different Error Orders) We consider

$$f(x) = -e^{1-\cos(\pi x)}$$

and approximate $f''(1) \approx 72.9270606$ with the central difference quotient for the second derivative

$$a(h) = \frac{f(x+h) - 2f(x) + f(x-h)}{h^2}.$$

We perform two steps of extrapolation each time. Initially, we choose the suboptimal order $q = 1$. The results can be seen in Table 9.4 and Fig. 9.5. A single step of extrapolation does not provide any advantage. This can be explained by the fact that the purely quadratic error development of the difference quotient cannot be approximated by a linear polynomial. We repeat the calculation and choose—adapted to the difference quotient—the extrapolation order $q = 2$. The results can be seen in Table 9.5 and Fig. 9.6. ◄

Table 9.5 Extrapolation of the central difference approximation with optimal order $q = 2$

h	$a(h_k) = p_{kk}$	$a(h_k) - f'(1)$	$p_{k,k+1}$	$p_{k,k+1} - f'(1)$	$p_{k,k+2}$	$p_{k,k+2} - f'(1)$
2^{-2}	60.03	1.29×10^1	72.41896	5.08×10^{-1}	72.922734	4.33×10^{-3}
2^{-3}	69.32	3.60×10^0	72.89125	3.58×10^{-2}	72.926985	7.56×10^{-5}
2^{-4}	72.00	9.28×10^{-1}	72.92475	2.31×10^{-3}	72.927059	4.22×10^{-6}
2^{-5}	72.69	2.34×10^{-1}	72.92692	1.45×10^{-4}	72.927060	1.91×10^{-8}
2^{-6}	72.87	5.85×10^{-2}	72.92705	9.11×10^{-6}		
2^{-7}	72.91	1.46×10^{-2}				

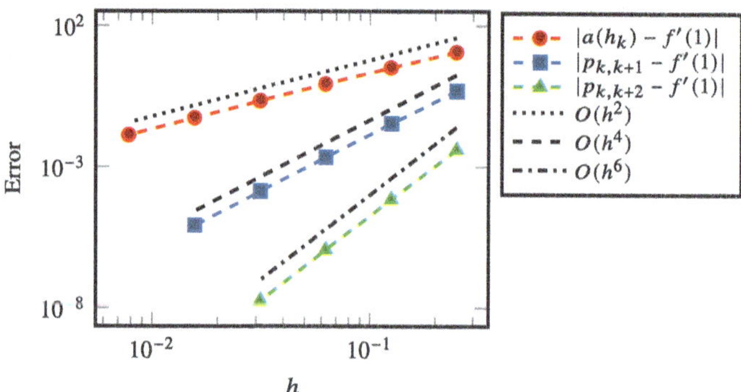

Fig. 9.6 Convergence of the extrapolation of the central difference approximation with optimal order $q = 2$

9.5 Best Approximation

Approximation is a concept generalizing the idea of interpolation. Again, we look for a simple function $p \in P$ (where the function space P is usually again the space of polynomials), which approximates data or a given function f as well as possible. However, instead of an exact approximation in points $y_i \in \mathbb{R}$, the measure for can be more general, and usually we understand the *best approximation* to be the minimal distance from p to f (or to the discrete data) in a norm

$$\|f - p\| = \min_{q \in P} \|f - q\|.$$

It turns out that the task is highly dependent on the choice of the norm. The Lagrange interpolation can also be formulated as a special type of best approximation:

Remark 9.36 (Lagrange Interpolation as Approximation) Let $x_0, x_1, \ldots, x_n$ be $n + 1$ pairwise different points. The best approximation in the "norm"

$$|p|_n := \max_{i=0,\dots,n} |p(x_i)|$$

is equivalent to the Lagrange interpolation. ($|\cdot|_n$ is indeed a norm on the polynomial space, but not on general spaces of continuous functions.) This example, however, directly shows the greater flexibility of the best approximation. Even if more than n pairwise different points are given, the best approximation problem $|f - p|_n \le |f - q|_n$ is meaningful.   ♦

In the following, we will consider the L^2-norm

$$\|f\|_{L^2((a,b))} := \sqrt{(f,f)} = \left(\int_a^b f(x)^2 \mathrm{d}x \right)^{\frac{1}{2}},$$

and the maximum norm

$$\|f\|_\infty := \max_{x \in [a,b]} |f(x)|.$$

The best approximation in the L^2-norm using trigonometric polynomials is the *Fourier series*. The approximation with respect to the maximum norm is of great importance in practical applications as it solves the problem of unwanted oscillations in parts of the interval which were typical for polynomial interpolation (see Example 9.18). If the error is under control in the maximum norm, it means that we have uniform approximation properties on the whole interval. The maximum-norm best approximation is called the Chebychev approximation, see Sect. 9.5.2.

9.5.1 Gauss Approximation—Best Approximation in the L^2-Norm

Definition 9.37 (*Gauss Approximation*) Let $I = (a,b)$ and $f \in L^2(I)$. Let $V \subset L^2(I)$ be a finite-dimensional subspace. The best approximation in the $L^2(I)$-norm

$$\|f - p\|_{L^2(I)} \le \min_{q \in V} \|f - q\|_{L^2(I)}$$

is called *Gauss Approximation*.

Here, we will consider the space of polynomials $V = P_n$. This space is a vector space with a scalar product, and such spaces are called *pre-Hilbert spaces*. The scalar product is used to describe orthogonality relationships. For the approximation with respect to the L^2-norm, the following characterization applies:

Theorem 9.38 (Approximation and Orthogonality) *Let $f \in L^2([a,b])$ and $S \subset L^2([a,b])$ be a finite-dimensional subspace. Let*

$$(f, g) := \int_a^b f(x)g(x)dx, \quad \|f\| = (f, f)^{\frac{1}{2}},$$

be the L^2 scalar product with associated norm. Then $p \in S$ is the best approximation to $f \in L^2([a, b])$, i.e.,

$$\|f - p\| = \min_{\phi \in S} \|f - \phi\|,$$

if and only if the error $f - p$ is orthogonal to the space S, i.e.,

$$(f - p, \phi) = 0 \quad \forall \phi \in S.$$

Proof (i) We first show that every best approximation $p \in S$ satisfies the orthogonality condition. For each fixed $\phi \in S$, the quadratic function

$$F_\phi(t) := \|f - p - t\phi\|^2 = (f - p - t\phi, f - p - t\phi), \quad t \in \mathbb{R}$$

has a minimum at $t = 0$. This means $t = 0$ must be a stationary point, i.e.,

$$\frac{d}{dt} F_\phi(t)\bigg|_{t=0} = -2(f - p - t\phi, \phi)\bigg|_{t=0} \overset{!}{=} 0.$$

The orthogonality relationship follows for all $\phi \in S$ from $t = 0$.
(ii) Conversely, let the relationship

$$(f - p, \phi) = 0 \quad \forall \phi \in S$$

hold for a $p \in S$. Then for arbitrary $\phi \in S$, it holds that

$$\|f - p\|^2 = (f - p, f - p) = (f - p, f - \phi) + \underbrace{(f - p, \overbrace{\phi - p}^{\in S})}_{=0}$$

$$\leq \|f - p\| \, \|f - \phi\|,$$

which shows the minimality condition. $\square$

We have shown that we can describe the best approximation $p \in S$ through an orthogonality relationship. This connection is the key to the analysis of the Gauss approximation and also to practical numerical implementations.

Theorem 9.39 (General Gauss Approximation) *Let H be a vector space with scalar product $(\cdot, \cdot)$ and $S \subset H$ be a finite-dimensional subspace. For every $f \in H$, there exists a uniquely determined best approximation $p \in S$ in the induced norm $\| \cdot \|$:*

$$\|f - p\| = \min_{\phi \in S} \|f - \phi\|$$

Proof We use the characterization of the best approximation using Theorem 9.38.

(i) Uniqueness: Let $p_1, p_2 \in S$ be two best approximations. Then necessarily, we have

$$(f - p_i, \phi) = 0 \quad \Rightarrow \quad (p_1 - p_2, \phi) = 0 \quad \forall \phi \in S.$$

For $\phi := p_1 - p_2$ it follows, $\|p_1 - p_2\| = 0$, i.e., $p_1 = p_2$.

(ii) Existence: The finite-dimensional subspace $S \subset H$ has a basis $\{\phi_1, \ldots, \phi_n\}$ with $n := \dim(S)$. We can represent the best approximation $p \in S$ in this basis as

$$p = \sum_{i=1}^{n} \alpha_i \phi_i$$

with uniquely determined coefficients $\alpha_k \in \mathbb{R}$. Then,

$$(f - p, \phi_j) = \left(f - \sum_{i=1}^{n} \alpha_i \phi_i, \phi_j\right) = (f, \phi_j) - \sum_{i=1}^{n} \alpha_i (\phi_i, \phi_j) = 0$$

for $j = 1, \ldots, n$. With

$$a_{ji} := (\phi_i, \phi_j), \quad b_j := (f, \phi_j)$$

and coefficient matrix $A = (a_{ij})_{i,j=1}^{n}$ and right-hand side $b = (b_j)_{j=1}^{n}$, this corresponds to the linear system

$$Ax = b$$

with the solution vector $x = (\alpha_i)_{i=1}^{n}$. The matrix A is the *Gram matrix* for the basis $\phi_1, \ldots, \phi_n$ and is always regular. Thus, the system is always uniquely solvable. $\square$

The proof immediately provides a construction principle to determine the best approximation $p \in S$. Given a basis $\{\phi_1, \ldots, \phi_n\}$, we can set up the linear $(n \times n)$ system

$$\sum_{j=1}^{n} a_{ij}\alpha_j = b_i, \quad i = 1, \ldots, n,$$

with $a_{ij} = (\phi_j, \phi_i)$, and compute the vector $x = (\alpha_j)_j$. Then

$$p = \sum_{i=1}^{n} \alpha_i \phi_i.$$

In this case, we would first need to assemble the matrix, i.e., compute the n^2 entries of the matrix through quadrature. The resulting system is dense, and solving it has the complexity $O(n^3)$. Suppose we start the construction with the monomial basis $\{1, x, \ldots, x^{n-1}\}$ on the interval $[0, 1]$. Then the matrix given by

$$a_{ij} = \int_0^1 x^i\,x^j\,\mathrm{d}x$$

is the so-called *Hilbert matrix*

$$H_n = \begin{pmatrix}
1 & \frac{1}{2} & \frac{1}{3} & \frac{1}{4} & \cdots & \frac{1}{n} \\
\frac{1}{2} & \frac{1}{3} & \frac{1}{4} & \frac{1}{5} & \cdots & \frac{1}{n+1} \\
\frac{1}{3} & \frac{1}{4} & \frac{1}{5} & & \ddots & \frac{1}{n+2} \\
\frac{1}{4} & \frac{1}{5} & \ddots & & & \vdots \\
\vdots & & & \ddots & & \vdots \\
\frac{1}{n} & \frac{1}{n+1} & \frac{1}{n+2} & \frac{1}{n+3} & \cdots & \frac{1}{2n-1}
\end{pmatrix}.$$

However, the Hilbert matrix is extremely ill-conditioned. Even for $n = 4$, the error amplification of the Hilbert matrix is 15 000. This matrix must therefore be avoided at all costs.
The direct approach to creating the best approximation is thus not numerically stable.

Even though the choice of basis vectors $\{\phi_1, \ldots, \phi_n\}$ does not influence the theoretical
result, it significantly determines the computational effort. Suppose the basis is an orthonormal system, i.e., $(\phi_i, \phi_j) = \delta_{ij}$ would hold for all $i, j = 1, \ldots, n$. Then, the system matrix
is the identity

$$a_{ji} = (\phi_i, \phi_j) = \delta_{ij} \quad \Rightarrow \quad A = I,$$

and the sought coefficients and the best approximation can be read off immediately as

$$\alpha_i = (f, \phi_i), \quad \text{and} \quad p(x) = \sum_{i=1}^n (f, \phi_i)\phi_i(x).$$

Orthonormalization can be performed using the Gram-Schmidt algorithm. When using the
monomial basis of $S = P_n$, this leads to the Legendre polynomials.

For the polynomials orthogonal with respect to $L^2(-1, 1)$, a closed formula can be given:

Theorem 9.40 (Legendre Polynomials) *Gram-Schmidt orthogonalization of the monomial basis $\{1, x, x^2, \ldots, x^n\}$ with respect to the $L^2(-1, 1)$-scalar product gives the basis
$\{p_0, \ldots, p_n\}$ which has the explicit representation*

$$p_k(x) := \frac{k!}{(2k)!} \frac{d^k}{dx^k}(x^2 - 1)^k. \tag{9.16}$$

For these polynomials, it holds that

$$\|p_k\| = \frac{k!^2}{(2k)!}\sqrt{\frac{2^{2k+1}}{2k+1}}, \quad p_k(1) = \frac{2^k k!^2}{(2k)!}. \tag{9.17}$$

They can be calculated by the two-step recursion formula

$$p_0(x) = 1$$
$$p_1(x) = x$$
$$k = 1, \ldots, n - 1: \quad p_{k+1}(x) = x p_k(x) - \frac{k^2}{4k^2 - 1} p_{k-1}(x). \tag{9.18}$$

By normalizing at $x = 1$, the Legendre polynomials $\ell_n \in P_n$ *are defined as*

$$\ell_k(x) := \frac{(2k)!}{2^k k!^2} p_k(x), \quad \ell_k(1) = 1.$$

Proof *(i)* We first show the orthogonality by integration by parts. It holds for $k \geq l$

$$\frac{(2k!)(2l)!}{k!l!}(p_k, p_l) = \left(\frac{d^k}{dx^k}(x^2 - 1)^k, \frac{d^l}{dx^l}(x^2 - 1)^l \right)$$

$$= \underbrace{\frac{d^{k-1}}{dx^{k-1}}(x^2 - 1)^k \cdot \frac{d^l}{dx^l}(x^2 - 1)^l \Big|_{-1}^{1}}_{=0} - \left(\frac{d^{k-1}}{dx^{k-1}}(x^2 - 1)^k, \frac{d^{l+1}}{dx^{l+1}}(x^2 - 1)^l \right),$$

since $\frac{d^s}{dx^s}(x^2 - 1)^k$ always contains the factor $x^2 - 1$ if $s < k$. After $l \leq k$-fold partial differentiation, we obtain

$$\frac{(2k!)(2l)!}{k!l!}(p_k, p_l) = (-1)^l \left(\frac{d^{k-l}}{dx^{k-l}}(x^2 - 1)^k, \frac{d^{2l}}{dx^{2l}}(x^2 - 1)^l \right). \tag{9.19}$$

For $k = l$, we deduce

$$\frac{(2k)!^2}{k!^2} \|p_k\|^2 = (-1)^k \left((x^2 - 1)^k, 1 \right)(2k)! = (-1)^k (2k)! \int_{-1}^{1} (x^2 - 1)^k dx, \tag{9.20}$$

and therefore

$$\|p_k\|^2 = (-1)^k \frac{k!^2}{(2k)!} \int_{-1}^{1} (x^2 - 1)^k dx. \tag{9.21}$$

The integral is calculated as

$$\int_{-1}^{1} (x^2 - 1)^k dx = (-1)^k \frac{\Gamma(k + 1)\sqrt{\pi}}{\Gamma\left(\frac{3}{2} + k\right)} = (-1)^k \frac{k!(k + 1)! 2^{2k+2}}{(2(k + 1))!}.$$

Where $\Gamma(\cdot)$ is the Gamma function, see [2]. Using (9.21), the normalization (9.16) can be shown

$$\|p_k\|^2 = \frac{k!^4 2^{2k+1}}{(2k)!^2 (2k+1)}$$

and hence, the representation (9.17) holds.

For $k > l$, further partial differentiation shows orthogonality

$$\frac{(2k!)(2l)!}{k!l!}(p_k, p_l) = (-1)^l \underbrace{\frac{d^{k-l-1}}{dx^{k-l-1}}(x^2-1)^k \cdot \frac{d^{2l}}{dx^{2l}}(x^2-1)^l \bigg|_{-1}^{1}}_{=0}$$
$$- (-1)^l \left(\frac{d^{k-l-1}}{dx^{k-l-1}}(x^2-1)^k, \underbrace{\frac{d^{2l+1}}{dx^{2l+1}}(x^2-1)^l}_{=0} \right).$$

(ii) Normalization at $x = 1$ is straightforward for $k = 0$ and $k = 1$. Recursively, it follows:

$$\begin{aligned}
p_{k+1}(1) &= 1 \cdot \frac{2^k k!^2}{(2k)!} - \frac{k^2}{4k^2-1} \frac{2^{k-1}(k-1)!^2}{(2(k-1))!} \\
&= \frac{2^k k!^2}{(2k)!} - \frac{2^{k-1}k!^2}{(2k-1)(2k+1)(2(k-1))!} \\
&= \frac{2^k k!^2 (2k+1)(2k+2)}{(2(k+1))!} - \frac{2^{k-1}k!^2 2k(2k+2)}{(2(k+1))!} \\
&= \frac{2^{k+1}(k+1)!^2}{(2(k+1))!} \left(\frac{(2k+1)}{k+1} - \frac{k}{k+1} \right) = \frac{2^{k+1}(k+1)!^2}{(2(k+1))!}
\end{aligned}$$

(iii) The leading coefficient is one, so

$$p_k(x) = x^k \pm \ldots.$$

With the normalization of the leading coefficient, the polynomials orthogonalized from the monomial basis are uniquely determined.

(iv) It remains to prove the two-step recursion formula. We will postpone this to Theorem 9.41, which will introduce a general principle. $\qquad\square$

9.5.1.1 Two-Step Orthogonalization

In Sect. 6.2, we have already used a two-step orthogonalization in the context of the CG method. We will generalize this principle. For the following, let V be a general, finite or countably infinite-dimensional vector space, e.g., the space of polynomials (up to a degree n). Further, let $A : V \to V$ be a mapping on V. For $d \in V$, the Krylov subspace can then be defined as

$$K_l(d, A) := \text{span}\{d, Ad, \ldots, A^{l-1}d\} \subset V.$$

For example, in the case of polynomials, we choose $d = 1$, i.e., the constant monomial, and the mapping $A : P_k \to P_{k+1}$ as

$$p(z) \mapsto zp(z).$$

Then the Krylov space $K_k(1, z)$ is just the polynomial space P_{k-1}.

Theorem 9.41 (Two-Step Orthogonalization) *Let V be a vector space with scalar product $(\cdot, \cdot)$ and induced norm $\| \cdot \|$. Let $A : V \to V$ be such that*

$$(Ax, y) = (x, Ay) \quad \forall x, y \in V. \tag{9.22}$$

Let $d \in V$. An orthonormalization $\{q_1, \ldots, q_l\}$ of the Krylov space $K_l(d, A)$ can be calculated with the following two-step scheme:

$$q_1 := \frac{d}{\|d\|},$$

$$\tilde{q}_2 := Aq_1 - (Aq_1, q_1)q_1, \quad q_2 := \frac{\tilde{q}_2}{\|\tilde{q}_2\|},$$

$$i = 3, \ldots, l : \quad \tilde{q}_i := Aq_{i-1} - (Aq_{i-1}, q_{i-1})q_{i-1} - (Aq_{i-1}, q_{i-2})q_{i-2},$$

$$q_i := \frac{\tilde{q}_i}{\|\tilde{q}_i\|}.$$

Proof We prove by induction over l that $q_1, \ldots, q_l$ forms an orthonormal basis of $K_l(d, A)$ and further that either $K_{l+1}(d, A) \subset K_l(d, A)$ or $Aq_l \in K_{l+1}(d, A) \setminus K_l(d, A)$. The latter means that the Krylov space is enriched by considering Aq_l as the next vector to be orthogonalized.

For $l = 1$ it is

$$q_1 := \frac{d}{\|d\|}.$$

Assuming $K_2(d, A) := \text{span}\{d, Ad\} \not\subset K_1(d, A) := \text{span}\{d\}$, implies that $Aq_1 \notin K_1(d, A)$.

Assume that an orthonormal basis of $K_l(d, A)$ is given by $q_1, \ldots, q_l$. We consider two cases:

Case 1: It holds $K_{l+1}(d, A) \subset K_l(d, A)$, i.e., the Krylov space does not get larger and the procedure stops, as the orthonormal basis $\{q_1, \ldots, q_l\}$ is already given.

Case 2: The Krylov space $K_{l+1}(d, A)$ is larger than $K_l(d, A)$, and, by induction, we can assume that $Aq_l \in K_{l+1}(d, A) \setminus K_l(d, A)$.

(i) We orthogonalize $Aq_l \in K_{l+1}(d, A)$ with respect to $q_1, \ldots, q_l$ according to

$$\tilde{q}_{l+1} := Aq_l - \sum_{i=1}^{l} \beta_i q_i.$$

This yields

$$0 \overset{!}{=} (\tilde{q}_{l+1}, q_j) = (Aq_l, q_j) - \beta_j \quad j = 1, \ldots, l,$$

since $q_1, \ldots, q_l$ form an orthonormal basis. Furthermore,

$$\beta_j = (Aq_l, q_j) = (q_l, \underbrace{Aq_j}_{\in K_{j+1}(d,A)}) = 0 \quad \text{for } j = 1, \ldots, l-2.$$

From this, we deduce the iteration

$$\tilde{q}_2 = Aq_1 - (Aq_1, q_1)q_1,$$

and

$$\tilde{q}_{l+1} = Aq_l - (Aq_l, q_l)q_l - (Aq_l, q_{l-1})q_{l-1}$$

for $l > 1$.

(ii) Assume it holds $K_{l+2}(d, A) \not\subset K_{l+1}(d, A)$ which implies that $A^{l+1}d \notin K_{l+1}(d, A)$. We must show that $Aq_{l+1} \notin K_{l+1}(d, A)$. For $q_{l+1} \in K_{l+1}(d, A)$, there exist $\alpha_1, \ldots, \alpha_{l+1} \in \mathbb{R}$ with $\alpha_{l+1} \neq 0$. Since $q_{e+1} \in K_{e+1}(d, A) \setminus K_e(d, A)$, such that

$$q_{l+1} = \sum_{i=1}^{l+1} \alpha_i A^{i-1} d.$$

Then,

$$Aq_{l+1} = \sum_{i=1}^{l+1} \alpha_i A^i d = \underbrace{\alpha_{l+1} A^{l+1} d}_{\in K_{e+2}(d,A) \setminus k_{e+1}(d,A)} + \underbrace{\sum_{i=1}^{l} \alpha_i A^i d}_{\in K_{l+1}(d,A)} \in K_{l+2}(d, A),$$

which implies $Aq_{e+1} \in K_{e+2}(d, A) \setminus k_{e+1}(d, A)$. $\qquad\square$

In order to orthogonalize a sequence in this shortened procedure, we need two ingredients: The initial sequence must be generated in the sense of a Krylov space with a mapping $A : V \to V$. The scalar product must preserve the symmetry $(Ax, y) = (x, Ay)$ with respect to A.

In the following, we will apply this result to the Legendre polynomials. First, we have

$$p_{k+1}(x) = x p_k(x) - \big(t p_k(t),\, p_k(t)\big) p_k(x) - \big(t p_k(t),\, p_{k-1}(t)\big) p_{k-1}(x).$$

From (9.16), we read that p_k is either an even or odd function. Because $(x^2 - 1)^k$ is even and the k-th derivative is even when k is even, otherwise odd. Thus, p_k^2 is an even and $x p_k^2$ an odd function with zero mean. Therefore, the two-step recursion holds

$$p_{k+1}(x) = x p_k(x) - \big(t p_k(t),\, p_{k-1}(t)\big) p_{k-1}(x).$$

We calculate the factor $\gamma_k = \big(t p_k(t),\, p_{k-1}(t)\big)$ indirectly. With (9.17) we have

$$\frac{2^{k+1}(k+1)!^2}{(2(k+1))!} = \frac{2^k k!^2}{(2k)!} - \gamma_k \frac{2^{k-1}(k-1)!^2}{(2(k-1))!},$$

thus

$$\frac{4k^2(k+1)^2}{(2k-1)2k(2k+1)(2k+2)} = \frac{2k^2}{(2k-1)2k} - \gamma_k$$

and finally

$$\gamma_k = \frac{k}{2k-1} - \frac{k(k+1)}{(2k-1)(2k+1)} = \frac{k^2}{4k^2 - 1}.$$

With this, the proof of Theorem 9.40 is now completed, and we return to the Gauss approximation.

9.5.1.2 Gauss Approximation Using Legendre Polynomials

The first Legendre polynomials and their roots are

$$
\begin{aligned}
\ell_0(x) &= 1, \\
\ell_1(x) &= x, & x_0 &= 0, \\
\ell_2(x) &= \tfrac{1}{2}(3x^2 - 1), & x_{0/1} &= \pm\sqrt{\tfrac{1}{3}}, \\
\ell_3(x) &= \tfrac{1}{2}(5x^3 - 3x), & x_0 &= 0,\ x_{1/2} = \pm\sqrt{\tfrac{3}{5}}, \\
\ell_4(x) &= \tfrac{1}{8}(35x^4 - 30x^2 + 3), & x_{0/1} &\approx \pm 0.861136,\ x_{2/3} \approx \pm 0.339981.
\end{aligned}
\tag{9.23}
$$

The Legendre polynomials help us to solve the Gauss approximation problem.

Corollary 9.42 (Gauss Approximation With Polynomials) *Let $f \in C[-1, 1]$. The best approximation $p \in P_n$ with respect to the L^2-norm in the space of polynomials of degree n is uniquely determined by*

$$p(x) = \sum_{i=0}^{n} \frac{1}{\|\ell_i\|_{L^2[-1,1]}^2} (\ell_i, f)\ell_i(x),$$

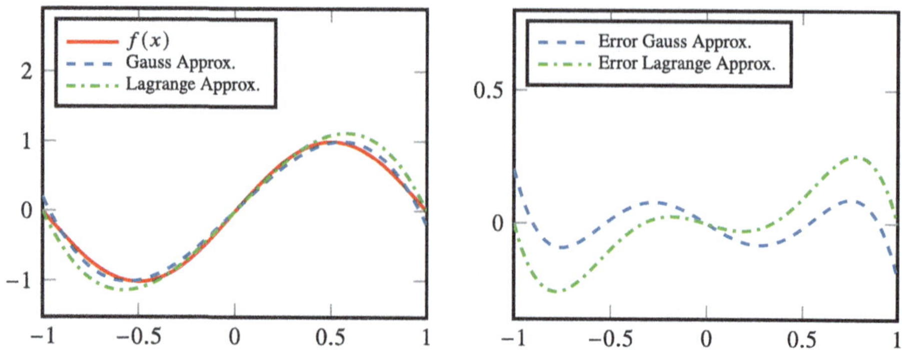

Fig. 9.7 Gauss approximation and Lagrange interpolation (both cubic) of the function $f(x) = \sin(\pi x)$. Right: Errors of the approximations

where the $\ell_i(x)$ are the normalized Legendre polynomials

$$\ell_i(x) := \frac{1}{2^i i!} \frac{d^i}{dx^i} (x^2 - 1)^i.$$

The normalization with $\|\ell_n\|^2$ is necessary, since the Legendre polynomials are not normalized with respect to the L^2-norm, c.f. Theorem 9.40. The property of $p \in P_n$ being the best approximation to f in the space of polynomials also means that p is a better approximation with respect to the L^2-norm than any Lagrange interpolation for any choice of nodes $x_0, \ldots, x_n$ in $[-1, 1]$. From this, we can trivially derive a simple error estimate. Let $f \in C^{n+1}([a, b])$ and $p \in P_n$ be the best approximation. It holds

$$\|f - p\|_{L^2[a,b]} \leq \|f - i_h f\|_{L^2[a,b]},$$

where $i_h f$ is any interpolation to f at $n + 1$ nodes. With the (pointwise) error estimate of the Lagrange interpolation, Theorem 9.13, it follows that

$$\|f - p\|_{L^2([a,b])} \leq \frac{\|f^{(n+1)}\|_\infty}{(n + 1)!} \min_{x_0,\ldots,x_n \in [a,b]} \left(\int_{-1}^{1} \prod_{j=0}^{n} (x - x_j)^2 dx \right)^{\frac{1}{2}}. \tag{9.24}$$

The minimum of the product over the nodes is not easy to compute, as all nodes in the interval can be freely varied. Theorem 9.54 will narrow this gap and give a good general choice for the nodes.

Example 9.43 (Gauss Approximation Versus Lagrange Interpolation) We approximate the function $f(x) = \sin(\pi x)$ on the interval $I = [-1, 1]$ with polynomials of degree three. First, we create the corresponding Lagrange interpolation polynomial at the four equidistantly distributed nodes

$$x_i = -1 + \frac{2i}{3}, \quad i = 0, \ldots, 3.$$

Due to $f(x_0) = f(x_3) = 0$, it holds

$$p_L(x) = \sin(\pi x_1) L_1^{(3)}(x) + \sin(\pi x_2) L_2^{(3)}(x) = \frac{27\sqrt{3}}{16}(x - x^3).$$

As a second approximation, we determine the Gauss approximation in the L^2-norm. For this, we use the representation over the normalized Legendre polynomials $\ell_n(x)$ on $[-1, 1]$. It holds

$$\int_{-1}^{1} \ell_0(x) f(x)\mathrm{d}x = 0, \quad \int_{-1}^{1} \ell_2(x) f(x)\mathrm{d}x = 0,$$

and

$$\int_{-1}^{1} \ell_1(x) f(x)\mathrm{d}x = \frac{\sqrt{6}}{\pi}, \quad \int_{-1}^{1} \ell_3(x) f(x)\mathrm{d}x = \frac{\sqrt{14}(\pi^2 - 15)}{\pi^3}.$$

From this, we obtain the Gauss approximation

$$p_G(x) = \frac{\sqrt{6}}{\pi}\ell_1(x) + \frac{\sqrt{14}(\pi^2 - 15)}{\pi^3}\ell_3(x)$$

$$= \frac{5x(7\pi^2 x^2 - 105x^2 - 3\pi^2 + 63)}{2\pi^3}.$$

For the two approximations, it holds

$$\|f - p_L\|_{L^2([-1,1])} \approx 0.2, \quad \|f - p_G\|_{L^2([-1,1])} \approx 0.1,$$

i.e., the Gauss approximation provides twice as good a result in the L^2-norm. In Fig. 9.7, we show both approximations, the function $f(x)$ as well as the errors $f(x) - p_L(x)$ and $f(x) - p_G(x)$. The maximum error in the interval is slightly lower for the Gauss approximation. ◀

9.5.1.3 Discrete Gauss Approximation

The Gauss approximation has a discrete analog: we consider the best approximation of discrete measurement values $(x_i, y_i), i = 1, 2, \ldots, m$, where we do not require that the nodes x_i are necessarily pairwise distinct. We define

$$|y|_2 = \left(\sum_{i=1}^{m} y_i^2\right)^{\frac{1}{2}}.$$

Let S be a finite-dimensional function space, e.g., the space of polynomials P_n, and we assume that $|\cdot|_2$ is a norm on P_n. This is achieved if the set of nodes x_i contains at least $n + 1$ pairwise distinct ones. Therefore, we will always assume that $m > n$. We are looking for the best approximation $p \in S$ with respect to the point norm:

$$|p - y|_2 = \left(\sum_{i=1}^{m} |p(x_i) - y_i|^2 \right)^{\frac{1}{2}} = \min_{\phi \in S} |\phi - y|_2.$$

If $m = n + 1$ and if all nodes are pairwise distinct, the result is the Lagrange interpolation. Here, we are interested in the case $m > n$, often $m \gg n$. We cannot demand equality at the points x_i, and we also allow that specific points x_i may appear multiple times. An example is the determination of the linear regression $p_1(x) = \alpha_0 + \alpha_1 x$ for many (possibly thousands) of measurement values. The Euclidean vector norm is derived from the Euclidean scalar product by

$$|x|_2 := (x, x)_2^{\frac{1}{2}},$$

therefore, the previously developed theory can be applied. The key to determining the best approximation is the characterization via orthogonality according to Theorem 9.38:

$$|p - y|_2 = \min_{\phi \in S} |\phi - y|_2 \quad \Leftrightarrow \quad (p - y, \phi)_2 = 0 \quad \forall \phi \in \mathbb{R}^n$$

All properties of the continuous Gauss approximation carry over, and we can immediately formulate the existence theorem:

Theorem 9.44 (Discrete Gauss Approximation) *Let m discrete data values be given by (x_i, y_i) for $i = 1, \ldots, m$. Furthermore, let S be a finite-dimensional function space with basis $\{\phi_1, \ldots, \phi_n\}$. The discrete Gauss approximation $p \in S$, with*

$$|p - y|_2 := \left(\sum_{i=1}^{m} |p(x_i) - y_i|^2 \right)^{\frac{1}{2}} = \min_{\phi \in S} |\phi - y|_2,$$

is uniquely determined if $|\cdot|_2$ is a norm on S.

Proof Through the basis representation, S is equivalent to the Euclidean space $\mathbb{R}^n$. Uniqueness and existence follow in $\mathbb{R}^n$ analogously to the proof of Theorem 9.39. $\qquad\square$

The construction of the discrete Gauss approximation follows by using the representation

$$p(x) = \sum_{i=1}^{n} \alpha_i \phi_i(x)$$

and the orthogonality relation

$$(p - y, \phi_k)_2 = \sum_{i=1}^{m} (p(x_i) - y_i)\phi_k(x_i) = 0, \quad k = 1, \ldots, n.$$

Then, a linear system of n equations in n unknowns determines the solution:

$$\sum_{j=1}^{n} \alpha_j \underbrace{\sum_{i=1}^{m} \phi_j(x_i)\phi_k(x_i)}_{=(\phi_j,\phi_k)_2} = \underbrace{\sum_{i=1}^{m} y_i\phi_k(x_i)}_{=(y,\phi_k)_2}, \quad k = 1, \ldots, n$$

The sought-after coefficient vector $\alpha = (\alpha_k)_{k=1}^{n}$ is given as the solution of the system

$$\begin{pmatrix} (\phi_0, \phi_0)_2 & (\phi_0, \phi_1)_2 & \cdots & (\phi_0, \phi_n)_2 \\ (\phi_1, \phi_0)_2 & \ddots & & \vdots \\ \vdots & & \ddots & \vdots \\ (\phi_n, \phi_0)_2 & \cdots & \cdots & (\phi_n, \phi_n)_2 \end{pmatrix} \begin{pmatrix} \alpha_0 \\ \alpha_1 \\ \vdots \\ \alpha_n \end{pmatrix} = \begin{pmatrix} (y, \phi_0)_2 \\ (y, \phi_1)_2 \\ \vdots \\ (y, \phi_n)_2 \end{pmatrix}.$$

Example 9.45 (Linear Regression) We aim at the linear best approximation $p(x) \in P_1$

$$p(x) = \alpha_0 + \alpha_1 x$$

for data (x_i, y_i). Let $\phi_0(x) \equiv 1, \phi_1(x) = x$. This results in the following linear system

$$\begin{pmatrix} (\phi_0, \phi_0)_2 & (\phi_0, \phi_1)_2 \\ (\phi_1, \phi_0)_2 & (\phi_1, \phi_1)_2 \end{pmatrix} \begin{pmatrix} \alpha_0 \\ \alpha_1 \end{pmatrix} = \begin{pmatrix} (y, \phi_0)_2 \\ (y, \phi_1)_2 \end{pmatrix} \quad \Rightarrow \quad \begin{pmatrix} m & \sum x_i \\ \sum x_i & \sum x_i^2 \end{pmatrix} \begin{pmatrix} \alpha_0 \\ \alpha_1 \end{pmatrix} = \begin{pmatrix} \sum y_i \\ \sum x_i y_i \end{pmatrix}$$

This system is regular if $m \geq 2$ and there are at least two support values $x_i \neq x_j$. ◀

Generalization of the Gauss Approximation

The discrete Gauss approximation and the approximation of functions can be extended to weighted scalar products and weighted induced norms.

Weights can be introduced to improve the accuracy of the Gauss approximation at the interval ends. For each integrable and positive $\omega(x) > 0$,

$$(f, g)_\omega := \int_a^b f(x)g(x)\omega(x)\mathrm{d}x$$

defines a scalar product with induced norm

$$\|f\|_\omega = (f, f)_\omega^{\frac{1}{2}}.$$

The weight function $w(x) = \frac{1}{\sqrt{1-x^2}}$ on $(-1, 1)$, for instance, is a good choice to put greater focus on the ends. The corresponding orthonormal polynomials are the Chebyshev polynomials.

Theorem 9.46 (Chebyshev Polynomials) *The* Chebyshev polynomials

$$T_n(x) := \cos(n \arccos x), \quad -1 \le x \le 1, \quad n \in \mathbb{N}$$

are the orthonormal polynomials $T_n \in P_n$ with respect to the $(\cdot, \cdot)_\omega$ scalar product with respect to the the weight function

$$\omega(x) = \frac{1}{\sqrt{1 - x^2}}.$$

The n pairwise different zeros of the n-th Chebyshev polynomial $T_n(x)$ are given by

$$x_k = \cos\left(\frac{2k + 1}{2n}\pi\right), \quad k = 0, \dots, n - 1.$$

On the interval $\mathbb{R} \setminus [-1, 1]$ the Chebyshev polynomials have the representation

$$T_n(x) = \frac{1}{2}\left[(x + \sqrt{x^2 - 1})^n + (x - \sqrt{x^2 - 1})^n\right].$$

Proof *(i)* We first have to prove that the functions $T_n(x)$ are polynomials. We prove this inductively by deriving an iterative formula for T_n. It holds that

$$T_0(x) = 1, \quad T_1(x) = x,$$

so in particular $T_0 \in P_0$ and $T_1 \in P_1$. From the addition theorem for the cosine function

$$\cos((n + 1)y) + \cos((n - 1)y) = 2\cos(ny)\cos(y),$$

we obtain with $y = \arccos(x)$ the two-step recursion formula

$$T_{n+1}(x) + T_{n-1}(x) = 2xT_n(x) \quad \Rightarrow \quad T_{n+1}(x) = 2xT_n(x) - T_{n-1}(x).$$

From this, we conclude inductively $T_n \in P_n$ with

$$T_n(x) = 2^{n-1}x^n + \dots.$$

(ii) We prove the orthogonality of the Chebyshev polynomials with respect to the weighted scalar product on $[-1, 1]$. With $x = \cos(t)$ and using the orthogonality of trigonometric polynomials, we have

$$\int_{-1}^{1} \frac{T_n(x) T_m(x)}{\sqrt{1 - x^2}} \, dx = \int_{0}^{\pi} \cos(nt) \, \cos(mt) dt = \begin{cases} \pi, & n = m = 0, \\ \frac{\pi}{2}, & n = m > 0, \\ 0, & n \neq m. \end{cases}$$

(iii) The zeros of the Chebyshev polynomials can be calculated directly from the explicit representation.

(iv) The second representation of the Chebyshev polynomials can also be traced back to the iteration formula. $\qquad\square$

The discrete Gauss approximation can be generalized in this respect. For this purpose, let $\omega \in \mathbb{R}^n$ be a vector with $\omega_i > 0$. Then for $x, y \in \mathbb{R}^n$,

$$(x, y)_\omega = \sum_{i=1}^{n} x_i \, y_i \, \omega_i, \quad |x|_\omega = (x, x)_\omega^{\frac{1}{2}},$$

gives a weighted scalar product with induced norm. Especially for the discrete approximation, the weights play a major role in applications: Suppose that estimates for the measurement error are known. Then support values with smaller measurement errors can be weighted more heavily.

All theorems on Gauss approximation were directly shown for arbitrary norms induced by scalar products. Therefore, all properties transfer to the weighted norms.

9.5.2   Chebyshev Approximation

If we choose the maximum norm instead of the $L^2[a, b]$-norm

$$\|f\|_\infty := \max_{[a,b]} |f(x)|,$$

the best approximation task cannot be approached in the same way as the Gauss approximation. The maximum norm is not induced by any scalar product. Thus, there is no connection to orthogonality.

Definition 9.47 *(Chebyshev Approximation)* Let $f \in C[a, b]$ and $S \subset C[a, b]$ be a finite-dimensional subspace. The function $p \in S$ with

$$\|f - p\|_\infty \leq \|f - g\|_\infty \quad \forall g \in S$$

is called the *Chebyshev approximation* to f.

Example 9.48 (Best Approximation in the Maximum Norm) We consider $f(x) = \cos(\pi x)$ on $I = [0, 1]$. We first look for the best approximation with the constant function $p \in P_0$, i.e., $p(x) = c$:

$$\min_{c \in \mathbb{R}} \| f - c \|_\infty$$

Because

$$-1 = \min_I f(x) < \max_I f(x) = 1,$$

it must also hold that $-1 \le c \le 1$. For this choice, we get

$$\| f - c \|_\infty = \max\{1 - c, 1 + c\}$$

and the optimal solution for $c = 0$.

In the case of the linear best approximation $p(x) = c + \alpha x$, this task is already more difficult to solve. ◀

The existence of the Chebyshev approximation cannot be shown using an orthogonality relationship. However, we can use a very general approximation result:

Theorem 9.49 (Existence of the Chebyshev Approximation) *Let E be a vector space with norm $\| \cdot \|$ and $S \subset V$ be a finite-dimensional subspace. Then for every $f \in V$ there exists a best approximation $p \in S$ with*

$$\| f - p \| = \min_{q \in S} \| f - q \|.$$

Proof We first show that a best approximation must lie in the set

$$S_f := \{\phi \in S, \ \|\phi\| \le 2\|f\|\}.$$

This set is not empty, since the zero function $g \equiv 0 \in S_f$ is in this set.

Assume, $g \in S$ with $\|g\| > 2\|f\|$. Then it follows with the reverse triangle inequality

$$\|g - f\| \ge \|g\| - \|f\| > \|f\| = \|f - 0\| \ge \inf_{q \in S} \|f - q\|,$$

i.e., $g \in S$ is not a best approximation. The subset $S_f \subset S$ is closed, bounded, finite-dimensional, and therefore compact. On S_f the function

$$F(\phi) := \| f - \phi \|$$

is continuous and has a minimum $p \in S_f$ as a continuous function on the compact set S_f. This minimum is the best approximation. □

To characterize the best approximation, we need a verifiable criterion. Many of the following theorems can be transferred to general vector spaces V and finite-dimensional subspaces S. However, we assume that $V = C[a, b]$ is the space of continuous functions and $S = P_n$ is the space of polynomials of degree up to n. Suppose, $p \in P_n$ is a best approximation to $f \in C[a, b]$. Then, we define

$$e(x) = f(x) - p(x), \quad \mathbf{e} := \max_{[a,b]} |e(x)|,$$

and the set of all points $x \in [a, b]$, where the error function reaches its maximum:

$$E(f, p) := \{x \in [a, b] : |e(x)| = \mathbf{e}\}.$$

Theorem 9.50 (Kolmogorov's Theorem) *The function $p \in P_n$ is the best approximation to $f \in C[a, b]$, if and only if the* Kolmogorov criterion

$$\forall q \in P_n : \quad \min\{(f(x) - p(x))q(x) : \forall x \in E(f, p)\} \le 0$$

holds.

Proof *(i)* Assume, p is the best approximation and the Kolmogorov criterion does not hold. Then there exists a $q \in P_n$ and an $\epsilon > 0$, such that for $e(x) := f(x) - p(x)$ it holds that

$$\forall x \in E(f, p) : e(x)q(x) > 2\epsilon.$$

The set $E(f, p)$ is compact. Therefore, there exists an open set U with $E(f, p) \subset U$ and

$$\forall x \in U : (f(x) - p(x))q(x) > \epsilon.$$

We define $M := \|q\|_\infty$ and $p_1 := p + \lambda q$. For $\lambda > 0$ it follows for $x \in U$

$$(f(x) - p_1(x))^2 = \left(e(x) - \lambda q(x)\right)^2 = e(x)^2 - 2\lambda e(x)q(x) + \lambda^2 q(x)^2$$
$$< \mathbf{e}^2 - 2\lambda\epsilon + \lambda^2 M^2.$$

Then, for all

$$0 < \lambda < \frac{\epsilon}{M^2},$$

it holds

$$(f(x) - p_1(x))^2 < \mathbf{e}^2 - \lambda\epsilon.$$

So there exists a function p_1, which is a better approximation to f in the vicinity of U than p. We consider $x \in [a, b] \setminus U$. This set is compact (U is open). Therefore, there exists a $\delta > 0$ with

$$|e(x)| < \mathbf{e} - \delta \quad \forall x \in [a, b] \setminus U,$$

because the maxima are just taken within the open vicinity U. For

$$\lambda < \frac{\delta}{2M},$$

it then holds

$$|f(x) - p_1(x)| \le |f(x) - p(x)| + \lambda |q(x)| \le \mathbf{e} - \delta + \frac{\delta}{2M} M = \mathbf{e} - \frac{\delta}{2}.$$

Hence, the function p_1 is a better approximation to f on the entire interval $[a, b]$. This is a contradiction, and the Kolmogorov criterion must hold.

(ii) Let's assume that the Kolmogorov criterion holds for p. Let $p_1 \in P_n$ be arbitrary and $q := p_1 - p$. Thus, there exists at least one $t \in E(f, p)$ with

$$e(t)q(t) \le 0.$$

From this, it follows

$$(f(t) - p_1(t))^2 = (e(t) - q(t))^2 = e(t)^2 - 2e(t)q(t) + q(t)^2 \ge e(t)^2 = \mathbf{e}^2,$$

i.e.,

$$\|f - p_1\|_\infty \ge \|f - p\|_\infty \quad \forall p_1 \in P_n,$$

so, p is the best approximation. $\qquad\Box$

The uniqueness of the best approximation in polynomial spaces can be proven with this criterion:

Theorem 9.51 (Haar's Uniqueness Theorem) *Let $f \in C[a, b] \setminus P_n$, P_n be the space of polynomials up to degree n, and $p \in P_n$ be the best approximation to f. Then the set $E(f, p)$ has at least $n + 2$ points and the best approximation is uniquely determined.*

Proof *(i)* Suppose the set $E(f, p)$ has only $n + 1$ points $x_0, x_1, \ldots, x_n \in [a, b]$. For these $n + 1$ pairwise different points, there exists a unique Lagrange interpolation $q \in P_n$ with

$$q(x_j) = e(x_j) = f(x_j) - p(x_j), \quad j = 0, \ldots, n.$$

We check the Kolmogorov criterion

$$e(x_j)q(x_j) = e(x_j)^2 = \mathbf{e}^2 > 0.$$

The criterion is not fulfilled, i.e., p cannot be the best approximation, so there exist at least $n + 2$ points in $E(f, p)$.

(ii) Suppose p_1, p_2 are two best approximations to $f \in C[a, b]$. Then

$$\left\| f - \frac{1}{2}(p_1 + p_2) \right\|_\infty \le \frac{1}{2}\|f - p_1\|_\infty + \frac{1}{2}\|f - p_2\|_\infty,$$

i.e., $p := (p_1 + p_2)/2$ is also a best approximation. According to part *(i)*, there exist at least $n + 2$ points $x_0, \ldots, x_{n+1} \in E(f, p)$ with

$$|f(x_j) - p(x_j)| = \mathbf{e}, \quad j = 0, \ldots, n + 1,$$

so

$$\left| \frac{1}{2}(f(x_j) - p_1(x_j)) + \frac{1}{2}(f(x_j) - p_2(x_j)) \right| = \mathbf{e}, \quad j = 0, \ldots, n + 1.$$

At the same time, for the two polynomials p_1, p_2, we have

$$|f(x_j) - p_i(x_j)| \leq \mathbf{e}, \quad i = 1, 2, \quad j = 0, \ldots, n + 1.$$

From this, it follows

$$f(x_j) - p_1(x_j) = f(x_j) - p_2(x_j) \quad \Rightarrow \quad p_1(x_j) = p_2(x_j),$$

i.e., the two polynomials $p_i \in P_n$ agree in $n + 2$ points and are therefore identical. The best approximation is unique. $\qquad\square$

To determine the best approximation and to further characterize it, we cite the following sharpening of the Kolmogorov criterion. The Chebyshev Alternation Theorem states that the maximum errors are realized with alternating signs:

Theorem 9.52 (Chebyshev's Alternation Theorem) *Let $f \in C[a, b]$ and $p \in P_n$ be the best approximation to f. Then there exists a sequence of $n + 2$ pairwise different and ascending sorted nodes $E(f, p) := (x_0, \ldots, x_{n+1})$, the so-called* alternant, *such that for the error $e(x_i) = f(x_i) - p(x_i)$ it holds*

$$e(x_i) = -e(x_{i+1}), \quad i = 0, \ldots, n.$$

In particular $|e(x_i)| = \mathbf{e} \in \mathbb{R}_+$ for all $i = 0, \ldots, n + 1$.

Proof The Haar Uniqueness Theorem already states that the set $E(f, p)$ has at least $n + 2$ pairwise different points, at which:

$$|e(x_j)| = \mathbf{e}, \quad j = 0, \ldots, n + 1$$

Furthermore, the Kolmogorov criterion holds in these $n + 2$ points. We now have to show that the sign of $e(x_j)$ alternates from point to point. Suppose there were only n sign changes, e.g.,

$$\begin{aligned}
e(x_j) &= (-1)^j \mathbf{e}, \quad j = 0, \ldots, i, \\
e(x_i) = e(x_{i+1}) &= (-1)^i \mathbf{e}, \\
e(x_j) &= (-1)^{j+1} \mathbf{e}, \quad j = i + 2, \ldots, n + 1.
\end{aligned} \quad (9.25)$$

We try to bring this to a contradiction by constructing a function $q_\alpha \in P_n$ that does not satisfy the Kolmogorov criterion. For this, we choose the uniquely determined polynomial $q_\alpha \in P_n$ with

$$\begin{aligned}
q_\alpha(x_j) &= e(x_j), \quad j = 0, \ldots, i-1 \\
q_\alpha(x_i) &= \alpha e(x_i), \quad j = i \\
q_\alpha(x_j) &= e(x_j), \quad j = i+2, \ldots, n+1,
\end{aligned}$$

with an arbitrary $\alpha > 0$. These are $n+1$ conditions for the $n+1$ unknowns of the polynomial space P_n, so q_α is uniquely determined. The point x_{i+1} is omitted. Let $L_j^{(n)}$ for $j = 0, \ldots, n$ be the Lagrange polynomials for the $n+1$ points

$$x_0, x_1, \ldots, x_i, x_{i+2}, \ldots, x_{n+1}.$$

Then q can be written as

$$q_\alpha(x) = \sum_{j=0}^{i-1} e(x_j) L_j^{(n)}(x) + \alpha e(x_i) L_i^{(n)}(x) + \sum_{j=i+2}^{n+1} e(x_j) L_{j-1}^{(n)}(x). \tag{9.26}$$

With this, we have

$$e(x_j) q_\alpha(x_j) = \begin{cases}
\mathbf{e}^2 & j = 0, \ldots, i-1 \\
\alpha \mathbf{e}^2 & j = i, \\
\mathbf{X_{i+1}} & j = i+1, \\
\mathbf{e}^2 & j = i+2, \ldots, n+1.
\end{cases}$$

At the point $j = i+1$, we have inserted $\mathbf{X}_{i+1}$ as a placeholder. We still need to calculate this. Using representation (9.26), we have

$$q_\alpha(x_{i+1}) = \tilde{q}_\alpha(x_{i+1}) + \alpha e(x_i) L_i^{(n)}(x_{i+1}),$$

where $\tilde{q}(x) = q_\alpha(x) - \alpha e(x_i) L_i^n(x)$. The Lagrange basis polynomial $L_i^{(n)}(x)$ cannot change sign between x_i and x_{i+2}. Otherwise, it would have $n+1$ zeros and would itself be the zero polynomial. Due to the assumption $e(x_i)e(x_{i+1}) > 0$, see (9.25), we have

$$\mathbf{X_{i+1}} = e(x_{i+1}) q_{\alpha(x_{i+1})} = e(x_{i+1}) \tilde{q}_{\alpha(x_{i+1})} + \alpha \underbrace{e(x_{i+1}) e(x_i)}_{=\mathbf{e}^2 > 0} \underbrace{L_i^{(n)}(x_{i+1})}_{> 0}.$$

For α large enough, it follows that

$$\mathbf{X_{i+1}} = e(x_{i+1}) q_\alpha(x_{i+1}) > 0,$$

in contradiction to the Kolmogorov theorem. Therefore, an alternant must exist. $\qquad\square$

The Remez Algorithm

The *Remez algorithm* constructs an alternant. If an alternant $x_0, \ldots, x_{n+1}$ is known, the coefficients $\alpha_0, \ldots, \alpha_n$ of the best approximation

$$p(x) = \sum_{i=0}^{n} \alpha_i x^i$$

can be determined. To this end, we have

$$\sum_{i=0}^{n} \alpha_i x_k^i + (-1)^k \alpha_{n+1} = f(x_k), \quad k = 0, \ldots, n+1, \tag{9.27}$$

where the coefficient α_{n+1} indicates the maximum error:

$$|\alpha_{n+1}| = \|f - p_n\| = \mathbf{e}$$

Through (9.27), a system of $n+2$ linear equations in the $n+2$ unknowns $\alpha_0, \ldots, \alpha_{n+1}$ is given.

The Remez algorithm tries to determine the alternant iteratively. In the following, we outline the procedure: First, an initial estimate for the alternant $x_0^{(0)}, \ldots, x_{n+1}^{(0)}$ must be made. Possible choices are equally distributed nodes in the interval or the zeros of the corresponding Chebyshev polynomial.

We assume that an estimate $A^{(l)} := \{x_0^{(l)}, \ldots, x_{n+1}^{(l)}\}$ for the alternant is given. The step $l \mapsto l+1$ consists of four parts:

1. Using $A^{(l)}$, the linear system (9.27) is solved. The solution $\alpha_0^{(l)}, \ldots, \alpha_{n+1}^{(l)}$ determines the polynomial

$$p^{(l)}(x) = \sum_{j=0}^{n} \alpha_j^{(l)} x^j.$$

2. Determine the error function

$$e^{(l)}(x) = f(x) - p^{(l)}(x)$$

and the set of extreme points $y_0^{(l)}, \ldots, y_m^{(l)}$ of $e^{(l)}$.

3. If $n+2$ of these extreme points approximately fulfill the alternant property, the iteration stops. The alternant property is fulfilled if

$$\frac{\left| |e^{(l)}(y_i)| - \|e^{(l)}\|_\infty \right|}{\|e^{(l)}\|_\infty} < tol, \quad e^{(l)}(y_i) \cdot e^{(l)}(y_{i+1}) < 0.$$

4. If no alternant is present, a new estimate is obtained from the set of extreme points $y_0^{(l)}, \ldots, y_m^{(l)}$ and the set of previous points $x_0^{(l)}, \ldots, x_{n+1}^{(l)}$.

The new point $x_j^{(l+1)}$ is chosen as one of the extreme points $y_k^{(l)}$ that is as close as possible to $x_j^{(l)}$ and satisfies the sign condition

$$e^{(l)}(x_j^{(l+1)}) \cdot e^{(l)}(x_{j-1}^{(l+1)}) < 0$$

If this is not possible, the old points $x_j^{(l)}$ are reused.

Example 9.53 (Chebyshev Approximation) We consider the function $f(x) = \sqrt{x}$ on the interval $[0, 1]$ and search for the quadratic Chebyshev approximations $p \in P_2$ using the Remez algorithm. For $n = 2$, the desired alternant has 4 points, and we start with

$$A^{(0)} = \{0, 0.3, 0.7, 1\}.$$

We set up the linear system (9.27) (rounded to three significant digits) and get

$$\begin{pmatrix} 1 & 0 & 0 & 1 \\ 1 & 0.3 & 0.09 & -1 \\ 1 & 0.7 & 0.49 & 1 \\ 1 & 1 & 1 & 1 \end{pmatrix} \begin{pmatrix} \alpha_0^{(0)} \\ \alpha_1^{(0)} \\ \alpha_2^{(0)} \\ \alpha_3^{(0)} \end{pmatrix} = \begin{pmatrix} 0 \\ 0.548 \\ 0.837 \\ 1 \end{pmatrix}$$

with the solution

$$\alpha^{(0)} = \begin{pmatrix} 0.0396 \\ 1.84 \\ -0.917 \\ -0.0397 \end{pmatrix} \quad \Rightarrow \quad p^{(0)}(x) = 0.0397 + 1.84x - 0.917x^2.$$

In Fig. 9.8, we show the approximation on the left and the error function on the right.

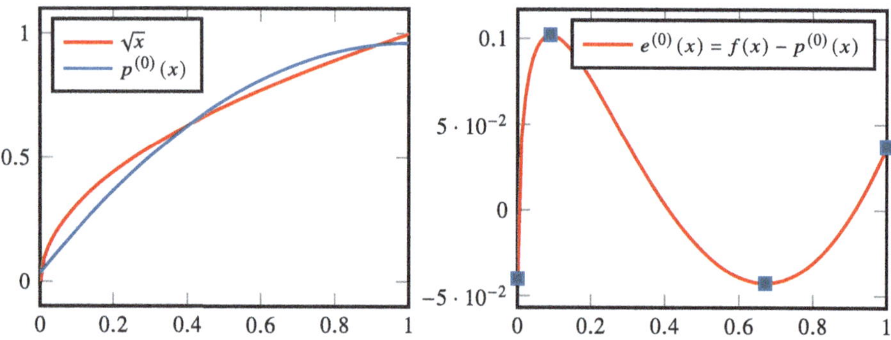

Fig. 9.8 First approximation and error function from the Remez algorithm for the Chebyshev approximation of $\sqrt{x}$

In addition to the two boundary points, the error function has two inner extreme points

$$\xi_0^{(0)} = 0, \quad \xi_1^{(0)} \approx 0.0889, \quad \xi_2^{(0)} \approx 0.671, \quad \xi_3^{(0)} = 1.$$

At the extreme points, we have

$$f(\xi_0^{(0)}) - p^{(0)}(\xi_0^{(0)}) \approx -0.0397, \quad f(\xi_1^{(0)}) - p^{(0)}(\xi_1^{(0)}) \approx 0.102,$$
$$f(\xi_2^{(0)}) - p^{(0)}(\xi_2^{(0)}) \approx -0.0423, \quad f(\xi_3^{(0)}) - p^{(0)}(\xi_3^{(0)}) \approx 0.0373.$$

The condition $e^{(0)}(\chi_i)e^{(0)}(\chi_{i+1}) < 0$ is satisfied, but $|e^{(0)}(\chi_i)| \approx |e^{(0)}(\chi_{i+1})| \approx |\alpha_3^{(0)}|$ does not hold, so we do not yet accept the alternant. In the next iteration, we therefore choose

$$A^{(1)} = \{\xi_0^{(0)}, \xi_1^{(0)}, \xi_2^{(0)}, \xi_3^{(0)} = 1\} = \{0, 0.0894, 0.669, 1\}.$$

We again set up the linear system

$$\begin{pmatrix} 1 & 0 & 0 & 1 \\ 1 & 0.0889 & 0.00790 & -1 \\ 1 & 0.671 & 0.450 & 1 \\ 1 & 1 & 1 & -1 \end{pmatrix} \begin{pmatrix} \alpha_0^{(1)} \\ \alpha_1^{(1)} \\ \alpha_2^{(1)} \\ \alpha_3^{(1)} \end{pmatrix} = \begin{pmatrix} 0 \\ 0.298 \\ 0.819 \\ 1 \end{pmatrix}$$

with the solution

$$\alpha^{(0)} = \begin{pmatrix} 0.0669 \\ 1.94 \\ -1.08 \\ -0.0669 \end{pmatrix} \quad \Rightarrow \quad p^{(1)}(x) = 0.0669 + 1.94x - 1.08x^2$$

which we show on the left in Fig. 9.9, while the error function is given on the right.

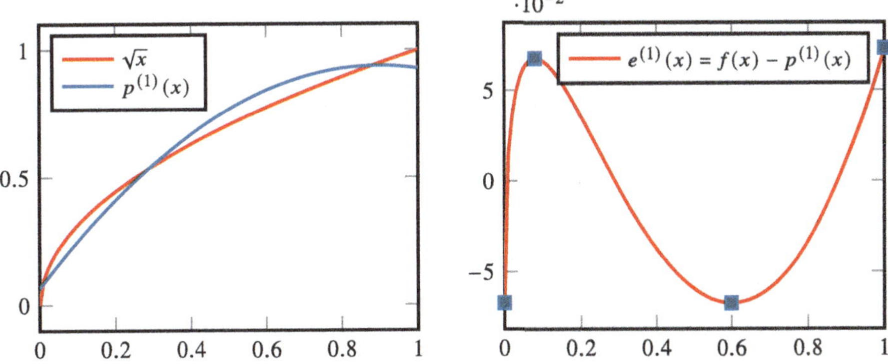

Fig. 9.9 Second approximation and error function from the Remez algorithm for the Chebyshev approximation of $\sqrt{x}$

The error function has the extreme points

$$\xi_0^{(1)} = 0, \quad \xi_1^{(1)} \approx 0.0801, \quad \xi_2^{(1)} \approx 0.599, \quad \xi_3^{(1)} = 1,$$

and here it holds

$$f(\xi_0^{(1)}) - p^{(1)}(\xi_0^{(1)}) \approx -0.0669, \quad f(\xi_1^{(1)}) - p^{(1)}(\xi_1^{(1)}) \approx 0.0677,$$
$$f(\xi_2^{(1)}) - p^{(1)}(\xi_2^{(1)}) \approx -0.0675, \quad f(\xi_3^{(1)}) - p^{(1)}(\xi_3^{(1)}) \approx 0.0731.$$

The alternation property is already very well fulfilled, and we accept $p^{(1)}(x) = 0.0669 + 1.94x - 1.08x^2 \in P_2$ as the quadratic Chebyshev approximation. ◄

9.5.3　Optimal Lagrange Interpolation

Finally, we want to use the Chebyshev approximation to create the optimal Lagrange interpolation, i.e., an optimal choice of nodes, for a function $f \in C^{n+1}[a, b]$. According to Theorem 9.13, the general Lagrange interpolation to the pairwise different nodes $x_0, \ldots, x_n$, satisfies

$$f(x) - p_n(x) = \frac{f^{(n+1)}(\xi)}{(n + 1)!} \prod_{i=0}^{n}(x - x_i)$$

with an intermediate value ξ. The goal is to determine the nodes $x_0, \ldots, x_n$, such that

$$\max_{x \in [a,b]} \|L(x)\|, \quad L(x) = \prod_{i=0}^{n}(x - x_i).$$

This choice of nodes is then optimal for general functions $f \in C^{n+1}[a, b]$ and does not depend on the specific function. This task can be described as a best approximation task. It holds

$$L(x) = x^{n+1} - g(x)$$

with a polynomial $g \in P_n$. So, the best approximation $g \in P_n$ to the function $f(x) = x^{n+1}$ in the interval $I = [a, b]$ is sought. From Theorem 9.52, we know that the error function $e(x) = f(x) - g(x)$ has at least $n + 1$ zeros (between the $n + 2$ extreme points).

Theorem 9.54 (Optimal Nodes of Lagrange Interpolation) *On the interval $[-1, 1]$, the Chebyshev approximation $g \in P_n$ to $f(x) = x^{n+1}$ is given by*

$$g(x) = x^{n+1} - 2^{-n}T_{n+1}(x)$$

with the Chebyshev polynomial (see Theorem 9.46)

$$T_{n+1}(x) = \cos\left((n + 1)\arccos(x)\right).$$

The zeros of $T_{n+1}(x)$

$$x_k = \cos\left(\frac{\pi(2k+1)}{2(n+1)}\right), \quad k = 0, \ldots, n,$$

are the sought optimal nodes *of the Lagrange interpolation on* $[-1, 1]$. *On the general interval* $[a, b]$, *the optimal nodes are given by*

$$y_k = a + \frac{b-a}{2}(1 + x_k).$$

Proof The zeros of the Chebyshev polynomials are easy to evaluate, see also Theorem 9.46. As also shown in the latter theorem, the following recursive relationship

$$T_0(x) = 1, \quad T_1(x) = x, \quad T_{n+1}(x) = 2x T_n(x) - T_{n-1}(x)$$

holds. From this, it follows

$$T_{n+1}(x) = 2^n x^{n+1} + q(x)$$

with a $q \in P_n$. On the other hand, T_{n+1} can be written in terms of its linear factors as

$$2^{-n} T_{n+1}(x) = \prod_{k=0}^{n}(x - x_k) = L(x).$$

From this, it further follows

$$\max_{[-1,1]} |L(x)| = \max_{[-1,1]} \prod_{k=0}^{n} |x - x_k| = 2^{-n} \max_{[-1,1]} |T_{n+1}(x)| = 2^{-n}.$$

The function $T_{n+1}(x)$ takes exactly $n + 2$ times the extreme values ± 1. These $n + 2$ extreme points form an alternant for the unique best approximation to $f(x) = x^{n+1}$

$$g(x) = x^{n+1} - 2^{-n} T_{n+1}(x) \in P_n.$$

It remains to show that this best approximation really minimizes the product $|L(x)|$ in $[-1, 1]$:

$$\max_{[-1,1]} |L(x)| = 2^{-n} \max_{[-1,1]} |T_{n+1}(x)|$$

$$= \max_{[-1,1]} \left|x^{n+1} - (x^{n+1} - 2^{-n} T_{n+1}(x))\right| = \max_{[-1,1]} \left|x^{n+1} - g(x)\right|$$

$$\leq \min_{p \in P_n} \max_{[-1,1]} \left|x^{n+1} - p(x)\right|$$

$$\leq \min_{-1 \leq y_0 < \cdots < y_n \leq 1} \max_{[-1,1]} \left|\prod_{k=0}^{n}(x - y_k)\right|,$$

i.e., the choice of nodes is indeed optimal. $\qquad\square$

9.6   Trigonometric Interpolation

So far, we have exclusively dealt with interpolation in polynomial spaces. In this section, we consider the interpolation with trigonometric polynomials, i.e., functions of the type

$$t_n(x) = \frac{1}{2}a_0 + \sum_{k=1}^{m}\left(a_k \cos\left(\frac{2\pi kx}{\omega}\right) + b_k \sin\left(\frac{2\pi kx}{\omega}\right)\right),$$

where $n = 2m$ and $a_k, b_k \in \mathbb{R}$ are the real coefficients are. For these *trigonometric polynomials* it holds

$$t_n(x) = t_n(x + \omega),$$

i.e., $\omega > 0$ is the period of the function $t_n(x)$. Trigonometric interpolation is thus suitable for the interpolation of periodic functions or data. Fourier analysis, known from analysis, is not interpolation, but can be considered as the best approximation (see Sect. 9.5) in the space of trigonometric functions.

In the following, we consider without loss of generality the period $\omega = 2\pi$, so that the trigonometric basis polynomials can be simplified as

$$t_n(x) = \frac{1}{2}a_0 + \sum_{j=1}^{m}\left\{a_j \cos(jx) + b_j \sin(jx)\right\}. \tag{9.28}$$

We investigate the interpolation task

$$t_n(x_k) = y_k, \quad k = 0, \ldots, n.$$

Here, we consider equidistantly nodes in the interval $[0, 2\pi]$, i.e.,

$$x_k = \frac{2\pi k}{n+1}, \quad k = 0, \ldots, n. \tag{9.29}$$

In the case of trigonometric functions, the consideration of complex numbers is often simpler. For coefficients $c_0, \ldots, c_n \in \mathbb{C}$, we define the 2π-periodic function

$$t_n^*(x) = \sum_{k=0}^{n} c_k e^{ikx}$$

with the complex unit $i^2 = -1$.

Theorem 9.55 (Complex Trigonometric Interpolation) *Let $x_0, \ldots, x_n \in [0, \pi)$ be the $n + 1$ pairwise different equidistant nodes and $y_0, \ldots, y_n \in \mathbb{C}$ support values. Then the trigonometric interpolation polynomial*

$$t_n^*(x_i) = y_i, \quad i = 0, \ldots, n$$

is uniquely determined. The coefficients are determined as

$$c_k = \frac{1}{n+1} \sum_{j=0}^{n} y_j e^{-ijx_k}.$$

Proof *(i)* We define

$$w := e^{ix}, \quad w_k := e^{ix_k}.$$

Then the polynomial t_n^* is written as

$$t_n^*(w) = \sum_{j=0}^{n} c_j w^j$$

and the interpolation task as

$$t_n^*(w_k) = y_k, \quad k = 0, \ldots, n.$$

The interpolation polynomial is then uniquely determined in the sense of Lagrange interpolation.

(ii) To calculate the coefficients, we observe that the w_k are just the $n+1$-th roots of unity, because

$$w_k^{n+1} = \exp(2\pi ik) = 1 \quad \forall k \in \mathbb{N}.$$

Further, it holds

$$w_k^j = \exp\left(\frac{2\pi ijk}{n+1}\right) = w_j^k.$$

Then, we have

$$\sum_{j=0}^{n} y_j w_k^{-j} = \sum_{j=0}^{n} t_n^*(x_j) w_k^{-j} = \sum_{j,l=0}^{n} c_l w_j^l w_k^{-j}$$

$$= \sum_{l=0}^{n} c_l \sum_{j=0}^{n} w_j^{l-k} = \sum_{l=0}^{n} c_l \sum_{j=0}^{n} w_{l-k}^j. \tag{9.30}$$

For an $(n+1)$-th root of unity, it holds

$$0 = w^{n+1} - 1 = (w-1)(w^n + w^{n-1} + \cdots + 1).$$

Therefore,

$$\sum_{j=0}^{n} w_j^{l-k} = \begin{cases} 0 & l \neq k \\ n+1 & l = k, \end{cases}$$

since $w_0 = 1$. Then, using (9.30), we obtain the representation

$$\sum_{j=0}^{n} y_j w_k^{-j} = (n+1)c_k \quad \Leftrightarrow \quad c_k = \frac{1}{n+1} \sum_{j=0}^{n} y_j w_k^{-j}.$$

$\square$

Due to the periodic structure of the basis polynomials and the equidistant distribution of the nodes, we have:

$$c_{n+1-k} = \frac{1}{n+1} \sum_{j=0}^{n} y_j e^{-ijx_{n+1-k}} = \frac{1}{n+1} \sum_{j=0}^{n} y_j e^{-\frac{2i\pi j(n+1-k)}{n+1}}$$

$$= \frac{1}{n+1} \sum_{j=0}^{n} y_j e^{\frac{2i\pi jk}{n+1}} =: c_{-k} \qquad (9.31)$$

The relationship between the complex polynomials $t_n^*(x)$ and a desired real trigonometric interpolation $t_n(x)$ is achieved on the basis of Euler's formula

$$e^{ix} = \cos(x) + i\sin(x).$$

Theorem 9.56 (Real Trigonometric Interpolation) *For $n \in \mathbb{N}$ let $x_0, \ldots, x_n \in [0, \pi)$ and $y_0, \ldots, y_n \in \mathbb{R}$. There is exactly one trigonometric polynomial of the form*

$$t_n(x) = \frac{1}{2}a_0 + \sum_{k=1}^{m} \left(a_k \cos(kx) + b_k \sin(kx) \right) + \frac{\theta}{2} a_{m+1} \cos\big((m+1)x\big),$$

with $t_n(x_j) = y_j$ for $j = 0, \ldots, n$ and

$$m = \begin{cases} \frac{n}{2} & n \text{ even} \\ \frac{n-1}{2} & n \text{ odd} \end{cases}, \quad \theta = \begin{cases} 0 & n \text{ even} \\ 1 & n \text{ odd} \end{cases}.$$

The real coefficients are determined as

$$a_k = \frac{2}{n+1} \sum_{j=0}^{n} y_j \cos(jx_k), \quad b_k = \frac{2}{n+1} \sum_{j=0}^{n} y_j \sin(jx_k).$$

Proof *(i)* Let t_n^* be the complex interpolation polynomial determined according to Theorem 9.55 with

$$t_n^*(x_j) = y_j, \quad j = 0, \ldots, n.$$

We define the coefficients using (9.31) as

$$a_k := c_k + c_{-k}, \quad b_k = i(c_k - c_{-k}), \quad k = 1, \ldots, m,$$
$$a_{m+1} := 2c_{m+1}.$$

The coefficients are real, since

$$c_k \pm c_{-k} = \frac{1}{n+1} \sum_{j=0}^{n} y_j \left(\left(\cos(jx_k) - i\sin(jx_k) \right) \pm \left(\cos(jx_k) + i\sin(jx_k) \right) \right)$$

$$\Rightarrow \quad a_k = \frac{2}{n+1} \sum_{j=0}^{n} y_j \cos(jx_k),$$

$$b_k = \frac{2}{n+1} \sum_{j=0}^{n} y_j \sin(jx_k).$$

For $n = 2m + 1$ odd, we obtain

$$m + 1 = \frac{n+1}{2} = (n+1) - \frac{n+1}{2} = (n+1) - (m+1),$$

so $a_{m+1} = c_{m+1} + c_{-(m+1)} \in \mathbb{R}$. We define the polynomial

$$t_n(x) = \frac{a_0}{2} + \sum_{k=1}^{m} \left(a_k \cos(kx) + b_k \sin(kx) \right) + \frac{\theta}{2} a_{m+1} \cos\left((m+1)x \right).$$

(ii) We show that

$$t_n(x_j) = t_n^*(x_j), \quad j = 0, \ldots, n.$$

It holds

$$t_n(x_j) = c_0 + \sum_{k=1}^{m} \left((c_k + c_{-k}) \cos(kx_j) + i(c_k - c_{-k}) \sin(kx_j) \right)$$
$$+ \theta c_{m+1} \cos\left((m+1)x_j \right)$$
$$= c_0 + \sum_{k=1}^{m} \left(c_k \left(\cos(kx_j) + i\sin(kx_j) \right) + c_{-k} \left(\cos(kx_j) - i\sin(kx_j) \right) \right)$$
$$+ \theta c_{m+1} \left(\cos((m+1)x_j) + \underbrace{i\sin((m+1)x_j)}_{=0 \text{ for } n=2m+1 \text{ odd}} \right)$$
$$= c_0 + \sum_{k=1}^{m} \left(c_k e^{ikx_j} + c_{-k} e^{-ikx_j} \right) + \theta c_{m+1} e^{i(m+1)x_j},$$

since $\cos(x) = \cos(-x)$ and $-\sin(x) = \sin(-x)$. In the case $n = 2m + 1$, odd, we could insert the term

$$\sin\left((m+1)x_j\right) = \sin\left(\frac{n+1}{2}\frac{2\pi j}{n+1}\right) = \sin(\pi j) = 0, \quad j = 0, \ldots, n.$$

Using (9.31), i.e., $c_{-k} = c_{n+1-k}$ and

$$e^{-ikx_j} = e^{i\frac{2\pi kj}{n+1}} = e^{i\frac{2\pi(n+1-k)j}{n+1}} = e^{i(n+1-k)x_j},$$

it follows

$$t_n(x_j) = \sum_{k=0}^{n} c_k e^{ikx_j} y_j = 0,$$

i.e., the interpolation condition is fulfilled at all points.

(iii) It remains to show the uniqueness of this interpolation. We can proceed as usual: To determine the $n + 1$ coefficients $a_0, \ldots, a_m$ (and a_{m+1}, if n is odd) and $b_1, \ldots, b_m$, $n + 1$ conditions $t_n(x_j) = y_j$ must be fulfilled. We have shown that this system is solvable for any values $y_0, \ldots, y_n$. The corresponding $(n + 1) \times (n + 1)$-matrix thus has full rank and is regular. The interpolation is therefore uniquely determined. $\qquad\square$

Trigonometric interpolation is suitable for approximating periodic functions. Often, functions are not periodic or are only defined on an interval $[a, b]$. In this case, the function can first be transformed to the interval $[0, \pi]$ and then continued periodically:

$$\bar{f}(x) = \begin{cases} f(x) & x \in [0, 2\pi] \\ f(x - 2j\pi) & x \in (2\pi j, 2\pi(j + 1)) \end{cases}$$

In the case $f(0) \neq f(2\pi)$, the periodically continued function has discontinuities at the points $2\pi j$. We define

$$f(2\pi j) = \frac{1}{2}(f(0) + f(2\pi)).$$

This averaging is motivated by the Fourier expansion of functions, which converges at discontinuities against the average of the two one-sided limits.

We assume that a continuous function $f \in C[0, \pi]$ with $f(0) = f(\pi) = 0$ is given. This can be continued in two ways. The *even continuation* is defined as

$$\bar{f}(x) = \begin{cases} f(x) & x \in [0, \pi] \\ f(2\pi - x) & x \in [\pi, 2\pi] \\ 2\pi - \text{periodic} & \text{elsewhere,} \end{cases} \tag{9.32}$$

the *odd continuation* as

$$\bar{f}(x) = \begin{cases} f(x) & x \in [0, \pi] \\ -f(2\pi - x) & x \in [\pi, 2\pi] \\ 2\pi - \text{periodic} & \text{elsewhere.} \end{cases} \tag{9.33}$$

For these two, special cases arise in the calculation of the coefficients a_k and b_k, n = 2m + 1 odd:

Theorem 9.57 (Cosine and Sine Transformation) *Let $f \in C[0, \pi]$ be a function with $f(0) = f(\pi) = 0$. For the trigonometric interpolation of the even continuation*

$$x_k = \frac{2\pi k}{n + 1}, \quad y_k = \begin{cases} f(x_k) & k = 0, \ldots, m, \\ f(2\pi - x_k) & k = m + 1, \ldots, n, \end{cases}$$

it holds

$$c(x) = \frac{1}{2}a_0 + \sum_{k=1}^{m+1} a_k \cos(kx), \quad a_k = \frac{2}{m + 1} \sum_{j=0}^{m} f(x_j) \cos(jx_k).$$

For the trigonometric interpolation of the odd continuation

$$x_k = \frac{2\pi k}{n + 1}, \quad y_k = \begin{cases} f(x_k) & k = 0, \ldots, m, \\ -f(2\pi - x_k) & k = m + 1, \ldots, n, \end{cases}$$

it holds

$$s(x) = \sum_{k=1}^{m} b_k \sin(kx), \quad b_k = \frac{2}{m + 1} \sum_{j=0}^{m} f(x_j) \sin(jx_k).$$

Proof We first consider the complex coefficients. In the case of the even continuation, with $y_k = y_{n+1-k}$, we have

$$c_k = \frac{1}{n + 1} \sum_{j=0}^{n} y_j e^{-ijx_k} = \frac{1}{n + 1} \sum_{j=0}^{n} y_{n+1-j} e^{-ijx_k}$$

$$= \frac{1}{n + 1} \sum_{j=0}^{n} y_j e^{-i(n+1-j)x_k} = c_{n+1-k} = c_{-k}.$$

So it follows

$$b_k = i(c_k - c_{-k}) = 0$$

and

$$a_k = c_k + c_{-k} = 2c_k = \frac{2}{n+1} \sum_{j=0}^{m} y_j e^{-ijx_k} + \frac{2}{n+1} \sum_{j=m+1}^{n} y_j e^{-ijx_k}$$

$$= \frac{2}{n+1} \sum_{j=0}^{m} y_j e^{-ijx_k} + \frac{2}{n+1} \sum_{j=1}^{m} \underbrace{y_{n+1-j}}_{=y_j} e^{-i(n+1-j)x_k}$$

$$= \frac{2}{n+1} \sum_{j=0}^{m} y_j e^{-ijx_k} + \frac{2}{n+1} \sum_{j=1}^{m} y_j e^{ijx_k},$$

since

$$e^{-i(n+1-j)x_k} = e^{-i2\pi\left(1-\frac{j}{n+1}\right)k} = e^{-i2\pi k} e^{ijx_k} = e^{ijx_k}.$$

Here, we have used $e^{-i2\pi k} = 1$ in the last step. Taking into account $y_0 = y_{m+1} = 0$, this implies

$$a_k = \frac{4}{n+1} \sum_{j=1}^{m} y_j \frac{e^{ijx_k} + e^{-ijx_k}}{2} = \frac{2}{m+1} \sum_{j=1}^{m} y_j \cos(jx_k).$$

In the case of the odd continuation, the proof follows analogously. $\qquad\square$

We conclude this section with an example in which we approximate a 2π-periodic function using trigonometric interpolation.

Example 9.58 Let f be a 2π-periodic function with

$$f(x) = \begin{cases} -1 & x \in [0, \pi), \\ +1 & x \in [\pi, 2\pi). \end{cases}$$

The function is shown on the left of Fig. 9.10.

We apply Theorem 9.56 with equidistant choice of the nodes x_j according to (9.29). We calculate the trigonometric polynomials $t_n(x)$ up to $n = 3$. We list our results in Table 9.6 and the corresponding visualization can be found on the right of Fig. 9.10. ◀

9.7   Discrete Fourier Transformation

The transformation of support values $y_0, \ldots, y_n$ to coefficients a_k, b_k is called *discrete Fourier transformation* or in the case of an even or odd continuation *discrete sine transformation* and *discrete cosine transformation*. In this section, we will derive with the *fast Fourier transformation* (FFT) a very fast algorithm to calculate the coefficients a_k and b_k of the transformation.

We have seen in the previous section that this process is reversible, i.e., there exist bijective mappings

Table 9.6 Trigonometric polynomial interpolation for Example 9.58

n	m	θ	x_j	y_j	$t_n(x)$	a_k	b_k
0	0	0	0	-1	$\frac{1}{2}a_0$	$a_0 - 2$	
1	0	1	0	-1		$a_0 = 0$	
			π	$+1$	$\frac{1}{2}a_0 + \frac{1}{2}a_1\cos(x)$	$a_1 = -2$	
2	1	0	0	-1		$a_0 = -\frac{2}{3}$	
			$\frac{2}{3}\pi$	-1		$a_1 = -\frac{2}{3}$	
			$\frac{4}{3}\pi$	$+1$	$\frac{1}{2}a_0 + a_1\cos(x) +$ $b_1\sin(x)$		$b_1 = -1.1547$
3	1	1	0	-1		$a_0 = 0$	
			$\frac{1}{2}\pi$	-1		$a_1 = -1$	
			π	$+1$			$b_1 = -1$
			$\frac{3}{2}\pi$	$+1$	$\frac{1}{2}a_0 + a_1\cos(x) +$ $b_1\sin(x) +$ $\frac{1}{2}a_2\cos(2x)$	$a_2 = 0$	

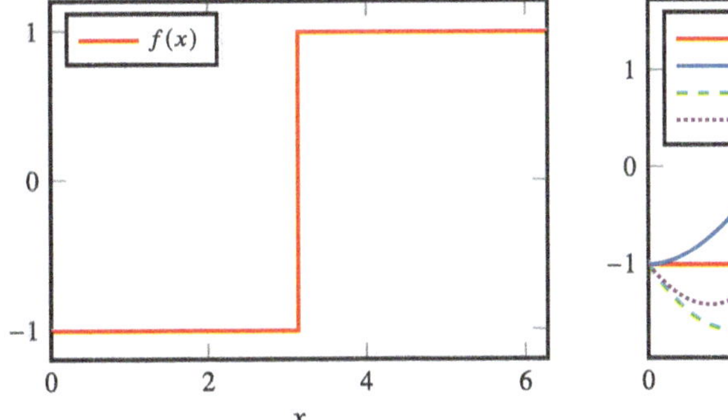

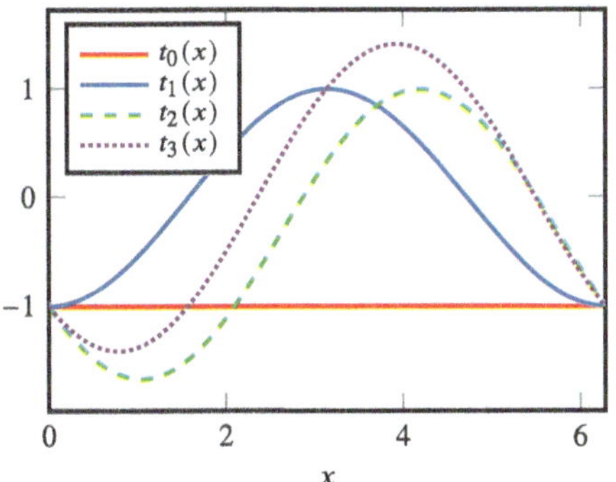

Fig. 9.10 Example 9.58. Left: The function f is to be represented using trigonometric interpolation. Right: Approximation of the function f using the trigonometric polynomials $t_0(x), \ldots, t_3(x)$, which were calculated in Table 9.6. We can immediately see that the interpolation conditions $t_n(x_j) = y_j$ are fulfilled in each case

$$\{y_0, \ldots, y_n\} \quad \leftrightarrow \quad \{a_0, \ldots, a_{m+1}, b_1, \ldots, b_m\}$$

and

$$\{y_0, \ldots, y_m\} \quad \leftrightarrow \quad \begin{cases} \{a_0, \ldots, a_{m+1}\} & \text{even extension} \\ \{b_1, \ldots, b_m\} & \text{odd extension.} \end{cases}$$

The basis polynomials $\cos(kx)$ and $\sin(kx)$ can be seen as the essential oscillations of the periodic function $f(x)$. The coefficients a_k, b_k indicate the respective dominance of the frequencies. The discrete Fourier transformation helps to analyze data. For example, from measurements, the dominant oscillations of mechanical structures can be determined. On the other hand, the Fourier transformation is the basis for many compression methods. For example, the *JPEG* format for images is based on the cosine transformation, see Excursus 9.8.

To derive the *fast Fourier transformation*, we choose the complex notation, in which complex values

$$c_k = \sum_{j=0}^{n} \bar{y}_j w^{jk}, \quad k = 0, \ldots, n, \tag{9.34}$$

with the abbreviations

$$\bar{y}_j := \frac{1}{n+1} y_j, \quad w := e^{-i\frac{2\pi}{n+1}},$$

are to be calculated. From these, a_k and b_k are determined as

$$a_k = c_k + c_{-k}, \quad b_k = i(c_k - c_{-k}). \tag{9.35}$$

The calculation of the $n+1$ values c_k with the Horner scheme would require $n^2 + O(n)$ operations. A more efficient method can be achieved by a hierarchical division of the sums. For the following, let $n+1 = 2^p$ for $p \in \mathbb{N}$. Then

$$\begin{aligned}
c_k &= \sum_{j=0}^{2^p-1} \bar{y}_j w^{jk} \\
&= \sum_{j=0}^{2^{p-1}-1} \left(\bar{y}_{2j}(w^2)^{jk} + \bar{y}_{2j+1}(w^2)^{jk} w^k \right) \\
&= \sum_{j=0}^{2^{p-1}-1} \bar{y}_{2j}(w^2)^{jk} + \sum_{j=0}^{2^{p-1}-1} \bar{y}_{2j+1}(w^2)^{jk} w^k.
\end{aligned} \tag{9.36}$$

Now, let $k_1 \in \{0, 1, \ldots, 2^{p-1} - 1\}$ be the integer remainder when dividing k by 2^{p-1}, i.e.,

$$k \equiv k_1 \bmod 2^{p-1}.$$

Because $w^{n+1} = w^{2^p} = 1$, it follows for a $\lambda \in \mathbb{N}$ that

$$(w^2)^{jk} = (w^2)^{\lambda 2^{p-1} j}(w^2)^{jk_1} = (w^{2^p})^{j}(w^2)^{jk_1} = (w^2)^{jk_1}.$$

With this, (9.36) can be written as

$$c_k = \sum_{j=0}^{2^{p-1}-1} \bar{y}_{2j}(w^2)^{jk_1} + w^k \sum_{j=0}^{2^{p-1}-1} \bar{y}_{2j+1}(w^2)^{jk_1},$$

or

$$c_k = \tilde{c}_{k_1} + w^k \bar{c}_{k_1}, \quad k = 0, \ldots, 2^p - 1, \quad k \equiv k_1 \bmod 2^{p-1},$$

where

$$\tilde{c}_{k_1} = \sum_{j=0}^{2^{p-1}-1} \bar{y}_{2j}(w^2)^{jk_1}, \quad \bar{c}_{k_1} = \sum_{j=0}^{2^{p-1}-1} \bar{y}_{2j+1}(w^2)^{jk_1}, \quad k_1 = 0, \ldots, 2^{p-1} - 1.$$

To calculate the $n + 1 = 2^p$ quantities $c_0, \ldots, c_n$ it is therefore sufficient to calculate the two times 2^{p-1} quantities $\tilde{c}_0, \ldots, \tilde{c}_{2^{p-1}-1}$ and $\bar{c}_0, \ldots, \bar{c}_{2^{p-1}-1}$. The $n + 1 = 2^p$-dimensional Fourier transformation was replaced by two Fourier transformations of dimension 2^{p-1}. If we were to determine these coefficients each with the Horner scheme, we would have already halved the effort, since

$$2 \cdot \left(\frac{n}{2}\right)^2 = \frac{n^2}{2}.$$

The FFT algorithm is based on the recursive application of this technique. The two Fourier transformations of dimension 2^{p-1} are replaced by four transformations of dimension 2^{p-2}. Finally, there are 2^p Fourier transformations of trivial dimension 1. These can be "solved" without further computational effort by assigning the function values $\bar{y}_j$.

Theorem 9.59 (Fast Fourier Transformation) *In the case of $n + 1 = 2^p$, the* fast Fourier transformation *calculates the $n + 1$ coefficients $c_0, \ldots, c_n$ with*

$$2(n + 1) \log_2(n + 1)$$

complex operations.

Proof Assume that the effort to calculate the $n + 1 = 2^p$ components is given by r_p. If $\tilde{c}_{k_1}, \bar{c}_{k_1}$ are known, the powers w^k and the sums

$$c_k = \tilde{c}_{k_1} + w^k \bar{c}_{k_1}, \quad k = 0, \ldots, n$$

need to be calculated. For this, $(n + 1) + (n - 1) = 2n \leq 2 \cdot 2^p$ complex operations are necessary. Iteratively, we get

$$r_p \leq 2 \cdot r_{p-1} + 2 \cdot 2^p, \quad p = 1, 2, \ldots.$$

We show by induction that $r_p \leq 2p \cdot 2^p$. The initial case with $r_0 = 0$ is trivial. Then, using the induction hypothesis, we get

$$r_p \leq 2 \cdot r_{p-1} + 2 \cdot 2^p = 2 \cdot (2(p - 1) \cdot 2^{p-1}) + 2 \cdot 2^p$$
$$= 2p \cdot 2^p = 2(n + 1) \log_2(n + 1).$$

$\square$

In a practical implementation of the fast Fourier transform, we start with the 2^p atomic coefficients, which are given by the support values y_j. Subsequently, the recursive formulas

$$c_k = \tilde{c}_{k_1} + \bar{c}_{k_1}, \quad k = 0, \ldots, 2^p - 1, \ k \equiv k_1 \bmod 2^{p-1},$$

are executed. For $n + 1 = 1\,024$, the fast Fourier transform requires $20\,480$ complex operations. Based on the Horner scheme, $1\,000\,000$ complex operations would be necessary. The fast Fourier transform is one of the few methods that, if applicable, only has advantages. This is rare in numerics. Most "advanced" numerical methods have certain disadvantages, e.g., higher regularity requirements (in the Gauss approximation) or less robustness (in Newton's method with poor initial values). Another method of this kind is the multigrid method to solve sparse systems. If it can be applied, a fixed error reduction in linear effort $O(n)$ is achieved.

9.8   Excursus: The JPEG Format for Image Compression

One of the most common formats for efficient storage of digital images is the *JPEG*[2] *file format* [89]. The mathematical background of the file format is the *discrete cosine transformation*, see Theorem 9.57 and Sect. 9.7. The image data is transformed into its frequency components using the fast cosine transformation. For compression, not all frequencies are stored, but—depending on the desired quality—only the dominant components are stored. The JPEG image is then obtained as the inverse transformation of this reduced data. The JPEG format is a *lossy compression method*. In Fig. 9.11, we show an example image and a small detail of the size 256×256 pixels. The detail is shown in four different quality levels: 100% (this corresponds to a lossless representation), 50, 12, and 6%. Increasing artifacts in the representation can be seen. In particular, a block structure of the image is recognizable. We will later recognize the causes.

Even though the mathematical core of the JPEG format is the cosine transformation, many intermediate steps are necessary to create an efficient, i.e., sparse representation, that simultaneously has high quality. We present the individual steps in detail below:

1. Adjustment of the color space and *subsampling*. The image is divided into three channels for the representation of brightness and color. The individual channels are then treated like black-and-white images.
2. Block formation and cosine transformation. The individual channels are divided into blocks of size 8×8. In each of these blocks, the discrete cosine transformation is performed.
3. Quantization. This is the lossy compression; not all frequencies are stored equally.
4. Further compression by encoding. Further encoding methods compress the result, but with lossless storage.

Color Spaces

---

[2] *Joint Photographic Expert Group* was introduced in 1992 in the standard ISO/IEC 10918-1.

The representation of b/w images (black and white) is straightforward on a computer. An image of size $x \times y$ consists of $x \cdot y$ values for the brightness of each pixel. The *color depth* describes how many levels are used to represent the brightness of each pixel. Typical values are 8-bit or 16-bit. Here, we limit ourselves to a color depth of 8 bits. Thus, for each pixel, 256 different brightness levels can be specified, see Sect. 1.4. For example, the value 0 can stand for black, the value 255 for white. The quality of an image thus depends on the one hand on the resolution, i.e., the number of pixels $X \cdot Y$, and on the other hand on the color depth.

To display colored images, various color spaces exist. The most common model is the *RGB color space*. Here, intensity components (corresponding to the brightness) are stored for the three colors red, green, and blue. The actual image is then obtained by additive mixing of these three components. This color model is common for display on monitors. An RGB image with resolution 800×800 and color depth 8 bit can thus be described in the form of three fields:

```
int8 R[640000], G[640000], B[640000];
```

Here, `int8` denotes a data type for an integer number with 8 bits. Without compression, 1 920 000 bytes, about 2 megabytes, are needed for storage. The JPEG compression is based on the analysis of *monochrome* images. One variant for encoding would thus be a separate application of the procedure to the three color channels red, green, and blue. However, the image is first transformed into the $Y'CbCr$ color space. This model also relies on three channels, the Y' channel, which indicates the brightness (luminance), and the Cb and Cr channels, which indicate the chrominance, i.e., the colorfulness with respect to the blue-yellow and red-green values. In Fig. 9.12, we show the decomposition of an RGB image into these three channels. The individual channels are again monochrome, so the image can be stored (losslessly) in the following form:

```
int8 Y[640000], Cb[640000], Cr[640000];
```

If the red/green/blue values are in the range $\{0, 255\}$, then the image can be transformed into the $Y'CbCr$ color space by a simple linear transformation:

$$\begin{bmatrix} Y' \\ Cb \\ Cr \end{bmatrix} \approx \begin{bmatrix} 0 \\ 128 \\ 128 \end{bmatrix} + \begin{bmatrix} 0.299 & 0.587 & 0.114 \\ -0.168736 & -0.331264 & 0.5 \\ 0.5 & -0.418688 & -0.081312 \end{bmatrix} \cdot \begin{bmatrix} R \\ G \\ B \end{bmatrix} \tag{9.37}$$

Except for rounding, this transformation is lossless. The reason for this transformation lies in the perception of the human eye: The resolution for brightness differences is much more pronounced in the eye than the resolution for the perception of color differences. For a good representation of an image, the Y' channel is, therefore, more important than the Cb and Cr channels.

At this point, the first step of JPEG compression, the so-called *subsampling*, comes into play: In the fields for Cb and Cr, only every second piece of information is stored.[3] With $4:2:0$-*subsampling*, far less information needs to be stored:

```
int8 YY[640000], Cb[160000], Cr[160000];
```

In total, the storage requirement is reduced by a factor of 2 to about 1 megabyte. The human eye usually cannot perceive these differences.

These three *monochrome* fields separately serve as the basis for the following actual JPEG compression. We will therefore describe this only using black-and-white fields, regardless of whether the intensity indicates a brightness or a color value.

Block Formation and Cosine Transformation

We continue to assume that a monochrome intensity field with an 8-bit color depth and resolution $x \times y$ is present. In the first step, this field is divided into blocks of size 8×8. We assume that both image width x and image height y are multiples of 8. In the case of color channels Cb and Cr, this means that due to the *subsampling*, the image width and image height of the original image must be multiples of 16. The artifacts seen in Fig. 9.11 are exactly this blocking when storing the image at low quality. If the original image cannot be divided into blocks of size 8×8, further adjustments are required at the boundary, see [89]. In Fig. 9.13, we show such a section of size 8×8 from the Y channel, i.e., the brightness of the image. We also provide the intensities in the range $\{-128, 127\}$ as a matrix $\mathbf{B}$. The shift of the discrete values around the center 0 later facilitates the discrete cosine transformation.

In the following step, a discrete cosine transformation of the intensity values is performed. See Theorem 9.57 and the discrete variant in Sect. 9.7. We want to represent the matrix $\mathbf{B} \in \mathbb{R}^{8 \times 8}$ using a two-dimensional cosine transformation. For this, we choose the form

$$\mathbf{B}_{ij} = \frac{1}{4} \sum_{u,v=1}^{8} \alpha(u)\alpha(v)\mathbf{G}_{uv} \cos\left(\frac{2i-1}{16}u\pi\right) \cos\left(\frac{2j-1}{16}v\pi\right) \tag{9.38}$$

with the factors

$$\alpha(u) = \begin{cases} \frac{1}{\sqrt{2}} & u = 1 \\ 1 & 1 < u \le 8. \end{cases}$$

The special form of the basis (9.38) results from the chosen symmetry of the even continuation at the pixels $\frac{1}{2}$ and $8 + \frac{1}{2}$, so the endpoints are doubled in the continuation. In Fig. 9.14 we show the 64 different basis functions of the discrete cosine transformation.

---

[3] The JPEG standard provides for various rates for subsampling. The rate $4:4:4$ means no subsampling, i.e., full information retention, the usual rate $4:2:0$ means horizontal and vertical reduction of the Cb and Cr information by a factor of 2, and the rate $4:2:2$ means a horizontal reduction of the color rates by a factor of 2, but full information retention in the vertical direction. This compression is common in various video systems, see [89].

Fig. 9.11 Original image (left) and detail (right) with different compression strengths. File sizes 100, 26, 18, and 14 kB (from top left to bottom right)

Theorem 9.60 (Discrete Two-Dimensional Cosine Transformation) *Let* $\mathbf{B} \in \mathbb{R}^{8\times8}$. *For the coefficient matrix* $\mathbf{G} \in \mathbb{R}^{8\times8}$ *of the discrete two-dimensional cosine transformation, it holds*

$$\mathbf{G}_{uv} = \frac{1}{4}\alpha(u-1)\alpha(v-1)\sum_{i=1}^{8}\sum_{j=1}^{8}\mathbf{B}_{ij}\cos\left(\frac{2i-1}{16}u - 1\pi\right)\cos\left(\frac{2j-1}{16}v - 1\pi\right).$$

Proof The proof is first reduced to the one-dimensional case due to the tensor product form. Inserting the coefficients yields

$$\mathbf{B}_{ij} \overset{!}{=} \sum_{k,l=1}^{8}\mathbf{B}_{kl}\left(\sum_{u=1}^{8}\frac{\alpha(u-1)^2}{4}\cos\left(\frac{2k-1}{16}u - 1\pi\right)\cos\left(\frac{2i-1}{16}u - 1\pi\right)\right)$$
$$\cdot\left(\sum_{v=1}^{8}\frac{\alpha(v-1)^2}{4}\cos\left(\frac{2l-1}{16}v - 1\pi\right)\cos\left(\frac{2j-1}{16}v - 1\pi\right)\right).$$

The result then follows with the orthogonality of the cosine function. $\square$

Fig. 9.12 Original image (top left) and representation in the $Y'CbCr$ color space. The brightness (luminance) Y' is shown at the top right, the blue-difference chrominance (colorfulness) Cb at the bottom left, and the red-difference chrominance Cr at the bottom right

For our example, the transformation of **B** is given by

$$
\mathbf{G} \approx \begin{bmatrix}
-110 & -203 & 214 & 45.8 & 4.87 & 33.5 & 6.42 & -2.63 \\
365 & 187 & -72.5 & 26.1 & 25.1 & -44.9 & 10.6 & -2.22 \\
58.6 & 139 & 24.9 & -60.5 & 24.9 & 12.7 & -60.1 & 35.5 \\
-16.7 & -63.2 & -0.407 & 10.9 & -65.5 & 17 & 14.5 & -22.7 \\
-1.38 & -30 & -17.8 & 12.6 & -7.38 & -10.9 & 6.41 & -0.757 \\
21.4 & 1.96 & -24.8 & -4.62 & 24.1 & -13.8 & -1.85 & 15.6 \\
21 & 13.9 & -11.8 & 9.18 & 7.43 & -2.24 & 13.1 & -3.74 \\
-2.13 & -0.791 & 7.58 & 2.12 & -2.35 & 5.35 & 8.92 & -3.74
\end{bmatrix},
\tag{9.39}
$$

$$\begin{bmatrix} 105 & 90 & 59 & 42 & 50 & 36 & 50 & 48 \\ 97 & 69 & 22 & 14 & 9 & 3 & 41 & 47 \\ 79 & 13 & -7 & 1 & 25 & 22 & 11 & 57 \\ -20 & -59 & -66 & -51 & 5 & 44 & 27 & 63 \\ -82 & -112 & -112 & -112 & -95 & 63 & 55 & 54 \\ -84 & -112 & -112 & -112 & -112 & 14 & 50 & 42 \\ -72 & -104 & -112 & -112 & -112 & -76 & 37 & 41 \\ -47 & -88 & -111 & -107 & -109 & -111 & 18 & 29 \end{bmatrix}$$

Brightness values of individual pixels in the 8 × 8 block on the scale from −128 (dark) to 127 (bright).

Fig. 9.13 Block of size 8 × 8 for brightness, i.e., the Y value. Excerpt from Image 9.11. On the right are the values for the individual pixels of the 8 × 8 block in the range {−128, 127} given

Fig. 9.14 The 8 × 8 basis functions of the two-dimensional Cosine Transformation. Each field of 8 × 8 values can be linearly combined from these base elements. The JPEG format is based on the principle that the human eye can distinguish different base functions to varying degrees

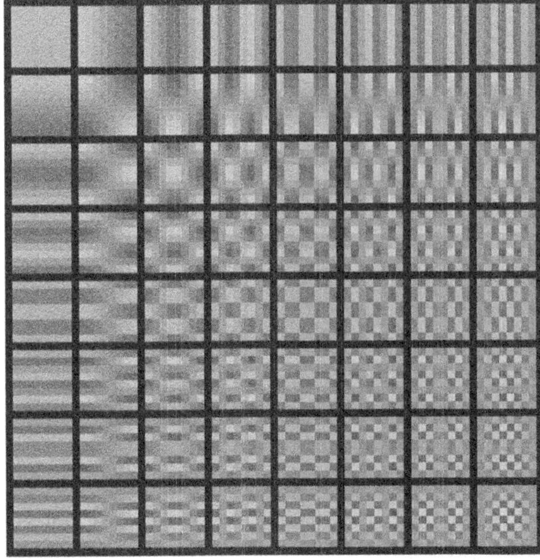

and a reverse transformation according to Theorem 9.60 would yield the original matrix **B**, i.e., the intensities of the image section of size 8 × 8. Up to this point, no data compression has been achieved. Storing the coefficient matrix **G** is just as complex as storing the intensities **B**. Since the coefficients are not integers, even greater storage effort can be expected.

Quantization

Another essential idea of the JPEG compression is the observation that the human eye has difficulty perceiving high-frequency differences. Therefore, the determined coefficients, in our case the matrix (9.39), are weighted by a different factor depending on the frequency, which is called *quantization*. A common quantization matrix is:

$$
\mathbf{Q} = \begin{bmatrix}
10 & 15 & 25 & 37 & 51 & 66 & 82 & 100 \\
15 & 19 & 28 & 39 & 52 & 67 & 83 & 101 \\
25 & 28 & 35 & 45 & 58 & 72 & 88 & 105 \\
37 & 39 & 45 & 54 & 66 & 79 & 94 & 111 \\
51 & 52 & 58 & 66 & 76 & 89 & 103 & 119 \\
66 & 67 & 72 & 79 & 89 & 101 & 114 & 130 \\
82 & 83 & 88 & 94 & 103 & 114 & 127 & 142 \\
100 & 101 & 105 & 111 & 119 & 130 & 142 & 156
\end{bmatrix}
$$

Depending on the desired quality level of the JPEG format, different matrices $\mathbf{Q}$ are chosen. The actual quantization is then done by element-wise division of the entries of $\mathbf{G}$ by $\mathbf{Q}$ and rounding to the nearest whole number, i.e.,

$$
\mathbf{G}^q_{ij} = \left[\frac{\mathbf{G}_{ij}}{\mathbf{Q}_{ij}} \right].
$$

In our example, we get:

$$
\mathbf{G}^q = \begin{bmatrix}
-11 & -17 & 9 & 1 & 0 & 1 & 0 & 0 \\
24 & 10 & -3 & 1 & 0 & -1 & 0 & 0 \\
2 & 5 & 1 & -1 & 0 & 0 & -1 & 0 \\
0 & -2 & 0 & 0 & -1 & 0 & 0 & 0 \\
0 & -1 & 0 & 0 & 0 & 0 & 0 & 0 \\
0 & 0 & 0 & 0 & 0 & 0 & 0 & 0 \\
0 & 0 & 0 & 0 & 0 & 0 & 0 & 0 \\
0 & 0 & 0 & 0 & 0 & 0 & 0 & 0
\end{bmatrix}
\tag{9.40}
$$

The rounding in the quantization is the most important approximation and thus compression in the JPEG format.

Encoding

Finally, the matrices $\mathbf{G}^q$ must be stored efficiently. This takes advantage of the fact that the entries are small (due to the quantization), integers, and that many entries vanish. Typically, most of the non-zero entries are in the top left of the matrix, thus belonging to low frequencies. The further encoding is lossless. The basic idea is a renumbering of the entries of $\mathbf{G}^q$, arranged in a zijzey pattern

$$
\mathbf{G}^q = \{-11, -17, 24, 2, 10, 9, 1, -3, 5, 0, 0, -2, 1, 1, 0, 1, 0, -1, \ldots\}.
$$

By this type of sorting, the entries become smaller and typically only zeros appear from a certain point on. Stored for each 8×8 block is the number of entries up to the point where only zeros appear. These non-vanishing entries are represented using *Huffman encoding*.[4] For details, we refer to [89].

Decoding

When displaying JPEG images, these must first be decoded. To do this, the steps described above must be reversed. We substitute the matrix $\mathbf{G}^q$ into (9.40) and first reverse the quantization step, thus calculating the matrix $\bar{\mathbf{G}}$ as

$$\bar{\mathbf{G}}_{ij} = \mathbf{Q}_{ij}\mathbf{G}^q_{ij}.$$

This results in:

$$\bar{\mathbf{G}} = \begin{bmatrix} -110 & -204 & 225 & 37 & 0 & 66 & 0 & 0 \\ 360 & 190 & -84 & 39 & 0 & -67 & 0 & 0 \\ 50 & 140 & 35 & -45 & 0 & 0 & -88 & 0 \\ 0 & -78 & 0 & 0 & -66 & 0 & 0 & 0 \\ 0 & -52 & 0 & 0 & 0 & 0 & 0 & 0 \\ 0 & 0 & 0 & 0 & 0 & 0 & 0 & 0 \\ 0 & 0 & 0 & 0 & 0 & 0 & 0 & 0 \\ 0 & 0 & 0 & 0 & 0 & 0 & 0 & 0 \end{bmatrix}$$

The error of the compression can be indicated by the expression

$$\frac{\|\mathbf{G} - \bar{\mathbf{G}}\|}{\|\mathbf{G}\|}.$$

In the Frobenius norm (2.1), it holds

$$\frac{\|\mathbf{G} - \bar{\mathbf{G}}\|_F}{\|\mathbf{G}\|_F} \approx 0.19.$$

This corresponds to a representation error of just under 20%. The actual recoding is done using the basis representation from (9.38), i.e., in the form

$$\bar{\mathbf{B}}_{ij} = \frac{1}{4} \sum_{u,v=1}^{8} \alpha(u-1)\alpha(v-1)\bar{\mathbf{G}}_{uv} \cos\left(\frac{2i-1}{16}u - 1\pi\right) \cos\left(\frac{2j-1}{16}v - 1\pi\right).$$

---

[4] Encoding method developed by David Huffman in 1952. In the sense of a dictionary, numbers or chains of numbers are combined into words. Words that occur very frequently are given a short "name", rare words a longer one. In our example, for instance, the words "0", "1" or "−1" occur frequently. See [56].

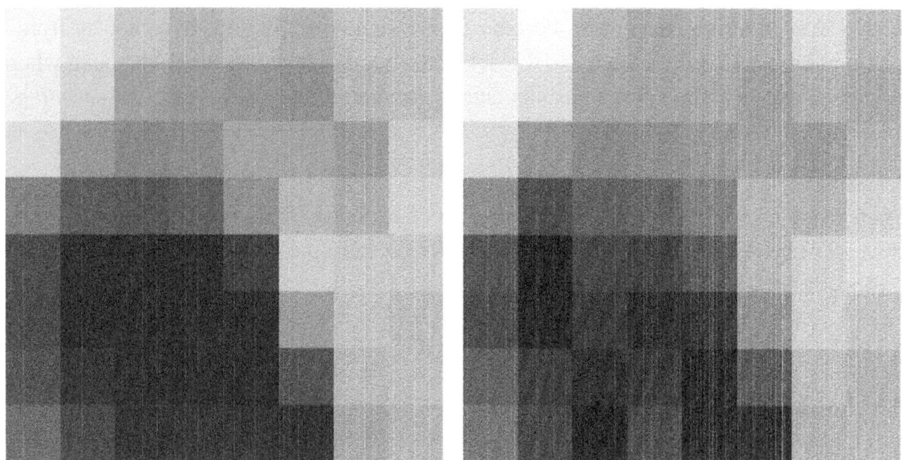

Fig. 9.15 Left: Block of size 8 × 8 from the original image. Right: Recoding of the JPEG compression with the method given here. Instead of 64 intensity levels (one byte each) only 38 non-zero frequencies need to be stored be stored. Using Huffman coding, this can be efficiently done with about 10 bytes of storage

In Fig. 9.15, we show the recoding as well as the original 8 × 8 block. For color images, the three channels Y', Cb, and Cr are treated separately. Subsequently, the *subsampling* step must be reversed by enlarging the two chrominance channels Cb and Cr. Finally, the linear transformation from (9.37) can be easily reversed.

9.9   Numerical Building Blocks for Solving Initial Value Problems

Initial value problems (see, e.g., [26, 76, 80, 86, 99]) play a major role in numerical mathematics, as they appear in the mathematical modeling of many applications. A time-dependent process, e.g., growth of a species, can often be described by *differential equations*. Further examples of differential equations in this book are oscillations of a spring pendulum in Excursus 5.6 or the minimal surface Excursus 3.8.

Definition 9.61 (*Ordinary Differential Equation*) Let $y(t)$ be a function and $t \in \mathbb{R}$ the independent variable. A differential equation is a mathematical equation that relates the function $y(t)$ with at least one of its derivatives $y'(t)$, $y''(t)$, ..., so that both the function itself and the derivative(s) satisfy the differential equation.

Derivation of the Initial Value Problem
Initial value problems form such a class of differential equations. Here, starting from a given initial value, one is interested in the future development of the process. Let $y := y(t)$ be the

number of a species at time t. At a starting point t_0, the number is y_0. We are interested in the development of this. Very often, local changes are of interest. In an infinitesimal time unit dt, the local relative change dy/y is described by the proportional relation

$$\frac{dy}{y} = (g - m)dt,$$

where g describes the birth rate and m the mortality rate. If $g - m > 0$, then the species will grow. For $g - m = 0$, the number of members of the species will not change and we should remain at the initial status y_0. For $g - m < 0$, the species should become extinct. In the limit $dt \to 0$, after rearrangement $dy/dt = y'(t)$, which is just the difference quotient (Excursus 3.8 and Chap. 7). With this, we have the differential equation:

$$y'(t) = (g - m)y(t).$$

It is clear that the solution $y(t)$ is not yet uniquely determined and must be pinned down with the help of an additional condition. This is precisely the initial condition $y(t_0) = y_0$.

In the following, we generalize the concrete derivation. Let $I = [t_0, t_0 + T]$ be a time interval with starting time t_0 and ending time $T > t_0$. On the interval $I = [t_0, t_0 + T]$ a function $y \in C^1(I)$ is sought, such that the differential equation and the initial value

$$y'(t) = f(t, y(t)) \quad t \in I, \quad y(t_0) = y_0$$

are satisfied. Here, $f(t, y(t))$ is a given right-hand side. For example, $f(t, y) := (g - m)y(t)$.

Numerical Treatment
The approximation of initial value problems is based on a *discretization* of the interval $I = [0, T]$ into discrete time steps with the time points

$$0 = t_0 < t_1 < \cdots < t_N = t_0 + T \tag{9.41}$$

and in an *approximation* of the sought solution $y(t)$ in these discrete points

$$y_n \approx y(t_n).$$

Instead of a function $y \in C^1(I)$, the discrete function values

$$(t_0, y_0), \ldots, (t_N, y_N)$$

are now sought. These points can then be connected, for example, by linear interpolation (Sect. 9.1) to obtain a solution curve. The simplest methods for approximation are the explicit and the implicit Euler methods. Both are based on the approximation of the derivative $y'(t_n)$ with a first-order difference quotient and the iteration rules are

$$y_{n+1} = y_n + (t_{n+1} - t_n) f(t_n, y_n), \quad n = 0, \ldots, N-1 \qquad (9.42)$$

for the explicit Euler method and

$$y_{n+1} = y_n + (t_{n+1} - t_n) f(t_{n+1}, y_{n+1}), \quad n = 0, \ldots, N-1 \qquad (9.43)$$

in the case of the implicit Euler method. Details on discretization methods can be found, for example, in the works [26, 76, 80, 86, 99]. The major differences between the two approaches are as follows.

- *Explicit methods* can be implemented and executed very quickly. They can be written as an iterative process of the type

$$y_{n+1} = F(y_n)$$

  and in each step the function $F(\cdot)$ must be evaluated (which depends on the right-hand side $f(t, y(t))$). A disadvantage of *explicit methods* is often the lack of stability. Explicit methods often require a vast number of iterations, i.e., tiny time steps $t_n - t_{n-1}$, so this advantage can be lost again.
- *Implicit methods* can be written as iterative processes of the type

$$y_{n+1} = G(y_{n+1}; y_n).$$

In each step, a system must be solved. If the relationship in $G(\cdot)$ is nonlinear, even a nonlinear system must be solved. The advantage of *implicit methods* is that they are often very stable. An approximation of the sought solution can be determined with far fewer iterations than is the case with *explicit methods*. However, each iteration is significantly more complex.

In this excursus, we are specifically interested in the numerical analysis of an *implicit method*. Due to the (generally nonlinear) relationship in the function $G(\cdot)$, other numerical methods such as the Newton method (Sect. 8.3) must often be used. The aim of this excursus is the derivation of such constructions. In particular, the nesting of different numerical methods is a central aspect in most practice-oriented applications. Specifically we shall use the difference quotient (Sect. 9.3), the polynomial interpolation (for the final graphical representation) from Sect. 9.1 or piecewise polynomial interpolation (Sect. 9.2), and, in the case of nonlinear differential equations, a fixed point iteration or Newton's methods.

A Linear Initial Value Problem

In the following, we consider the growth of a species as modeled above. This is a simple linear growth model, given by

$$y'(t) = a y(t) \text{ for } t \geq t_0 \text{ and } y(t_0) = y_0, \qquad (9.44)$$

where $y := y(t)$ denotes the number of species and t is the time variable. Furthermore, $a \in \mathbb{R}$ is the growth parameter.[5] The initial condition at time t_0 is y_0. For $a < 0$, the species will eventually die out, and for $a > 0$, the species will multiply exponentially. In fact, the solution to this model can even be given explicitly using the *separation of variables* as

$$y(t) = y_0 \exp(a(t - t_0)). \tag{9.45}$$

Example 9.62 Using a specific case study, we calculate the development of a small unit of such a species. In the year $t_0 = 2011$, there are two individuals, i.e., $y(2011) = 2$. We assume a growth rate of $a = 0.25$. Using the above equations, we can estimate (or, in this case, calculate exactly) that, for example, in the year $t = 2014$

$$y(2014) = 2 \exp(0.25 \cdot (2014 - 2011)) = 4.117 \approx 4,$$

so there are four members, and in the year $t = 2022$ already

$$y(2022) = 2 \exp(0.25 \cdot (2022 - 2011)) = 31.285 \approx 31,$$

so 31 individuals are alive. From this, we can see that the model is fundamentally correct, but it seems somewhat unrealistic when applied to humans.   ◄

Since the exact solution is often unknown, we develop an appropriate algorithm that enables us to solve the problem numerically. To do this, we first decompose the interval $I = [t_0, t_0 + T]$ according to (9.41) and denote the *step size* by $k_n = t_n - t_{n-1}$. To approximate the differential equation, we must in particular replace the derivative $y'(t)$ with a discrete approximation, i.e., a difference quotient (see Sect. 9.3). With a Taylor expansion, we have at each discrete time point t_n

$$y'(t_n) = \frac{y(t_n) - y(t_{n-1})}{k_n} + \frac{y''(\xi_n)}{2} k_n$$

with an intermediate point $\xi_n \in [t_{n-1}, t_n]$. For a detailed analysis of difference quotients, we refer to Sect. 9.3.1. At the discrete points t_n, we denote the sought numerical approximation to the solution by $y_n \approx y(t_n)$. The goal is to construct a sequence step by step

$$y_0, y_1, \ldots, y_N, \quad N \in \mathbb{N}.$$

The following algorithm describes the *implicit Euler method* (see, for example, [26, 76, 80]).

---

[5] The parameter a is a problem-specific coefficient, which generally has to be determined by given data or experimental results. Often, such a measurement is also error-prone and ultimately an approximation.

Algorithm 9.66: Implicit Euler Method

> **Input:** Initial value y_0
> 1 **for** $n = 1$ **to** N **do**
> 2 　　$y_n = y_{n-1} + k_n f(t_n, y_n)$
> **Result:** Approximation $y_n \approx y(t^n)$.

In our above example, the right-hand side is given as

$$f(t_n, y_n) = a y_n,$$

and a step of the implicit Euler method can be written as

$$y_n = y_{n-1} + k_n a y_n \quad \Leftrightarrow \quad y_n = (1 - k_n a)^{-1} y_{n-1}.$$

Whenever $1 - k_n a \neq 0$ holds, we can perform the method.

A fundamental problem of the implicit Euler method is the fact that the sought solution y_n appears both on the left and on the right side f. For a general function $f(t, y(t))$, the equation cannot be solved explicitly in each step of the method. This is where the Newton method comes into play. We write each step for $n = 1, 2, \ldots, N$

$$\frac{y_n - y_{n-1}}{k_n} = f(t_n, y_n) \tag{9.46}$$

as a root-finding problem. To this end, we bring all terms to one side and define the function $g(\cdot)$ by

$$g(y) := y - y_{n-1} - k_n f(t_n, y).$$

In each step, we seek $y_n \in \mathbb{R}$ such that

$$g(y_n) = 0. \tag{9.47}$$

As we can easily see, task (9.47) is equivalent to (9.46). Starting from an initial value $y_n^{(0)} \in \mathbb{R}$, we can apply the Newton method from Algorithm 8.13 and obtain the iteration

$$y_n^{(l+1)} = y_n^{(l)} - \frac{g(y_n^{(l)})}{g'(y_n^{(l)})}, \quad l = 0, 1, 2, \ldots,$$

where l indicates the Newton iteration. The derivative here is given as

$$g'(y_n^{(l)}) := 1 - k_n f_y'(t_n, y_n^{(l)}),$$

where f_y' is the derivative with respect to the variable y. In our specific example, we get

$$g'(y_n^{(l)}) = 1 - k_n a,$$

since $f(t_n, y_n^{(l)}) = a y_n^{(l)}$. The iteration is stopped when the Newton termination criterion

$$|g(y_n^{(l+1)})| < \epsilon$$

is satisfied. For instance, $\epsilon = 10^{-12}$.

Remark 9.67 A good initial value for a Newton method within a time step method is the previous solution y_{n-1}, i.e., $y_n^{(0)} := y_{n-1}$. For this initial value, we can even provide an a priori error estimate. Assuming y_n is the exact solution. Then, using the procedure of the implicit Euler method, we have

$$y_n - y_n^{(0)} = y_n - y_{n-1} = k_n f(t_n, y_n).$$

For the error, the estimate $|y_n - y_n^{(0)}| = O(k_n)$ holds. The smaller we choose the step size, the smaller the initial error of Newton's method. ♦

With this iteration, we obtain an approximation $y_n^{(l)}$ for the sought value y_n (which itself is an approximation of the actual value $y(t_n)$). This means that we are dealing with a nesting of two numerical methods here.

For our initial value problem above, each step of Newton's method is given as

$$y_n^{(l+1)} = y_n^{(l)} - \frac{y_n^{(l)} - y_{n-1} - k_n a y_n^{(l)}}{1 - k_n a}.$$

We have already discussed that this problem is linear. A use of the Newton method is therefore not necessary. For more complex problems, the main effort will be in the calculation of the derivative $g'(y_n^{(l)})$. In higher-dimensional problems, which we discussed in Sect. 8.4.1, the derivative is a matrix, the *Jacobian*. Then, a significant numerical effort is the solution of a linear system with the Jacobian.

We calculate the numerical solution of the growth problem for various N and $a = 0.25$, and compile the results in Table 9.7. We show for various N the numerical approximation $y_N \approx y(T)$ as well as the relative and absolute errors compared to the exact solution $y(T) \approx 31.28526$.

We see that for larger N, we get, as expected, an increasingly better approximation to the exact solution. We also illustrate this in Fig. 9.16.

Furthermore, we observe that the error at the end time T halves when twice as many intervals are used to calculate the numerical solution. Finally, we return to our original motivation to use the Newton method within the time step method. We observe that we always need a single Newton step. This is clear, since the function $g(y_n)$ is linear in y_n so Newton's converges in one step.

A Nonlinear Initial Value Problem

The species from the previous section has increased from two to 31 individuals in eleven years. We now consider extinction (due to lack of food), which follows this (nonlinear) law:

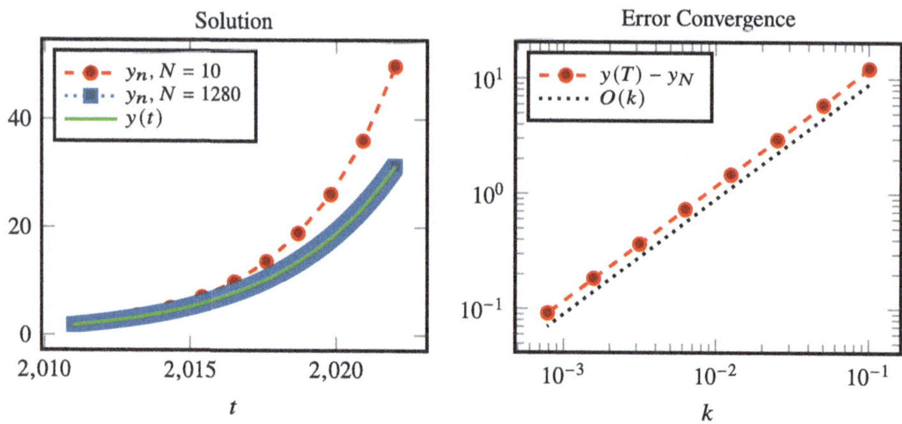

Fig. 9.16 Linear problem $y'(t) = ay(t)$ with $a = 0.25$. Left: Comparison of numerical solution using the implicit Euler method and exact solution for $N = 10$ and $N = 1\,280$. Right: Convergence plot of solution error at $t = T$ depending on $k = 1/N$

Table 9.7 Numerical error of the implicit Euler method for the approximation of $y'(t) = ay(t)$ with $a = 0.25$

N	y_N	$\lvert y(T) - y_N \rvert / \lvert y(T) \rvert$	$\lvert y(T) - y_N \rvert$
10	49.848	0.52082	12.377
20	38.534	0.21894	5.9697
40	34.544	0.10144	2.9629
80	32.835	0.04893	1.4791
160	32.042	0.02404	0.7394
320	31.659	0.01192	0.3697
640	31.471	0.00593	0.1848
1 280	31.378	0.00296	0.0924

$$y'(t) = ay^2(t) \quad \text{for } t \in [t_0, T] \quad \text{and} \quad y(t_0) = y_0.$$

In particular, a decline is characterized by a negative parameter $a < 0$. We choose $a = -0.1$, $t_0 = 2022$, $T = 2050$ and $y_0 = 31$. Based on our previous considerations, we are again interested in a formulation as a root-finding problem

$$g(y_n) = 0$$

to apply Newton's method. The function $g(y_n)$ and its derivative are given by

Fig. 9.17 Numerical solutions of the nonlinear problem $y'(t) = ay(t)^2$ with negative growth $a = -0.1$ for $N = 10$ and $N = 1\,280$. As before, we recognize significant differences between the two approximations. This is particularly significant at the beginning, when the function drops steeply, i.e., the first derivative is huge

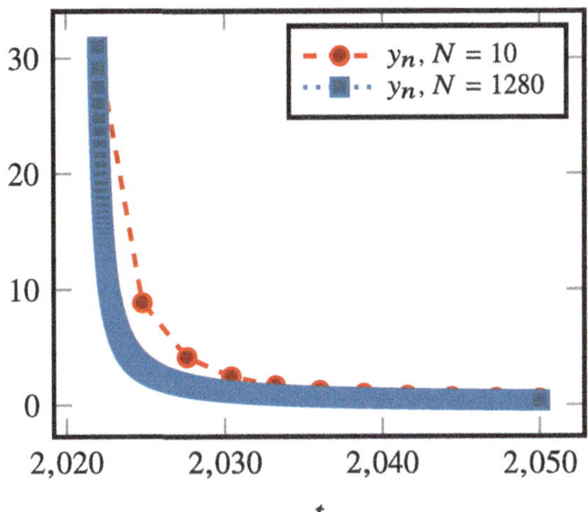

$$g(y_n) = y_n - y_{n-1} - k_n a y_n^2,$$
$$g'(y_n) = 1 - 2ka y_n.$$

In time step n, the rule for the $l + 1$-th iteration is

$$y_n^{(l+1)} = y_n^{(l)} - \frac{g(y_n^{(l)})}{g'(y_n^{(l)})} = y_n^{(l)} - \frac{y_n^{(l)} - y_{n-1} - k_n a (y_n^{(l)})^2}{1 - 2k_n a y_n^{(l)}}.$$

Due to the nonlinearity of the right-hand side $f(y_n, t) = ay_n^2$, we now expect more than one Newton iteration per time step solution n.

At time $Z = 2050$, the exact solution is $y(T) \approx 0.35$ (see Table 9.8 with the finer discretization $N = 1\,280$). This means that in the year 2050 all individuals are likely to be extinct according to the model used here. In fact, we see a very rapid drop in the solution curve in Fig. 9.17. At time $t = 2026.75$, only $y_{2026.75} = 1.995961$, i.e., less than two, individuals remain. This makes reproduction impossible, assuming the exclusion of asexual reproduction. Finally, we note that at $t = 2031.73$, we already have $y_{2031.73} = 0.9995613$.

Let us now take a closer look at the performance of Newton's method. With a tolerance of $\mathtt{tol} = 10^{-10}$, we need six iterations in the first time step ($n = 1$) and three iterations at $n = 10$ for $N = 10$. In total, 39 Newton iterations are needed. For $N = 1\,280$, we observe three iterations in the first time step and two iterations at $n = 1\,280$. The total sum of the required Newton steps is summarized in Table 9.8.

First, we observe that we indeed need more than one Newton iteration for this problem. Then, we identify a dependency on the interval length (i.e., N) of the time-stepping method. This is clear, since we choose the previous time step as the initial value for the Newton method in each time step. As we can easily see in Fig. 9.17, the function values of $y(t)$ changes quickly, at the beginning. This means that the previous time step solution y_{n-1}

Table 9.8 Accuracy of the final time value $y_n(T)$ and summed number of Newton iterations as well as average number of Newton iterations per time step n for different N and two different Newton tolerances

tol	N	y_N	#Newton steps	#Newton steps/N
10^{-4}	10	0.5054	28	2.80
10^{-4}	1 280	0.3543	1312	1.02
10^{-10}	10	0.5054	39	3.90
10^{-10}	1 280	0.3543	2583	2.01

is not a good initial value for the Newton method. This is particularly the case when only a few points (i.e., small N) are used for time discretization. Since the solution decreases exponentially and the curve becomes flatter towards the end time $T = 2050$, the previous solutions y_{n-1} are better initial values for the Newton iteration. For larger N, we then also see a significant improvement of the Newton iteration with an average of two iterations per time step. At the same time, we saw in the previous section that the accuracy of the solution itself also improves significantly with larger N. With larger N, we thus achieve two goals: higher efficiency of the inner solver, here Newton, and also higher accuracy of the outer method, here the implicit Euler method. Nevertheless, there is a higher overall computational effort, as more time steps have to be solved.

The computational costs can be further reduced if we raise the Newton tolerance from $\mathtt{tol} = 10^{-10}$ to $\mathtt{tol} = 10^{-4}$. The corresponding values are in Table 9.8. With such changes in accuracy, however, care must be taken that the accuracy of the solution curve or the target functional, i.e., here the final time value y_N, does not deteriorate significantly. This is not the case here or with the coarse discretization with $N = 10$, the value changes in the 10th sigificant digit, and with $N = 1\,280$ in the 12th sigificant digit. These differences are obviously negligible. Therefore, in this example, it is justified to choose the Newton tolerance relatively large, which almost halves the computational effort.

The numerical approximation of integrals is called *quadrature* or *numerical integration*. This can be necessary for various reasons:

- The antiderivative cannot be expressed by an elementary function, such as in

$$\int_0^\infty \exp(-x^2)\,dx, \quad \int_0^\infty \frac{\sin(x)}{x}\,dx.$$

- An antiderivative exists in closed form, but the calculation is costly, and numerical methods are more efficient.
- The integrand is only known at discrete points, for example, in measured data.

The fundamental ideas for quadrature are all based on interpolation methods. Hence, we will call these methods *interpolatory quadrature rules* or *piecewise interpolatory quadrature rules* (also often denoted as *summed quadrature*) when they are based on the spline interpolation.

Definition 10.1 (Quadrature Rule) Let $f \in C[a, b]$. A *numerical quadrature rule* for the approximation of $I(f) := \int_a^b f(x)\,dx$ has the form

$$I^n(f) := \sum_{i=0}^n \alpha_i f(x_i)$$

with $n + 1$ pairwise different *quadrature points* $x_0, \ldots, x_n$ and $n + 1$ *quadrature weights* $\alpha_0, \ldots, \alpha_n$. The weights do not depend on the function f.

© The Author(s), under exclusive license to Springer-Verlag GmbH, DE, part of Springer
Nature 2026

T. Richter et al., *Introduction to Numerical Mathematics*, Mathematics Study
Resources 25, https://doi.org/10.1007/978-3-662-72546-7_10

Some simple examples are well known:

Definition 10.2 (Box Rule) The box rule for integration on $[a, b]$ is the simplest quadrature rule. It is based on interpolation of $f(x)$ with a constant polynomial $p(x) = f(a)$ and integration of $p(x)$ (see the top left of Fig. 10.1):

$$I^0(f) = (b - a) f(a)$$

In addition to this *left-sided box rule*, there is also the *right-sided box rule* with $I^0(f) = (b - a) f(b)$. See the top right of Fig. 10.1.

A quadrature rule with better approximation properties is obtained by exploiting symmetry.

Definition 10.3 (Midpoint Rule) The midpoint rule is based on the integration of the constant interpolating polynomial at the interval midpoint. It is given by

$$I^0(f) = (b - a) f\left(\frac{a + b}{2}\right).$$

See the bottom left of Fig. 10.1.

The box rule and midpoint rule are based on an interpolation of the function f with a constant function. The trapezoidal rule interpolates $f(x)$ linearly at the two interval ends, see bottom right of Fig. 10.1.

Definition 10.4 (Trapezoidal Rule) The trapezoidal rule is formed by integrating the linear interpolation in $(a, f(a))$ and $(b, f(b))$:

$$I^1(f) = \frac{b - a}{2}(f(a) + f(b))$$

10.1   Interpolatory Quadrature

The interpolatory quadrature rules are constructed from a suitable interpolant polynomial. For the given quadrature points $a \le x_0 < \ldots < x_n \le b$, the Lagrange interpolat is formed as an approximation of the function f. Using the notation introduced in Sect. 9.1, this reads as

$$p_n(x) = \sum_{i=0}^{n} f(x_i) L_i^{(n)}(x).$$

Integrating this polynomial gives

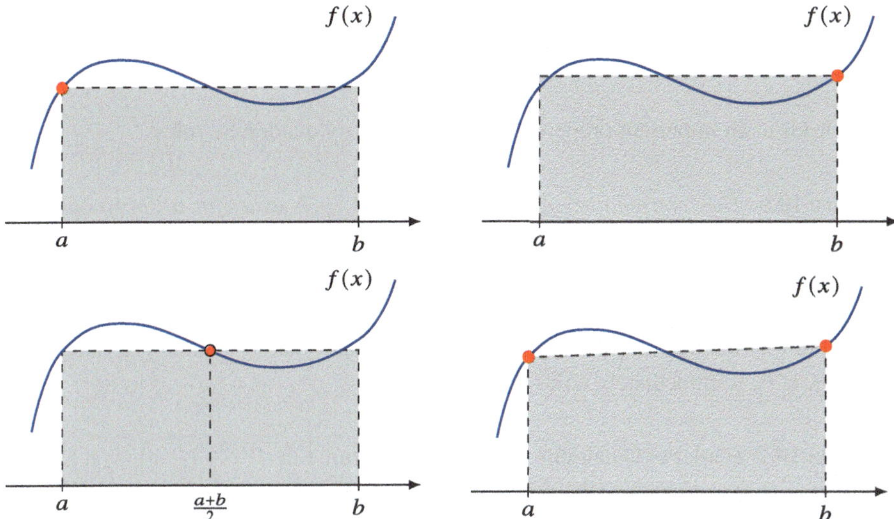

Fig. 10.1 Basic quadrature rules for approximating the integral $\int_a^b f(x)\,dx$. Top: Left- and right-sided box rule. Bottom left: Midpoint rule with the quadrature point $x_m = (a+b)/2$. Bottom right: Trapezoidal rule

$$I^{(n)}(f) := \int_a^b p_n(x)\,dx = \sum_{i=0}^n f(x_i) \underbrace{\int_a^b L_i^{(n)}(x)\,dx}_{=:\alpha_i} = \sum_{i=0}^n \alpha_i f(x_i)$$

and the quadrature weights

$$\alpha_i = \int_a^b L_i^{(n)}(x)\,dx. \tag{10.1}$$

The weights clearly do not depend on the function $f(x)$ to be integrated, but they do depend on the interval $[a, b]$ and the quadrature points $x_0, \ldots, x_n$. This raises the question of whether the quality of the weights can be improved by clever distribution of the quadrature points.

Before we analyze individual quadrature rules and search for the most powerful ones, we can derive a general yet straightforward result.

Theorem 10.5 (Lagrange Quadrature) *For the interpolatory quadrature rules $I^n(f)$ with $n+1$ pairwise different quadrature points $x_0, x_1, \ldots, x_n$, the error representation for the approximation of the integral $I(f) = \int_a^b f(x)\,dx$ is given by*

$$I(f) - I^n(f) = \int_a^b f[x_0, \ldots, x_n, x] \prod_{j=0}^n (x - x_j)\,dx.$$

Proof The proof follows by integrating the error estimate of the Lagrange interpolation, see Theorem 9.15.   □

This leads to an important property of the interpolatory quadrature rules:

Corollary 10.6 *The interpolatory quadrature rule* $I^{(n)}(\cdot)$ *is exact for all polynomials of degree n. This result is independent of the specific choice of the* $n+1$ *pairwise different quadrature points.*

Proof Follows directly from the construction of the interpolatory quadrature rules, since for each $f \in P_n$ it immediately holds that $p = f$.   □

Definition 10.7 (Order of Quadrature Rules) A quadrature rule $I^{(n)}(\cdot)$ is called (at least) of order m, if it integrates exactly all polynomials in P_{m-1}. That is, the interpolatory quadrature rules $I^{(n)}(\cdot)$ for $n+1$ quadrature points are at last of order $n+1$.

In the following, we will analyze the simple quadrature rules introduced earlier in detail and determine their error bounds and order. For this, we will use the Newtonian integral representation of the interpolation error from Theorem 10.5.

Theorem 10.8 (Box Rule) *Let* $f \in C^1([a, b])$. *The box rule (Definition 10.2) is of first order, and the error estimate is given by*

$$I^0(f) := (b-a)f(a), \quad I(f) - I^0(f) = \frac{(b-a)^2}{2} f'(\xi)$$

with an intermediate point $\xi \in (a, b)$.

Proof The box rule is based on interpolation with a constant polynomial. Therefore, it is first order. From Theorem 10.5 it follows with $x_0 = a$

$$I(f) - I^0(f) = \int_a^b f[a, x](x - a)\, dx,$$

where with Theorem 9.15 it further holds

$$I(f) - I^0(f) = \int_a^b \int_0^1 f'(a + t(x - a))(x - a)\, dt\, dx.$$

We apply the mean value theorem of integral calculus twice and obtain

$$I(f) - I^0(f) = f'(\xi) \int_a^b \int_0^1 (x - a)\, dt\, dx = \frac{1}{2} f'(\xi)(b - a)^2$$

with an intermediate value $\xi \in (a, b)$.   □

The box rule is the simplest quadrature rule. Analogous to the discussion on difference quotients, we can expect a better approximation order for the midpoint rule by exploiting symmetry. To achieve this better order, superapproximation properties must be used. This generally requires a little more effort:

Theorem 10.9 (Midpoint Rule) *Let $f \in C^2([a, b])$. The midpoint rule (Definition 10.3) is of order 2, and the error estimate is given by*

$$I_m^0(f) := (b - a) f\left(\frac{a+b}{2}\right), \quad I(f) - I_m^0(f) = \frac{(b-a)^3}{24} f''(\xi),$$

with an intermediate point $\xi \in (a, b)$.

Proof Due to the interpolation with a constant polynomial, initially only first order is expected. It holds with Theorem 9.15

$$I(f) - I_m^0(f) = \int_a^b f\left[\frac{a+b}{2}, x\right]\left(x - \frac{a+b}{2}\right) dx.$$

For simplification, we set $x_m := (a + b)/2$ and let $\bar{f} \in \mathbb{R}$ be any fixed value. Then we have

$$I(f) - I_m^0(f) = \int_a^b (f[x_m, x] - \bar{f})(x - x_m)\, dx + \bar{f} \underbrace{\int_a^b x - x_m\, dx}_{=0}. \tag{10.2}$$

With the integral representation of the divided differences from Theorem 9.15 and a Taylor expansion at x_m, we have

$$f[x_m, x] = \int_0^1 f'(x_m + t(x - x_m))\, dt = \int_0^1 f'(x_m) + t(x - x_m) f''(\xi)\, dt$$

for some $\xi \in [x_m, x]$. Choosing $\bar{f} = f'(x_m)$ and with (10.2) the error is

$$I(f) - I_m^0(f) = f''(\xi) \int_a^b \int_0^1 t(x - x_m)^2\, dt\, dx = \frac{|b-a|^3}{24} f''(\xi).$$

$\square$

With the same effort as the box rule, i.e., with one evaluation of the function $f(x)$, the midpoint rule achieves second order.

Theorem 10.10 (Trapezoidal Rule) *Let $f \in C^2([a, b])$. The trapezoidal rule (Definition 10.4) is of order two and the error estimate*

$$I^1(f) := \frac{b-a}{2}(f(a) + f(b)), \quad I(f) - I^1(f) = -\frac{(b-a)^3}{12}f''(\xi),$$

holds with an intermediate point $\xi \in (a, b)$.

Proof An optimal error estimate for the trapezoidal rule can be derived with a trick. We insert a function $g'' \equiv 1$, and using integration by parts, we obtain

$$\begin{aligned}
\int_a^b f(x)\mathrm{d}x &= \int_a^b f(x)g''(x)\mathrm{d}x = fg'\Big|_a^b - \int_a^b f'(x)g'(x)\mathrm{d}x \\
&= fg'\Big|_a^b - f'g\Big|_a^b + \int_a^b f''(x)g(x)\mathrm{d}x.
\end{aligned} \tag{10.3}$$

We must determine a function $g \in C^2([a, b])$ such that the boundary term $f'g\Big|_a^b$ vanishes and the term $fg'\Big|_a^b$ exactly takes the form of the trapezoidal rule. From $g'' \equiv 1$ it follows

$$g(x) = \frac{1}{2}(x - c_1)(x - c_2)$$

and from $g(a) = g(b) = 0$, so that $f'g\Big|_a^b = 0$, the function $g(x)$ must be

$$g(x) = \frac{1}{2}(x - a)(x - b),$$

thus

$$g'(x) = x - \frac{a+b}{2} \quad \Rightarrow \quad g'(a) = -\frac{b-a}{2}, \quad g'(b) = \frac{b-a}{2}.$$

We insert this function into (10.3), and obtain

$$\int_a^b f(x)\mathrm{d}x = \frac{b-a}{2}(f(a) + f(b)) + \frac{1}{2}\int_a^b f''(x)(x - a)(x - b)\mathrm{d}x.$$

Since the function $g(x)$ does not change sign in the interval $[a, b]$, we conclude with the mean value theorem of integral calculus

$$\begin{aligned}
\int_a^b f(x)\mathrm{d}x - \frac{b-a}{2}(f(a) + f(b)) &= f''(\xi)\frac{1}{2}\int_a^b (x - a)(x - b)\mathrm{d}x \\
&= -f''(\xi)\frac{(b-a)^3}{12},
\end{aligned}$$

with an intermediate value $\xi \in [a, b]$. $\qquad\square$

The use of a quadratic interpolation polynomial leads to Simpson's rule.

Theorem 10.11 (Simpson's Rule) *Let $f \in C^4([a, b])$. Simpson's rule, based on interpolation with a quadratic polynomial, is*

$$I^2(f) = \frac{b-a}{6}\left(f(a) + 4f\left(\frac{a+b}{2}\right) + f(b)\right),$$

and is of order four. It holds

$$I(f) - I^2(f) = -\frac{f^{(4)}(\xi)}{2880}(b-a)^5,$$

where $\xi \in (a, b)$.

Proof The proof follows analogously to the proofs for the midpoint and trapezoidal rules and is left as an exercise for the reader. $\qquad\square$

Remark 10.12 As for the midpoint rule, the order of Simpson's rule is a power higher than one would expect. Suppose we continue this game and construct even higher quadrature rules. In that case, we recognize a general principle: With quadrature using *even* interpolation polynomials, the error order is increased by one power. $\qquad\blacklozenge$

Definition 10.13 (Newton-Cotes Rules) Quadrature rules with equidistantly distributed quadrature points in the interval $[a, b]$ are called *Newton-Cotes rules*. If the Interval ends a and b are quadrature points, the rules are called *closed*. If the interval ends are not included, they are called *open*.

In Table 10.1, we summarize some common Newton-Cotes rules. Closed forms with $n \geq 8$ and open ones for $n \geq 2$ have some negative quadrature weights α_i. Cancellation can occur even with purely positive integrands. Therefore, these quadrature rules are not useful from a numerical perspective. The weights of a quadrature rule always satisfy

$$\sum_{i=0}^{n} \alpha_i = b - a.$$

In the case of negative weights, it holds

$$\sum_{i=0}^{n} |\alpha_i| > (b - a).$$

A general theorem shows that for the Newton-Cotes rules, it holds

$$\sum_{i=0}^{n} |\alpha_i| \to \infty \quad (n \to \infty).$$

Table 10.1 Properties and weights of some Newton-Cotes rules

n	Weights α_i		Name	Order
0	1	Open	Box rule	1
1	$\frac{1}{2}, \frac{1}{2}$	Closed	Trapezoidal rule	2
2	$\frac{1}{6}, \frac{4}{6}, \frac{1}{6}$	Closed	Simpson's rule	4
3	$\frac{1}{8}, \frac{3}{8}, \frac{3}{8}, \frac{1}{8}$	Closed	Newton-3/8 rule	4
4	$\frac{7}{90}, \frac{32}{90}, \frac{12}{90}, \frac{32}{90}, \frac{7}{90}$	Closed	Milne's rule	6
6	$\frac{41}{840}, \frac{216}{840}, \frac{27}{840}, \frac{272}{840}, \frac{27}{840}, \frac{216}{840}, \frac{41}{840}$	Closed	Weddle's rule	8

This is shown in [74], where it is also discussed that Newton-Cotes rules do not even necessarily converge for analytic functions. The weights tend to oscillate and therefore, quadrature rules of higher degree are numerically unstable.

10.2 Piecewise Interpolatory Quadrature Formulas

Interpolatory quadrature rules from the previous section are based on the integration of an interpolation polynomial p. For this, it holds due to Theorem 9.13, that

$$f(x) - p(x) = \frac{f^{(n+1)}(\xi)}{(n+1)!} \prod_{i=0}^{n}(x - x_i).$$

The accuracy of the quadrature can, in principle, be improved in two ways: The prefactor can be kept small by choosing more quadrature points, because $1/(n+1)! \to 0$. However, this only leads to convergence of the error if the derivatives $f^{(n+1)}$ do not grow too quickly, see Example 9.18. The second factor is the product of the quadrature points. A rough estimate shows

$$\left| \prod_{i=0}^{n}(x - x_i) \right| \le (b - a)^{n+1}.$$

Powers of the interval size can bound interpolation errors (and thus also quadrature errors). This is used to develop stable and convergent quadrature rules that keep the degree of polynomial interpolation low but reduce the interval size by splitting it into subintervals, exactly in the spirit of piecewise interpolation (Sect. 9.2). Let

$$a = y_0 < y_1 < \cdots < y_m = b, \quad h_i := y_i - y_{i-1},$$

be a partition of the interval $[a, b]$ into m subintervals with step size h_i. On each of these m subintervals, the function f is approximated using a quadrature rule:

$$I_h^n(f) = \sum_{i=1}^m I_{[y_{i-1}, y_i]}^n(f)$$

Theorem 10.14 (Summed Quadrature) *Let $f \in C^{n+1}([a, b])$ and $a = y_0 < \cdots < y_m = b$ be a partition of the interval with step sizes $h_i := y_i - y_{i-1}$ and $h := \max h_i$. For the summed quadrature rule $I_h^n(f)$, the error estimate*

$$I(f) - I_h^n(f) \le c \max_{[a,b]} |f^{(n+1)}| h^{n+1}$$

holds, with a constant $c(n) > 0$.

Proof The proof is a combination of Theorem 10.5 and the differential residual representation of the Lagrange interpolation. □

An important conclusion can be drawn from this result: For $h \to 0$, i.e., for smaller interval partitions, summed quadrature rules converge for any n. With the help of summed quadrature rules, functions with rapidly growing derivatives can thus be integrated. Furthermore, summed rules are also suitable for integrating functions that only have low regularity, such as $f \in C^1([a, b])$. Since it is possible to keep the order n small, summed quadrature rules are numerically stable. The summed trapezoidal rule is shown in Fig. 10.2.

In the following, we specify some simple summed quadrature rules. For this, we consider an equidistant division of the interval $[a, b]$:

$$a = y_0 < \cdots < y_m = b, \quad h := \frac{b - a}{m}, \quad y_i := a + ih, \quad i = 0, \ldots, m \qquad (10.4)$$

Theorem 10.15 (Summed Box Rule) *Let $f \in C^1([a, b])$ and an equidistant division of the interval be given by (10.4). For the summed box rule*

$$I_h^0(f) = h \sum_{i=1}^m f(y_i),$$

the quadrature error is bounded by

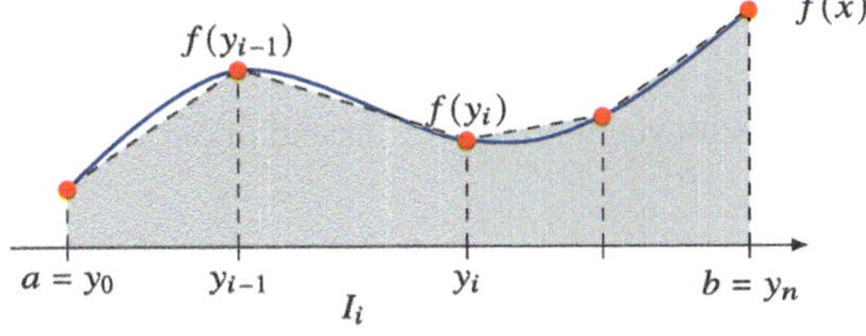

Fig. 10.2 Summed trapezoidal rule for integral approximation

$$|I(f) - I_h^0(f)| \leq \frac{b-a}{2} h \max_{[a,b]} |f'|.$$

Proof Applying Theorem 10.8 on each subinterval and summing over all intervals leads to

$$I(f) - I_h^0(f) = \sum_{i=1}^{m} \frac{h^2}{2} f'(\xi_i)$$

with intermediate points $\xi_i \in [y_{i-1}, y_i]$. Noting that $(b-a) = \sum_i h_i$ and by passing to the maximum, the error estimate is shown. $\qquad\qquad\square$

Theorem 10.16 (Summed Trapezoidal Rule) *Let $f \in C^2([a,b])$ and an equidistant division of the interval be given by (10.4). For the summed trapezoidal rule*

$$I_h^1(f) = \frac{h}{2} \sum_{i=1}^{m} (f(y_{i-1}) + f(y_i)) = \frac{h}{2} f(a) + h \sum_{i=1}^{m-1} f(y_i) + \frac{h}{2} f(b)$$

it holds

$$|I(f) - I_h^1(f)| \leq \frac{b-a}{12} h^2 \max_{[a,b]} |f''|.$$

Proof Exercise. $\qquad\qquad\square$

The summed trapezoidal rule is particularly attractive because the quadrature points overlap: The point y_i is a support point in the subinterval $[y_{i-1}, y_i]$ as well as $[y_i, y_{i+1}]$, and $f(y_i)$ only needs to be determined once. In addition, the summed trapezoidal rule is suitable as a basis for *adaptive quadrature methods*, where the accuracy is adjusted step by step to the problem: If the approximation $I_h^1(f)$ was determined on a division with step size h, the accuracy can be easily increased by halving the interval $I_{h/2}^1(f)$. All quadrature points $f(y_i)$ can be reused, and the function $f(x)$ only needs to be calculated in the new interval centers. The summed Simpson rule has a similar property:

$$I_h^2(f) = \sum_{i=1}^{m} \frac{h}{6} \left(f(y_{i-1}) + 4f\left(\frac{y_{i-1} + y_i}{2}\right) + f(y_i) \right)$$

$$= \frac{h}{6} f(a) + \sum_{i=1}^{m} \frac{h}{3} f(y_i) + \sum_{i=1}^{m-1} \frac{2h}{3} f\left(\frac{y_{i-1} + y_i}{2}\right) + \frac{h}{6} f(b)$$

As seen above, only $2m+1$ instead of $3m$ function evaluations are necessary.

10.3   Romberg Quadrature

The idea of Richardson extrapolation in Sect. 9.4 is to achieve a higher-order approximation of general limit processes $a(h) \to 0$ (for $h \to 0$) by interpolating $a(h_i)$ for a sequence $h_i \to 0$ and evaluating the interpolation polynomial at zero. The summed trapezoidal rule for approximating $\int_a^b f \, dx$ is given by

$$I_h(f) = h \left(\frac{1}{2} f(a) + \sum_{j=1}^{N-1} f(x_j) + \frac{1}{2} f(b) \right), \quad x_j := a + jh,$$

and given $f \in C^2([a, b])$ the error estimate

$$I(f) - I_h(f) = \frac{b-a}{12} h^2 f^{(2)}(\xi)$$

with an intermediate point $\xi \in [a, b]$ holds. The idea of Rhomberg quadrature is to construct a sequence approximating the integral using summed quadrature and to apply Richardson extrapolation to realize a more accurate approximation (sequence) of the integral.

In order to successfully apply the extrapolation method, we must know higher orders of the error estimate. The basis for this is the Euler-Maclaurin summation formula, which shows an expansion of the error in even powers of h:

Theorem 10.17 (Euler-Maclaurin Summation Formula) *Let $f \in C^{2m+2}([0, 1])$. Then the trapezoidal rule has the error expansion*

$$\int_0^1 f(x)dx - \frac{1}{2}\Big(f(0) + f(1)\Big) = \sum_{k=1}^{m} \frac{B_{2k}}{(2k)!} \left(f^{(2k-1)}(1) - f^{(2k-1)}(0) \right)$$

$$+ h^{2m+2} \frac{B_{2m+2}}{(2m+2)!} B_{2m+1} f^{(2m+2)}(\xi)$$

with $\xi \in [0, 1]$ and the Bernoulli numbers B_{2k}.

Proof The proof of this error estimate closely follows the approach to the simple error estimate of the trapezoidal rule in Theorem 10.10. We only sketch the idea and refer to the literature [36].
(i) We define a sequence of polynomials $b_k \in P$ with the properties

$$b_0 \equiv 1, \quad b'_{k+1} = (k+1)b_k. \tag{10.5}$$

The polynomials are not uniquely determined, as a constant may shift the antiderivatives. However, it holds that $b_k \in P_k$ and the coefficient in front of the highest monomial is always one. Now, we have

$$\int_0^1 f(x)\mathrm{d}x = \int_0^1 f(x)b_0(x)\mathrm{d}x = fb_0\Big|_0^1 - \int_0^1 f'(x)b_1(x)\mathrm{d}x$$

$$= f(1)(1+c_1) - f(0)c_1 - \int_0^1 f'(x)b_1(x)\mathrm{d}x.$$

We choose $c_1 = -\frac{1}{2}$ and obtain

$$\int_0^1 f(x)\mathrm{d}x - \frac{1}{2}\Big(f(0) + f(1)\Big) = -\int_0^1 f'(x)b_1(x)\mathrm{d}x. \tag{10.6}$$

For $k = 1, 2, \ldots$, we have

$$\int_0^1 f^{(k)}b_k(x)\mathrm{d}x = \frac{1}{k}f^{(k)}b_{k+1}\Big|_0^1 - \frac{1}{k}\int_0^1 f^{(k+1)}(x)b_{k+1}(x)\mathrm{d}x. \tag{10.7}$$

The polynomials $b_k = b'_{k+1}$ are given up to a constant. We can choose these constants so that for odd b_{2k+1} it holds:

$$b_{2k+1}(0) = b_{2k+1}(1) = 0, \quad k = 1, 2, \ldots$$

The polynomials defined in this way are the *Bernoulli polynomials* [61]. The corresponding *Bernoulli numbers* are given by $B_k := b_k(0)$. Odd Bernoulli numbers are zero, i.e., $B_{2k+1} = 0$. The Bernoulli polynomials have the symmetry property

$$b_k(x) = (-1)^k b_k(1-x), \quad k = 0, 1, \ldots,$$

such that $B_k = b_k(0) = b_k(1)$ (as $B_k = 0$ for k odd).

(ii) Starting from (10.6) and (10.7), it holds

$$\int_0^1 f(x)\mathrm{d}x - \frac{1}{2}\Big(f(0) + f(1)\Big) = \sum_{k=2}^{2m} \frac{(-1)^k}{k!}f^{(k-1)}b_k\Big|_0^1$$

$$+ \frac{1}{(2m+1)!}\int_0^1 f^{(2m+1)}(x)b_{2m+1}(x)\mathrm{d}x.$$

For all odd summands, with $B_{2k+1} = b_{2k+1}(0) = b_{2k+1}(1)$, we have

$$f^{(2k)}b_{2k+1}\Big|_0^1 = f^{(2k)}(1)b_{2k+1}(1) - f^{(2k)}(0)b_{2k+1}(0) = 0.$$

For the even ones, with $B_{2k} = b_{2k}(0) = b_{2k}(1)$, we get

$$f^{(2k-1)}b_{2k}\Big|_0^1 = B_{2k}\Big(f^{(2k-1)}(1) - f^{(2k-1)}(0)\Big).$$

It follows that

$$\int_0^1 f(x)\mathrm{d}x - \frac{1}{2}\Big(f(0) + f(1)\Big) = \sum_{k=1}^{m} \frac{B_{2k}}{(2k)!}\Big(f^{(2k-1)}(1) - f^{(2k-1)}(0)\Big)$$

$$+ \frac{1}{(2m+1)!} \int_0^1 f^{(2m+1)}(x)b_{2m+1}(x)\mathrm{d}x.$$

To estimate the remainder, we perform integration by parts

$$\int_0^1 f^{(2m+1)}(x)b_{2m+1}(x)\mathrm{d}x = \frac{1}{2m+2} f^{(2m+1)}(b_{2m+2} - B_{2m+2})\Big|_0^1$$

$$- \frac{1}{2m+2} \int_0^1 f^{(2m+2)}(x)\Big(b_{2m+2}(x) - B_{2m+2}\Big)\mathrm{d}x$$

We used that both $b_{2m+2}(x)$ and $b_{2m+2}(x) - B_{2m+2}$ are antiderivatives of $(2m+2)b_{2m+1}$. The boundary term vanishes, since $B_{2m+2} = b_{2m+2}(0) = b_{2m+2}(1)$. It can be shown that the function

$$b_{2m+2}(x) - B_{2m+2}$$

does not change its sign within the interval $[0, 1]$, see [36]. Then, the mean value theorem of integral calculus gives

$$\int_0^1 f^{(2m+1)}(x)b_{2m+1}(x)\mathrm{d}x = \frac{1}{2m+2} f^{(2m+2)}(\xi) \int_0^1 B_{2m+2}\mathrm{d}x,$$

as $\int_0^1 b_{2m+2}(x)\mathrm{d}x = 0$. This completes the proof. $\square$

The Euler-Maclaurin formula can be applied to the summed version of the trapezoidal rule.

Theorem 10.18 (Euler-Maclaurin Summation Formula for the Summed Trapezoidal Rule) *Let $I = [a, b]$ and $h = (b - a)/N$ for $N \in \mathbb{N}$. For the function $f^{2m+2}([a, b])$ it holds*

$$\int_a^b f(x)dx - h\left(\frac{f(a)}{2} + f(x_1) + \cdots f(x_{n-1}) + \frac{f(b)}{2}\right)$$

$$= \sum_{k=1}^{m} h^{2k} \frac{B_{2k}}{(2k)!}\Big(f^{(2k-1)}(b) - f^{(2k-1)}(a)\Big)$$

$$+ h^{2m+2} \frac{b-a}{(2m+2)!} B_{2m+2} f^{(2m+2)}(\xi)$$

with an intermediate value $\xi \in [a, b]$, the Bernoulli numbers B_{2k} and the quadrature points

$$x_n = a + hn.$$

Proof This estimate follows by transforming the subintervals $[x_{n-1}, x_n]$ to a reference interval $[0, 1]$. On each of the subintervals, Theorem 10.17 is applied. Due to

$$\lim_{s\downarrow 0} f^{(k)}(x_n + s) = \lim_{s\downarrow 0} f^{(k)}(x_n - s), \quad n = 1, \ldots, N - 1,$$

all derivatives in the inner points vanish. $\qquad\qquad\qquad\qquad\qquad\qquad\qquad\qquad\square$

The Bernoulli numbers can be defined as the coefficients of the Taylor series representation of

$$\frac{x}{e^x - 1} = \sum_{k=0}^{\infty} \frac{B_k}{k!} x^k = 1 - \frac{1}{2}x + \frac{1}{6}\frac{x^2}{2!} - \frac{1}{30}\frac{x^4}{4!} + \frac{1}{42}\frac{x^6}{6!} + \cdots,$$

and satisfy the recursion

$$B_0 = 0, \quad B_k = -\sum_{j=0}^{k-1} \frac{k!}{j!(k-j+1)!} B_j, \quad k = 1, 2, \ldots.$$

The first Bernoulli numbers are given as

$$1, -\frac{1}{2}, \frac{1}{6}, 0, -\frac{1}{30}, 0, \frac{1}{42}, 0, -\frac{1}{30}, \ldots.$$

Except for $B_1 = -\frac{1}{2}$, every second (odd index) Bernoulli number $B_{2k+1} = 0$. Otherwise, they do not follow any recognizable law. For large k, the Bernoulli numbers grow very quickly and behave asymptotically as (see e.g. [65])

$$|B_{2k}| \sim 2(2k)!(2\pi)^{-2k}, \quad k \to \infty.$$

The Euler-Maclaurin summation formula states that the trapezoidal rule has an expansion in even powers of h. Therefore, it is optimally suited for extrapolation in two ways: Due to the quadratic error expansion, we gain two orders of accuracy in each extrapolation step and due to the choice of quadrature points at the ends of the subintervals x_{j-1} and x_j already calculated values $f(x_j)$ for a step size h can be reused in the next finer approximation to $h/2$. To approximate

$$a(0) \approx \int_a^b f(x)\,dx, \quad a(h) := I_h(f) = h\left(\frac{1}{2}f(a) + \sum_{j=1}^{N-1} f(x_j) + \frac{1}{2}f(b)\right)$$

we use the following algorithm:

Algorithm 10.19: Romberg Quadrature

> **Input:** Sequence of step sizes h_k for $k = 1, 2, \ldots$ with $\frac{h_{k+1}}{h_k} \leq \rho < 1$.
> 1 **for** $k = 0$ **to** m **do**
> 2   Evaluate summed quadrature rule $a_k := a(h_k)$
> 3 **for** $k = 0$ **to** m **do**
> 4   Extrapolate the pairs $(h_k^2, a(h_k))$ with polynomials in h^2
> **Result:** Approximation of $a(0) = I_0(f)$

The exponential sequence

$$h_k = 2^{-k} h$$

is one common option. This sequence is called *Romberg sequence* and has the advantage that already calculated quadrature points can be reused. The disadvantage of this sequence is the rapid growth of the number of quadrature points. The extrapolation itself is carried out with the Neville scheme from Algorithm 9.9.

The diagonal elements $a_{k,k}$ are the approximations to $a(0)$. Based on Theorem 9.33 (Richardson extrapolation), we obtain the error estimate:

Theorem 10.20 (Romberg Quadrature) *Let $f^{2m+2}([a, b])$ and $h > 0$ be given. m steps of the Romberg method applied to the sequence $h_k = 2^{-k} h, \ k = 0, \ldots, m$ yield the approximation*

$$I(f) - a_{m,m} = O(h^{2m+2}).$$

Proof The proof follows by using the special form of the error remainder given by the Euler-Maclaurin summation formula for extrapolation. $\qquad\square$

Remark 10.21 An alternative to the Romberg sequence $h_k = 2^{-k} h$ is the Burlisch sequence. This alternated between $2^{-k} h$ and $2^{-k} h/3$. It requires significantly fewer function evaluations. For $h = 1$, the first steps are

$$\frac{1}{2}, \frac{1}{3}, \frac{1}{4}, \frac{1}{6}, \frac{1}{8}, \frac{1}{12}, \ldots$$

$\blacklozenge$

Example 10.22 (Romberg Quadrature Method) We consider the integral

$$I(f) = \int_0^1 \exp\left(1 + \sin^2(x)\right) \, \mathrm{d}x \approx 3.662415705380723.$$

It holds $f \in C^\infty(\mathbb{R})$ and the Romberg method should therefore be well applicable. In the following table, we carry out the Romberg method with the summed trapezoidal rule. The convergence is shown in Fig. 10.3.

h	$I_h(f) = a(h)$	$I(f) - I_h(f)$	$a^{(1)}(h)$	$I(f) - a^{(1)}(h)$	$a^{(2)}(h)$	$I(f) - a^{(2)}(h)$
2^{-0}	4.11830	-4.56×10^{-1}	3.65324	9.17×10^{-3}	3.6623410	7.47×10^{-5}
2^{-1}	3.76951	-1.07×10^{-1}	3.66177	6.43×10^{-4}	3.6624169	1.24×10^{-6}
2^{-2}	3.68871	-2.63×10^{-2}	3.66238	3.91×10^{-5}	3.6624157	2.30×10^{-8}
2^{-3}	3.66896	-6.54×10^{-3}	3.66241	2.42×10^{-6}	3.6624157	3.65×10^{-10}
2^{-4}	3.66405	-1.63×10^{-3}	3.66242	1.51×10^{-7}	3.6624157	5.72×10^{-12}
2^{-5}	3.66282	-4.08×10^{-4}	3.66242	9.42×10^{-9}	3.6624157	8.93×10^{-14}
2^{-6}	3.66252	-1.02×10^{-4}	3.66242	5.89×10^{-10}		
2^{-7}	3.66244	-2.55×10^{-5}				
Order		$O(h^2)$		$O(h^4)$		$O(h^6)$

In this example, the Romberg quadrature quickly reaches machine precision with only two extrapolation steps.

As a second example, we consider the function

$$f(x) = |\sin(x) - 0.75|, \quad I(f) = \int_0^1 f(x)\,\mathrm{d}x \approx 0.304666468.$$

Here, extrapolation of the trapezoidal rule does not result in a significant improvement:

h	$I_h(f) = a(h)$	$I(f) - I_h(f)$	$a^{(1)}(h)$	$I(f) - a^{(1)}(h)$	$a^{(2)}(h)$	$I(f) - a^{(2)}(h)$
2^{-0}	0.42074	-1.16×10^{-1}	0.32063	1.60×10^{-2}	0.3045314	1.35×10^{-4}
2^{-1}	0.34565	-4.10×10^{-2}	0.30554	8.71×10^{-4}	0.3036544	1.01×10^{-3}
2^{-2}	0.31557	-1.09×10^{-2}	0.30377	8.94×10^{-4}	0.3049991	3.33×10^{-4}
2^{-3}	0.30672	-2.05×10^{-3}	0.30492	2.56×10^{-4}	0.3045325	1.34×10^{-4}
2^{-4}	0.30537	-7.06×10^{-4}	0.30456	1.10×10^{-4}	0.3046924	2.59×10^{-5}
2^{-5}	0.30476	-9.42×10^{-5}	0.30468	1.75×10^{-5}	0.3046680	1.58×10^{-6}
2^{-6}	0.30470	-3.66×10^{-5}	0.30467	2.57×10^{-6}		
2^{-7}	0.30468	-1.11×10^{-5}				

A convergence order cannot be clearly determined in this example. The reason for this is the lack of regularity of the function f. Although the function is integrable, its derivatives have singularities at the point $\arcsin(0.75) \approx 0.86$. ◀

Integration of Periodic Functions

We consider periodic functions $f \in C^{2m+2}(\mathbb{R})$ satisfying

$$f(a) = f(b), \quad f^{(2k-1)}(a) = f^{(2k-1)}(b), \quad k = 1, 2, \ldots,$$

for $a, b \in \mathbb{R}$. The Euler-Maclaurin summation formula is *significantly* improved. From

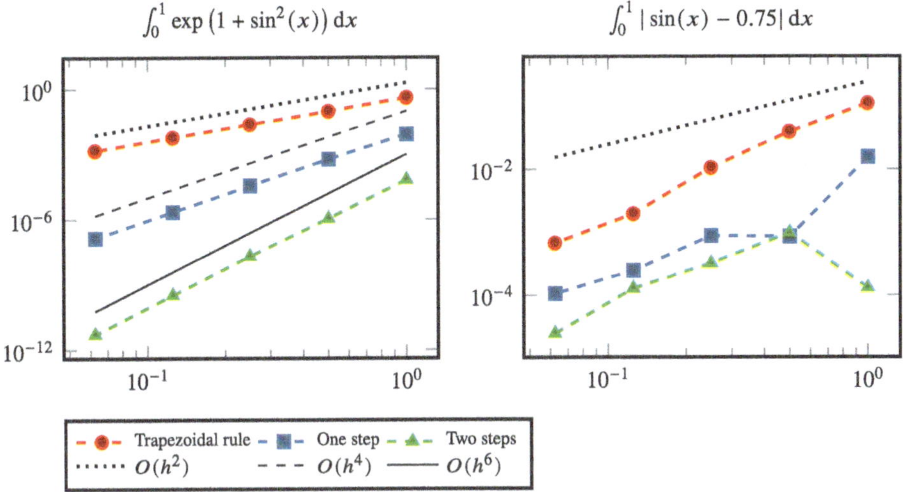

Fig. 10.3 Convergence of the Romberg method with one and two extrapolation steps. Left: integration of a smooth function. Right: integration of a function with limited regularity, where the extrapolation principle does not increase the order of convergence

$$I(f) - I_h(f) = \sum_{k=1}^{m} h^{2k} \frac{B_{2k}}{(2k)!} \left(f^{(2k-1)}(b) - f^{(2k-1)}(a) \right)$$

$$+ h^{2m+2} \frac{b-a}{(2m+2)!} B_{2m+2} f^{(2m+2)}(\xi)$$

with

$$I_h(f) = h \left(\frac{1}{2} f(a) + \sum_{j=1}^{N-1} f(x_j) + \frac{1}{2} f(b) \right), \quad x_j := a + jh,$$

it follows that

$$I(f) - I_h(f) = h^{2m+2} \frac{b-a}{(2m+2)!} B_{2m+2} f^{(2m+2)}(\xi) = O(h^{2m+2})$$

with

$$I_h(f) = h \left(\frac{1}{2} f(a) + \sum_{j=1}^{N-1} f(x_j) + \frac{1}{2} f(b) \right) = h \left(f(a) + \sum_{j=1}^{N-1} f(x_j) \right) = h \left(\sum_{j=0}^{N-1} f(x_j) \right).$$

The summed trapezoidal rule *simplifies* to the summed left-sided box rule.

If $f \in C^\infty(\mathbb{R})$ is fully periodic on $[a, b]$, the summed box rule converges faster to the integral value for $h \to 0$ than any other quadrature rule.

Example 10.23 (Quadrature of Periodic Functions) As above, we consider the function

$$f(x) = \exp\left(1 + \sin^2(x)\right),$$

this time integrating from 0 to π.

$$I(f) = \int_0^\pi f(x)\,dx \approx 14.97346455769737.$$

The summed trapezoidal rule yields:

h	$I_h(f) = a(h)$	$I(f) - I_h(f)$
$2^{-0}\pi$	8.53973	6.43×10^0
$2^{-1}\pi$	15.87657	-9.03×10^{-1}
$2^{-2}\pi$	14.97811	-4.64×10^{-3}
$2^{-3}\pi$	14.97346	-1.07×10^{-8}
$2^{-4}\pi$	14.97346	$< eps$

10.4 Gauss Quadrature

The interpolatory quadrature rules

$$I^{(n)}(f) = \sum_{i=0}^n \alpha_i f(x_i)$$

for $x_0, \ldots, x_n \in [a, b]$ are constructed to be at least of order $n + 1$, i.e.,

$$I(p) = I^{(n)}(p) \quad \forall p \in P_n.$$

This result is valid for all choices of (pairwise different) quadrature points. Furthermore, we have already learned that specific rules, such as the midpoint rule $n = 0$ or the Simpson rule, have even higher orders.

In this section, we investigate the question of whether the order can be further increased by appropriately selecting the quadrature points. So far, the freedom was only in the choice of the polynomial degree and thus in the number of quadrature points. Here, we will construct the *Gauss quadrature rules*, which are interpolatory quadrature rules with order $2n + 2$.

Theorem 10.24 (Order Barrier of Quadrature) *An interpolatory quadrature rule for $n + 1$ quadrature points can have at most the order $2n + 2$.*

Proof Assume that $I^{(n)}(\cdot)$ with the quadrature points $x_0, \ldots, x_n$ would be of higher order, i.e., it would be exact for polynomials of degree $2n + 2$. Hence, the polynomial

$$p(x) = \prod_{j=0}^{n}(x - x_j)^2 \in P_{2n+2}$$

would be integrated exactly. The unique interpolation $p_n \in P_n$ with $p_n(x_i) = p(x_i) = 0$ has $n + 1$ zeros and thus represents the zero function. This results in the contradiction

$$0 < \int_a^b p(x)dx = I^{(n)}(p) = I^{(n)}(p_n) = 0.$$

$\square$

In the examine interpolatory quadrature rules to derive conditions under which the highest possible order $2n + 2$ can be realized. We must find quadrature rules with $n + 1$ quadrature points that are exact for all polynomials from P_{2n+1}. We start with a formula consisting of the $2n + 2$ quadrature points $x_0, x_1, \ldots, x_n, x_{n+1}, \ldots, x_{2n+1}$. This rule has the order $2n + 2$. In Newton's representation of the interpolation polynomial, the error is

$$I(f) - I^{2n+1}(f) = I(f) - \sum_{i=0}^{2n+1} f[x_0, \ldots, x_i] \int_a^b \prod_{j=0}^{i-1}(x - x_j)\,dx$$

$$= I(f) - I^n(f) - \sum_{i=n+1}^{2n+1} f[x_0, \ldots, x_i] \int_a^b \prod_{j=0}^{i-1}(x - x_j)\,dx,$$

where $I^n(\cdot)$ is the quadrature rule given by the first $n + 1$ quadrature points $x_0, \ldots, x_n$. We now try to determine the quadrature points $x_0, \ldots, x_{2n+1}$ so that the residual term

$$\sum_{i=n+1}^{2n+1} f[x_0, \ldots, x_i] \int_a^b \prod_{j=0}^{i-1}(x - x_j)\,dx = 0$$

vanishes for all functions $f \in C^{2n+2}([a, b])$. Consequently the addition of the quadrature points $x_{n+1}, \ldots, x_{2n+1}$ is not necessary, i.e.,

$$I(f) - I^{2n+1}(f) = I(f) - I^n(f) \quad \Rightarrow \quad I^{2n+1}(f) = I^n(f).$$

The formula $I^n(f)$ with $n + 1$ quadrature points then already has the order $2n + 1$. The Newton basis polynomials for $i = n + 1, \ldots, 2n + 1$ must fulfill the following condition:

$$\underbrace{\int_a^b \prod_{j=0}^{i-1}(x-x_j)\,\mathrm{d}x}_{\in P_{2n+1}} = \int_a^b \underbrace{\prod_{j=0}^{n}(x-x_j)}_{\in P_{n+1}} \underbrace{\prod_{j=n+1}^{i-1}(x-x_j)}_{\in P_n}\,\mathrm{d}x \overset{!}{=} 0$$

The goal is to determine the first $n+1$ quadrature points $x_0, \ldots, x_n$ such that the integral

$$\int_a^b \underbrace{\prod_{j=0}^{n}(x-x_j)\,q(x)}_{=:p_{n+1}(x)}\,\mathrm{d}x = 0, \quad \forall q \in P_n, \tag{10.8}$$

vanishes for all $q \in P_n$. Choosing a basis of P_n, we obtain a system of $n+1$ equations for the $n+1$ unknowns $x_0, \ldots, x_n$. Geometrically, this is an orthogonalization problem with respect to the L^2 scalar product

$$(f, g)_{L^2[a,b]} := \int_a^b f(x)g(x)\,\mathrm{d}x.$$

In short: Find $x_0, \ldots, x_n \in \mathbb{R}$ such that $p_{n+1}(x) := \prod_{i=0}^{n}(x-x_i)$ satisfies

$$(p_{n+1}, q) = 0 \quad \forall q \in P_n. \tag{10.9}$$

The space of polynomials P_n is a linear vector space. A scalar product is given by $(\cdot, \cdot)$. Using the Gram-Schmidt orthogonalization method, Theorem 2.10, an orthogonal polynomial $p_{n+1} \in P_{n+1}$ can be generated from the monomial basis $\{1, x, \ldots, x^n\}$ satisfying (10.9). To finally find a Gaussian quadrature rule, we must still show that this polynomial p_{n+1} has $n+1$ different zeros $x_0, \ldots, x_n \in [a, b]$.

Before we consider the general solvability of this task, we examine the simple cases $n = 0$ and $n = 1$.

Example 10.25 (Optimal Quadrature Points Through Orthogonalization) We want to determine the polynomials $p_1(x) = (x - x_0)$ and $p_2(x) = (x - x_0)(x - x_1)$ such that they are orthogonal in the L^2 scalar product to all polynomials of degree 0 and 1, respectively. Without loss of generality, we consider the interval $[-1, 1]$:

For $n = 0$ with $q \equiv \alpha_0 \in P_0$

$$0 \overset{!}{=} (x - x_0, \alpha_0)_{L^2([-1,1])} = -2x_0\alpha_0 \quad \forall \alpha_0 \in \mathbb{R},$$

so, $x_0 = 0$ must hold.

In the case $n = 1$, setting $q(x) = \alpha_0 + \beta_0 x$, we get

$$0 \overset{!}{=} ((x - x_0)(x - x_1), \alpha_0 + \beta_0 x)_{L^2[-1,1]}$$
$$= \frac{-2(x_0 + x_1)}{3}\beta_0 + \frac{2(1 + 3x_0 x_1)}{3}\alpha_0 \quad \forall \alpha_0, \beta_0 \in \mathbb{R}.$$

From $\alpha_0 = 0$ and $\beta_0 = 1$ we conclude $x_0 = -x_1$ and from $\alpha_0 = 1$ and $\beta_0 = 0$, we get $x_{1/2} = \pm\frac{1}{\sqrt{3}}$. The corresponding weights are determined by

$$\alpha_i = \int_a^b L_i^{(n)}(x)\,dx,$$

and using (10.1), they are given as

$$\alpha_0^0 = \int_{-1}^1 1\,dx = 2, \quad \alpha_0^1 = \int_{-1}^1 \frac{x - \frac{-1}{\sqrt{3}}}{\frac{1}{\sqrt{3}} - \frac{-1}{\sqrt{3}}}\,dx = 1, \quad \alpha_1^1 = \int_{-1}^1 \frac{x - \frac{1}{\sqrt{3}}}{\frac{-1}{\sqrt{3}} - \frac{1}{\sqrt{3}}}\,dx = 1.$$

We have derived the first two Gauss quadrature rules

$$I_G^0(f) = 2f(0), \quad I_G^1(f) = f\left(-\frac{1}{\sqrt{3}}\right) + f\left(\frac{1}{\sqrt{3}}\right).$$

The first formula is just the midpoint rule, which we already know to be of order two (i.e., $2n + 2 = 2 \cdot 0 + 2$). The second formula has order 4, which can be easily verified by integrating the monomials $1, x, x^2, x^3$. ◀

10.4.1 Gauss-Legendre Quadrature

The Gauss rules based on the $L^2([-1, 1])$ orthogonal polynomials are called *Gauss-Legendre rules*, c.f. Sect. 9.5.1. We will examine general results on the existence and uniqueness of Gaussian quadrature rules $I_G^n(f)$. The interval $[-1, 1]$ is usually chosen to normalize the rules. An integral transformation obtains the transition to general integration domains.

Remark 10.26 (Transformation of Quadrature Rules) Let

$$\int_{-1}^1 f(x)\,dx \approx I_n(f) = \sum_{i=0}^n \alpha_i f(x_i)$$

be a numerical quadrature rule on the interval $[-1, 1]$ with quadrature points $x_i \in [-1, 1]$. The transition to a general interval $[a, b]$ is made by the linear transformation

$$z(x) = a + \frac{b - a}{2}(x + 1),$$

and the quadrature rule becomes

$$\int_a^b f(z)\,dz \approx \frac{b-a}{2} \sum_{i=1}^{n} \alpha_i\, f\!\left(a + \frac{b-a}{2}(x_i + 1)\right).$$

$\blacklozenge$

Definition 10.27 (Gauss Quadrature) A quadrature rule for the integration of a function $f : [a, b] \to \mathbb{R}$ with

$$I_G^n(f) = \sum_{k=0}^{n} \alpha_k f(x_k)$$

with $n + 1$ quadrature nodes, it is called a Gauss quadrature rule if all polynomials $p \in P_{2n+1}$ are integrated exactly, i.e., if it holds:

$$\int_{-1}^{1} p(x)dx = \sum_{k=0}^{n} \alpha_k p(x_k) \quad \forall p \in P_{2n+1}$$

The two quadrature rules derived in Example 10.25 are thus Gaussian quadrature rules. We want to investigate the existence of Gaussian quadrature rules of general order, i.e., for any number of nodes $n \in \mathbb{N}$. First, we prove that the nodes of Gauss quadrature must always be roots of orthogonal polynomials:

Theorem 10.28 *Let $x_0, \ldots, x_n$ be pairwise different quadrature nodes of a Gauss quadrature rule $I^n(\cdot)$. Then the orthogonality relationship holds*

$$\int_{-1}^{1} p_{n+1}(x)q(x)dx = 0 \quad \forall q \in P_n, \quad p_{n+1}(x) := (x - x_0)\cdots(x - x_n).$$

Proof This statement has already been proven in the derivation of the Gauss rules, compare (10.8) and (10.9). If the quadrature rule with $n + 1$ points has the order $2n + 2$, i.e., is a Gaussian rule, then the remainder of the possible points $x_{n+1}, \ldots, x_{2n+1}$ must vanish. This corresponds precisely to the orthogonality relationship in (10.8). $\qquad\square$

If an interpolatory quadrature rule has order $2n + 2$, then the $n + 1$ nodes form a polynomial $p_{n+1}(x) = \prod(x - x_i)$, which is orthogonal to P_n. It remains to show that the roots of an orthogonal polynomial always form the nodes of a Gauss rule, i.e., the reverse direction.

Theorem 10.29 *Let $p_{n+1} \in P_{n+1}$ be a polynomial*

$$p_{n+1}(x) = \prod_{j=0}^{n} (x - x_j), \tag{10.10}$$

with pairwise different and real roots $x_j \in (-1, 1)$ that is L^2-orthogonal to all polynomials $q \in P_n$, i.e.,

$$\int_{-1}^{1} p_{n+1}(x)q(x)\,dx = 0 \quad \forall q \in P_n.$$

Then, the nodes $x_0, \ldots, x_n$ and weights

$$\alpha_i = \int_{-1}^{1} L_i^{(n)}(x)\,dx = \int_{-1}^{1} \prod_{j=0, j \neq i}^{n} \frac{x - x_j}{x_i - x_j}\,dx$$

define a Gaussian quadrature rule of order $2n + 2$.

Proof Let p_{n+1} be given as in (10.10) and let $f \in C^{2n+2}([-1, 1])$ be arbitrary. Furthermore, let $p_n \in P_n$ be the interpolation polynomial through the nodes $x_0, \ldots, x_n$. In addition, let p_{2n+1} be a second interpolation polynomial through the nodes $x_0, \ldots, x_n$, as well as through further (pairwise different) nodes $x_{n+1}, \ldots, x_{2n+1}$. Thus, p_{2n+1}, with $2n + 2$ pairwise different nodes, generates at least one quadrature rule of order $2n + 2$. In Newton's representation, p_{2n+1} and p_n are given as

$$p_{2n+1}(x) = \sum_{i=0}^{2n+1} f[x_0, \ldots, x_i] \prod_{j=0}^{i-1} (x - x_j)$$

$$= p_n(x) + \sum_{i=n+1}^{2n+1} f[x_0, \ldots, x_i] \prod_{j=0}^{i-1} (x - x_j).$$

For the difference $p_{2n+1} - p_n$ it holds

$$p_{2n+1}(x) - p_n(x) = \sum_{i=n+1}^{2n+1} f[x_0, \ldots, x_i] \underbrace{\prod_{j=0}^{n} (x - x_j)}_{=p_{n+1}(x)} \underbrace{\prod_{j=n+1}^{i-1} (x - x_j)}_{=:q(x)}$$

with a polynomial $q \in P_n$. Since $p_{n+1} \perp q$, it follows

$$\int_{-1}^{1} p_n(x)\,dx = \int_{-1}^{1} p_{2n+1}(x)\,dx.$$

That is, the nodes $x_0, \ldots, x_n$ already generate a Gaussian quadrature rule with the order $2n + 2$. $\qquad\square$

These two theorems state that the characterization of Gaussian quadrature rules by roots of orthogonal polynomials is unique. We have already answered this question in Sect. 9.5.1 when discussing Gauss approximation and introduced the orthogonal Legendre polynomials

Table 10.2 Some Gauss-Legendre quadrature rules

# Points	$n-1$	x_i	α_i	Order	Exact
1	0	0	2	2	1
2	1	$\pm\sqrt{\frac{1}{3}}$	1	4	3
3	2	0	$\frac{8}{9}$	6	5
		$\pm\sqrt{\frac{3}{5}}$	$\frac{5}{9}$		
4	3	± 0.8611363116	0.347854845	8	7
		± 0.3399810436	0.652145154		
5	4	0	0.568888889		
		± 0.9061798459	0.236926885		
		± 0.5384693101	0.478628670	10	9

in Theorem 9.40. In that theorem, we also derived a fast, two-step recursion formula for calculating the orthogonal basis. In (9.23), we listed the first of these polynomials.

To set up Gauss rules, we need the roots of these polynomials for which closed formulae do not exist. For higher polynomial degrees, the roots must be found numerically, e.g., using Newton's method. The corresponding quadrature weights can then be calculated via (10.1):

$$\alpha_k = \int_{-1}^{1} \prod_{j=0,\, j\neq k}^{n} \frac{x - x_j}{x_k - x_j}\, dx$$

In Table 10.2, we summarize the quadrature points and weights of the first Gauss-Legendre rules.

Theorem 10.30 (Zeros of Orthogonal Polynomials) *The polynomials p_n orthogonal with respect to the $L^2([-1, 1])$ scalar product have real, simple zeros in the interval $(-1, 1)$.*

Proof The proof is conducted using contradiction arguments. The polynomials $p_n(x)$ are all real-valued.

(i) We first assume that p_n has a real zero λ that is not in the interval $(-1, 1)$. We define

$$q(x) := \frac{p_n(x)}{x - \lambda}.$$

The function q is a polynomial $q \in P_{n-1}$. Therefore, $q \perp p_n$. This implies

$$0 = (q, p_n) = \int_{-1}^{1} \frac{p_n^2(x)}{x - \lambda}\, dx.$$

But this is a contradiction, since $p_n(x)^2$ is always positive on $[-1, 1]$ (since p_n is real-valued). Further, $p_n(x)^2$ is not the zero function and moreover the factor $(x - \lambda)^{-1}$ cannot change its sign, since λ is outside of $[-1, 1]$. Therefore, p_n cannot have a zero outside of $(-1, 1)$.

(ii) In the second part, we show that only simple and real zeros lie in the interval $(-1, 1)$. We again work with a contradiction argument. We assume that p_n has a root λ, which is either a multiple or complex root. If λ is not real, then $\bar{\lambda}$ is also a zero. In both cases, we define

$$q(x) := \frac{p_n(x)}{(x - \lambda)(x - \bar{\lambda})} = \frac{p_n(x)}{|x - \lambda|^2} \in P_{n-2}.$$

It holds

$$0 = (p_n, q) = \int_{-1}^{1} \frac{p_n^2(x)}{|x - \lambda|^2} dx.$$

Again, a contradiction follows, since p_n^2 and $|x - \lambda|^2$ are both greater than or equal to zero, but not the zero function. $\square$

With these preliminaries, we are able to state the main theorem of this section:

Theorem 10.31 (Existence and Uniqueness of Gauss Quadrature) *For $n \in \mathbb{N}$ there exists a unique Gaussian quadrature rule of order $2n + 2$ with $n + 1$ pairwise different quadrature points in $(-1, 1)$, which are given as zeros of the orthogonal polynomial $p_{n+1} \in P_{n+1}$.*

Proof The proof consists of combining the previous results:

1. Theorem 9.40 provides the unique existence of the orthogonal polynomial $p_{n+1} \in P_{n+1}$.
2. Theorem 10.30 guarantees that this polynomial has $n + 1$ pairwise different, real zeros.
3. Theorem 10.29 states that by choosing the zeros as quadrature points with corresponding weights, a quadrature rule of order $2n + 2$, i.e., a Gauss quadrature, is given, see Definition 10.27.
4. Finally, Theorem 10.28 states that only by this choice of quadrature points a Gauss quadrature rule is generated, thus providing the uniqueness of the construction principle.

$\square$

A disadvantage of the Newton-Cotes rules is negative weights for $n \geq 8$ for closed quadrature rules. On the other hand, the Gauss quadrature weights are always positive.

Theorem 10.32 (Positivity of Gauss Quadrature Weights) *The weights of the Gaussian quadrature rules for the quadrature points $x_0, \ldots, x_n$ are always positive, and it holds*

$$\alpha_k = \int_{-1}^{1} L_k^{(n)}(x)\,dx = \int_{-1}^{1} L_k^{(n)}(x)^2\,dx > 0.$$

Proof Let the Gauss quadrature for the $n+1$ quadrature points $x_0, \ldots, x_n$ be given by $I_G^n(f) = \sum_k \alpha_k f(x_k)$. For the Lagrange basis polynomials, it holds

$$L_i^{(n)}(x) = \prod_{j \neq i} \frac{x - x_j}{x_i - x_j}, \quad L_i^{(n)}(x_k) = \delta_{ik}.$$

Since $L_i^{(n)} \in P_n$, $(L_i^{(n)})^2$ is also integrated exactly by the Gauss rule it holds

$$0 < \int_{-1}^{1} L_k^{(n)}(x)^2\,dx = \sum_{i=0}^{n} \alpha_i L_k^{(n)}(x_i)^2 = \alpha_k.$$

$\square$

Besides the high order, the Gaussian quadrature rules have the additional advantage of positive weights such that cancellation will not be a problem.

Moreover, we have seen that the Newton–Cotes rules suffer from the same deficiency as Lagrange interpolation: If the derivatives of the integrand increase too quickly, convergence may not occur as the degree of the polynomial increases. In the case of Gauss quadrature, we obtain convergence of the integral for $n \to \infty$ given sufficient regularity.

Theorem 10.33 (Convergence of Gauss Quadrature) *Let $f \in C^{\infty}([-1, 1])$. The sequence $(I^{(n)}(f))_n$, $n = 1, 2, \ldots$ of Gauss rules for the approximation of*

$$I(f) = \int_{-1}^{1} f(x)\,dx$$

converges:

$$I^{(n)}(f) \to I(f) \quad (n \to \infty)$$

Proof It holds

$$I^{(n)}(f) = \sum_{i=0}^{n} \alpha_i^{(n)} f(x_i^{(n)}), \quad \alpha_i^{(n)} > 0, \quad \sum_{i=0}^{n} \alpha_i^{(n)} = 2.$$

Let $\epsilon > 0$. According to the Weierstrass Approximation Theorem, Theorem 7.6, there exists a $p \in P_m$ (where m is sufficiently large), such that

$$\max_{-1 \leq x \leq 1} |f(x) - p(x)| \leq \frac{\epsilon}{4}.$$

For sufficiently large $2n + 2 > m$, it holds $I(p) - I_G^n(p) = 0$. From this, we infer

$$|I(f) - I^{(n)}(f)| \leq \underbrace{|I(f - p)|}_{\leq \frac{\epsilon}{4} \cdot 2} + \underbrace{|I(p) - I^{(n)}(p)|}_{=0} + \underbrace{|I^{(n)}(p - f)|}_{\leq \frac{\epsilon}{4} \cdot 2} \leq \epsilon.$$

Since $\epsilon > 0$ was chosen arbitrarily, $I^{(n)}(f) \to I(f)$ must converge for $n \to \infty$. $\square$

Theorem 10.34 (Gauss Quadrature Error Estimate) *Let $f \in C^{2n+2}([a, b])$. Then the remainder representation for a Gauss quadrature rule of order $2n + 2$ is*

$$R^{(n)}(f) = \frac{f^{(2n+2)}(\xi)}{(2n + 2)!} \int_a^b p_{n+1}^2(x)dx,$$

where $\xi \in [a, b]$ and with $p_{n+1} = \prod_{j=0}^n (x - \lambda_j)$.

Proof The proof is carried out with the help of Hermite interpolation (see Theorem 9.19). According to this, there exists a polynomial $h \in P_{2n+1}$ interpolating f with the conditions

$$h(x_i) = f(x_i), \quad h'(x_i) = f'(x_i), \quad i = 0, \dots, n.$$

The remainder is

$$f(x) - h(x) = f[x_0, x_0, \dots, x_n, x_n, x] \prod_{j=0}^n (x - x_j)^2.$$

If we now apply the Gaussian quadrature rule to $h(x)$, then it firstly holds $I_G^{(n)}(h) = I(h)$ and further by applying the mean value theorem of integral calculus, Theorem 7.4, we have

$$
\begin{aligned}
I(f) - I_G^{(n)}(f) &= I(f - h) - I_G^{(n)}(f - h) \\
&= \int_{-1}^1 f[x_0, x_0, \dots, x_n, x_n, x] \prod_{j=0}^n (x - x_j)^2 dx - \sum_{i=0}^n \alpha_i (f(x_i) - h(x_i)) \\
&= \frac{f^{(2n+2)}(\xi)}{(2n + 2)!} \int_{-1}^1 \prod_{j=0}^n (x - x_j)^2 dx.
\end{aligned}
$$

The mean value theorem can be applied here, since $\prod_{j=0}^n (x - x_j)^2 \geq 0$. The term $f(x_i) - h(x_i) = 0$ vanishes as h is the Hermite interpolation of f in x_i. This completes the proof. $\square$

Remark 10.35 (On the Regularity in the Error Formula) The same requirements apply to the Gauss quadrature as for all other integration methods: In order to achieve the full order $2n + 2$, the function $f(x)$ to be integrated must be correspondingly regular, i.e., $f \in C^{2n+2}([a, b])$. If this regularity is not present, Gauss quadrature rules cannot achieve their full order. In this case, a summed formula of lower order (possibly also a Gauss rule) is appropriate.   ♦

Example 10.36 (Gauss Rules with Legendre Polynomials) Finally, we discuss an example of Gauss quadrature. We integrate

$$\int_0^5 e^{-x^2}\,\mathrm{d}x \approx 0.8862269255.$$

We use Remark 10.26 to transform the integration domain and the quadrature points and weights from Table 10.2. For $n = 0$, we have

$$I^0(f) = \frac{5}{2}\alpha_0 f\left(\frac{5}{2}x_0 + \frac{5}{2}\right) = 5f\left(\frac{5}{2}\right) \approx 0.00966.$$

For $n = 1$, we get

$$I^1(f) = \frac{5}{2}\sum_{k=0}^{1}\alpha_k f\left(\frac{5}{2}x_k + \frac{5}{2}\right) = \frac{5}{2}\left(f\left(-\frac{1}{\sqrt{3}}\frac{5}{2} + \frac{5}{2}\right) + f\left(\frac{1}{\sqrt{3}}\frac{5}{2} + \frac{5}{2}\right)\right) \approx 0.81860.$$

For $n = 2$, we get

$$I^2(f) = \frac{5}{2}\sum_{k=0}^{2}\alpha_k f\left(\frac{5}{2}x_k + \frac{5}{2}\right)$$

$$= \frac{5}{2}\left(\frac{5}{9}f\left(-\sqrt{\frac{3}{5}}\frac{5}{2} + \frac{5}{2}\right) + \frac{8}{9}f\left(0\cdot\frac{5}{2} + \frac{5}{2}\right) + \frac{5}{9}f\left(\sqrt{\frac{3}{5}}\frac{5}{2} + \frac{5}{2}\right)\right)$$

$$= 1.01531.$$

Without further details, for $n = 3, 4, 5$, we get

$$I^3(f) \approx 0.87803,$$
$$I^4(f) \approx 0.87941,$$
$$I^5(f) \approx 0.88774.$$

                                                          ◀

Example 10.37 (Integration of the Double-Well Potential) We integrate the double-well potential, see Fig. 8.4, and examine up to which polynomial degree the Gauss rules integrate exactly:

$$I(f) = \int_{-3}^{3} f(x)\,\mathrm{d}x, \quad f(x) = (1 - x^2)^2$$

A Gauss rule with $n = 1$ (two-point rule) should integrate polynomials up to degree $2n + 1 = 3$ exactly and, therefore, not deliver the exact result:

$$I^1(f) = \frac{6}{2} \sum_{k=0}^{1} \alpha_k f\left(\frac{6}{2}x_k\right) = 3\left(f\left(-\frac{3}{\sqrt{3}}\right) + f\left(\frac{3}{\sqrt{3}}\right)\right) = 24$$

The exact value is $336/5 = 67.2$. This means we have a relative error of about 64%. The rule is therefore very inaccurate. The addition of a single additional support point leads to a three-point Gauss rule with $n = 2$ of order 6. It should integrate polynomials up to degree 5 exactly and therefore deliver the correct result. From Table 10.2, we read off the quadrature points and weights. A calculation shows

$$\begin{aligned}
I^2(f) &= \frac{6}{2} \sum_{k=0}^{2} \alpha_k f\left(3x_k\right) \\
&= 3\left(\alpha_0 f(3x_0) + \alpha_1 f(3x_1) + \alpha_2 f(3x_2)\right) \\
&= 3\left(\frac{5}{9}f(-3\sqrt{\frac{3}{5}}) + \frac{8}{9}f(0) + \frac{5}{9}f(3\sqrt{\frac{3}{5}})\right) = 67.2.
\end{aligned}$$

$\blacktriangleleft$

10.4.2 Gauss-Chebyshev Quadrature

The Gaussian quadrature rules are designed to integrate polynomials up to a certain degree exactly. Often, however, the functions to be integrated are not polynomials at all, e.g.,

$$f(x) = \sqrt{1 - x^2}$$

on the interval $[-1, 1]$. The derivatives of this function have singularities at the interval ends. The simple Gauss quadrature is not suitable for approximating $\int_{-1}^{1} f(x)\,dx$ since the required high regularity is not given.

The idea of Gauss quadrature can be generalized by finding optimal rules for functions that can be written as

$$f(x) = \omega(x)p(x)$$

where $p(x)$ is a polynomial and $\omega(x)$ is a *weight function*. Using this weight function, we define a modified scalar product:

Theorem 10.38 (Weighted Scalar Product) *Let the weight* $\omega \in L^\infty([-1, 1])$ *be positive*

$$\omega(x) > 0 \quad \text{almost everywhere.}$$

Then

$$(f, g)_\omega := \int_{-1}^{1} \omega(x)\, f(x)\, g(x)\, dx$$

defines a scalar product.

Proof Exercise.					□

We can now construct quadrature rules using the weighted scalar product. Suppose that $p_{n+1}^{\omega}(x)$ is a corresponding orthogonal polynomial

$$(p_{n+1}^{\omega}, q)_\omega = (\omega p_{n+1}^{\omega}, q) = 0 \quad \forall q \in P_n.$$

Theorem 10.30 can be directly transferred, i.e., p_{n+1}^{ω} has $n + 1$ pairwise different zeros in $(-1, 1)$. According to Theorem 10.29, which can also be immediately transferred, these zeros form a Gaussian quadrature rule for integrals of the type

$$I_\omega(f) = \int_{-1}^{1} \omega(x) f(x)\, dx.$$

We can apply the same argument as with the Gauss-Legendre rules.

The Gauss-Chebyshev quadrature chooses as a weight the function

$$\omega(x) = \frac{1}{\sqrt{1 - x^2}}.$$

This function places a strong focus on the two interval ends. The corresponding orthogonal polynomials are the Chebyshev polynomials, see Theorem 9.46, which play a special role in the optimal choice of the quadrature points for the Lagrange interpolation. To derive the Gauss-Chebyshev quadrature rules, the zeros of the orthogonal polynomials, i.e., the quadrature points, as well as the quadrature weights, must finally be determined. The zeros of the Chebyshev polynomials in $(-1, 1)$ can be read from the representation (compare Theorem 9.46)

$$T_n(x) := \cos(n \arccos x), \quad -1 \leq x \leq 1.$$

For the weights, a very simple relation holds.

Theorem 10.39 (Weights of the Gauss-Chebyshev Quadrature Rules) *The weights α_k of the Gauss-Chebyshev quadrature rule with n quadrature points are given by*

$$\alpha_k = \frac{\pi}{n}, \quad k = 0, \ldots, n - 1.$$

Proof The functions

$$\frac{T_m(x)}{\sqrt{1 - x^2}}, \quad m = 0, \ldots, n - 1,$$

are integrated exactly by the Chebyshev rule with n support points. That is, it holds

$$\sum_{k=0}^{n-1} \alpha_k T_m(x_k) = \int_{-1}^{1} \frac{T_m(x)}{\sqrt{1-x^2}}\, dx, \quad m = 0, \ldots, n-1.$$

With Theorem 9.46, it follows

$$\sum_{k=0}^{n-1} \alpha_k \cos\left(\frac{(2k+1)m}{2n}\pi\right) = \int_{-1}^{1} \frac{T_m(x)}{\sqrt{1-x^2}}\, dx, \quad m = 0, \ldots, n-1,$$

with

$$\int_{-1}^{1} \frac{T_m(x)}{\sqrt{1-x^2}}\, dx = \begin{cases} \pi & m = 0, \\ 0 & m = 1, \ldots, n-1. \end{cases}$$

From these equations, a linear system in the $\alpha_0, \ldots, \alpha_{n-1}$ can be derived with the solutions $\alpha_k = \pi/n$. $\qquad\square$

With this, we have collected all components to state the Gauss-Chebyshev rule, including the remainder:

Theorem 10.40 (Gauss-Chebyshev Rule) *The Gauss-Chebyshev rule of degree $2n$ with remainder is*

$$\int_{-1}^{1} \frac{f(x)}{\sqrt{1-x^2}} dx = \frac{\pi}{n} \sum_{k=0}^{n-1} f\left(\cos\left(\frac{2k+1}{2n}\pi\right)\right) + \frac{\pi}{2^{2n-1}(2n)!} f^{(2n)}(\xi)$$

with $\xi \in [-1, 1]$.

Proof This follows directly by inserting the quadrature points and weights into the general definition of a quadrature rule. The remainder is given by Theorem 10.34. $\qquad\square$

The Gauss-Chebyshev quadrature has the advantage that quadrature points and weights can be specified explicitly. They are naturally tailored to the special case of functions that have singularities at the edge of the integration range.

Example 10.41 We use the Gauss-Chebyshev quadrature to calculate the integral (half the area of a circle)

$$I(f) = \int_{-1}^{1} \sqrt{1-x^2}\, dx = \int_{-1}^{1} \omega(x) \underbrace{(1-x^2)}_{=:f(x)}\, dx$$

with the Gauss-Chebyshev weight $\omega(x)$ and $(1-x^2) \in P_2$. Since $f \in P_2$, we choose $n = 2$ in the method from Theorem 10.40 for exact calculation of the integral, i.e.,

$$\int_{-1}^{1} \omega(x) f(x)\mathrm{d}x \approx \frac{\pi}{2} \sum_{k=0}^{1} f\left(\cos\left(\frac{2k+1}{2\cdot 2}\pi\right)\right)$$

with $\xi \in (-1, 1)$ and the quadrature weights $\alpha_0 = \alpha_1 = \frac{\pi}{2}$. We obtain

$$\begin{aligned}
I(f) &= \int_{-1}^{1} \sqrt{1 - x^2}\,\mathrm{d}x = \int_{-1}^{1} \omega(x)(1 - x^2)\,\mathrm{d}x \\
&= \frac{\pi}{2} \sum_{k=0}^{1} \left(1 - \left(\cos\left(\frac{2k+1}{4}\pi\right)\right)^2\right) \\
&= \frac{\pi}{2} \left(\left(1 - \cos\left(\frac{1}{4}\pi\right)^2\right) + \left(1 - \cos\left(\frac{3}{4}\pi\right)^2\right)\right) \\
&= \frac{\pi}{2} \left((1 - 0.5) + (1 - 0.5)\right) = \frac{\pi}{2},
\end{aligned}$$

i.e., the integral is calculated exactly.　　　　　　　　　　　　　　　　　　◀

10.5　Excursus: Kepler's Barrel Rule

Kepler's Barrel Rule was developed by Johannes Kepler in the 17th century and is based on a special case of Simpson's Rule applied to rotational bodies. The background was that at Kepler's time, the price of a wine barrel was calculated according to its volume. In the 17th century, Kepler was tasked with defining new units of measurement for the city of Ulm, including those for the volumes of wine barrels. However, Kepler found that there were large deviations due to inaccurate calculations. Therefore, he developed a formula himself based on simple geometric considerations. This formula was published in 1615 [59] and is based on Simpson's Rule, discussed in Theorem 10.11. A quadratic polynomial approximates the shape of the outer wall, and the volume is approximated as

$$K_2 = \frac{b - a}{6}\left(f(a) + 4f(\frac{a+b}{2}) + f(b)\right). \tag{10.11}$$

This formula is of 4th order and, therefore, exact for polynomials up to degree three.

1. Attempt

First, we note that a barrel is a body of rotation with respect to the y-axis. The cross-sectional area perpendicular to the longitudinal axis (i.e., the y-axis) of a barrel is a circle with radius $r = f(x)$. The area is known to be given by $A := A(r) = \pi r^2$. The height of the barrel is h or more generally $b - a = h$, where $a = 0$ is the bottom and b is the lid. Kepler's Barrel Rule for volumes then reads:

$$V_K = \frac{b - a}{6}\left(A(a) + 4A(\frac{a+b}{2}) + A(b)\right) = \frac{h}{6}\left(A(0) + 4A(\frac{h}{2}) + A(h)\right) \tag{10.12}$$

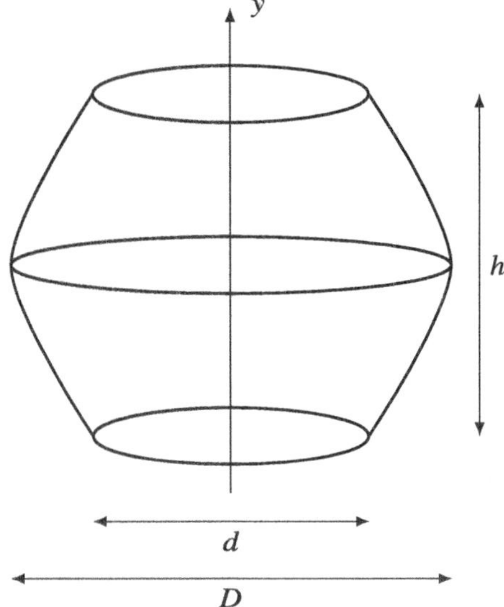

Fig. 10.4 Illustration of Kepler's Barrel rule

We illustrate the situation in Fig. 10.4.

We now have to determine the circumferences u of the bottom and lid, as well as the circumference U in the middle of the barrel, in order to be able to determine the volume approximately with

$$V_K = \frac{h}{6}\left(\pi \left(\frac{u}{2\pi}\right)^2 + 4\left(\frac{U}{2\pi}\right)^2 \pi + \pi \left(\frac{u}{2\pi}\right)^2\right) = \frac{h}{12\pi} \cdot (u^2 + 2U^2).$$

With this, we have all the data to determine the volume of a barrel quickly and relatively accurately. An oak wine barrel (French: Barrique) for red wine might have the following dimensions:

- Height: $h = 98$ cm
- Belly diameter: $D = 70$ cm
- Bottom and top diameter: $d = 58$ cm

Unfortunately, the manufacturer of the wine barrel could not provide any information about the volume at short notice. Using Kepler's barrel rule, we calculate the volume ourselves. First, we calculate the circumferences u and U:

$$u = 2 \cdot \pi \cdot \frac{d}{2} = 182.21$$

$$U = 2 \cdot \pi \cdot \frac{D}{2} = 219.91$$

With this, we get

$$V_K = \frac{h}{12\pi} \cdot (u^2 + 2U^2) = \frac{98}{12\pi}(182.21^2 + 2 \cdot 219.91^2) = 3.3773 \times 10^5 cm^3.$$

We know that $1\,dm^3 = 1000\,cm^3 = 1\,l$ corresponds. So we get a volume of $V_K = 338$ liters.

2. Attempt

The volume calculated above is a rough overestimate (i.e., an upper bound) and does not correspond to the actual fill volume. The reason is that the wine barrels have relatively thick walls (about $3\,cm$), lids, and bottoms that are set inwards by about $5\,cm$, and these are each $4\,cm$ thick. The modified data are therefore:

- Height: $h = 80$ cm
- Belly diameter: $D = 64$ cm
- Bottom diameter: $d = 52$ cm

For the circumferences, we get:

$$u = 2 \cdot \pi \cdot \frac{d}{2} = 163.36$$

$$U = 2 \cdot \pi \cdot \frac{D}{2} = 201.06$$

With this, we get

$$V_K = \frac{h}{12\pi} \cdot (u^2 + 2U^2) = \frac{80}{12\pi}(163.36^2 + 2 \cdot 201.06^2) = 2.28315 \times 10^5 cm^3,$$

which corresponds to 228 liters. This volume calculation is far more realistic than the result in attempt 1.

Mathematical Optimization and Artificial Neural Networks

A fundamental task of analysis is the search for the minimum or maximum of a function $f : \mathbb{R}^n \to \mathbb{R}$. For simple functions, e.g.,

$$f : \mathbb{R}^3 \to \mathbb{R}, \quad f(x) = \|x - x_0\|^2 + 5,$$

this task is straightforward, and the unique minimum is at $x = x_0$ where the function takes the value $f(x_0) = 5$. Theorem 7.13, the *theorem of the minimum and maximum* declares that continuous functions have a minimum and maximum on every closed interval. If we compare the proof of this theorem, for example, with that of the intermediate value theorem, we notice that the intermediate value theorem is based on a constructive argument, which can be directly implemented as a numerical method. The proof of the theorem of the minimum and maximum is based on the selection of a subsequence. It is a typical result of the type *there exists a point x, such that ...*, but the proof does not provide us with a clear way to find this point. Indeed, the task of finding minima and maxima is much more difficult than finding roots. This is particularly true when we know very little about the function $f : \mathbb{R}^n \to \mathbb{R}$. Theorem 7.13 requires the continuity of f. If we do not know more about f, we will not be able to derive efficient methods.

The key to success are characterizations of minima and maxima as extreme points, see Theorems 7.14 and 7.16. If a function is differentiable, then a minimum is an extreme point where the gradient vanishes. In this way, we can link the search for minima and maxima with the search for roots.

We will only provide a very small insight into basic methods here, as they have a direct connection to various topics in numerical mathematics. Optimization methods will also be used to "train neural networks", which we will discuss in Sect. 11.5. The Conjugate Gradient (CG) method introduced in Sect. 6.2 for solving linear systems can also be written

T. Richter et al., *Introduction to Numerical Mathematics*, Mathematics Study Resources 25, https://doi.org/10.1007/978-3-662-72546-7_11

as a minimization problem. This interpretation allows us to prove convergence of the method. For a broad introduction to the field of optimization, we refer to the literature [70].

11.1   Nonlinear Optimization

Given a continuous function

$$f : X \subset \mathbb{R}^n \to \mathbb{R},$$

we are looking for the point $x \in X$, which minimizes the function $f(\cdot)$:

$$f(x) = \min_{y \in X} f(y),$$

i.e., with

$$f(x) \leq f(y) \quad \forall y \in X$$

holds. In Definition 7.12, we introduced the concepts of local and global minima.

Figure 11.1 shows a function $f \in C(\mathbb{R})$, which has different types of minima:

$$f(x) = \begin{cases} \sin(4x) + \sin(x) + \frac{1}{4}x^2 & x < x_* \\ \sin(4x_*) + \sin(x_*) + \frac{1}{4}x_*^2 & x_* \leq x \leq x_* + \frac{1}{2} \\ \sin(4(x - \frac{1}{2})) + \sin(x - \frac{1}{2}) + \frac{1}{4}(x - \frac{1}{2})^2 & x > x_* + \frac{1}{2} \end{cases}$$

with $x_* = 2.72086$. Finding *global minima* is a very difficult task, comparable to finding *all roots* of a function.

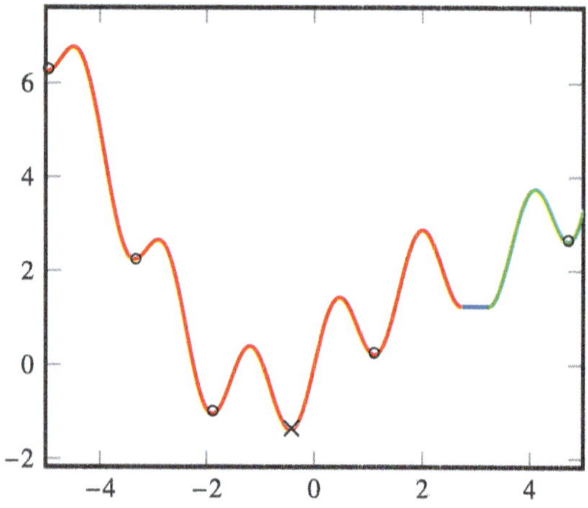

Fig. 11.1 Function with various minima. Local isolated minima are marked by o, the global (strict) minimum by ×, and the thickly marked area marks a non-isolated local minimum

11.1.1 Gradient Descent

The general descent method for minimizing a continuous function $f : X \subset \mathbb{R}^n \to \mathbb{R}$ starts with an initial value $x^0 \in X$ and seeks successive iterations $x^1, x^2, \ldots$ for which $f(x^k) < f(x^{k-1})$ holds. In each step k, we first choose the descent direction $d^k \in \mathbb{R}^n$ and then search for the next iteration only in this direction, i.e., we search for $x^k = x^{k-1} + s_k d^k \in X$, such that

$$f(x^{k-1} + s_k d^k) < f(x^k)$$

for some $s_k \in \mathbb{R}$. Consequently, we have transformed the n-dimensional minimization problem into a one-dimensional problem. This is also not easy to solve, but here we will learn efficient methods for approximation.

Definition 11.1 (Descent Direction) Let $f \in C(X)$ for $X \subset \mathbb{R}^d$ and $x \in X$. A vector $d \in \mathbb{R}^n$ is called *descent direction*, if there is a number $\alpha > 0$ such that

$$f(x + sd) < f(x) \quad \forall s \in (0, \alpha)$$

holds. We call $s \in \mathbb{R}$ the step size.

Definition 11.1 means that the function in direction d must be strictly monotonically decreasing locally. If the function f is additionally differentiable, it must hold that the derivative in direction d is negative, i.e.,

$$\nabla f(x) \cdot d < 0.$$

This relationship helps us in the search for the optimal descent direction:

Theorem 11.2 (Optimal Descent Direction) *Let $f : X \subset \mathbb{R}^n \to \mathbb{R}$ be a continuously differentiable function. Let $x \in X$ with $\nabla f(x) \neq 0$. The direction of the steepest descent in $x \in \mathbb{R}^n$ is uniquely given as the negative gradient*

$$d = -\frac{\nabla f(x)}{\|\nabla f(x)\|}.$$

Proof For arbitrary $d \in \mathbb{R}^n$ with $\|d\| = 1$, it holds

$$(\nabla f(x), d) \geq -\|\nabla f(x)\| \, \|d\| = -\|\nabla f(x)\|.$$

Equality only holds for $d = -\nabla f(x)/\|\nabla f(x)\|$. Thus, the unique normalized direction of the steepest descent is given by $-\nabla f(x)/\|\nabla f(x)\|$. $\qquad\square$

Algorithm 11.3　(Gradient Descent)

> **Input:** $f : X \subset \mathbb{R}^n \to \mathbb{R}$ continuously differentiable and starting value $x^0 \in X$.
> 1 **for** $k = 1, 2, \dots$ **do**
> 2 　　**if** $\nabla f(x^{k-1}) = 0$ **then**
> 3 　　　　Break
> 4 　　Determine descent direction $d^k = -\nabla f(x^{k-1})/\|\nabla f(x^{k-1})\|$
> 5 　　Choose a step size $s_k \in \mathbb{R}$
> 6 　　Calculate new approximation $x^k = x^{k-1} + s_k d^k$ with $f(x^k) < f(x^{k-1})$
> **Result:** Minimizer x^k of $f(x)$.

The gradient descent method is a concretization as it determines the search direction; however, the choice of step size is still open.

Definition 11.1 gives an idea for the basic procedure: If d^k is a descent direction, which is guaranteed by Theorem 11.2, then there must be a step size $s_k > 0$ such that the objective function is reduced. However, this step size may be close to zero. The methods generally work iteratively. In iteration $k \in \mathbb{N}$, we first choose an initial step size, e.g., $s_k^{(0)} = 1$, test whether a reduction is achieved by $x^{k-1} + s_k^{(0)} d^k$. If this is not the case, the step size is reduced by a factor $\beta \in (0, 1)$ to $s_k^{(1)} = \beta \cdot s_k^{(0)}$ and the test is repeated.

This is called *line search*. We have already come across this method in Sect. 8.4.2 as a globalization method of the Newton method.

Algorithm 11.4　(Line Search)

> **Input:** $\beta \in (0, 1)$, current step $x^{k-1} \in \mathbb{R}^n$ and search direction $d^k \in \mathbb{R}^n$.
> 1 $s_k^{(0)} = 1$
> 2 **for** $l = 0, 1, 2, \dots$ **do**
> 3 　　**if** $f(x^k + s_k^{(l)} d^k) < f(x^k)$ **then**
> 4 　　　　Break
> 5 　　$s_k^{(l+1)} = \beta \cdot s_k^{(l)}$
> **Result:** Step size $s_k = s_k^{(l+1)}$

The simple line search algorithm ensures that a descent is achieved, but cannot guarantee any properties of the descent direction beyond that. A modification is the *Armijo step size rule*. This uses the fact that the function f can be represented locally in search direction as a linear function with the slope $-\nabla f(x^{k-1})$. The line search is only terminated when the found direction achieves this linear descent.

Algorithm 11.5 (Armijo Step Size Rule)

> **Input:** Let $\beta, \gamma \in (0, 1)$, the current step by $x^k \in \mathbb{R}^n$ and the chosen search direction by
> $d^k \in \mathbb{R}^n$ be given.
>
> 1 $s_k^{(0)} = 1$
> 2   **for** $l = 0, 1, 2, \ldots$ **do**
> 3   **if** $f(x^k + s_k^{(l)} d^k) < f(x^k) + \gamma s_k^{(l)} \nabla f(x^k) \cdot d^k$ **then**
> 4   Break
> 5 $s_k^{(l+1)} = \beta \cdot s_k^{(l)}$
>   **Result:** Step size $s_k = s_k^{(l+1)}$

For both algorithms, it can be shown that they terminate in a finite number of steps, i.e., a suitable step size is indeed found. We demonstrate this for the Armijo rule. But any step size that satisfies the Armijo condition also naturally satisfies the simple line search condition.

Theorem 11.6 (Armijo Step Size Rule) *Let $f : X \to \mathbb{R}$ be continuously differentiable on the open set $X \subset \mathbb{R}^n$. Let $\gamma \in (0, 1)$ and $d \in \mathbb{R}^n$ be any descent direction at a point $x \in X$ with*

$$\nabla f(x) \cdot d < 0.$$

Then there exists a $k \in \mathbb{N}$ such that

$$s_k = \beta^k, \quad f(x + s_k d) - f(x) \le \gamma s_k (\nabla f(x), d).$$

Proof We have

$$\frac{f(x + sd) - f(x)}{s} - \gamma \nabla f(x) \cdot d \to (1 - \gamma) \nabla f(x) \cdot d < 0 \quad (s \to 0).$$

For sufficiently small s, due to the continuity of ∇f, there exists an ϵ such that for all $s < \epsilon$:

$$\frac{f(x + sd) - f(x)}{s} - \gamma \nabla f(x) \cdot d \le 0 \quad \forall |s| < \epsilon$$

For $\beta \in (0, 1)$, there exists a minimal $k \in \mathbb{N}$ such that $s_k := \beta^k < \epsilon$. $\qquad\square$

Gradient descent often converges very slowly. In Fig. 11.2, we show the convergence behavior using the so-called *Rosenbrock function*, a typical test case for optimization methods:

$$f(x, y) = 10(y - x^2)^2 + (1 - x)^2 \to \min.$$

We choose the initial values $x^0 = (-1, 1.5)$ and $x^0 = (2, -2)$. For the first case, we summarize the convergence behavior in Table 11.1.

There are many methods to increase the efficiency of gradient descent, and we refer to the literature [70, 94].

Table 11.1 Results for the approximation of the Rosenbrock function

k	x^k	y^k	$\|\nabla f(x^k)\|$	$f(x^k)$
0	−1	1.5	356	
1	−1.212	1.3675	4.9958	9.5597
2	−1.0898	1.3940	4.7929	6.3382
3	−1.1373	1.3533	4.6039	1.9585
4	−1.0384	1.2768	4.5494	5.7600
5	−1.0836	1.2337	4.3770	1.9861
...				
100	1.0050	1.0100	2.55e-05	0.012493
101	1.0048	1.0101	2.52e-05	0.012492
102	1.0049	1.0099	2.48e-05	0.012491
103	1.0047	1.0099	2.44e-05	0.012491
104	1.0049	1.0097	2.41e-05	0.012490
105	1.0046	1.0098	2.37e-05	0.012489

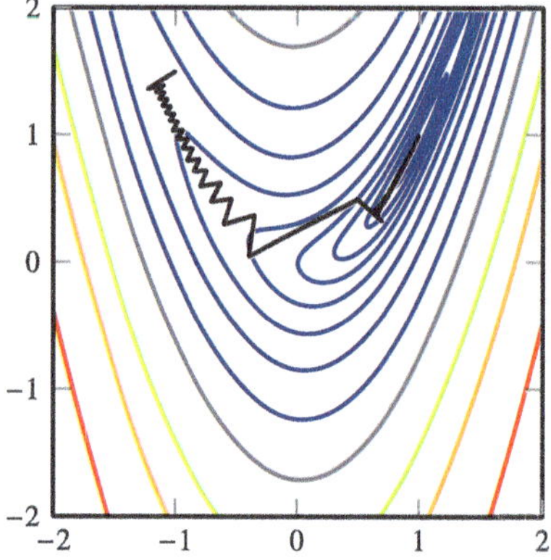
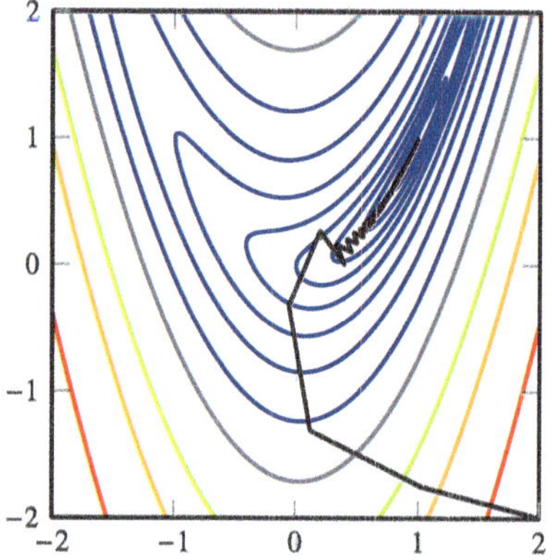

Fig. 11.2 Convergence behavior of gradient descent for the approximation of the minimum of the so-called *Rosenbrock function* $f(x, y) = 10(y - x^2)^2 + (1 - x)^2$ at various starting values

11.1.2 The Newton Method for Minimization

If the underlying function $f(x)$ is twice continuously differentiable, then the Newton method can be used to solve the minimization problem. As previously discussed, the necessary condition is given by

$$F(x) := \nabla f(x) \stackrel{!}{=} 0.$$

The general iteration, starting from an initial value $x^0 \in X$, is for $k = 1, 2, 3, \ldots$

Table 11.2 Convergence of the Newton method for the Rosenbrock function

k	x^k	y^k	$\|\nabla f(x^k)\|$	$f(x^k)$
0	-1	1.5	18.86	6.5
1	-1.222	1.444	6.92	4.963
2	-0.104	-1.239	26.08	16.84
3	-0.063	0.002	2.128	1.127
4	0.963	-0.123	45.51	11.03
5	0.965	0.931	7.05×10^{-2}	0.00124
6	0.999	0.999	5.56×10^{-2}	1.55×10^{-5}
7	1.000	1.000	9.67×10^{-8}	2.34×10^{-15}
0	2	-2	496.71	361
1	1.992	3.966	1.988	0.983
2	1.001	0.021	43.91	9.620
3	1.001	1.002	2.57×10^{-3}	1.65×10^{-6}
4	1.000	1.000	7.41×10^{-5}	2.74×10^{-11}

$$DF(x^k)(x^{k+1} - x^k) = -F(x^k).$$

Here,

$$DF(x^k) = H_f(x^k),$$

where $H_f(x^k)$ is the Hessian of the function f

$$H_f(x) := \begin{pmatrix} \frac{\mathrm{d}}{\mathrm{d}x_1} F_1 & \cdots & \frac{\mathrm{d}}{\mathrm{d}x_n} F_1 \\ \vdots & & \vdots \\ \frac{\mathrm{d}}{\mathrm{d}x_1} F_n & \cdots & \frac{\mathrm{d}}{\mathrm{d}x_n} F_n \end{pmatrix} = \begin{pmatrix} \frac{\mathrm{d}^2 f}{\mathrm{d}x_1 \mathrm{d}x_1} & \cdots & \frac{\mathrm{d}^2 f}{\mathrm{d}x_1 \mathrm{d}x_n} \\ \vdots & & \vdots \\ \frac{\mathrm{d}^2 f}{\mathrm{d}x_1 \mathrm{d}x_n} & \cdots & \frac{\mathrm{d}^2 f}{\mathrm{d}x_n \mathrm{d}x_n} \end{pmatrix}$$

To use Newton's method for optimization, the *objective function* $f(x)$ must be at least twice continuously differentiable. The Newton-Kantorovich theorem, Theorem 8.41, guarantees quadratic convergence, if the Hessian of f is Lipschitz. In Fig. 11.3, we show convergence of the Newton method for the Rosenbrock function. We summarize the numerical values in Table 11.2.

In both cases, the minima can be approximated very well in just a few iterations. The values in the table show that the Newton method is not a *descent method*. The iterates $f(x^k)$ do not converge monotonically towards the minimum.

Especially in optimization, the major weakness of the Newton method is often the very small convergence region. Here, *globalization methods* play a special role again. When applying the Newton method for minimization, efficient methods are available that combine gradient descent with the Newton method. Since we know that gradient descent converges

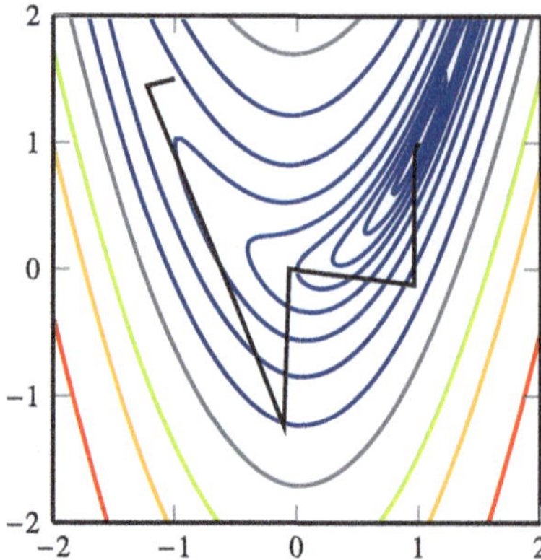 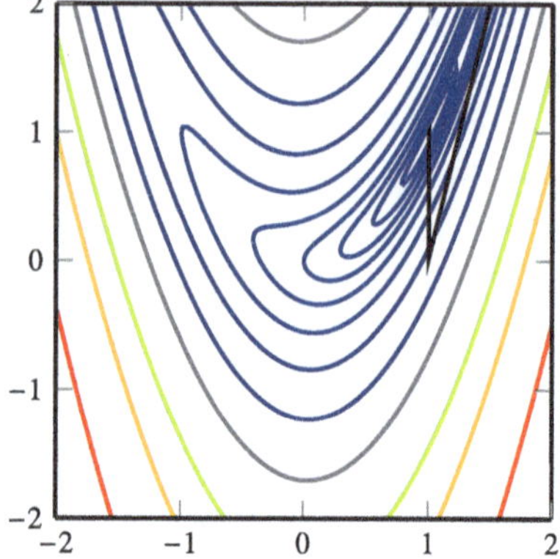

Fig. 11.3 Convergence of the Newton method for the approximation of the minimum of the Rosenbrock function $f(x, y) = 10(y - x^2)^2 + (1 - x)^2$

globally, it can be used when we are still far from the region of quadratic convergence. Near the minimum, a switch to the Newton method then takes place.

11.2   Steepest Descent Methods for Linear Systems

The idea of the gradient descent method can also be used to solve linear systems. This will lead us to another interpretation of the Conjugate Gradient method introduced in Chap. 6 as a Krylov subspace method. The starting point is the following theorem, which presents the solution of a linear system as a minimization problem.

Theorem 11.7 (Linear System and Minimization) *Let $A \in \mathbb{R}^{n \times n}$ be a symmetric positive definite matrix, $x, b \in \mathbb{R}^n$. The following conditions are equivalent:*

$$(i) \quad Ax = b,$$

$$(ii) \quad Q(x) \leq Q(y) \quad \forall y \in \mathbb{R}^n, \quad Q(y) = \frac{1}{2}(Ay, y)_2 - (b, y)_2.$$

Proof $(i) \Rightarrow (ii)$ Let x be the solution of the linear equation system, i.e., $Ax = b$. Then, for any $y \in \mathbb{R}^n$, we have:

$$\begin{aligned}
2Q(y) - 2Q(x) &= (Ay, y)_2 - 2(b, y)_2 - (Ax, x)_2 + 2(b, x) \\
&= (Ay, y)_2 - 2(Ax, y)_2 + (Ax, x)_2 \\
&= (A(y - x), y - x)_2 \geq 0,
\end{aligned}$$

where we use the symmetry of A with $(Ax, y)_2 = (Ay, x)_2$ to split $2(Ax, y)_2 = (Ax, y) + (x, Ay)$ from the penultimate to the last line. Finally, we have shown that $Q(y) \geq Q(x)$.

$(ii) \Rightarrow (i)$ Conversely, let $Q(x)$ now be the minimum. That is, $x \in \mathbb{R}^n$ is a stationary point of the quadratic form $Q(x)$, i.e.,

$$0 \overset{!}{=} \frac{\partial}{\partial x_i} Q(x) = \frac{\partial}{\partial x_i} \left\{ \frac{1}{2}(Ax, x)_2 - (b, x)_2 \right\} = (Ax)_i - b_i, \quad i = 1, \ldots, n. \tag{11.1}$$

That is, x is the solution of the linear system. $\qquad\square$

Instead of determining a solution of a linear system $Ax = b$, we look for the minimum of the so-called *energy functional* $Q(x)$. With this equivalent formulation, we have entered the realm of minimization problems. Instead of solving the system $Ax = b$, we minimize the function $Q : \mathbb{R}^n \to \mathbb{R}$. At first, this does not seem efficient, as solving a minimization problem is generally much more complex than solving or approximating a linear system. However, the objective functional $Q(x)$ is not a general function but has special structural properties.

Theorem 11.8 *Let $A \in \mathbb{R}^{n \times n}$ be symmetric positive definite. The functional*

$$Q(y) = \frac{1}{2}(Ay, y) - (b, y)$$

is quadratic and strictly convex.

Proof We have already determined the gradient of $Q(\cdot)$, see (11.1). It holds

$$\nabla Q(x) = \begin{pmatrix} (Ax)_1 - b_1 \\ (Ax)_2 - b_2 \\ \vdots \\ (Ax)_n - b_n \end{pmatrix} = Ax - b \in \mathbb{R}^n. \tag{11.2}$$

For the second derivatives, i.e., the Hessian matrix of $Q(\cdot)$, we have

$$H_Q(x) = A.$$

Since A is symmetric positive definite, it immediately follows that $Q(\cdot)$ is strictly convex. $\qquad\square$

Since A is symmetric positive definite, $\|x\|_A := \sqrt{(Ax, x)_2}$ is a norm, the so-called *energy norm*. The minimization of the energy functional $Q(\cdot)$ is equivalent to the minimization of the error $x^k - x$ in the energy norm. Let us suppose $x \in \mathbb{R}^n$ is the solution of the equation system and $y \in \mathbb{R}^n$ is any approximation to this solution. Then, we have

$$\|y - x\|_A^2 = (A(y - x), y - x)_2 = (Ay, y) - \underbrace{2(Ay, x)}_{=2(b,y)} + (Ax, x) = 2Q(y) + (Ax, x).$$

Minimization of $Q(y)$ also minimizes the error in the energy norm.

We can now apply the idea of gradient descent to the minimization of the objective functional $Q(x)$ and determine a sequence of approximations $x^k \in \mathbb{R}^n$. In each step of the method, the descent direction d^k is first chosen, then the approximation is improved in this direction: $x^{k+1} = x^k + \omega^k d^k$. The step size $\omega^k \in \mathbb{R}$ is chosen so that the resulting value of the energy functional is minimal. We summarize:

Algorithm 11.9 (Descent Method for Linear Systems)

> **Input:** $A \in \mathbb{R}^{n \times n}$ symmetric positive definite, $x^0, b \in \mathbb{R}^n$.
> 1   **for** $k = 0, 1, 2, \dots$ **do**
> 2   Choose descent direction $d^k \in \mathbb{R}^n$
> 3   Determine ω^k as the minimum of $\omega^k = \arg\min_{\omega \in \mathbb{R}} Q(x^k + \omega d^k)$
> 4   Update $x^{k+1} = x^k + \omega^k d^k$
> **Result:** Approximate solution x^{k+1} of $Ax = b$.

We assume that the search directions are given, and we are looking for the solution of the scalar minimization problem. In the general nonlinear minimization problem discussed in Sect. 11.1, the one-dimensional problem

$$\omega^k \in \mathbb{R} \quad Q(x^k + \omega^k d^k) \leq Q(x^k + s d^k) \quad \forall s \in \mathbb{R} \tag{11.3}$$

is too difficult to solve exactly. Instead, methods to determine the step sizes, like the Arimijo rule, were introduced as an approximation. In the case of the linear system, the situation is simpler, and we can determine the solution of the quadratic minimization problem as a stationary point

$$0 \overset{!}{=} \frac{\partial}{\partial \omega^k} Q(x^k + \omega^k d^k) = \omega^k (Ad^k, d^k) + (Ax^k, d^k) - (b, d^k),$$

so we get

$$\omega^k = \frac{(b - Ax^k, d^k)}{(Ad^k, d^k)}. \tag{11.4}$$

When choosing the search direction, we could orient ourselves on the simple fixed point iterations from Sect. 3.7 and determine d^k based on the Jacobi or Gauss-Seidel method, i.e.,

$$d_J^k = -D^{-1}(b - Ax^k), \quad d_{GS}^k = -(D + L)^{-1}(b - Ax^k).$$

Subsequently, the optimal step size is determined using (11.4). However, it turns out that these methods have hardly any advantage over the Jacobi and Gauss-Seidel methods themselves.

The *gradient method to solve linear systems* corresponds to the general gradient method from Sect. 11.1, by choosing the negative gradient as the search direction in step k, i.e.,

$$d^k = -\nabla Q(x^k) = b - Ax^k,$$

see (11.2). Then, Theorem 11.2 applies accordingly and $b - Ax^k$ is the direction of the *steepest descent*, in which the value of $Q(x)$ thus reduces the fastest. For a point $x \in \mathbb{R}^n$, this is the direction $d \in \mathbb{R}^n$, which is orthogonal (i.e., perpendicular) to the level set $N(x)$:

$$N(x) := \{y \in \mathbb{R}^n : Q(y) = Q(x)\}$$

At a point x, the level set is spanned by all directions $\delta x \in \mathbb{R}^n$ with

$$0 \stackrel{!}{=} Q'(x) \cdot \delta x = (\nabla Q(x), \delta x) = (b - Ax, \delta x).$$

The vectors δx, which span the level set, are orthogonal to the residual $b - Ax$. This, therefore, points in the direction of the strongest change of $Q(\cdot)$. We choose $d^k := b - Ax^k$. The search direction found in this way is then combined with the descent method, i.e., we iterate

$$d^k := b - Ax^k, \quad \omega^k := \frac{\|d^k\|_2^2}{(Ad^k, d^k)_2}, \quad x^{k+1} := x^k + \omega^k d^k.$$

We define the *gradient descent method*:

Algorithm 11.10 (Gradient Descent for Linear Systems)

> **Input:** $A \in \mathbb{R}^{n \times n}$ symmetric positive definite, $b \in \mathbb{R}^n$, and $x^0 \in \mathbb{R}^n$.
> 1 $d^0 := b - Ax^0$
> 2 **for** $k = 0, 1, 2, \ldots$ **do**
> 3 $\quad r^k := Ad^k$
> 4 $\quad \omega^k = \frac{\|d_k\|_2^2}{(r^k, d^k)_2}$
> 5 $\quad x^{k+1} = x^k + \omega^k d^k$
> 6 $\quad d^{k+1} = d^k - \omega^k r^k$
> **Result:** Approximate solution x^{k+1} of $Ax = b$.

By introducing a second auxiliary vector $r^k \in \mathbb{R}^n$, a matrix-vector product can be saved in each iteration. For matrices with diagonal part $D = \alpha I$, gradient descent is just the Jacobi method in combination with the descent method. Therefore, no improved convergence statement can generally be achieved for this method. We show:

Theorem 11.11 (Gradient Descent for Linear Systems) *Let $A \in \mathbb{R}^{n \times n}$ be symmetric positive definite. Then gradient descent converges for every starting vector $x^0 \in \mathbb{R}^n$ to the solution of the equation $Ax = b$.*

Proof Let $x^k \in \mathbb{R}^n$. Further, let $d := b - Ax^k$. Then, a step of gradient descent is calculated as

$$x^{k+1} = x^k + \frac{(d, d)}{(Ad, d)} d.$$

For the energy functional, it holds:

$$
\begin{aligned}
Q(x^{k+1}) &= \frac{1}{2}(Ax^{k+1}, x^{k+1}) - (b, x^{k+1}) \\
&= \frac{1}{2}(Ax^k, x^k) + \frac{1}{2}\frac{(d, d)^2}{(Ad, d)^2}(Ad, d) + \frac{(d, d)}{(Ad, d)}(Ax^k, d) \\
&\quad - (b, x^k) - \frac{(d, d)}{(Ad, d)}(b, d) \\
&= Q(x^k) + \frac{(d, d)}{(Ad, d)}\left\{\frac{1}{2}(d, d) + (Ax^k, d) - (b, d)\right\} \\
&= Q(x^k) + \frac{(d, d)}{(Ad, d)}\left\{\frac{1}{2}(d, d) + (\underbrace{Ax^k - b}_{=-d}, d)\right\}
\end{aligned}
$$

Therefore, it follows that

$$Q(x^{k+1}) = Q(x^k) - \frac{(d, d)^2}{2(Ad, d)}.$$

Due to the positive definiteness of A, it holds that

$$\lambda_{\min}(A)(d, d) \leq (Ad, d) \leq \lambda_{\max}(A)(d, d)$$

and finally, with

$$Q(x^{k+1}) \leq Q(x^k) - \underbrace{\frac{(d, d)}{2\lambda_{\max}}}_{>0}$$

the sequence $Q(x^k)$ is monotonically decreasing. As long as $d = b - Ax \neq 0$, the sequence strictly decreases. Furthermore, $Q(x^k)$ is bounded from below by $Q(x)$. Therefore, the sequence $Q(x^k) \to c \in \mathbb{R}^n$ converges. In the limit, it must hold that $0 = (d, d) = \|b - Ax\|^2$, i.e., $Ax = b$. $\qquad\square$

Finally, we show an estimate of the convergence speed of gradient descent:

Theorem 11.12 (Convergence of Gradient Descent (Simplified)) *Let $A \in \mathbb{R}^{n \times n}$ be a symmetric positive definite matrix. Then, for the gradient descent method to solve $Ax = b$, the error estimate*

$$\|x^0 - x\|_A \leq \left(1 - \frac{1}{\kappa}\right)^k, \quad \kappa := \mathrm{cond}_2(A) = \frac{\lambda_{max}(A)}{\lambda_{min}(A)}$$

holds.

Proof The matrix $A \in \mathbb{R}^{n \times n}$ is symmetric positive definite. Thus, there exists a system of n eigenvalues

$$0 < \lambda_{min} =: \lambda_1 \leq \cdots \leq \lambda_n =: \lambda_{max}$$

and orthonormal eigenvectors $w_1, \ldots, w_n \in \mathbb{R}^n$. Let

$$e^k = x^k - x = \sum_{i=1}^{n} e_i^k w_i \tag{11.5}$$

be an expansion of the error in the eigenvectors. For one step of the gradient descent method, with

$$d^k = b - Ax^k = -Ae^k,$$

the following error propagation holds

$$x^{k+1} = x^k + \frac{(d^k, d^k)}{(Ad^k, d^k)} d^k \quad \Rightarrow \quad e^{k+1} = e^k - \frac{(Ae^k, Ae^k)}{(A^2 e^k, Ae^k)} Ae^k. \tag{11.6}$$

With (11.5) and $Aw_i = \lambda_i w_i$, it follows that

$$e^{k+1} = \sum_{i=1}^{N} \left(1 - \frac{(Ae^k, Ae^k)}{(A^2 e^k, Ae^k)} \lambda_i \right) e_i^k w_i = \sum_{i=1}^{N} (1 - \mu_i) e_i^k w_i, \tag{11.7}$$

where

$$\mu_i = \lambda_i \frac{(Ae^k, Ae^k)}{(A^2 e^k, Ae^k)}.$$

Due to the orthonormality of the eigenvectors and since $\lambda_i > 0$, it follows

$$\mu_i = \lambda_i \frac{\left(\sum_{j=1}^{n} \lambda_j e_j^k w_j, \sum_{j=1}^{n} \lambda_j e_j^k w_j \right)}{\left(\sum_{j=1}^{n} \lambda_j^2 e_j^k w_j, \sum_{j=1}^{n} \lambda_j e_j^k w_j \right)} = \lambda_i \frac{\sum_{j=1}^{n} \lambda_j^2 (e_j^k)^2}{\sum_{j=1}^{n} \lambda_j^3 (e_j^k)^2}$$
$$\geq \lambda_{min} \frac{\sum_{j=1}^{n} \lambda_j^2 (e_j^k)^2}{\lambda_{max} \sum_{j=1}^{n} \lambda_j^2 (e_j^k)^2} = \frac{\lambda_{max}}{\lambda_{min}}. \tag{11.8}$$

We continue with (11.7) and the error in the A-norm is estimated as

$$\|e^{k+1}\|_A^2 = (Ae^{k+1}, e^{k+1}) = \left(A \sum_{i=1}^{N} (1 - \mu_i) e_i^k w_i, \sum_{i=1}^{N} (1 - \mu_i) e_i^k w_i \right)$$
$$= \sum_{i=1}^{N} (1 - \mu_i)^2 \lambda_i (e_i^k)^2.$$

With (11.8) and the relation

$$\|e^k\|_A^2 = (Ae^k, e^k) = \sum_{i=1}^{N} \lambda_i (e_i^k)^2,$$

it further follows

$$\|e^{k+1}\|_A^2 \leq \left(1 - \frac{1}{\kappa}\right)^2 \|e^k\|_A^2, \quad \kappa := \mathrm{cond}_2(A) = \frac{\lambda_{max}}{\lambda_{min}}.$$

The repeated application of this inequality gives

$$\|e^k\|_A \leq \left(1 - \frac{1}{\kappa}\right)^k \|x^0 - x\|_A.$$

$\square$

This error estimate is not optimal. Greater effort leads to an improved error estimate that holds under the same assumptions and that shows

$$\|x^k - x\|_A \leq \left(\frac{1 - 1/\kappa}{1 + 1/\kappa}\right)^k, \quad \kappa := \mathrm{cond}_2(A).$$

Asymptotically, it holds

$$\frac{1 - \frac{1}{\kappa}}{1 + \frac{1}{\kappa}} = 1 - \frac{2}{\kappa} + O\left(\frac{1}{\kappa^2}\right),$$

i.e., the method converges twice as fast. The key to this optimal proof is the Kantorovich theorem, an estimate for the eigenvalues of a positive definite matrix. For the proof, we refer to the literature [39].

The asymptotic convergence factor of the gradient descent method is related to the condition number of the matrix. For the model matrix with $\kappa = O(n^2)$, see Example 3.72, this is

$$\rho = \frac{1 - \frac{1}{n^2}}{1 + \frac{1}{n^2}} = 1 - \frac{2}{n^2} + O\left(\frac{1}{n^4}\right).$$

The convergence is therefore as slow as that of the Jacobi method. For the search directions of the gradient method, the following relationship holds:

Theorem 11.13 (Descent Directions in the Gradient Method) *Let $A \in \mathbb{R}^{n \times n}$ be symmetric positive definite. Then any two consecutive descent directions d^k and d^{k+1} of the gradient method are orthogonal to each other, i.e., $(d^k, d^{k+1}) = 0$.*

Proof We consider Algorithm 11.10. It holds

$$d^{k+1} = d^k - \omega^k r^k = d^k - \frac{(d^k, d^k)}{(Ad^k, d^k)} Ad^k.$$

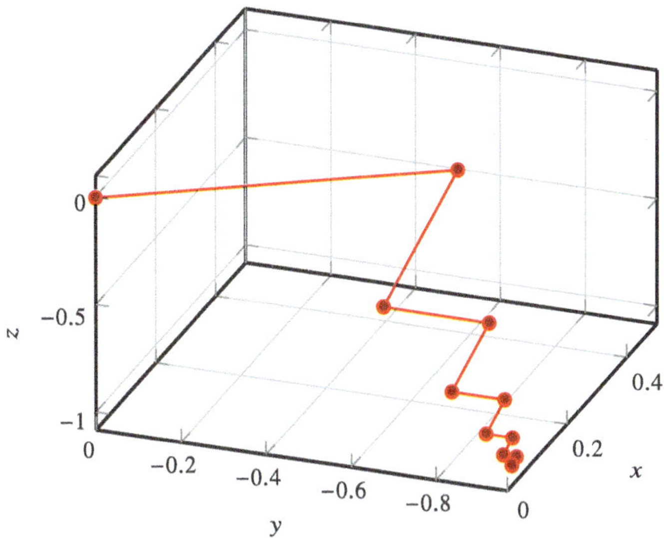

Fig. 11.4 Convergence of the gradient method for approximating a linear system $Ax = b$ with $A \in \mathbb{R}^{n \times n}$. Consecutive search directions are orthogonal, i.e., $d^k \perp d^{k+1}$. However, for d^k and d^{k+2}, it can be the case that these directions run almost parallel

Thus, it holds

$$(d^{k+1}, d^k) = (d^k, d^k) - \frac{(d^k, d^k)}{(Ad^k, d^k)}(Ad^k, d^k) = (d^k, d^k) - (d^k, d^k) = 0.$$

$\square$

In Fig. 11.4, we show the first 10 steps of the gradient method for approximating the solution of the linear system

$$\begin{pmatrix} 2 & -1 & 0 \\ -1 & 2 & -1 \\ 0 & -1 & 2 \end{pmatrix} \begin{pmatrix} x_1 \\ x_2 \\ x_3 \end{pmatrix} = \begin{pmatrix} 1 \\ -1 \\ -1 \end{pmatrix}$$

with the exact solution $x = (0, -1, -1)$. Two consecutive directions always lie parallel. The approximation proceeds in a zig-zag line and the convergence is very slow.

The CG method described in Sect. 6.2 can be derived as a further improvement of the gradient descent method. Instead of the one-dimensional minimization problems (11.3), which are solved in each step of the gradient descent method, we consider a minimization problem of increasing complexity. In the k-th step, the approximation $x^k = x^0 + \sum_{i=0}^{k-1} \alpha_i d^i$ is sought as the minimum over all $\alpha = (\alpha_0, \ldots, \alpha_{k-1})$ with respect to $Q(x^k)$:

$$\min_{\alpha \in \mathbb{R}^k} Q\left(x^0 + \sum_{i=0}^{k-1} \alpha_i d^i\right)$$

$$= \min_{\alpha \in \mathbb{R}^k} \left\{ \frac{1}{2}\left(Ax^0 + \sum_{i=0}^{k-1} \alpha_i Ad^i, x^0 + \sum_{i=0}^{k-1} \alpha_i d^i\right) - \left(b, x^0 + \sum_{i=0}^{k-1} \alpha_i d^i\right)\right\}$$

The stationary point is determined by

$$0 \overset{!}{=} \frac{\partial}{\partial \alpha_j} Q(x^k) = \left(Ax^0 + \sum_{i=0}^{k-1} \alpha_i Ad^i, d^j\right) - (b, d^j)$$

$$= -\left(b - Ax^k, d^j\right), \quad j = 0, \ldots, k-1.$$

This means that the new residual $b - Ax^k$ is orthogonal to all search directions d^j for $j = 0, \ldots, k-1$. The resulting system of k equations for the k unknowns $\alpha_0, \ldots, \alpha_{k-1}$

$$(b - Ax^k, d^j) = 0 \quad \forall j = 0, \ldots, k-1, \tag{11.9}$$

where $x^k = x^0 + \sum_{i=0}^{k-1} \alpha_i d^i$ holds, is exactly the *Galerkin equation*, which we discovered as (6.3) in Sect. 6.2. In the latter, the equation was the starting point of our derivation of the CG method as a Krylov subspace method.

11.3   Convergence Proof for the Conjugate Gradient Method

The convergence analysis of the CG method from Sect. 6.2 turns out to be rather involved and requires techniques from the Chebyshev Approximation presented in Sect. 9.5.2. The key is the following characterization of an iteration $x^k = x^0 + K_k$ as

$$x^k = x^0 + p_{k-1}(A)d^0,$$

where $p_{k-1} \in P_{k-1}$ is a polynomial in A:

$$p_{k-1}(A) = \sum_{i=0}^{k-1} \alpha_i A^i$$

With this, the minimization property from Theorem 6.7 can be written as

$$\|b - Ax^k\|_{A^{-1}} = \min_{y \in x^0 + K_k} \|b - Ay\|_{A^{-1}} = \min_{q \in P_{k-1}} \|b - Ax^0 - Aq(A)d^0\|_{A^{-1}}.$$

If we switch to the $\|\cdot\|_A$-norm, with $d^0 = b - Ax^0 = A(x - x^0)$, we get

$$\|b - Ax^k\|_{A^{-1}} = \|x - x^k\|_A = \min_{q \in P_{k-1}} \|(x - x^0) - q(A)A(x - x^0)\|_A,$$

so

$$\|x - x^k\|_A = \min_{q \in P_{k-1}} \|[I - q(A)A](x - x^0)\|_A.$$

In terms of a best approximation, we can write this task as

$$p \in P_{k-1}: \quad \|[I - p(A)A](x - x^0)\|_A = \min_{q \in P_{k-1}} \|[I + q(A)A](x - x^0)\|_A. \qquad (11.10)$$

The proof of convergence to the CG method, therefore, builds on Sect. 9.5 and in particular Sect. 9.5.2.

As $q(A)A \in P_k(A)$, we are looking for a polynomial $q \in P_k$ with the property $q(0) = 1$, such that

$$\|x^k - x\|_A \le \min_{q \in P_k,\, q(0)=1} \|q(A)\|_A \|x - x^0\|_A. \qquad (11.11)$$

The convergence of the CG method depends on whether we succeed in finding a polynomial $q \in P_k$ with the property $q(0) = 1$ with an as small as possible A-norm.

Lemma 11.14 (Bound for Matrix Polynomials) *Let $A \in \mathbb{R}^{n \times n}$ be symmetric positive definite with eigenvalues $0 < \lambda_1 \le \cdots \le \lambda_n$ and $p \in P_k$ a polynomial with $p(0) = 1$. It holds*

$$\|p(A)\|_A \le M, \quad M := \min_{p \in P_k,\, p(0)=1} \sup_{\lambda \in [\lambda_1, \lambda_n]} |p(\lambda)|.$$

Proof Let $\{q_1, \ldots, q_n\}$ be an orthogonal basis of eigenvectors. Any $y \in \mathbb{R}^n$ has the representation

$$y = \sum_{i=1}^{n} \gamma_i q_i, \quad \gamma_i \in \mathbb{R}$$

With $Q = [q_1, \ldots, q_n]$, we have

$$A = Q^T D Q, \quad D = \mathrm{diag}(\lambda_1, \ldots, \lambda_n),$$

such that for polynomials $p \in P_{k-1}$ it follows:

$$p(A) = \sum_{i=0}^{k-1} \alpha_i A^i = \sum_{i=0}^{k-1} \alpha_i (Q^T D Q)^i = Q^T \left(\sum_{i=0}^{k-1} \alpha_i D^i \right) Q = Q^T p(D) Q$$

Then,

$$\|p(A)y\|_A^2 = \sum_{i=1}^{n} \lambda_i p(\lambda_i)^2 \gamma_i^2 \le \underbrace{\sup_{\lambda \in [\lambda_1, \lambda_n]} |p(\lambda)|^2}_{=:M^2} \sum_{i=1}^{n} \lambda_i \gamma_i^2 =: M^2 \|y\|_A^2$$

and we get

$$\|p(A)\|_A = \sup_{y\in\mathbb{R}^n,\, y\neq 0} \frac{\|p(A)y\|_A}{\|y\|_A} = M.$$

$\square$

With this result and the error estimate (11.11), we can now derive a convergence estimate for the CG method.

Theorem 11.15 (Convergence of the CG Method) *Let $A \in \mathbb{R}^{n\times n}$ be symmetric positive definite, $b \in \mathbb{R}^n$ be the right-hand side, and $x^0 \in \mathbb{R}^n$ be any initial value. It holds*

$$\|x^k - x\|_A \leq 2\left(\frac{1 - 1/\sqrt{\kappa}}{1 + 1/\sqrt{\kappa}}\right)^k \|x^0 - x\|_A, \quad k \geq 0,$$

with the spectral condition $\kappa = \mathrm{cond}_2(A)$ of the matrix A.

Proof From the auxiliary theorem and the estimate (11.11) it follows

$$\|x^k - x\|_A \leq M\|x^0 - x\|_A$$

with

$$M = \min_{q\in P_k,\, q(0)=1} \max_{\lambda\in[\lambda_1,\lambda_n]} |q(\lambda)|.$$

The goal is to find the sharpest possible estimate for the size of M. We are looking for a polynomial $q \in P_k$ that takes the value one at the origin, $q(0) = 1$, and is as close as possible (in the maximum norm) to 0 on the interval $[\lambda_1, \lambda_n]$.

This problem can be tackled as a Chebyshev approximation. We are looking for the best approximation $p \in P_k$ to the zero function on $[\lambda_1, \lambda_n]$. This polynomial should also have the normalization property $p(0) = 1$. Therefore, the trivial solution $p = 0$ is excluded. The Chebyshev polynomials (see Sect. 9.5.2 and Theorem 9.46)

$$T_k(x) = \cos\left(k \arccos(x)\right)$$

have the property

$$2^{-k-1} \max_{[-1,1]} |T_k(x)| = \min_{\alpha_0,\dots,\alpha_{k-1}} \max_{[-1,1]} \left|x^k + \sum_{i=0}^{k-1} \alpha_i x^k\right|.$$

Normalized, it is the polynomial whose maximum on $[-1, 1]$ is minimal. We now choose the transformation

$$x \mapsto \frac{\lambda_n + \lambda_1 - 2t}{\lambda_n - \lambda_1},$$

and obtain with

$$p(t) = T_k\left(\frac{\lambda_n + \lambda_1 - 2t}{\lambda_n - \lambda_1}\right) T_k\left(\frac{\lambda_n + \lambda_1}{\lambda_n - \lambda_1}\right)^{-1}$$

the polynomial of degree k, which is minimal on $[\lambda_1, \lambda_n]$ and satisfies the normalization

$$p(0) = 1.$$

It holds

$$\sup_{t\in[\lambda_1,\lambda_n]} |p(t)| = T_k\left(\frac{\lambda_n + \lambda_1}{\lambda_n - \lambda_1}\right)^{-1} = T_k\left(\frac{\kappa + 1}{\kappa - 1}\right)^{-1} \tag{11.12}$$

with the spectral condition

$$\kappa := \frac{\lambda_n}{\lambda_1}.$$

We use the representation of the Chebyshev polynomials outside of $[-1, 1]$, taken from Theorem 9.46:

$$T_n(x) = \frac{1}{2}\left[(x + \sqrt{x^2 - 1})^n + (x - \sqrt{x^2 - 1})^n\right]$$

For $x = \frac{\kappa+1}{\kappa-1}$ it holds

$$\frac{\kappa + 1}{\kappa - 1} + \sqrt{\left(\frac{\kappa + 1}{\kappa - 1}\right)^2 - 1} = \frac{\kappa + 2\sqrt{\kappa} + 1}{\kappa - 1} = \frac{\sqrt{\kappa} + 1}{\sqrt{\kappa} - 1}$$

and correspondingly

$$\frac{\kappa + 1}{\kappa - 1} - \sqrt{\left(\frac{\kappa + 1}{\kappa - 1}\right)^2 - 1} = \frac{\sqrt{\kappa} - 1}{\sqrt{\kappa} + 1}.$$

With this, (11.12) can be estimated by

$$T_k\left(\frac{\kappa + 1}{\kappa - 1}\right) = \frac{1}{2}\left[\left(\frac{\sqrt{\kappa} + 1}{\sqrt{\kappa} - 1}\right)^k + \left(\frac{\sqrt{\kappa} - 1}{\sqrt{\kappa} + 1}\right)^k\right] \geq \frac{1}{2}\left(\frac{\sqrt{\kappa} + 1}{\sqrt{\kappa} - 1}\right)^k.$$

Therefore, it follows

$$\sup_{t\in[\lambda_1,\lambda_n]} T_k\left(\frac{\kappa + 1}{\kappa - 1}\right)^{-1} \leq 2\left(\frac{\sqrt{\kappa} - 1}{\sqrt{\kappa} + 1}\right)^k = 2\left(\frac{1 - \frac{1}{\sqrt{\kappa}}}{1 + \frac{1}{\sqrt{\kappa}}}\right)^k.$$

$\square$

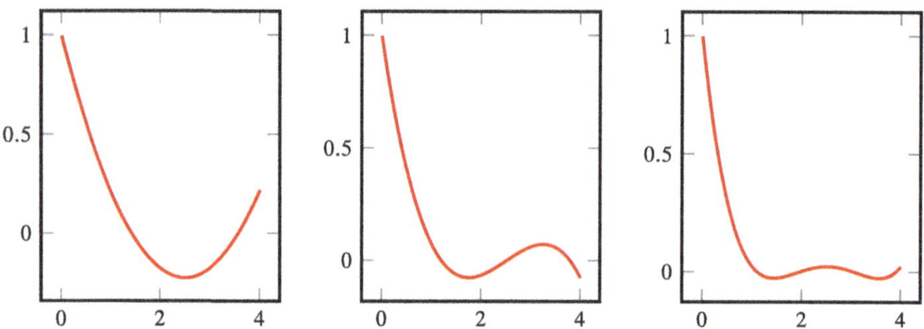

Fig. 11.5 Optimal polynomials for limiting CG convergence. It holds $p(0) = 1$ and a minimization on the interval $[1, 4]$. On the left $n = 2$, in the middle $n = 3$ and on the right $n = 4$

In Fig. 11.5, we show the corresponding polynomials in the case $n = 2, 3, 4$ for $\lambda_1 = 1$ and $\lambda_n = 4$.

The convergence of the CG method is twice as fast as that of the gradient descent method. It holds

$$\rho := \frac{1 - 1/\sqrt{\kappa}}{1 + 1/\sqrt{\kappa}} = 1 - 2\sqrt{\kappa} + O(\kappa).$$

For the model matrix, we get $\rho = 1 - \frac{1}{n}$. The CG method is one of the most efficient iterative methods for sparse symmetric systems. The convergence depends significantly on the condition $\mathrm{cond}_2(A)$ of the matrix A. For $\mathrm{cond}_2(A) \approx 1$ the method is optimal: To reduce the error by a given factor ϵ, a fixed number of steps is necessary.

11.4 Excursus: Honey Pricing

In this excursus, we illustrate the solution of the least squares problem (see also Sect. 4.5) using the gradient descent method (Sect. 11.2). Dominik, Thomas's brother, aims to optimize honey sales at fairs, see Fig. 11.6. First, he wants to understand better how increasing or reducing the price for a jar (0.5 kg) of honey affects his sales numbers and revenue. After that, he wants to determine his personal profit maximum. That is, at what price does he achieve sales numbers so that the revenue or profit becomes maximum? Furthermore, we ask ourselves whether these sales numbers are identical. For this purpose, he works with a sample of N measurements, denoted as $(x_1, p_1), \ldots, (x_N, p_N)$, to test how many jars x_k are sold at the price p_k.

In economics, such problems are dealt with using the so-called price-sales curve (PSC) or price-sales function, see, e.g., [73]. The profit maximum is the point of the PSC, at which the maximum profit is achieved. To answer the two questions mentioned above, Dominik conducts a sample on 14 different fairs in Germany and Austria to varying prices per jar of honey. From this, he then constructs a regression line that represents the PSC. Based on this,

Fig. 11.6 Beehives with bees in search of honey and the finished product

Table 11.3 Sample of $N = 14$ measurements at different prices with the respective sales numbers per jar of honey

Numberof jars sold	50	30	21	22	27	30	26	32	28	26	21	16	8	4
Priceper jar		3.5	4.0	4.5	5.0	5.5	6.0	6.5	7.0	7.5	8.0	8.5	9.0	9.5	10.0

we can then determine the revenue function and the profit function to answer the second question.

The sample on $N = 14$ different fairs leads to the measurement series given in Table 11.3.

Linear Regression

For the mathematical analysis, we assume a statistical model and a linear relationship. The latter ultimately leads to determining a linear regression line

$$p(x) = b + mx,$$

where $p(x)$ stands for the price (per jar), b for the maximum price, m for the slope and x for the sales quantity ($x = 1$ jar). The maximum price (also called the prohibitive price) is reached at $x = 0$ and is too high for a buyer to accept. The slope m is negative, as the quantity x sold decreases with increasing price. The zero of p characterizes the maximum quantity that can be sold, also known as the saturation quantity. The latter is easy to understand, because if Dominik gives away his honey, he will achieve the maximum sales. Ultimately, we need to determine two unknown parameters b and m of the above equation.

Determining the Parameters Using Least Squares

To determine the two unknowns, we refer back to Table 11.3 and use the model of least squares (see also Sect. 4.5), leading to

$$S := S(m, b) = \frac{1}{N} \sum_{k=1}^{N} (p_k - (b + mx_k))^2,$$

where (x_k, p_k) represent the sales-price pairs from Table 11.3. The number of measurements is $N = 14$. Our goal is to minimize the error S, i.e., min S.

Numerical Solution Using Gradient Descent

We are interested in the minimum of S while simultaneously calculating the two unknown parameters m and b. This task can be classified as a nonlinear optimization problem for which we use an iterative method to approximate its solution. For this, we use the gradient descent method from Sect. 11.2. This means we need

- initial values m_0 and p_0,
- a step size ρ,
- the gradient of the function $S := S(m, b)$.

The gradient is given as

$$\frac{\partial S}{\partial m} = \frac{2}{N} \sum_{k=1}^{N} -x_k(p_k - (mx_k + b))$$

$$\frac{\partial S}{\partial b} = \frac{2}{N} \sum_{k=1}^{N} -(p_k - (mx_k + b))$$

The final algorithm is given as follows: For $l = 1, 2, 3, \ldots, N_{max}$ iterate

$$\begin{pmatrix} m_{l+1} \\ b_{l+1} \end{pmatrix} = \begin{pmatrix} m_l \\ b_l \end{pmatrix} - \rho \begin{pmatrix} \frac{\partial S(m_l)}{\partial m} \\ \frac{\partial S(b_l)}{\partial b} \end{pmatrix}.$$

If we take as initial values $b_0 = 10.5$ and $m_0 = -2$ and the step size is $\rho = 10^{-4}$, then after $N_{max} = 200$ iterations, we get the values

$$b = 10.5613025417, \quad m = -0.154072634203.$$

Results and Interpretation

Now that we have determined m and b, we obtain the following regression line for the price

$$p(x) = \underbrace{10.561}_{=b} - \underbrace{0.1541}_{=m}\, x. \tag{11.13}$$

The prohibitive price is therefore around $10.50€$ per jar. If we form the demand function (i.e., the inverse of (11.13)), namely

$$x(p) = 68.548 - 6.4904\,p,$$

then we can also easily calculate how much the average sale will decrease if we increase the price by 1 EUR per jar: This is on average 6.5 jars. The saturation quantity is therefore at 68.548 jars (at this price, Dominik would give away each jar for free). The two regression lines and the sample data are illustrated in Fig. 11.7.

We now return to the second of our original questions, namely at which point a maximum profit would be achieved. The profit G is known to be composed of revenue U minus costs K. First, we examine the revenue function, which is calculated from the price multiplied by the quantity x:

$$U := U(x) = p(x) \cdot x = 10.561x - 0.1541x^2$$

The revenue U is thus a quadratic function, see Fig. 11.8. The maximum of U then represents the maximum revenue:

$$U'(x) = 0 \quad \Leftrightarrow \quad 10.5613 - 0.30815x = 0 \quad \Rightarrow x = 34.273$$

This means that at (rounded) 34 sold jars at a price of $5.28€$ per jar (substituting $x = 34$ into (11.13)), Dominik will achieve the maximum revenue (mathematically this is also justified,

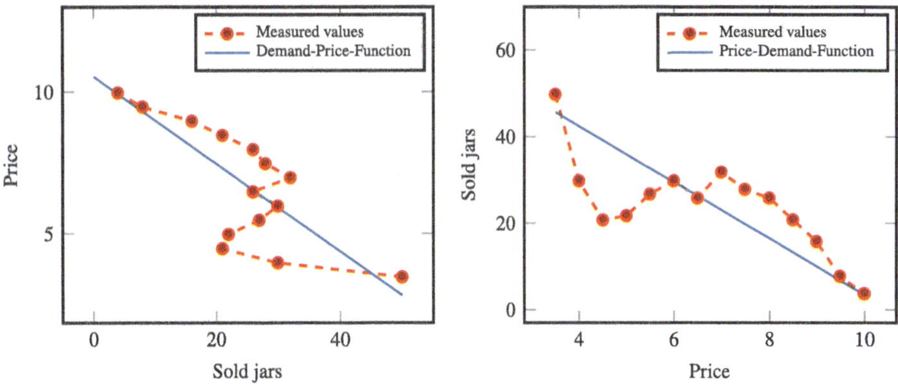

Fig. 11.7 The demand function (left) and the PAK (right)

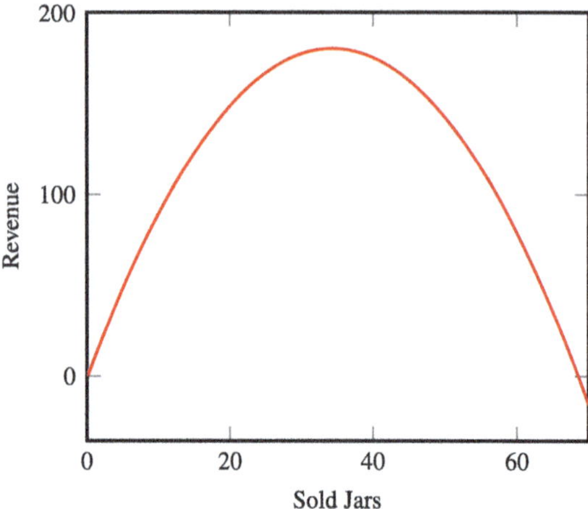

Fig. 11.8 The revenue function U in terms of EUR. The highest revenue is achieved at 34.273 jars sold. No revenue is made at no sales (left root) and at the saturation quantity $x = 68.548$ (right root), when every jar is given away for free

since the second derivative of U is negative ($U''(x) = -0.30815 < 0$), and, therefore, indeed a maximum is present). In light of these evaluations, the price per jar should, thus, be 5.30€ to achieve the maximum revenue.

Finally, we consider the profit curve: $G = U - K$. For the costs, we assume an average value (at 20 bee colonies) of 2.59€ per jar [17]. This average value is classified as variable costs per jar (i.e., any global fixed costs are already included):

$$G(x) = U(x) - K(x) = p(x) \cdot x - K(x)$$
$$= 10.561x - 0.1541x^2 - 2.59x$$
$$= 7.971x - 0.1541x^2$$

From this, we can immediately determine the marginal profit, $G'(p)$, which indicates by how much the profit changes, when one more jar of honey is sold[1]:

$$G'(x) = 7.971 - 0.30815x$$

From $G'(x)$ we also calculate the maximum of $G(x)$, which here is at $25.87 \approx 26$ jars. Based on our mathematical model, the underlying sample and the numerical solution, the maximum profit is achieved at a sale of 26 jars at a price of 5.28€ per jar.

---

[1] This statement is identical to the definition of the marginal tax rate in Example 1.6 in the introduction.

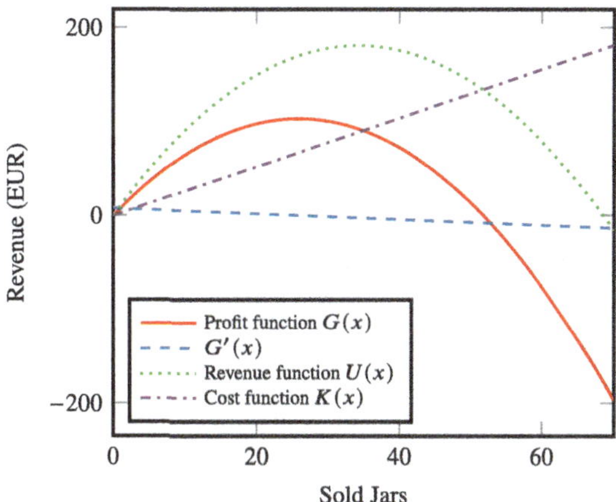

Fig. 11.9 The profit function $G(x)$ and its derivative $G'(x)$ compared to the revenue function $U(x)$ and the cost function $K(x)$

In Fig. 11.9, we also recognize the typical behavior of the profit maximum: This lies to the left of the revenue maximum. To achieve maximum profit, Dominik sells a smaller quantity of honey than for maximal revenue.

11.5   Artificial Neural Networks

Mathematically speaking, *artificial neural networks* (ANN) are functions that are composed of elementary building blocks and have a vast number of free parameters. These parameters can be determined so that the artificial neural network performs a specific mathematical task, which means nothing more than approximating a function that determines the result of the task. *Machine learning*, and especially working with artificial neural networks, can, therefore, be seen as a kind of approximation. However, unlike polynomial interpolation or trigonometric interpolation, a completely different type of function is used in *machine learning* with *artificial neural networks*.[2] Here, we do not choose a structured vector space for the approximation of a function, such as the space of polynomials or trigonometric polynomials, but instead compose functions from elementary components, the artificial neurons.

Initially, mathematical research on artificial neural networks was genuinely inspired by natural models, i.e., the functioning of neurons in the brain. Today, they are a powerful tool that can be used very efficiently on modern computers. Many long-standing problems have been successfully solved with the help of artificial neural networks. Particularly noteworthy is the prediction of protein structures by AlphaFold. More commonly known are advances in language processing, such as automatic translation (used to translate this book), speech

---

[2] These developments have been awarded in 2024 with the Nobel Prize: 'The Nobel Prize in Physics 2024 was awarded jointly to John J. Hopfield and Geoffrey Hinton "for foundational discoveries and inventions that enable machine learning with artificial neural networks"; see https://www.nobelprize.org/prizes/physics/2024/summary/.

recognition, and generative language models like ChatGPT, as well as the great successes in image and video processing.

Mathematically, we can consider artificial neural networks as a method for approximating functions. However, their analysis is more difficult, and many tools are missing: Lagrange interpolation was based mainly on the Taylor expansion and the representation of the interpolation polynomial with basis polynomials. Neural networks, however, are often not differentiable and usually do not represent a vector space with a basis. The best approximation according to Gauss is a minimization problem with a quadratic functional, which can be solved exactly using the scalar product. Neural networks are also determined using minimization problems, but these minimization tasks are of a complicated nonlinear structure, not quadratic, and usually not uniquely solvable. Thus, critical structural properties for the analysis are missing.

11.5.1 Structure of Artificial Neural Networks

The basic element is the *artificial neuron* and the name is modeled after the behavior of neurons in the brain:

Definition 11.16 (Artificial Neuron) Let $w \in \mathbb{R}^n$ and $b \in \mathbb{R}$ and $f : \mathbb{R}^n \to \mathbb{R}$. Then the function

$$f : \mathbb{R}^n \to \mathbb{R}, \quad f(x) = \sigma\Big(\sum_{i=1}^{n} w_i x_i + b \Big),$$

is called an *artificial neuron*. The vector $w \in \mathbb{R}^n$ is called weight and $b \in \mathbb{R}$ is the bias. The function $\sigma : \mathbb{R} \to \mathbb{R}$ is called *activation function*.

Like many things in the field of artificial neural networks, *activation functions* are not clearly defined. The simplest variant is the Heaviside step function

$$\sigma_{\mathrm{HS}}(s) = H(s) := \begin{cases} 0 & s < 0, \\ 1 & s \geq 0. \end{cases} \tag{11.14}$$

A continuous activation function is the *Rectified Linear Unit (ReLU)*

$$\sigma_{\mathrm{ReLU}}(s) = \mathrm{ReLU}(s) := \begin{cases} 0 & s \leq 0, \\ s & s > 0. \end{cases} \tag{11.15}$$

Continuously differentiable (arbitrarily often) are the *Sigmoid activation function*

$$\sigma(s) = \frac{1}{1 + \exp(-s)}, \tag{11.16}$$

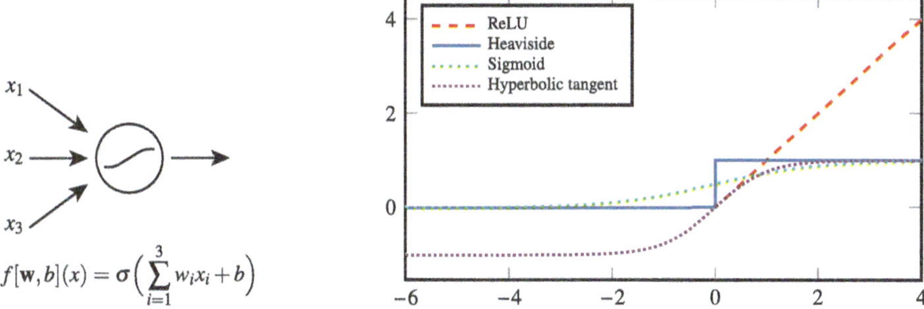

Fig. 11.10 Left: Sketch of an artificial neuron with 3 input signals, the linear layer, and the nonlinear activation. Right: Progression of some typical activation functions

or the hyperbolic tangent

$$\sigma_{\text{tanh}}(s) = \tanh(s). \tag{11.17}$$

The latter two are closely related; it holds

$$\sigma(s) = \frac{1}{2}\tanh\left(\frac{1}{2}s\right) + \frac{1}{2}.$$

In Fig. 11.10, we show the course of these activation functions as well as the schematic structure of an artificial neuron.

Example 11.17 (Classification With Artificial Neurons) An artificial neuron with Heaviside activation

$$f(x) = H\big((\mathbf{w}, x)_2 + b\big),$$

separates the space at a hyperplane, because every plane with normal vector $\mathbf{n}$ and selected point x_0 is given by

$$0 = n^T(x - x_0) = n^T x - n^T x_0.$$

The weight thus corresponds to the normal vector, and the bias is the scalar product with the selected point. A neuron thus separates the space into the two halves, which is one of the fundamental tasks for which artificial neural networks can be used. Given are $x_1, \ldots, x_N$, each assigned a label y_i, for example $y_i \in \{0, 1\}$. We are looking for the weights $\mathbf{w}$ and the bias b of the artificial neuron, so that $f[\mathbf{w}, b](x_i) = y_i$ holds. That is, we are looking for a model that performs the classification of the data with respect to the labels.

A model consisting of a single neuron is very limited, and it must be assumed that the task cannot be solved exactly. Instead, we try to find the best model, i.e., we consider the minimization task analogous to regression in Sect. 4.5 and the Excursus 11.4

$$\min_{(\mathbf{w},b)} \sum_{i=1}^{N} \|y_i - f[\mathbf{w}, b](x_i)\|^2.$$

Artificial neural networks are composed of individual artificial neurons. The simplest ANNs are so-called *single-layer networks*

Definition 11.18 (Single-Layer Network) Let $W \in \mathbb{R}^{n \times m}$ and $b \in \mathbb{R}^n$ and $\sigma : \mathbb{R} \to \mathbb{R}$ be an activation function. Then the function

$$f(x) = \sigma\left(Wx + b\right),$$

is called an *artificial neural network with one layer* and m artificial neurons. The application of the activation function is component-wise, i.e., for $y \in \mathbb{R}^m$ it is

$$\sigma(y) = \left(\sigma(y_i)\right)_{i=1,\ldots,m}.$$

From this definition, we quickly move to more general artificial neural networks, so-called *deep neural networks*. To do this, we connect several single-layer networks in series. The output of layer l is the input of layer $l + 1$.

Definition 11.19 (Deep Neural Network) Let $L \in \mathbb{N}$ be the number of layers. For $l = 1, \ldots, L$ let $W^l \in \mathbb{N}^{n_l \times n_{l-1}}$ and $b^l \in \mathbb{R}^{n_l}$, and let $\sigma : \mathbb{R} \to \mathbb{R}$ be an activation function. Then $f : \mathbb{R}^{n_0} \to \mathbb{R}^{n_L}$, given by

$$\begin{aligned}
x^0 &= x \\
z^l &= W^l x^{l-1} + b^l, \quad x^l = \sigma(z^l), \quad l = 1, \ldots, L - 1, \\
f(x) &= x^L = W^L x^{L-1} + b^L,
\end{aligned}$$

(11.18)

is called a *deep neural network*. The first layer x^0 is called the *input layer*, the last layer $x^L = f^L(\cdot)$ the *output layer*, and generally no activation function is added here. The inner layers are called *hidden layers*. We denote the architecture of a deep neural network with $\mathcal{N}(\sigma; n_0, n_1, \ldots, n_L)$, or briefly with $\mathcal{N}(L, n)$ with $n = \max_l n_l$. We denote the associated parameter space with

$$\mathcal{P}_\mathcal{N} = \{\mathbf{W} = (W_1, \ldots, W_L), \ \mathbf{b} = (b_1, \ldots, b_L), \ W_l \in \mathbb{R}^{n_l \times n_{l-1}}, \ b_l \in \mathbb{R}^{n_l}\}.$$

For a parameter choice $(\mathbf{W}, \mathbf{b}) \in \mathcal{P}_\mathcal{N}$ we denote with $f[\mathbf{W}, \mathbf{b}] : \mathbb{R}^{n_0} \to \mathbb{R}^{n_L}$ the resulting function according to (11.18). The set of *realizations*, i.e., the set of functions that can be represented in the architecture, we denote with

$$\mathcal{F}_\mathcal{N} = \{f[\mathbf{W}, \mathbf{b}] \ : \ (\mathbf{W}, \mathbf{b}) \in \mathcal{P}_\mathcal{N}\}.$$

Corollary 11.20 (Number of Free Parameters) *Let $N(\sigma; n_0, \ldots, n_L)$ be an artificial neural network with L layers and n_l artificial neurons each. Then the parameter space $\mathcal{P}_N$ has a total of*

$$\#\mathcal{P}_N = \sum_{l=1}^{L} n_l(n_{l-1} + 1)$$

free parameters, of which $\sum_{l=1}^{L} n_l n_{l-1}$ are weights and $\sum_{l=1}^{L} n_l$ are biases.

A basic operation in machine learning is the evaluation of a given network $N(\sigma; n_0, \ldots, n_L)$ with given parameters $(\mathbf{W}, \mathbf{b}) \in \mathcal{P}_N$ for a data point $x \in \mathbb{R}^{n_0}$ or, more often, for a whole list of (possibly very many) data points $x_i \in \mathbb{R}^{n_0}$ for $i = 1, \ldots, N_{tr}$.

Algorithm 11.21 (Evaluation of an Artificial Neural Network)

> **Input:** Artificial neural network $N(\sigma; n_0, \ldots, n_L)$, parameters $(\mathbf{W}, \mathbf{b}) \in \mathcal{P}$ and input
> $\quad\quad x \in \mathbb{R}^{n_0}$
> 1 $x^0 = x$
> 2 **for** $l = 1, 2, \ldots, L$ **do**
> 3 $\quad\quad z^l = W^l x^{l-1} + b^l$
> 4 $\quad\quad$ **if** $l < L$ **then**
> 5 $\quad\quad\quad\quad x^l = \sigma(z^l)$
> 6 $\quad\quad$ **else**
> 7 $\quad\quad\quad\quad x^L = z^L$
> **Result:** Network output $x^L = f[\mathbf{W}, \mathbf{b}](x)$

Remark 11.22 (Network Architectures) The above definition introduces only the simplest type of artificial neural networks: all outputs from layer $l - 1$ are inputs for all artificial neurons of layer l. These networks are also called *Multilayer perceptrons*. They belong to the category of *Feedforward neural networks*, i.e., the information only goes in one direction, from layer 1 to layer 2, and so on. Neural networks in which the output of layer l can also serve as input for earlier layers $l' < l$ are called *recurrent neural networks*.

In principle, a different activation function can be used in each layer. If an activation function is applied in the last layer, this restricts the image space, with the sigmoid activation to $(0, 1)$, with the ReLU activation to the non-negative numbers.

So-called *skip connections* skip individual layers. That is, the output of layer l can be used directly as input, e.g., for layer $l + 2$. Another simple modification is to consider *residual networks*, where a layer has the form

$$f(x) = x + \sigma(Wx + b),$$

i.e., the result of the artificial neuron is added to the input.

In image processing, *convolutional neural networks* play a major role. Here, each layer is a *convolution*. The special thing about convolution networks is that the size of the input x does not necessarily have to be linked to the number of weights. With the convolution weights $W \in \mathbb{R}^{n_f}$, we define

$$f_{conv}(x) = \sigma(W * x), \quad (W * x)_i = \sum_{j=1}^{n_f} W_j x_{i-j},$$

where $x_0 = x_{-1} = \cdots = x_{1-n_f} = 0$ is set. The same weights are always applied only to a small part of the inputs. In this way, functions in arbitrarily high-dimensional spaces can be represented with only a few free parameters (the weights).

Modern architectures, such as those used in language models like ChatGPT,[3] are much more complex, but also use and combine the models introduced here. Artificial neural networks used in practice are enormous, for example, GPT-4 is said to have nearly 1.8 trillion, i.e., $2 \cdot 10^{12}$ free parameters.

In Fig. 11.11, we give sketches of some typical artificial neural networks. A comprehensive overview is provided in the literature [12, 18, 44, 53, 69].   ◆

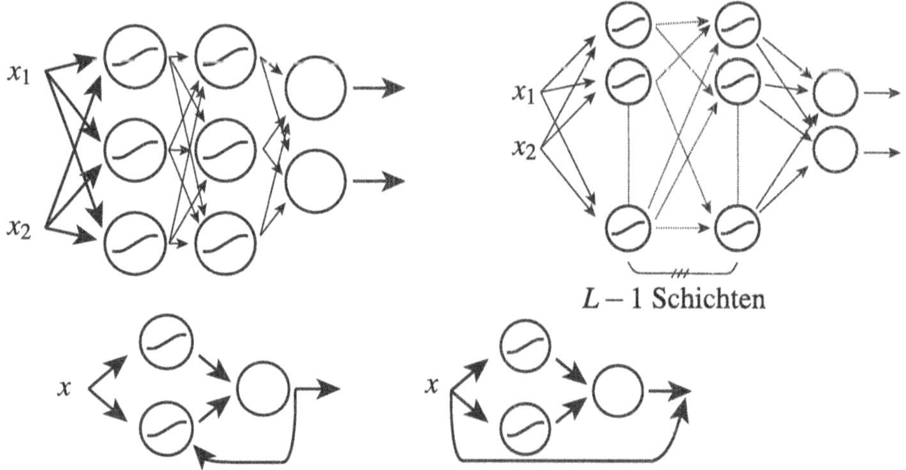

Fig. 11.11 Examples of artificial neural networks. Top left: simple multilayer network $\mathcal{N}(\sigma; 2, 3, 3, 2)$ with sigmoid activation (except for the output layer). Top right: this network represents functions $f : \mathbb{R}^2 \to \mathbb{R}^2$. Deep neural network with 2 inputs and outputs and $L - 1$ hidden layers, each with n neurons. This network also represents functions $f : \mathbb{R}^2 \to \mathbb{R}^2$. Bottom right: Artificial neural *residual network*. The input is added to the output. This network represents functions $f : \mathbb{R} \to \mathbb{R}$ with the structure $f(x) = x + f[\mathbf{W}, \mathbf{b}](x)$. Bottom left: *recurrent neural network*, where outputs can also be inputs for earlier layers

---

[3] Chat Generative Pre-trained Transformer, a language model based on neural networks, see https://en.wikipedia.org/wiki/ChatGPT.

11.5.2 Approximation with Neural Networks

We exclusively consider deep artificial neural networks of the type of Definition 11.19 and consider a network with L layers as a function $f : \mathbb{R}^{n_0} \to \mathbb{R}^{n_L}$ with the weight matrices W_l and bias vectors b_l as free parameters. First, we establish a simple but important relationship.

Theorem 11.23 *Let* $\mathcal{N} = \mathcal{N}(\sigma, n_0, \ldots, n_L)$ *be a network architecture with L layers and parameters* $W_l \in \mathbb{R}^{n_l \times n_{l-1}}$ *and* $b_l \in \mathbb{R}^{n_l}$ *for* $l = 1, \ldots, L$. *The set of realizations*

$$\mathcal{F}_{\mathcal{N}} := \{ f[\mathbf{W}, \mathbf{b}] : \mathbb{R}^{n_0} \to \mathbb{R}^{n_L}, \quad (\mathbf{W}, \mathbf{b}) \in \mathcal{P}_{\mathcal{N}} \},$$

is generally not a vector space. In particular, $\mathcal{F}_{\mathcal{N}}$ *is not closed with respect to addition or scalar multiplication.*

Proof We construct a simple counter example:

$$\mathcal{F}_{\mathcal{N}} := \{ f : \mathbb{R} \to \mathbb{R}, \ \mathrm{ReLU}(wx + b), \ w, b \in \mathbb{R} \}$$

and choose the two realizations $f_1, f_2 \in \mathcal{F}_{\mathcal{N}}$

$$f_1(x) = \mathrm{ReLU}(x) = \max\{x, 0\}, \quad f_2(x) = \mathrm{ReLU}(-x) = \max\{-x, 0\}.$$

It holds
$$f_1(x) + f_2(x) = |x| \notin \mathcal{F}_{\mathcal{N}},$$

since the absolute value function cannot be written in the form $\mathrm{ReLU}(wx + b)$. $\square$

The connection may be clear, but it has significant consequences: There is no basis for $\mathcal{F}_{\mathcal{N}}$ and we cannot use any familiar structural properties such as scalar products or linearity. These tools were crucial for the efficient implementation of best approximations or Lagrange interpolations. Therefore, we need to find new ways to search for the optimal parameters and coefficients.

Already in the 1980s, it was proven that neural networks have excellent approximation properties, more precisely, that the set $\mathcal{F}_{\mathcal{N}(L,n)}$ for $L \to \infty$ or $n \to \infty$ is dense in the set of continuous functions.

Theorem 11.24 (Universal Approximation Theorem) *Let* $I = [a, b] \subset \mathbb{R}$, *and* σ *be a continuous activation function with the property*

$$\sigma(s) \xrightarrow[s \to -\infty]{} 0, \quad \sigma(s) \xrightarrow[s \to \infty]{} 1.$$

Then the set of neural networks with one hidden layer $\mathcal{F}_{\mathcal{N}(\sigma; N, 1)}$, *i.e.,*

$$\mathcal{F} = \bigcup_{N \in \mathbb{N}} \{ \sum_{i=1}^{N} \omega_i^2 \sigma(\omega_i^1 x + b_i^1), \quad (\omega_1^1, b_0^1), \dots, (\omega_N^1, b_N^1), \omega_1^2 \in \mathbb{R} \},$$

is dense in $C(I)$.

Proof Proofs are given, for example, by Cybenko [21] or Hornik [54]. Some deep results of functional analysis are required. $\qquad\square$

The theorem means that for every continuous function $f \in C(I)$ and any positive $\epsilon > 0$ there is a neural network with architecture $\mathcal{N}(\sigma; 1, N, 1)$ and a realization $f[\mathbf{W}, \mathbf{b}] \in \mathcal{F}_N$, such that $\max_{x \in I} \| f(x) - f[\mathbf{W}, \mathbf{b}](x) \| \leq \epsilon$ holds. However, the result does not provide any quantitative estimates, i.e., it does not say anything about how fast N must grow when $\epsilon \to 0$ becomes smaller.

The original result was shown for activation functions of the sigmoid type. Still, it can be extended to almost any function, and the property remains as long as the activation function is not a polynomial. The density also extends to functions $C(X)$ for $X \subset \mathbb{R}^n$ and also to vector-valued functions.

ReLU networks are easier to understand.

Theorem 11.25 (ReLU Networks) *Let $\mathcal{N}$ be a neural network architecture with ReLU activation. Then every $f[\mathbf{W}, \mathbf{b}] \in \mathcal{F}_N$ is a piecewise linear function.*

Proof The network is built as a concatenation of single-layer networks

$$f = f_L \circ f_{L-1} \circ \cdots \circ f_1.$$

Each $f_l(x) = \mathrm{ReLU}(W_l x_l + b_l)$ is piecewise linear as a composition of the affine linear mapping $W_l x_l + b_l$ and the piecewise linear mapping $\mathrm{ReLU}(\cdot)$. Therefore, f is also piecewise linear. $\qquad\square$

In the simple scalar case, we can construct a corresponding neural network for each polygonal line:

Theorem 11.26 *Let $I = [a, b]$. Every piecewise linear function with $N \in \mathbb{N}$ pairwise different support points can be represented as an ReLU network with one layer and N artificial neurons.*

Proof Let $(x_i, y_i) \in [a, b] \times \mathbb{R}$ for $i = 1, \dots, N$ be support point pairs of the piecewise linear function p_h, $p_h(x_i) = y_i$. Without loss of generality, we assume that $x_{i-1} < x_i$ holds. Then we construct the network as

$$f(x) = y_1 + \sum_{i=2}^{N} \frac{y_i - y_{i-1}}{x_i - x_{i-1}} \text{ReLU}(x - x_{i-1}) - \sum_{i=3}^{N} \frac{y_{i-1} - y_{i-2}}{x_{i-1} - x_{i-2}} \text{ReLU}(x - x_{i-1}).$$

$$(11.19)$$

On each subinterval (x_{k-1}, x_k), $f(x)$ is a linear function, since

$$\text{ReLU}(x - x_{i-1}) = \begin{cases} 0 & x < x_{i-1}, \\ x & x \geq x_{i-1}, \end{cases}$$

i.e., $\text{ReLU}(x - x_{i-1})$ is continuous and piecewise linear. Thus, the sum is also piecewise linear.

Let $x \in [x_{k-1}, x_k]$. Then $\text{ReLU}(x - x_{i-1}) = x - x_{i-1}$ holds for $x \geq x_{i-1}$, i.e., $i \leq k$ and $\text{ReLU}(x - x_{i-1}) = 0$ for $x \leq x_{i-1}$ i.e., for $k < i$. From this follows

$$\begin{aligned}
f(x)\Big|_{[x_{k-1},x_k]} &= y_1 + \sum_{i=2}^{k} \frac{y_i - y_{i-1}}{x_i - x_{i-1}}(x - x_{i-1}) - \sum_{i=3}^{k} \frac{y_{i-1} - y_{i-2}}{x_{i-1} - x_{i-2}}(x - x_{i-1}) \\
&= y_1 + \sum_{i=2}^{k} \frac{y_i - y_{i-1}}{x_i - x_{i-1}}x - \sum_{i=2}^{k-1} \frac{y_i - y_{i-1}}{x_i - x_{i-1}}x \\
&\quad - \sum_{i=2}^{k} \frac{y_i - y_{i-1}}{x_i - x_{i-1}}(x_{i-1}) + \sum_{i=2}^{k-1} \frac{y_i - y_{i-1}}{x_i - x_{i-1}}(x_i) \\
&= y_1 + \underbrace{\sum_{i=2}^{k-1} y_i - y_{i-1}}_{=y_{k-1}} + \frac{y_k - y_{k-1}}{x_k - x_{k-1}}(x - x_{k-1}).
\end{aligned}$$

This corresponds exactly to the piecewise linear function on $[x_{k-1}, x_k]$. Then, (11.19) can be written as

$$f(x) = b_1 + \sum_{i=2}^{N} w_i \, \text{ReLU}(x - b_i),$$

with the $N - 1$ weights

$$w_2 = \frac{y_i - y_{i-1}}{x_i - x_{i-1}}, \quad w_i = \frac{y_i - y_{i-1}}{x_i - x_{i-1}} - \frac{y_{i-1} - y_{i-2}}{x_{i-1} - x_{i-2}}, \quad i = 3, \ldots, N,$$

and the N bias values

$$b_1 = y_1, \quad b_i = x_{i-1} \quad i = 2, \ldots, N.$$

$\square$

With this, we can derive a simple quantitative convergence result for neural networks:

Corollary 11.27 *Single-layer ReLU networks are exactly the set of piecewise linear functions. For a function $f \in C^2[a, b]$ there exists a sequence of neural network functions $f_N \in \mathcal{F}_{N(\text{ReLU}, N)}$ with*

$$\max_{x \in [a,b]} \| f(x) - f_N(x) \| \le C \frac{1}{N^2} \max_{x \in [a,b]} \| f''(x) \|$$

with some constant $C > 0$.

Proof Theorem 11.26 shows that f_N is a piecewise linear function with N intervals. With uniform partitioning of $[a, b]$, we have that $h = (b - a)/N$ is the interval size. The error estimate for piecewise linear functions (9.7) then immediately gives the error estimate. $\square$

Remark 11.28 (Approximation of High-Dimensional Functions) Artificial neural networks particularly excel at approximating high-dimensional functions $f : \mathbb{R}^n \to \mathbb{R}$ with $n \gg 1$. If we want to approximate such a function with traditional methods, e.g., with piecewise linear functions, the first step is always a partition of the domain $X \subset \mathbb{R}^n$ into a *grid*. As a simple example, we choose the n-dimensional unit cube

$$W_n = \{ x \in \mathbb{R}^n, \ 0 \le x_i \le 1, \ i = 1, \ldots, n \}.$$

A regular point grid of W_n with grid width $h = 1/M$ consists of the points

$$(x_i^1, x_i^2, \ldots, x_i^n), \quad x_i^j = \frac{i}{M}, \quad i = 0, \ldots, M.$$

In total, the grid has $(M + 1)^n$ points, resulting in exponentially increasing complexity. This means that the effort to achieve a certain accuracy increases exponentially with the dimension. This relationship is common in the approximation of all high-dimensional problems and is known as the *curse of dimensionality*.

In certain situations, artificial neural networks can approximate such functions much more efficiently. This is usually the case when the function to be approximated has an inherent low-dimensional structure. Let us make a simple example. The ten-dimensional function

$$f : \mathbb{R}^{10} \to \mathbb{R}, \quad f(x) = \sum_{i=1}^{10} x_i^2,$$

can be represented as a one-dimensional function via a coordinate transformation

$$f(r) = r^2, \quad r = \left(\sum_{i=1}^{10} x_i^2 \right)^{\frac{1}{2}}.$$

Here, we have exploited the symmetric structure of the function and represented it more simply. The effort would decrease from $O(M^{10})$ to $O(M)$. Such low-dimensional structures are found in many tasks, but they are usually not so obvious.

Artificial neural networks may be able to *learn* these relationships automatically and can then provide very efficient approximations. We refer to the literature [13].   ♦

11.5.3 Training of Artificial Neural Networks

In this section, we describe the search for parameters $(\mathbf{W}, \mathbf{b}) \in \mathcal{P}_N$, which represent the optimal realization $f[\mathbf{W}, \mathbf{b}] \in \mathcal{F}_N$ in a given network architecture $\mathcal{N}(\sigma; n_0, \dots, n_L)$, to approximate a given function $f : \mathbb{R}^{n_0} \to \mathbb{R}^{n_L}$ on a set $X \subset \mathbb{R}^{n_0}$, i.e., $f[\mathbf{W}, \mathbf{b}] \approx f$. For this, we choose a set of points $x_i \in X$ and values $y_i = f(x_i)$ for $i = 1, \dots, N_{tr}$. We call this set the *training data*

$$\mathcal{T} := \{(x_1, y_1), \dots, (x_{N_{tr}}, y_{N_{tr}})\}.$$

The interpolation task

$$f[\mathbf{W}, \mathbf{b}](x_i) = y_i, \quad i = 1, \dots, N_{tr}, \tag{11.20}$$

cannot be solved exactly for several reasons, primarily because the number of parameters and data points usually differ. More importantly, the set of neural networks is not a function space. Consequently, we do not have simple representations that allow us to directly formulate the interpolant. The key difference from polynomial interpolation is that the unknown coefficients, weights, and bias appear in a nonlinear form, and there is no basis for the set of functions available for interpolation. In the sense of best approximation (see Sect. 4.5), we will therefore define a functional (i.e., a mapping with image space $\mathbb{R}$) which measures the violation of the interpolation task (11.20)

$$J(\mathbf{W}, \mathbf{b}) := \frac{1}{N_{tr}} \sum_{i=1}^{N_{tr}} \| f[\mathbf{W}, \mathbf{b}](x_i) - y_i \|^2, \tag{11.21}$$

and try to determine the minimum of this functional, i.e.,

$$(\mathbf{W}_{\text{opt}}, \mathbf{b}_{\text{opt}}) = \arg \min_{\mathbf{W}, \mathbf{b} \in \mathcal{P}} J(\mathbf{W}, \mathbf{b}). \tag{11.22}$$

The functional $J(\cdot)$ is often called the *loss function*.

In general, we cannot solve this optimization problem exactly. Instead, we have to approach a minimum with an approximate method, such as the gradient method from Sect. 11.1. We restate Algorithm 11.3 in the context of artificial neural networks:

Algorithm 11.29 (Gradient Method to Train Artificial Neural Networks)

> **Input:** Continuously differentiable loss function $J : \mathcal{P}_N \to \mathbb{R}$ and initial value
> $\quad p^0 := (\mathbf{W}^0, \mathbf{b}^0) \in \mathcal{P}$.
> 1 **for** $k = 1, 2, \ldots$ **do**
> 2 $\quad$ **if** $\nabla J(p^{k-1}) = 0$ **then**
> 3 $\quad\quad$ Terminate
> 4 $\quad$ Determine descent direction $d^k = -\nabla J(p^{k-1})/\|\nabla J(p^{k-1})\|$
> 5 $\quad$ Choose a step size $s_k \in \mathbb{R}$
> 6 $\quad$ Compute new approximation $p^k = p^{k-1} + s_k d^k$ with $J(p^k) < J(p^{k-1})$
> **Result:** Minimizer $p^k = (\mathbf{W}^k, \mathbf{b}^k) \in \mathcal{P}$ of $J(x)$.

The application of the method to artificial neural networks brings some problems. Depending on the chosen activation function, the target function $J(\cdot)$ is not differentiable everywhere. This is the case, for example, with the ReLU function. Artificial neural networks are often heavily overparameterized, i.e., the number of free parameters can be very large. The target function is also not convex and usually has many local minima. We must therefore assume that it will hardly be possible to find a global minimum. Numerous heuristics must be applied for efficient optimization to find a good minimum robustly, rather than remaining in the first local minimum. Finally, the evaluation of the target function and its gradient is very complex when the number of data points is large. The work [38] provides a good overview of the research in this field.

11.5.3.1 Backpropagation–The Derivative of an Artificial Neural Network

As we already know, the most critical component in the gradient method is the determination of the descent direction, i.e., the gradient of the loss function $J(p) = J(\mathbf{W}, \mathbf{b})$. For this, all derivatives of $J(\cdot)$ with respect to all parameters of the artificial neural network must be determined, i.e., in the direction of the weight matrices W_l and the bias vectors b_l, each for $l = 1, \ldots, L$. Corollary 11.20 has shown that the number of parameters can increase very quickly. In addition, the loss function $J(\cdot)$ naturally depends on the result of the neural network $f[\mathbf{W}, \mathbf{b}](\cdot)$ and this is only given in algorithmic notation, of Definition 11.19. The evaluation of the derivative of the loss function is the most complex part in training the neural network.

In the following, we denote with "X" any derivative direction, e.g., the entry of a weight matrix $X = W_{rs}^l$ in layer l. Then with (11.21)

$$\frac{\mathrm{d}J(\mathbf{W}, \mathbf{b})}{\mathrm{d}X} = \frac{2}{N_{tr}} \sum_{i=1}^{N_{tr}} \left(y_i - f[\mathbf{W}, \mathbf{b}](x_i), \frac{\mathrm{d}f[\mathbf{W}, \mathbf{b}](x_i)}{\mathrm{d}X} \right), \tag{11.23}$$

where $(\cdot, \cdot)$ is the Euclidean scalar product. While this outer derivative is easy to set up, the calculation of the derivatives of the network function $f[\mathbf{W}, \mathbf{b}](x)$ is more involved. We will develop this derivative in the following.

Schematically, an artificial neural network can be understood as a nested application of activation functions and affine linear functions, i.e.,

$$f[\mathbf{W}, \mathbf{b}](x) = W_L(\cdots \sigma(W_2\sigma(W_1x + b_1) + b_2)\cdots) + b_L. \tag{11.24}$$

Derivatives are needed with respect to all weight matrices W_l and bias vectors b_l, and possibly with respect to the input x. The basic tool for the calculation is simply the chain rule, which we will formulate efficiently for exactly this application. In Definition 11.19, we introduced the notation

$$z^l = A_l(x^{l-1}) := W^l x^{l-1} + b^l, \; l = 1, \ldots, L, \quad x^l = \sigma(z^l), \quad l = 1, \ldots, L-1,$$

where $x^0 = x$ and $x^L = z^L = f[\mathbf{W}, \mathbf{b}](x)$ is the desired output. Then, we can write the entire network compactly in the form

$$f[\mathbf{W}, \mathbf{b}](x) = A_L \circ \sigma \circ A_{L-1} \circ \sigma \circ \cdots \circ \sigma \circ A_2 \circ \underbrace{\sigma \circ \overbrace{A_1(x)}^{z^1}}_{=x^1}. \tag{11.25}$$

Let us further combine x^l and z^l, i.e., we use the notation

$$x^l : \mathbb{R}^{n_{l-1}} \to \mathbb{R}^{n_l} : \quad \begin{aligned} x^l(x^{l-1}) &= \sigma(W^l x^{l-1} + b^l), \quad l = 1, \ldots, L-1, \\ x^L(x^{L-1}) &= W^L x^{L-1} + b^L, \end{aligned} \tag{11.26}$$

Then the network is given as

$$f[\mathbf{W}, \mathbf{b}](x) = x^L \circ x^{L-1} \circ \cdots \circ x^1(x). \tag{11.27}$$

With the repeated application of the chain rule, we write the derivative with respect to an abstract variable "X", which occurs in layer l or deeper, as

$$\begin{aligned} \frac{\mathrm{d}f[\mathbf{W}, \mathbf{b}]_{i_L}(x)}{\mathrm{d}X} &= \frac{\mathrm{d}X_{i_L}^L}{\mathrm{d}X} \\ &= \sum_{i_{L-1}=1}^{n_{L-1}} \sum_{i_{L-2}=1}^{n_{L-2}} \cdots \sum_{i_l=1}^{n_l} \frac{\mathrm{d}x_{i_L}^L}{\mathrm{d}x_{i_{L-1}}^{L-1}} \frac{\mathrm{d}x_{i_{L-1}}^{L-1}}{\mathrm{d}x_{i_{L-2}}^{L-2}} \cdots \frac{\mathrm{d}x_{i_{l+1}}^{l+1}}{\mathrm{d}x_{i_l}^l} \frac{\mathrm{d}x_{i_l}^l}{\mathrm{d}X}. \end{aligned} \tag{11.28}$$

Before we get more specific, let's further simplify the notation. The derivative $\mathrm{d}x^{l+1}/\mathrm{d}x^l$ corresponds to the gradient of the function $x^l(x)$ according to (11.26). That is, we can write (11.28) as

$$\frac{\mathrm{d}f[\mathbf{W},\mathbf{b}]_{i_L}(x)}{\mathrm{d}X} = \sum_{i_l=1}^{n_l}(\nabla x^L \cdot \nabla x^{L-1} \cdots \nabla x^{l+1})_{i_L i_l}\frac{\mathrm{d}x_{i_l}^l}{\mathrm{d}X}, \tag{11.29}$$

where "$\cdot$" is the matrix-matrix product. Even shorter, we write

$$\frac{\mathrm{d}f[\mathbf{W},\mathbf{b}]_{i_L}(x)}{\mathrm{d}X} = \sum_{i_l=1}^{n_l}(\nabla_{x^l}x^L)_{i_L i_l}\frac{\mathrm{d}x_{i_l}^l}{\mathrm{d}X}, \tag{11.30}$$

where $\nabla_{x^l}x^L$ is the gradient of the composite function $x^L \circ \cdots \circ x^{l+1}(x^l)$.

We now begin with the calculation of the gradient of a layer with respect to its input.

Theorem 11.30 (Derivative of Artificial Neural Networks With Respect to the Input) *Let $\mathcal{N}(\sigma; n_0, \ldots, n_L)$ be an artificial neural network with differentiable activation function $\sigma(\cdot)$. For $x^0 := x \in \mathbb{R}^{n_0}$, let x^l and z^l for $l = 1, \ldots, L$ be the intermediate results from Definition 11.19. Then for $l = 1, \ldots, L-1$*

$$\frac{dx^l}{dx^{l-1}} = \sigma'(z^l)\star W^l \quad and \quad \frac{dx^L}{dx^{L-1}} = W^L.$$

Here, for a vector $a \in \mathbb{R}^n$ and a matrix $A \in \mathbb{R}^{n\times m}$, the component-wise product $a\star A$ is defined as

$$(a\star A)_{ij} = a_i A_{ij}, \quad i = 1, \ldots, n, \quad j = 1, \ldots, m. \tag{11.31}$$

It holds

$$\nabla x^l = \frac{dx^l}{dx^{l-1}} \in \mathbb{R}^{n_l \times n_{l-1}}.$$

Proof The result can be calculated component-wise. Let $l < L$. Then for $i = 1, \ldots, n_l$ and $j = 1, \ldots, n_{l-1}$, using the chain rule applied to $x_i^l = \sigma(z_i^l)$, we have

$$\frac{\mathrm{d}}{\mathrm{d}x_j^{l-1}}\sigma(z_i^l) = \sigma'(z_i^l)\frac{\mathrm{d}}{\mathrm{d}x_j^{l-1}}z_i^l = \sigma'(z_i^l)\sum_{i_{l-1}=1}^{n_{l-1}}\frac{\mathrm{d}}{\mathrm{d}x_j^{l-1}}W_{i,i_{l-1}}^l x_{i_{l-1}}^{l-1} = \sigma'(z_i^l)W_{i,j}^l.$$

We continue by considering the activation function and its derivative as a scalar function, which is applied separately to each entry of the vector, i.e.,

$$\left(\sigma'(z^l)\right)_i = \sigma'(z_i^l), \quad i = 1, \ldots, n_l.$$

The result can be read off when using the component-wise product (11.31). It also immediately follows that $\nabla x^l \in \mathbb{R}^{n_l \times n_{l-1}}$. For the last element x^L, we have $x^L = z^L$, i.e., the gradient is the inner derivative, so $\nabla x^L = W^L$. $\qquad\square$

Remark 11.31 (Special Treatment of the Output Layer) In the calculation of the gradient, we treated the last layer x^L separately. This is because we generally do not apply the activation function to the output see also Remark 11.22. There are indeed applications where it is desirable for the result to lie in the interval $[0, 1]$, and then the activation function would also be used in the output layer, i.e.,

$$f[\mathbf{W}, \mathbf{b}](x) = \sigma\left(W^L \sigma(\cdots) + \cdots + b^L\right).$$

The first factor of the derivative would then correspondingly be $\sigma'(z^L)\star W^L$ instead of W^L.

With the multilayer perceptron, Definition 11.19, we are only considering a very simple case of artificial neural networks here. However, for more complex architectures, the building blocks we develop here can be combined. ◆

The derivatives of the stages depend on the derivatives of the activation function. It holds:

Theorem 11.32 (Derivative of the Activation Functions) *For the derivatives of the sigmoid function and the hyperbolic tangent, it holds*

$$\sigma'(s) = \sigma(s)\left(1 - \sigma(s)\right), \quad \tanh'(s) = 1 - \tanh^2(s).$$

The derivative of the ReLU activation is only piecewise defined:

$$\mathrm{ReLU}'(s) = \begin{cases} 0 & s < 0 \\ 1 & s \geq 0 \end{cases}$$

Proof Check by calculation. □

So far, we can calculate the derivative of the neural network with respect to the input.

Corollary 11.33 (Derivative of the Neural Network With Respect to the Input) *With the prerequisites of Theorem 11.30, it holds*

$$\nabla f[\mathbf{W}, \mathbf{b}](x) = W^L \cdot \left(\sigma'(z^{L-1})\star W^{L-1}\right) \cdots \left(\sigma'(z^1)\star W^1\right),$$

and for the gradient with respect to the output of layer l:

$$\nabla_{x^l} f[\mathbf{W}, \mathbf{b}](x) = W^L \cdot \left(\sigma'(z^{L-1})\star W^{L-1}\right) \cdots \left(\sigma'(z^{l+1})\star W^{l+1}\right).$$

It is $\nabla_{x^l} f[\mathbf{W}, \mathbf{b}](x) \in \mathbb{R}^{n_L \times n_l}$.

The next step is the calculation of the derivatives in the direction of the weights W^l and b^l in layer l, see the abstract form of the derivative (11.28). For this, it is sufficient only to consider the l-th layer of the network:

Theorem 11.34 (Derivative of a Layer With Respect to the Weights) *For $l = 1, \ldots, L$, let*

$$x^l : \mathbb{R}^{n_{l-1}} \to \mathbb{R}^{n_l}, \quad x^l(x) = \begin{cases} W^l x^{l-1} + b^l & l = L, \\ \sigma(W^l x^{l-1} + b^l) & l < L, \end{cases}$$

with $W^l \in \mathbb{R}^{n_l \times n_{l-1}}$, $b^l \in \mathbb{R}^{n_l}$ and a differentiable activation function $\sigma(\cdot)$. Then for $l = 1, \ldots, L - 1$ it holds

$$\frac{dx_i^l}{dW_{rs}^l} = \delta_{ir}\sigma'(z_i^l)x_s^{l-1}, \qquad i, r = 1, \ldots, n_l, \ s = 1, \ldots, n_{l-1},$$

$$\frac{dx_i^l}{db_r} = \delta_{ir}\sigma'(z_i^l), \qquad i, r = 1, \ldots, n_l,$$

(11.32)

and

$$\frac{dx_i^L}{dW_{rs}^L} = \delta_{ir}x_s^{L-1}, \qquad i, r - 1, \ldots, n_l, \ s = 1, \ldots, n_{l-1},$$

$$\frac{dx_i^L}{db_r} = \delta_{ir}, \qquad i, r = 1, \ldots, n_l.$$

(11.33)

Proof We calculate the derivative component-wise. For $i, r = 1, \ldots, n_l$ and $s = 1, \ldots, n_{l-1}$, it holds

$$\frac{dx_i^l}{dW_{rs}^l} = \frac{d}{dW_{rs}^l}\sigma(W^l x^{l-1} + b^l)_i = \sigma'(z_i^l)\sum_{j=1}^{n_{l-1}} \frac{d}{dW_{rs}^l}W_{ij}^l x_j^{l-1} = \delta_{ir}\sigma'(z_i^l)x_s^{l-1}.$$

Most derivatives, more precisely all derivatives for $i \neq r$, vanish. Instead of $n_l^2 n_{l-1}$, only $n_l n_{l-1}$ derivatives need to be calculated. Accordingly, it holds

$$\frac{dx_i^l}{db_r} = \frac{d}{db_r}\sigma(W x^{l-1} + b^l)_i = \sigma'(z_i^l)\frac{d}{db_r}b_i^l = \delta_{ir}\sigma'(z_i^l).$$

Here, only n_l of the actual n_l^2 derivatives remain, which do not vanish. $\square$

With this, we can now specify the derivatives of the artificial neural network in the direction of all parameters.

Corollary 11.35 (Derivative of the Neural Network With Respect to the Parameters) *With the prerequisites of Theorem 11.30, it holds for $l = 1, \ldots, L - 1$ that*

$$\frac{d f_i[\mathbf{W}, \mathbf{b}](x)}{dW_{rs}^l} = \left(\nabla_{x^l} f[\mathbf{W}, \mathbf{b}](x)\right)_{ir} \sigma'(z_r^l) x_s^{l-1},$$

$$\frac{d f_i[\mathbf{W}, \mathbf{b}](x)}{db_r^l} = \left(\nabla_{x^l} f[\mathbf{W}, \mathbf{b}](x)\right)_{ir} \sigma'(z_r^l),$$

as well as

$$\frac{d f_i[\mathbf{W}, \mathbf{b}](x)}{dW_{rs}^L} = \delta_{ir} x_s^{l-1}, \quad \frac{d f_i[\mathbf{W}, \mathbf{b}](x)}{db_r^L} = \delta_{ir}.$$

Proof Substituting the relations from Theorem 11.34 into (11.30) gives

$$\frac{\mathrm{d} f_i[\mathbf{W}, \mathbf{b}](x)}{\mathrm{d}W_{rs}^l} = \sum_{j=1}^{n_l} \left(\nabla_{x^l} f[\mathbf{W}, \mathbf{b}](x)\right)_{ij} \delta_{jr} \sigma'(z_j^l) x_s^{l-1}$$

$$\frac{\mathrm{d} f_i[\mathbf{W}, \mathbf{b}](x)}{\mathrm{d}b_r^l} = \sum_{j=1}^{n_l} \left(\nabla_{x^l} f[\mathbf{W}, \mathbf{b}](x)\right)_{ij} \delta_{jr} \sigma'(z_j^l),$$

i.e., in the sum, only the term for $j = r$ remains. For $l = L$, the calculation simplifies accordingly. $\square$

The derivative with respect to the bias can be written compactly as a gradient, i.e., $\nabla_{b^l} f[\mathbf{W}, \mathbf{b}](x) = \nabla_{x^l} x^L \star \sigma'(z^l)$, where we use the notation corresponding to (11.31) for $A \in \mathbb{R}^{n \times m}$ and $a \in \mathbb{R}^m$

$$(A \star a)_{ij} = A_{ij} a_j.$$

The derivatives with respect to the weights W_{rs}^l, on the other hand, form a third-order tensor. However, we only need the derivatives for the calculation of the gradient of the loss function and do not need any further compact notation for the individual derivatives.

Theorem 11.36 (Gradient of the Loss Function) *With the prerequisites of Theorem 11.34, it holds for $l = 1, \ldots, L - 1$*

$$\nabla_{W^l} J(\mathbf{W}, \mathbf{b}) = \frac{2}{N_{tr}} \sum_{i_{tr}=1}^{N_{tr}} \left(\nabla_{x^l} f[\mathbf{W}, \mathbf{b}](x_{i_{tr}})^T d_{i_{tr}} \circ \sigma'(z_{i_{tr}}^l)\right) \left(x_{i_{tr}}^{l-1}\right)^T,$$

$$\nabla_{b^l} J(\mathbf{W}, \mathbf{b}) = \frac{2}{N_{tr}} \sum_{i_{tr}=1}^{N_{tr}} \nabla_{x^l} f[\mathbf{W}, \mathbf{b}](x_{i_{tr}})^T d_{i_{tr}} \circ \sigma'(z_{i_{tr}}^l),$$

as well as

$$\nabla_{W^L} J(\mathbf{W}, \mathbf{b}) = \frac{2}{N_{tr}} \sum_{i_{tr}=1}^{N_{tr}} d_{i_{tr}} (x_{i_{tr}}^{L-1})^T,$$

$$\nabla_{b^l} J(\mathbf{W}, \mathbf{b}) = \frac{2}{N_{tr}} \sum_{i_{tr}=1}^{N_{tr}} d_{i_{tr}}.$$

Here,

$$d_{i_{tr}} = y_{i_{tr}} - f[\mathbf{W}, \mathbf{b}](x_{i_{tr}}), \quad i_{tr} = 1, \ldots, N_{tr},$$

is the defect of the training data and $A \circ B$ is the Hadamard product of two matrices $A, B \in \mathbb{R}^{n \times m}$ (or vectors), given by

$$(A \circ B)_{ij} = A_{ij} B_{ij}, \quad i = 1, \ldots, n, \quad j = 1, \ldots, m.$$

Proof We combine (11.23) with the results from Corollary 11.35 and summarize the occurring sums in a compact notation

$$\frac{dJ(\mathbf{W}, \mathbf{b})}{dW_{rs}^l} = \frac{2}{N_{tr}} \sum_{i_{tr}=1}^{N_{tr}} \sum_{i=1}^{n_L} d_{i_{tr},i} \left(\nabla_{x^l} f[\mathbf{W}, \mathbf{b}](x_{i_{tr}})\right)_{ir} \sigma'(z_{i_{tr},r}^l) x_{i_{tr},s}^{l-1}$$

$$= \frac{2}{N_{tr}} \sum_{i_{tr}=1}^{N_{tr}} \left(\nabla_{x^l} f[\mathbf{W}, \mathbf{b}](x_{i_{tr}})^T d_{i_{tr}}\right)_r \sigma'(z_{i_{tr}}^l)_r x_{i_{tr},s}^{l-1}$$

$$= \frac{2}{N_{tr}} \sum_{i_{tr}=1}^{N_{tr}} \left(\nabla_{x^l} f[\mathbf{W}, \mathbf{b}](x_{i_{tr}})^T d_{i_{tr}} \circ \sigma'(z_{i_{tr}}^l)\right)_r x_{i_{tr},s}^{l-1},$$

and correspondingly

$$\frac{dJ(\mathbf{W}, \mathbf{b})}{db_r^l} = \frac{2}{N_{tr}} \sum_{i_{tr}=1}^{N_{tr}} \sum_{i=1}^{n_L} d_{i_{tr},i} \left(\nabla_{x^l} f[\mathbf{W}, \mathbf{b}](x_{i_{tr}})\right)_{ir} \sigma'(z_{i_{tr},r}^l)$$

$$= \frac{2}{N_{tr}} \sum_{i_{tr}=1}^{N_{tr}} \left(\nabla_{x^l} f[\mathbf{W}, \mathbf{b}](x_{i_{tr}})^T d_{i_{tr}}\right)_r \sigma'(z_{i_{tr}}^l)_r$$

$$= \frac{2}{N_{tr}} \sum_{i_{tr}=1}^{N_{tr}} \left(\nabla_{x^l} f[\mathbf{W}, \mathbf{b}](x_{i_{tr}})^T d_{i_{tr}} \circ \sigma'(z_{i_{tr}}^l)\right)_r.$$

$\square$

With Theorem 11.36, we have completed the derivation of the gradient of the loss function with respect to all parameters. What remains is the efficient algorithmic implementation. The main part of the effort lies in the calculation of the gradient with respect to the l-th layer, i.e., $\nabla_{x^l} x^L$, which must be calculated for each element of the training data. In Corollary 11.33, we derived the relationship

$$\nabla_{x^l} x^L = W^L \cdot \left(\sigma'(z^{L-1}) \star W^{L-1}\right) \cdots \left(\sigma'(z^{l+1}) \star W^{l+1}\right),$$

for this. The derivative with respect to layer l requires the product of the weight matrices $W^L, W^{L-1}, \ldots, W^{l+1}$ as well as the linear intermediate results $z^{L-1}, z^{L-2}, \ldots, z^{l+1}$. To calculate these, the network must be run from the input $x = x^0$ onwards. To avoid re calculating these values for each derivative, the gradient is set up in two steps: first, the network is evaluated for an input $x = x_i$ and all intermediate results $z_i^1, \ldots, z_i^{L-1}$ are stored. This step is called the *forward mode*. Subsequently, the gradients are calculated, starting from layer L. This second step is correspondingly called the *backward mode*. The entire algorithm is called *backpropagation*.

Algorithm 11.37 (Backpropagation: Evaluation of the Derivatives of the Artificial Neural Network and the Loss Function)

> **Input:** Artificial neural network $\mathcal{N}(\sigma; n_0, \ldots, n_L)$, parameters $(\mathbf{W}, \mathbf{b}) \in \mathcal{P}$ and training
> data $(x_{i_{tr}}, y_{i_{tr}}) \in \mathbb{R}^{n_0} \times \mathbb{R}^{n_L}$
>   /* Initialization */
> 1 **for** $l = 1, \ldots, L$ **do**
> 2 $\nabla_{W^l} J = 0$
> 3 $\nabla_{b^l} J = 0$
> 4 **for** $i_{tr} = 1, 2, \ldots, N_{tr}$ **do**
>   /* Forward mode */
> 5 $x^0 = x_{i_{tr}}$
> 6   **for** $l = 1, \ldots, L-1$ **do**
> 7   Store $z_{i_{tr}}^l = W^l x_{i_{tr}}^{l-1} + b^l$
> 8   Store $x_{i_{tr}}^l = \sigma(z_{i_{tr}}^l)$
> 9 $x^L = W^L x^{L-1} + b^L$
> 10 $d = y_{i_{tr}} - x^L$
>   /* Backward mode */
> 11 $G_{W^L} = W^L$
> 12   **for** $l = L-1, \ldots, 1$ **do**
> 13 $G_{W^l} = G_{W^{l+1}} \cdot \left(\sigma'(z_{i_{tr}}^l) \star W^l\right)$
> 14   Compute intermediate result $X = (G_{W^l})^T \circ \sigma'(z_{i_{tr}}^l)$
> 15 $\nabla_{W^l} J = \nabla_{W^l} J + \frac{2}{N_{tr}} X (x_{i_{tr}}^{l-1})^T$
> 16 $\nabla_{b^l} J = \nabla_{b^l} J + \frac{2}{N_{tr}} X$
> **Result:** Gradients $\nabla_{W^l} J$ and $\nabla_{b^l} J$

Theorem 11.38 (Backpropagation) *The backpropagation algorithm calculates the gradients of the loss function with respect to the weights. For the architecture $\mathcal{N}(\sigma; n_0, \ldots, n_L)$, the gradients can be calculated with the effort*

$$O(N_{tr} n^3 L).$$

To estimate the effort, all operations, particularly the matrix-matrix multiplications must be counted.

Remark 11.39 (Efficient Implementations) Efficient implementations of neural networks such as *PyTorch* [72] or *Tensorflow* [1] achieve their performance through extreme optimization of the basic matrix operations. We have formulated Algorithm 11.37 as a loop over all training data pairs. However, the networks and derivatives can be calculated much more efficiently if all data is evaluated simultaneously. This, however, requires dealing with tensors of higher order, i.e., with multilinear mappings $X_{i,j,k}$ with three or even more indices.

In addition, libraries like *PyTorch* or *Tensorflow* use special hardware like graphics cards (GPUs) or accelerator cards like TPUs (Tensor-Processing-Units), which are highly optimized for these tasks. ♦

Having the gradients of the loss function with respect to the weights and the bias, the artificial neural network can be *trained*. In other words, we can use the gradient method to search for the optimal parameters $(\mathbf{W}, \mathbf{b}) \in \mathcal{P}$ that minimize the loss function.

Example 11.40 (Function Approximation With Artificial Neural Networks)
We consider the function

$$f(x) = x(1 + \exp(x)) + 10\sin(3 + \log(x^2 + 1)), \tag{11.34}$$

which we have already analyzed in the search for roots, and try to represent it with a simple artificial neural network. We start with the architecture $\mathcal{N}(\sigma; 1, 8, 1)$, i.e., the set of functions

$$f[\mathbf{W}, \mathbf{b}](x) = W_2 \tanh\left(W_1 x + b_1\right) + b_2, \quad W_1 \in \mathbb{R}^{8\times 1}, \quad W_2 \in \mathbb{R}^{1\times 8}, \quad b_1 \in \mathbb{R}^8, \quad b_2 \in \mathbb{R}.$$

From the interval $I = [-10, 2]$ we choose $N_{tr} = 20$ random points $x_1, \ldots, x_{N_{tr}}$ and corresponding values $y_i = f(x_i)$. In Fig. 11.12, we show on the left the course of the loss function in 10000 steps of optimization. On the right, we give the course of the function $f(x)$ as well as the trained neural network $f[\mathbf{W}, \mathbf{b}](x)$. The network approximates most training points. Beyond the training data, especially in the range $[1, 2]$, however, the network does not provide any approximation to the function $f(x)$. This is not to be expected, as the network has not learned the function during training, but only some discrete values.

If we distribute more training points in the interval, the approximation improves. In Fig. 11.13, we show on the left the approximation for $N_{tr} = 100$ random of points. The network we have chosen is very small and only has 25 free parameters. In the right of Fig. 11.13, we show the approximation for the same $N_{tr} = 100$ random training points, but using a much larger network with the architecture $\mathcal{N}(\tanh; 1, 16, 16, 16, 1)$. This network can represent the function very well at all points. ◄

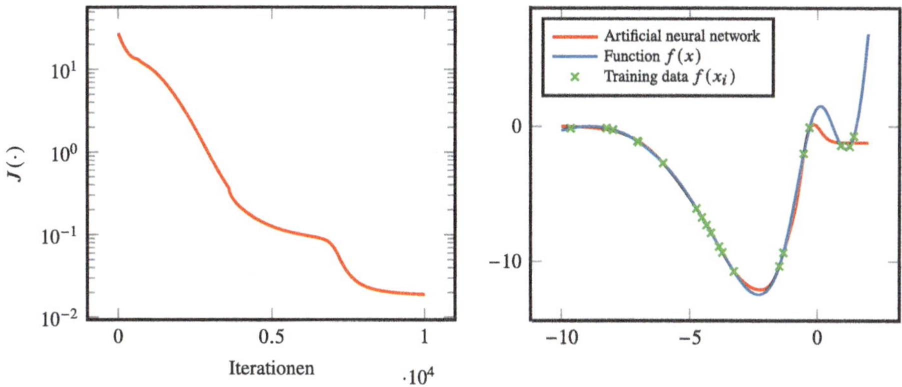

Fig. 11.12 Approximation of a function with artificial neural networks. On the left is the course of the training, i.e., the minimization of the loss function using the gradient method. On the right is the approximation property of the network as well as the 20 training points

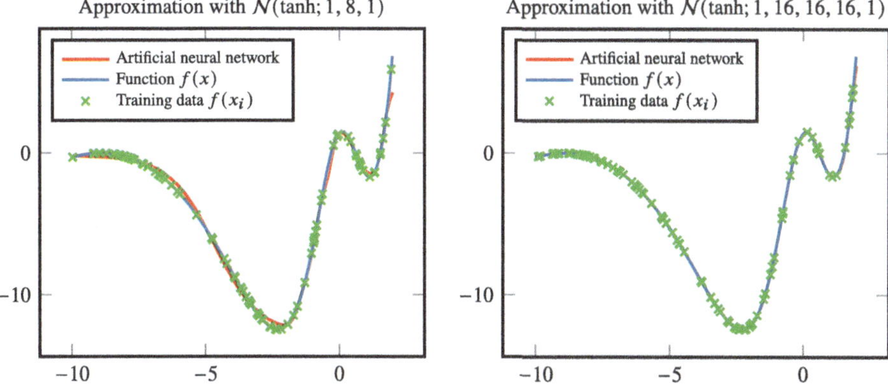

Fig. 11.13 Approximation of a function with artificial neural networks. Left: small network $\mathcal{N}(\tanh; 1, 8, 1)$ with $N_{tr} = 100$ training data and right: large network $\mathcal{N}(\tanh; 1, 16, 16, 16, 1)$ again with $N_{tr} = 100$ training data

11.5.3.2 Overfitting, Testing and Validation, and Generalization

Artificial Neural Networks are often designed with a huge number of free parameters. This makes them very powerful and flexible, but the overparametrization carries the risk of *overfitting*: The data pairs are well mapped $y_i \approx f[\mathbf{W}_{\mathrm{opt}}, \mathbf{b}_{\mathrm{opt}}](x_i)$, but in other points the network has not learned the function course. We consider the approximation of the function $f(x) = 0$, but assume that the function is only known with errors at training support points $x_i \in [-10, 2]\, y_i = 0 + N(0.1)$. Here, $N(0.1)$ is a normally distributed random number with standard deviation 0.1. Figure 11.14 shows the training data and the approximation with a (very richly parameterized) network.

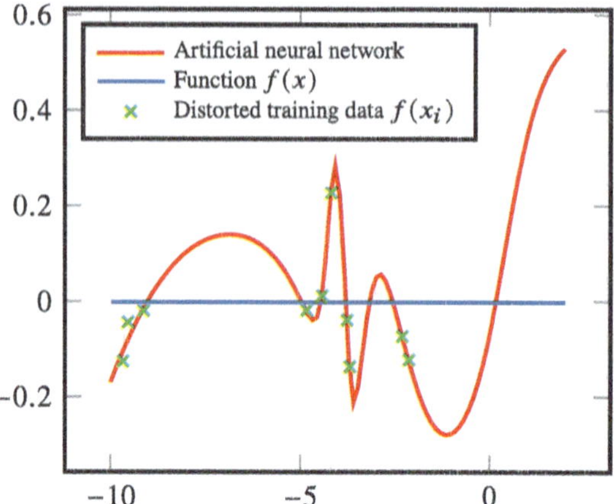

Fig. 11.14 Overfitting: the network approximates the $N_{tr} = 10$ training data very well, but otherwise does not reproduce the course of the function $f(x) = 0$

There are various strategies to minimize *overfitting*. The usual procedure is to divide the data into *training data* and *validation data*: The *training data* (perhaps 70% of the total data) are used to train the artificial neural network. That is, they are the basis for evaluating the gradient in optimization.

During optimization, the loss function is evaluated not only using the training data but also based on the *validation data*. These have never been included in the calculation of the gradient, but we expect the neural network also to provide good predictions on these validation data, and that the loss function of the validation data decreases during optimization. If there is a discrepancy, i.e., we reach a point where the loss function of the training data is further reduced, but the loss function of the validation data stagnates or even increases again, this is a sign of *overfitting* and the optimization should be terminated.

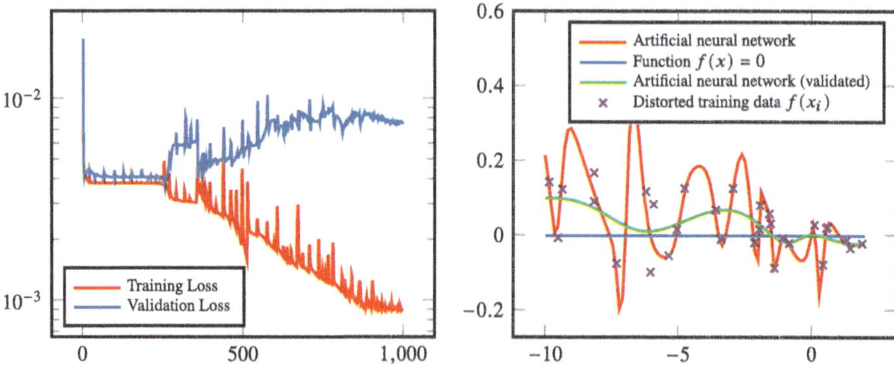

Fig. 11.15 Left: Optimization and course of the loss functions applied to training and validation data. Right: Result of the artificial neural networks after 10000 optimization steps and with premature termination after 2500 steps, before training and validation loss diverge

In Fig. 11.15, we show on the left the course of the loss function applied to $N_{tr} = 30$ training data and to $N_{va} = 10$ validation data. On the right, we show the network that was trained with 10000 steps and also network after termination of the minimization at 2500 steps, where training and validation loss still match well. The training points are no longer approximated as well, but the overall course of the network is closer to the underlying function $f(x) = 0$.

Another technique to avoid *overfitting* and to improve the training process is the *regularization* of the loss function. In the simplest case, this means that the loss function (11.21) is extended by a quadratic term:

$$J_\gamma(\mathbf{W}, \mathbf{b}) = \frac{1}{N_{tr}} \sum_{i=1}^{N_{tr}} \left(f[\mathbf{W}, \mathbf{b}](x_i) - y_i \right)^2 + \gamma \left(\sum_{l=1}^{L} \|W^l\|_F^2 + \|b^l\|_2^2 \right), \qquad (11.35)$$

with the regularization parameter $\gamma > 0$. In mathematical optimization, this method is called Tikhonov regularization.

11.6 Excursus: Eigenvalue Determination with Artificial Neural Networks

We will develop an artificial neural network that receives a symmetric matrix as input and returns all eigenvalues of the matrix as output. As a base set, we consider symmetric matrices that are constructed according to the following principle

$$X_d = \{A \in \mathbb{R}^{d \times d}, \ A = B + B^T, \ B \in \mathcal{U}(0, 1)^{d \times d}\}, \qquad (11.36)$$

where $\mathcal{U}(0, 1)$ are uniformly distributed numbers in the interval $[0, 1]$. In Chap. 5, we examined the conditioning of the eigenvalue problem and already found that the eigenvalues depend continuously on the matrix entries, see Remark 5.6. The problem should therefore be well-suited for approximation with neural networks. On the other hand, the determination of eigenvalues is sufficiently difficult. We have learned about various algorithms in Chap. 5. Direct methods are not suitable due to their lack of stability, and the iterative methods all involve a high effort.

First, we create the training and validation data. For this purpose, we choose $N_{tr} \in \mathbb{N}$ and $N_{va} \in \mathbb{N}$ random matrices $A_1^{tr}, \ldots, A_{N_{tr}}^{tr} \in X_d$ as well as $A_1^{va}, \ldots, A_{N_{va}}^{va} \in X_d$ and determine the eigenvalues $\Lambda_i^{tr}, \Lambda_i^{va} \in \mathbb{R}^d$ in each case. Since the matrices are symmetric, all eigenvalues are real, and we arrange them in ascending order, i.e.,

$$\Lambda_i^{tr} = (\lambda_{i,1}^{tr}, \ldots, \lambda_{i,d}^{tr}), \ \text{with} \ \lambda_{i,1}^{tr} \le \lambda_{i,2}^{tr} \le \cdots \le \lambda_{i,d}^{tr},$$

and

$$\Lambda_i^{va} = (\lambda_{i,1}^{va}, \ldots, \lambda_{i,d}^{va}), \ \text{with} \ \lambda_{i,1}^{va} \le \lambda_{i,2}^{va} \le \cdots \le \lambda_{i,d}^{va}.$$

The artificial neural network receives the matrix elements as inputs and the eigenvalues as the output. Thus, $n_0 = d^2$ and $n_L = d$. In the inner layers, we choose a fixed number of $n_l = n$ neurons. The hyperbolic tangent is used as the activation function. With the number of layers still free, the architecture is given as

$$\mathcal{N}(\tanh; d^2, \underbrace{n, \ldots, n}_{L \text{ layers}}, d).$$

The neural network thus has $Ln + d$ artificial neurons, and the number of free weights and bias is

$$\left(nd^2 + (L-1)n^2 + dn\right) + \left(Ln + d\right).$$

Assuming that the training data is given by $\mathbf{A} = (A_1, \ldots, A_{N_{tr}})$ and $\mathbf{\Lambda} = (\Lambda_1, \ldots, \Lambda_{N_{tr}})$, we choose the loss function

$$l(\mathbf{A}, \mathbf{\Lambda}) = \frac{1}{N_{tr}} \sum_{i=1}^{N_{tr}} \| f[\mathbf{W}, \mathbf{b}](A_i) - \Lambda_i \|_2^2. \tag{11.37}$$

Training the Network

We must now determine the number of layers L and their width n. This choice can often only be made experimentally. We start with a simple example and $d = 8$, i.e., symmetric matrices $A \in \mathbb{R}^{8 \times 8}$. Furthermore, we choose $L = 4$ layers with $n = 64$ artificial neurons each. For training, $N_{tr} = 5000$ random matrices are determined. Furthermore, we choose $N_{va} = 1\,250$ random matrices for validation. In Fig. 11.16 on the left, we show the course of the loss function for training and validation data. *Overfitting* is clearly visible: the network makes far better predictions on the training data than on the validation data. At the same time, it is evident that the optimization is not yet complete, as the curve of the training data continues to decrease.

Although we do not expect the resulting network to provide good predictions, we list the eigenvalues Λ and the eigenvalues Λ_N approximated by the network for two randomly selected matrices (which are not included in the training data) in Table 11.4. The results are each rounded to three decimal places.

The network trains too specifically on the given data. One way to avoid *overfitting* is to add a regularization term to the loss function (11.37). The simplest form of regularization is *Tikhonov regularization*, which adds the parameters to be determined as quadratic terms, i.e.,

$$l(\mathbf{A}, \mathbf{\Lambda}) = \frac{1}{N_{tr}} \sum_{i=1}^{N_{tr}} \| f[\mathbf{W}, \mathbf{b}](A_i) - \Lambda_i \|_2^2 + \gamma \sum_{l=1}^{L} \left(\|W^l\|_F^2 + \|b^l\|_2^2 \right), \tag{11.38}$$

Table 11.4 Eigenvalues and neural network approximation for two randomly selected matrices not included in the training data. The relative error between eigenvalues λ_i and network approximation λ_i^N is given as $(\lambda_i - \lambda_i^N)/\max_i |\lambda_i|$

Reference	−1.945	−1.337	−0.998	−0.016	0.355	1.105	1.444	7.278
Network	−2.015	−1.040	−0.648	−0.414	−0.187	0.725	2.105	7.218
Rel. Error	0.010	−0.041	−0.048	0.055	0.074	0.052	−0.091	0.008
Reference	−1.631	−1.246	−0.222	0.206	0.344	1.130	1.773	8.607
Network	−1.739	−0.853	−0.361	0.120	0.445	0.955	1.680	8.734
Rel. Error	0.013	−0.046	0.016	0.010	−0.012	0.020	0.011	−0.015

where $\gamma > 0$ is a small parameter. We choose $\gamma = 10^{-4}$ and in Fig. 11.16 on the right, the progression of the two loss functions is shown. Overfitting can thus be avoided. However, it also shows that the loss function can no longer be reduced accordingly. The choice of the parameter γ is not easy. For $\gamma = 10^{-5}$, the regularization loses its effect in this example. The choice $\gamma = 10^{-3}$ leads to a loss of the approximation property.

Table 11.5 shows the results for two test matrices using regularization. All eigenvalues can be predicted with a maximum error of 5%. The normalization in the relative error always refers to the largest eigenvalue of the matrix in absolute value.

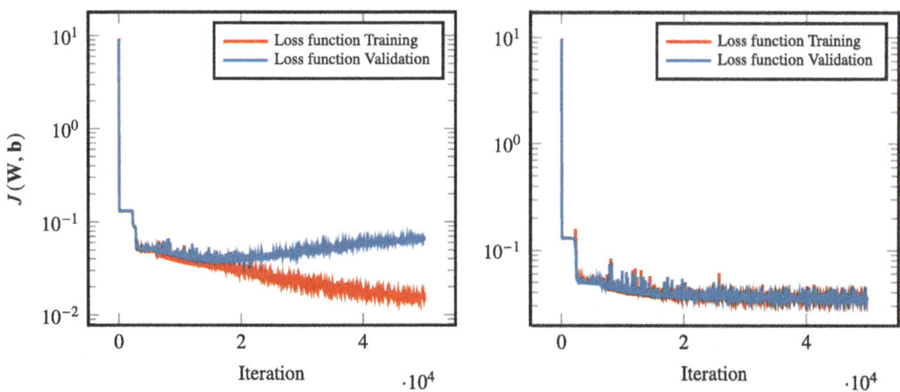

Fig. 11.16 Left: Progression of the loss function for training and validation data for eigenvalue determination of symmetric 8×8 matrices. Clear *overfitting* can be seen. Right: Training of the same network with *regularization*. The loss function is supplemented by the term $\gamma \sum_l \|W^l\|_F^2$

Table 11.5 Training with regularization. Eigenvalues and neural network approximation for two randomly selected matrices not included in the training data

Reference	−1.594	−1.106	−0.958	0.337	0.641	1.352	1.494	8.638
Network	−1.746	−1.074	−0.498	0.035	0.569	1.121	1.758	8.655
Rel. Error	0.018	−0.004	−0.053	0.035	0.008	0.027	−0.031	−0.002
Reference	−1.932	−0.958	−0.215	0.557	0.651	1.435	2.037	7.843
Network	−1.698	−0.936	−0.323	0.261	0.845	1.436	2.088	7.839
Rel. Error	−0.030	−0.003	0.014	0.038	−0.025	−0.000	−0.007	0.001

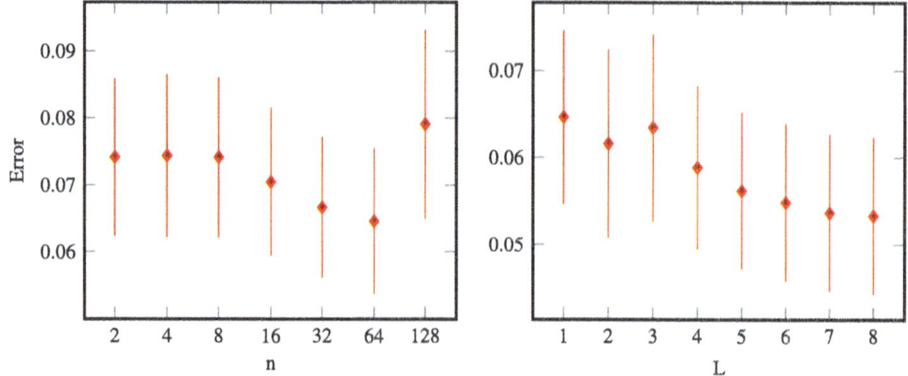

Fig. 11.17 Left: Mean error and standard deviation with increasing number of artificial neurons n per layer. $L = 4$ always applies. Right: Mean error and standard deviation when varying the number of layers L. Here, $n = 64$ always applies

Influence of the Network Size

The theoretical approximation quality of a neural network is the fundamental factor that determines how well a problem can be solved at all. This is called the *expressivity of the network*. We first examine the influence of the number of neurons n per layer. That is, we stick with the architecture $\mathcal{N}(\tanh; d^2, n, n, n, n, d)$ and vary the neurons per layer. For $n \in \{2, 4, 8, 16, 32, 64, 128, 256\}$ we train a network each and then apply it to 1000 random matrices. For these, we determine the relative errors and plot the results in Fig. 11.17 on the left. We show both the mean and the standard deviation of the errors.

As expected, larger networks are better able to approximate the eigenvalues. However, the convergence is slow. This may be due to other influences, such as the number of layers or the choice of the regularization parameter γ, which we have not adjusted here. Finally, we also keep the number of training data constant. For all cases, $N_{tr} = 10000$ applies.

Corresponding to this test, we now fix $n = 64$ and vary the number of layers L from $L = 1$ to $L = 8$. That is, the smallest network has the architecture $\mathcal{N}(\tanh; d^2, n, d)$ and the largest

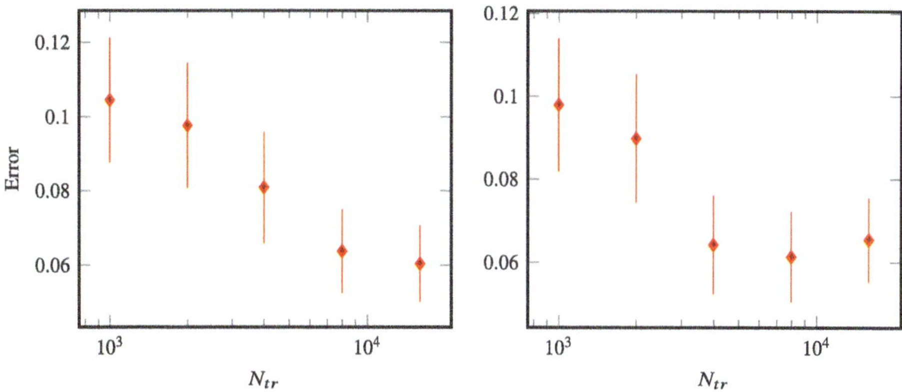

Fig. 11.18 Validation error (mean and standard deviation) with increasing number of training data. On the left, we show the result for the network $\mathcal{N}(\tanh; d^2, 64, 64, 64, 64, d)$ and on the right for a network with twice the number of neurons per layer, i.e., $\mathcal{N}(\tanh; d^2, 128, 128, 128, 128, d)$

has 8 inner layers. We show the results in Fig. 11.17 on the right. Deeper networks have better approximation properties. But here too, it is shown that the improvement stagnates beyond a certain depth. This does not mean that the network then has no better approximation property, but rather that we either need to increase the number of training data, possibly adjust the regularization parameter γ, or intervene in the optimization process in some other way. The improvement of the whole setup, i.e., the choice of the network, the data, the method of optimization, and the regularization, is called *hyperparameter optimization*. This usually requires a lot of trial and error and manual work.

Influence of the Training Data

We have seen that it only makes sense to increase the complexity of the neural network up to a certain limit, as further steps do not lead to an improvement in the application to test data. This is because we have always kept the number of training data the same. In Fig. 11.18, we now show the dependence of the network prediction on an increasing number of training data. The network itself has $L = 4$ layers with $n = 64$ artificial neurons each (left) and $n = 128$ (right).

It turns out that the approximation quality increases with the number of artificial neurons. However, a limit is quickly reached. Another effect, however, is the significantly reduced variance of the results. This is naturally due to the fact that the generalization error is lower. This is given by the distance between a validation matrix A and the test data, i.e.,

$$\min_{i=1,\ldots,N_{train}} \|A - A_i\|^2.$$

Table 11.6 Generalization of the artificial neural network for predicting the eigenvalues of a matrix. For 1000 random matrices of the training and test data, as well as for 1000 additional, this time normally distributed matrices, we give the average prediction error and the standard deviation of the errors

Dataset	Mean	Standard deviation
Training data	0.058	0.020
Test data	0.056	0.021
Generalization data	0.537	0.288
Generalization data (normalized)	0.042	0.013

On the right of the figure, we repeat the experiment with twice the number of artificial neurons.

Generalization

So far, we have only tested the networks with matrices that originate from the same X_d. In particular, the entries of the matrices have so far all been normally distributed numbers with $A_{ij} \in [0, 2]$, of (11.36) (the entries of B are in $B_{ij} \in [0, 1]$). We first apply an already trained network to matrices whose entries are normally distributed

$$Y_d = \{A \in \mathbb{R}^{d \times d}, \ A = B + B^T, \ B \in \mathcal{N}(0.5)^{d \times d}\},$$

where $\mathcal{N}(0.5)$ is a normally distributed number with standard deviation 1 and mean 0.5. This is called the *generalization* of an artificial neural network: *How well can the method be applied to data that did not occur at all during training?*

For this purpose, we train a network with the architecture

$$\mathcal{N}(\tanh; 64, 128, 128, 128, 128, 8)$$

with $N_{tr} = 16000$ training matrices. The training is stopped after about 40000 iterations when the training and validation errors are still close to each other.

In Table 11.6, we show the average error for predicting the eigenvalues of 1000 random matrices each. In addition, we give the standard deviation of the 1000 random matrices as a measure of the dispersion of the errors. We compare elements of the training data with elements of the validation data (i.e., random matrices that have uniformly distributed entries but were not part of the training) and generalization data from Y_d. On this dataset, the network does not provide useful predictions, see Table 11.6. The eigenvalues of the generalization data have an average prediction error of about 50%. This is not surprising, as the matrices

$A \in Y_d$ have entries that do not have to lie in the interval $[0, 2]$, as is the case with all training data. Normally distributed numbers can be larger or smaller, and our network has never seen entries of this magnitude. Therefore, it is not to be expected that the method generalizes well. We, therefore, modify the example somewhat: for a matrix $A \in Y_d$, we first determine the absolute maximum deviation from the mean 1:

$$a_{max} = \max_{i,j} |A_{ij} - 1|.$$

With this element, the matrix is scaled

$$\tilde{A} = \frac{1}{a_{max}}(A - 1) + 1$$

and subsequently, $\tilde{A}$ is shifted back to the mean 1. For an eigenvector x and eigenvalue λ of A, it holds

$$\tilde{A}x = \frac{1}{a_{max}}(A - 1)x = \left(\frac{\lambda}{a_{max}} - 1\right)x.$$

The network can be applied to $\tilde{A}$, and the resulting eigenvalue $\tilde{\lambda}$ can be scaled according to

$$\lambda = a_{max}\left(\tilde{\lambda} + 1\right)$$

to eigenvalues of A. The results in the last line of Table 11.6 show that in this way, a very good generalization to matrices with normally distributed values can be achieved.

Finally, we investigate what happens when we apply the method to matrices that are no longer symmetric at all. For this purpose, we choose matrices of the type

$$Y_{d,\alpha} = \{A \in \mathbb{R}^{d \times d}, \; A = A_x + \alpha C, \; A_x \in X_d, \; C \in N^{d \times d}\}.$$

where α is a small parameter. The actually symmetric matrix $A_x \in X_d$ is disturbed with normally distributed numbers. The results are shown in Fig. 11.19 for $\alpha = 0, 0.05, 0.1, \ldots, 0.5$. In Table 11.7, we give for $\alpha = 0, 0.2$ and $\alpha = 0.4$ the eight eigenvalues for a single randomly chosen matrix each.

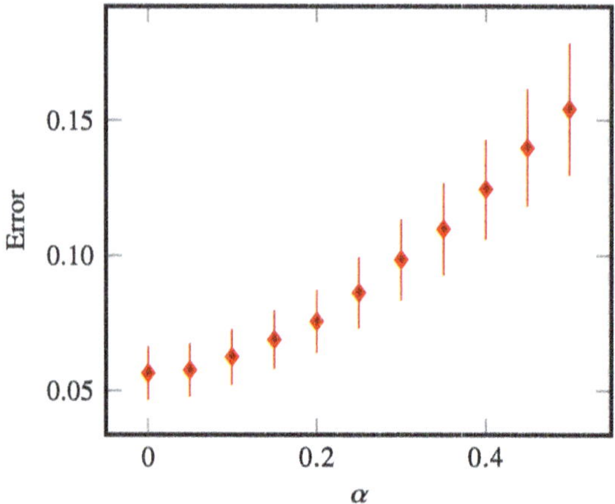

Fig. 11.19 Generalization error for an artificial neural network when applied to matrices with a normally distributed disturbance. The parameter α indicates the standard deviation of the disturbance

Table 11.7 Eigenvalues for three disturbed matrices of the type $A^T A + \alpha N$, where N are standard normally distributed numbers

	α	λ_1	λ_2	λ_3	λ_4	λ_5	λ_6	λ_7	λ_8
Exact	0.0	−1.80	−0.72	−0.37	0.15	0.51	1.32	1.99	8.26
Network	0.0	−1.62	−0.77	−0.38	0.15	0.63	1.31	1.80	8.20
Exact	0.2	−2.07	−1.33	−0.88	−0.38	0.67	0.84	1.60	7.76
Network	0.2	−2.08	−1.16	−0.83	−0.34	0.43	0.95	1.54	7.43
Exact	0.4	−3.34	−1.76	−0.42	0.14	0.70	1.36	2.57	6.59
Network	0.4	−2.77	−1.57	−0.76	0.24	0.85	1.73	1.89	6.21

List of Excursus

Python-Programs and Jupyter-Notebooks

Numerous algorithms were implemented as Python programs. These can be found online in the form of Jupyter notebooks and are intended for experimentation. The structure of the book is mirrored in the notebooks, and programs are provided for most sections:

https://github.com/sn-code-inside/EinfNumMath-RWW

The original German notebooks are found in the main directory of the above repository, while the english translation is available in the subdirectory EN/ of the same repository.

All implementations in the book, which can also be found on the GitHub page, have been highlighted with the Python logo.

References

1. M. Abadi, A. Agarwal, P. Barham, E. Brevdo, Z. Chen, C. Citro, G. S. Corrado, A. Davis, J. Dean, M. Devin, S. Ghemawat, I. Goodfellow, A. Harp, G. Irving, M. Isard, Yangqing Jia, R. Jozefowicz, L. Kaiser, M. Kudlur, J. Levenberg, D. Mané, R. Monga, S. Moore, D. Murray, C. Olah, M. Schuster, J. Shlens, B. Steiner, I. Sutskever, K. Talwar, P. Tucker, V. Vanhoucke, V. Vasudevan, F. Viégas, O. Vinyals, P. Warden, M. Wattenberg, M. Wicke, Y. Yu, and X. Zheng. TensorFlow: Large-scale machine learning on heterogeneous systems, 2015. Software available from tensorflow.org.
2. L. V. Ahlfors. *Complex Analysis. An Introduction to the Theory of Analytic Functions of One Complex Variable*. McGraw-Hill, Inc., 3rd edition, 1979.
3. G. P. Akilov and L. V. Kantorovich. *Functional Analysis in Normed Spaces*. Pergamon Press, 1964.
4. G. S. Almasi and A. Gottlieb. *Highly parallel computing*. Benjamin-Cummings Publishing Co., 2 edition, 1994.
5. H. Amann and J. Escher. *Analysis I*. Birkhäuser, 2006.
6. K. Bach and F. Otto, editors. *Seifenblasen. Eine Forschungsarbeit des Instituts für leichte Flächenwerke über Minimalflächen = Forming bubbles*. Mitteilungen des Instituts für Leichte Flächentragwerke. Kämer, Stuttgart, 1980. ISBN 3-7828-2018-5.
7. C. Bär. *Elementare Differentialgeometrie*. De Gruyter, Berlin, Boston, 2 edition, 2010.
8. R. Barrett, M. Berry, T. F. Chan, J. Demmel, J. Donato, J. Dongarra, V. Eijkhout, R. Pozo, C. Romine, and H. van der Vorst. *Templates for the Solution of Linear Systems: Building Blocks for Iterative Methods*. SIAM, Philadelphia, PA, 1994.
9. U. Becciani, E. Sciacca, M. Bandieramonte, A. Vecchiato, B. Bucciarelli, and M. G. Lattanzi. Solving a very large-scale sparse linear system with a parallel algorithm in the gaia mission. In *2014 International Conference on High Performance Computing Simulation (HPCS)*, pages 104–111, July 2014.
10. P. Benner, A. Cohen, M. Ohlberger, and K. Willcox. *Model Reduction and Approximation: Theory and Algorithms*. SIAM Philadelphia, 2015.
11. J. Bewersdorff. *Algebra für Einsteiger: Von der Gleichungsauflösung zur Galois-Theorie*. Vieweg+Teubner, Wiesbaden, 2007.
12. C. M. Bishop. *Pattern recognition and machine learning*. Springer, 2006.

T. Richter et al., *Introduction to Numerical Mathematics*, Mathematics Study Resources 25, https://doi.org/10.1007/978-3-662-72546-7

13. C. M. Bishop and H. Bishop. *Deep Learning. Foundations and Concepts.* Springer Cham, 2024.

14. M. Bonnet. Lecture notes on numerical linear algebra. Méthodes numériques matricielles avancées: Analyse et expérimentaiton, 2022.

15. Bronstein. Formelsammlung. auf CD.

16. S. L. Brunton and J. N. Kutz. *Daten-driven science and engineering: Machine learning, Dynamical Systems, and Control.* Cambridge University Press, 2019.

17. W. Burkhart. *Steuerrecht in der Imkerei.* Steuerbuero Burkhart, 2015.

18. A. Burkov. *The hundred-page machine learning book.* Andriy Burkov, 2019.

19. T. F. Chan. An improved algorithm for computing the singular value decomposition. *ACM Transactions on Mathematical Software*, 8(1):72–83, 1982.

20. F. Cucker and A. G. Corbolan. An alternate proof of the continuity of the roots of a polynomial. *The American Mathematical Monthly*, 96(4):342–345, 1989.

21. G. Cybenko. Approximation by superpositions of a sigmoidal function. *Mathematics of Control, Signals, and Systems*, 2:303–314, 1989.

22. W. Dahmen and A. Reusken. *Numerik für Ingenieure und Naturwissenschaftler.* Springer, 2008.

23. J. Demmel, M. Gu, S. Eisenstat, I. Slapničar, K. Veselić, and Z. Drmač. Computing the singular value decomposition with high relative accuracy. *Linear Algebra and its Applications*, 299:21–80, 1999.

24. Bundesministerium der Justiz und Verbraucherschutz. Einkommensteuergesetz. 2002.

25. P. Deuflhard. *Newton Methods for Nonlinear Problems. Affine Invariance and Adaptive Algorithms*, volume 35 of *Computational Mathematics*. Springer, 2011.

26. P. Deuflhard and F. Bornemann. *Gewöhnliche Differentialgleichungen.* De Gruyter, 2008.

27. Duden. *Rechnen und Mathematik: Das Lexikon für Schule und Praxis.* Dudenverlag, 5. ueberarbeitete Auflage, 1994.

28. C. Eck, H. Garcke, and P. Knabner. *Mathematische Modellierung.* Springer, 2008.

29. J. F. Epperson. *An introduction to numerical methods and analysis.* John Wiley & Sons, 2007.

30. D. Etling. *Theoretische Meteorologie.* Springer, 2008.

31. L. C. Evans. *Partial Differential Equations.* Graduate Studies in Mathematics. American Mathematical Society, 2010.

32. J. D. Faires and R. L. Burdon. *Numerische Methoden.* Spektrum Akademischer Verlag, 1994.

33. G. Fischer. *Lineare Algebra.* Vieweg + Teubner, 2010.

34. A. Fog. Instruction tables: Lists of instruction latencies, throughputs and micro-operation breakdowns for intel, amd and via cpus. Technical report, 2016. Accessed on 22.05.2016.

35. E. Freitag and R. Busam. *Funktionentheorie 1.* Springer, 2006.

36. R. Freund and R. Hoppe. *Stoer/Bulirsch: Numerische Mathematik 1.* Springer, 2007.

37. S. Froeba and A. Wassermann. *Die bedeutendsten Mathematiker.* Marixverlag, 2007.

38. C. Gambella, B. Ghaddar, and J. Naoum-Sawaya. Optimization problems for machine learning: A survey. *European Journal of Operational Research*, 290(3):807–828, 2021.

39. C. Geiger and C. Kanzow. *Numerische Verfahren zur Lösung unrestringierter Optimierungsaufgaben.* Springer, 1999.

40. C. Geiger and C. Kanzow. *Theorie und Numerik restringierter Optimierungsaufgaben.* Springer, 2002.

41. G. Golub and W. Kahan. Calculating the singular values and pseudo-inverse of a matrix. *SIAM Journal of Numerical Analysis*, 2(2):205–224, 1965.

42. G. H. Golub and C. Reinsch. Singular value decomposition and least squares solutions. *Numer. Math.*, 14:403–420, 1970.

43. G. H. Golub and C. F. van Loan. *Matrix computations.* Johns Hopkins University Press, 3rd edition, 1996.

44. I. Goodfellow, Y. Bengio, and A. Courville. *Deep Learning.* MIT Press, 2016.

45. A. Grama, A. Gupta, G. Karypis, and V. Kumar. *Introduction to Parallel Computing*. Addison-Wesley, 2003. 2nd edition.

46. Benedikt Großer and Bruno Lang. Efficient parallel reduction to bidiagonal form. *Parallel Computing*, 25(8):969–986, 1999.

47. C. Großmann and H. G. Roos. *Numerische Behandlung partieller Differentialgleichungen*. Teubner, 2005.

48. W. Hackbusch. *Iterative Lösung großer schwachbesetzter Gleichungssysteme*. Number 69 in Leitfäden der angewandten Mathematik und Mechanik. Springer, 1993.

49. G. Hämmerlin and K.-H. Hoffmann. *Numerische Mathematik*. Springer Verlag, 1992.

50. M. Hanke-Bourgeois. *Grundlagen der numerischen Mathematik und des Wissenschaftlichen Rechnens*. Vieweg-Teubner Verlag, 2009.

51. ...C. R. Harris, K. J. Millman, S. J. van der Walt, R. Gommers, P. Virtanen, D. Cournapeau, E. Wieser, J. Taylor, S. Berg, N. J. Smith, R. Kern, M. Picus, S. Hoyer, M. H. van Kerkwijk, M. Brett, A. Haldane, J. F. del Río, M. Wiebe, P. Peterson, P. Gérard-Marchant, K. Sheppard, T. Reddy, W. Weckesser, H. Abbasi, C. Gohlke, and T. E. Oliphant. Array programming with NumPy. *Nature*, 585(7825):357–362, 2020.

52. H. Heuser. *Lehrbuch der Analysis Teil 2*. Vieweg+Teubner, 14 edition, 2012.

53. C. F. Higham and D. J. Higham. Deep learning: An introduction for applied mathematicians. *SIAM Review*, 61(4):860–891, 2019.

54. K. Hornik. Multilayer feedforward networks are universal approximators. *Neural Networks*, 2:359–366, 1989.

55. M. Huber, B. Gmeiner, U. Rüde, and B. Wohlmuth. Resilience for massively parallel multigrid solvers. *SIAM Journal on Scientific Computing*, 38(5):S217–S239, 2016.

56. D. A. Huffman. A method for the construction of minimum-redundancy codes. *Proceedings of the I.R.E.*, pages 1098–1101, 1952. Aufgerufen 09/2016.

57. IEEE. IEEE 754-2008: Standard for floating-point arithmetic. Technical report, 2008.

58. W. Keiper, A. Milde, and S. Volkwein. *Reduced-Order Modeling (ROM) for Simulation and Optimization*. Springer Verlag, 2018.

59. J. Kepler. *Nova stereometria doliorum vinariorum*. online by courtesy of Carnegie Mellon University Libraries, 1615.

60. H. Kielhöfer. *Variationsrechnung: Eine Einführung in die Theorie einer unabhängigen Variablen mit Beispielen und Anwendungen*. Vieweg+Teubner Verlag, 2010.

61. M. Koecher. *Klassische elementare Analysis*. Springer, 1987.

62. K. Königsberger. *Analysis 1*. Springer, 6 edition, 2004.

63. R. Kress. *Numerical Analysis*. Springer Verlag, 1998.

64. W. Ma, Y. Ao, C. Yang, and S. Williams. Solving a trillion unknowns per second with hpgmg on sunway taihulight. *Cluster Computing*, 23(2):493–507, 2020.

65. G. Maess. *Vorlesungen über numerische Mathematik II, Analysis*. Birkhäuser Verlag, 1988.

66. S. Mandal, A. Ouazzi, and S. Turek. Modified Newton solver for yield stress fluids. In *Proceedings of ENUMATH 2015, the 11th European Conference on Numerical Mathematics and Advanced Applications*, volume 112 of *Lecture Notes in Computational Science and Engineering*. Springer, 2016.

67. Marvin Marcus and Henryk Minc. *Introduction to Linear Algebra*. Dover Publications Inc., 1988.

68. C. McEniry. The mathematics behind the fast inverse square root funciton code. Aufgerufen am 22.05.2016, August 2007.

69. M. A. Nielsen. Neural networks and deep learning, 2018.

70. J. Nocedal and S. J. Wright. *Numerical optimization*. Springer Ser. Oper. Res. Financial Engrg., 2006.

71. B. N. Parlett. *The Symmetric Eigenvalue Problem*. Classics in Applied Mathematics. SIAM, 1998.

72. A. Paszke, S. Gross, F. Massa, A. Lerer, J. Bradbury, G. Chanan, T. Killeen, Z. Lin, N. Gimelshein, L. Antiga, A. Desmaison, K. A., E. Yang, Z. DeVito, M. Raison, A. Tejani, S. Chilamkurthy, B. Steiner, L. Fang, J. Bai, and Soumith Chintala. Pytorch: An imperative style, high-performance deep learning library. In *Advances in Neural Information Processing Systems 32*, pages 8024–8035. Curran Associates, Inc., 2019.

73. D. Piekenbrock. *Einfuehrung in die Volkswirtschaftslehre und Mikrooekonomie*. Physica-Verlag HD, 2008.

74. G. Pólya. Über die konvergenz von quadraturverfahren. *Mathematische Zeitschrift*, 37:264–286, 1933.

75. R. Rannacher. *Einführung in die Numerische Mathematik*. Heidelberg University Publishing, 2017.

76. R. Rannacher. *Numerik gewöhnlicher Differentialgleichungen*. Heidelberg University Publishing, 2017.

77. R. Rannacher. *Numerik partieller Differentialgleichungen*. Heidelberg University Publishing, 2017.

78. R. Rannacher. *Probleme der Kontinuumsmechanik und ihre numerische Behandlung*. Heidelberg University Publishing, 2017.

79. T. Rauber and G. Rünger. *Parallele Programmierung*. Springer, 2 edition, 2007.

80. H.-J. Reinhardt. *Numerik gewöhnlicher Differentialgleichungen. Anfangs- und Randwertprobleme*. Berlin, Boston: De Gruyter, 2008.

81. C. Reinsch and M. Richter. Singular value decomposition in extended double precision arithmetic. *Numer. Algorithms*, 93(3):1137–1155, 2023.

82. Y. Saad. *Iterative Methods for Sparse Linear Systems*. SIAM, Philadelphia, PA, 2 edition, 2003.

83. Y. Saad. *Numerical methods for large eigenvalue problems*, volume 66 of *Classics in Applied Mathematics*. SIAM publications, 2011.

84. R. Schaback and H. Wendland. *Numerische Mathematik*. Springer, 2005.

85. J. Schüle. *Paralleles Rechnen*. de Gruyter, 2010.

86. H. R. Schwarz and N. Köckler. *Numerische Mathematik*. Vieweg-Teubner Verlag, 2011.

87. G. W. Stewart. *Matrix Algorithms. Volume II: Eigensystems*. SIAM publications, 2001.

88. Gilbert Strang. *Introduction to linear algebra*. Wellesley-Cambridge Press, Wellesley, MA, sixth edition, 2023.

89. T. Strutz. *Bilddatenkompression. Grundlagen, Codierung, Wavelets, JPEG, MPEG, H.264*. Praxiswissen. Vieweg+Teubner Verlag, 2005.

90. Terence Tao. *Analysis I*. Springer, 2022.

91. Terence Tao. *Analysis II*. Springer, 2022.

92. L. N. Trefethen and D. Bau. *Numerical Linear Algebra*. SIAM, 1997.

93. E. E. Tyrtyshnikov. *A Brief introduction to Numerical Analysis*. Birkhäuser, 1997.

94. M. Ulbrich and S. Ulbrich. *Nichtlineare Optimierung*. Mathematik Kompakt. Birkhäuser, 2011.

95. D. Werner. *Funktionalanalysis*. Springer, 2004.

96. Wikipedia. Fast inverse square root. https://en.wikipedia.org/wiki/Fast_inverse_square_root. Accessed on 24.12.2016.

97. J. H. Wilkinson. *The algebraic eigenvalue problem*. Numerical Mathematics and Scientific Computation. Oxford Science Publications, 1965.

98. J. Wloka. *Partielle Differentialgleichungen*. Teubner, Stuttgart, 1982.

99. W. Zulehner. *Numerische Mathematik: Ein Einfuehrung anhand von Differentialgleichungsproblemen; Band 2: Instationaere Probleme*. Mathematik Kompakt. Birkhaeuser Verlag, 2011.

Index

© The Editor(s) (if applicable) and The Author(s), under exclusive license to
Springer-Verlag GmbH, DE, part of Springer Nature 2025
T. Richter et al., *Introduction to Numerical Mathematics*, Mathematics Study Resources
25, https://doi.org/10.1007/978-3-662-72546-7

The manufacturer's authorised representative in the EU is Springer
Nature Customer Service Centre GmbH, Europaplatz 3, 69115 Heidelberg,
Germany. If you have any concerns regarding our products, please
contact ProductSafety@springernature.com

Printed and bound by CPI Group (UK) Ltd, Croydon, CR0 4YY
05/06/2026
02126662-0012